RESEARCH HANDBOOK ON BIODIVERSITY AND LAW

RESEARCH HANDBOOKS IN ENVIRONMENTAL LAW

This highly topical series addresses some of the most important questions and areas of research in Environmental Law. Each volume is designed by a leading expert to appraise the current state of thinking and probe the key questions for future research on a particular topic. The series encompasses some of the most pressing issues in the field, ranging from climate change, biodiversity and the marine environment through to the impacts of trade, regulation, and sustainable development.

Each *Research Handbook* comprises specially-commissioned chapters from leading academics, and sometimes practitioners, as well as those with an emerging reputation and is written with a global readership in mind. Equally useful as reference tools or high-level introductions to specific topics, issues and debates, these *Handbooks* will be used by academic researchers, post-graduate students, practising lawyers and lawyers in policy circles.

Titles in the series include:

Research Handbook on Climate Change Adaptation Law
Edited by Jonathan Verschuuren

Research Handbook on Climate Change Mitigation Law
Edited by Geert Van Calster, Wim Vandenberghe and Leonie Reins

Handbook of Chinese Environmental Law
Edited by Qin Tianbao

Research Handbook on International Marine Environmental Law
Edited by Rosemary Rayfuse

Research Handbook on Biodiversity and Law
Edited by Michael Bowman, Peter Davies and Edward Goodwin

Research Handbook on Biodiversity and Law

Edited by

Michael Bowman
University of Nottingham, UK

Peter Davies
University of Nottingham, UK

Edward Goodwin
University of Nottingham, UK

RESEARCH HANDBOOKS IN ENVIRONMENTAL LAW

Edward Elgar
PUBLISHING

Cheltenham, UK • Northampton, MA, USA

© The Editors and Contributors Severally 2016

All rights reserved. No part of this publication may be reproduced, stored in a retrieval system or transmitted in any form or by any means, electronic, mechanical or photocopying, recording, or otherwise without the prior permission of the publisher.

Published by
Edward Elgar Publishing Limited
The Lypiatts
15 Lansdown Road
Cheltenham
Glos GL50 2JA
UK

Edward Elgar Publishing, Inc.
William Pratt House
9 Dewey Court
Northampton
Massachusetts 01060
USA

A catalogue record for this book
is available from the British Library

Library of Congress Control Number: 2015954315

This book is available electronically in the Elgaronline
Law subject collection
DOI 10.4337/9781781004791

ISBN 978 1 78100 478 4 (cased)
ISBN 978 1 78100 479 1 (eBook)

Typeset by Columns Design XML Ltd, Reading
Printed and bound in Great Britain by TJ International Ltd, Padstow

Contents

List of contributors vii
Preface xii
List of abbreviations xiv

PART I VISIONS, VALUES AND VOICES

1. Law, legal scholarship and the conservation of biological diversity: 2020 vision and beyond 3
   Michael Bowman
2. In whose interest? Instrumental and intrinsic value in biodiversity law 55
   Mattia Fosci and Tom West
3. Participatory resource management: a Caribbean case study 78
   Nicole Mohammed
4. The role of non-state actors in treaty regimes for the protection of marine biodiversity 95
   Elizabeth A. Kirk

PART II SIGNIFICANT THREATS TO BIODIVERSITY

5. Climate change, marine biodiversity and international law 123
   Rosemary Rayfuse
6. Broad-spectrum efforts to enhance the conservation of vulnerable marine ecosystems 146
   Edward J. Goodwin
7. Alien invasive species: is the EU's strategy fit for purpose? 184
   Peter Davies
8. Countering fragmentation of habitats under international wildlife regimes 219
   Arie Trouwborst
9. Armed conflict and biodiversity 245
   Karen Hulme

PART III GENERAL PRINCIPLES OF INTERNATIONAL ENVIRONMENTAL LAW

10. The Convention on Biological Diversity and the concept of sustainable development: the extent and manner of the Convention's application of components of the concept 273
    Veit Koester
11. Whaling and inter- and intra-generational equity 297
    Malgosia Fitzmaurice

vi *Research handbook on biodiversity and law*

12. Common concern, common heritage and other global(-ising) concepts: rhetorical devices, legal principles or a fundamental challenge? 334
    Duncan French

PART IV REGULATORY CHALLENGES AND RESPONSES

13. Biodiversity, knowledge and the making of rights: reviewing the debates on bioprospecting and ownership 361
    Emilie Cloatre
14. Ecological restoration in international biodiversity law: a promising strategy to address our failure to prevent? 387
    Kees Bastmeijer
15. Non-compliance procedures and the implementation of commitments under wildlife treaties 414
    Karen N. Scott
16. 'Only connect'? Regime interaction and global biodiversity conservation 437
    Richard Caddell

Index 473

Contributors

Kees Bastmeijer is Professor of Nature Conservation and Water Law at Tilburg University, the Netherlands, and a visiting professor at the School of Business, Economics and Law at the University of Gothenburg, Sweden (2016–19). His research focuses on the role of international, European and domestic law in protecting nature, with a particular interest in nature conservation in the Polar Regions, relationships between law and philosophical human-nature attitudes, property rights and nature, and the role of law in protecting wilderness. His latest publication is the edited volume *Wilderness Protection in Europe. The Role of International, European and National Law* (CUP, forthcoming).

Michael Bowman is Associate Professor in Law at the University of Nottingham, and Director of its long-established Treaty Centre. He has served as both *rapporteur* and chair of the International Environmental Law Committee of the International Law Association's British Branch, and has provided extensive information, advice and assistance to governments, NGOs and international organisations on questions of international treaty law, conservation and animal welfare. He is the author of numerous essays and journal articles on these topics, and co-author (with Peter Davies and Catherine Redgwell) of the second edition of *Lyster's International Wildlife Law*, published in 2010 to mark the International Year of Biodiversity. He also co-edited *International Law and the Conservation of Biological Diversity* (with Catherine Redgwell, 1996) and *Environmental Damage in International and Comparative Law* (with Alan Boyle, 2002).

Richard Caddell is a Nippon Foundation Senior Nereus Fellow and a Senior Research Associate at the Netherlands Institute for the Law of the Sea, Utrecht University and a Lecturer in Law at Cardiff University. His main research interests engage international and EU environmental law, the law of the sea, biodiversity conservation, fisheries and Polar law. Dr Caddell is also an academic member of Francis Taylor Building, Inner Temple, London, the UK's foremost environmental and planning law barristers.

Emilie Cloatre is Senior Lecturer in Law at the University of Kent. She has held visiting positions at the Centre for the Study of Law and Society, University of California at Berkeley; the Genomics Forum, University of Edinburgh; the School of Law, University of Singapore; and the Ghana Institute of Management and Public Administration. She is principal investigator for the AHRC Network Technoscience, Law and Society: Interrogating the Nexus. Her main research interests lie in the intersection between law and contemporary 'science and society' issues – including pharmaceutical flows, access to health, and the politics of climate change regulation. Her approach to law is influenced by insights from Science and Technology Studies, and in particular by Actor-Network Theory. Her publications include *Pills for the*

Poorest: an Exploration of TRIPS and access to Medicines in sub-Saharan Africa (Palgrave Macmillan, 2013 – awarded the 2014 Hart Socio-Legal Book prize) and *Knowledge, Technology and Law* (Routledge, 2014, with Martyn Pickersgill).

Peter Davies is Associate Professor in the School of Law, University of Nottingham. He is co-author (with Michael Bowman and Catherine Redgwell) of *Lyster's International Wildlife Law* (CUP, 2010), and the author of *European Union Environmental Law* (Ashgate, 2004). He has a particular interest in wildlife conservation law and has published in a number of leading journals and books. He is a member of the Editorial Advisory Board of the *New Zealand Yearbook of International Law*.

Malgosia Fitzmaurice is Professor of Public International Law at the Department of Law, Queen Mary University of London. She specialises in international environmental law, treaties, indigenous peoples, and Arctic law, and has published widely on these subjects. Professor Fitzmaurice currently is one of the Queen Mary Investigators, as Principal Investigator, on a multinational interdisciplinary research project to fight environmental crime, funded by the European Commission under the 7th Framework Programme for Research. In July 2001 she delivered a lecture on 'International Protection of the Environment' at the Hague Academy of International Law, and in 1996 presented a paper at the 50th anniversary of the International Court of Justice in The Hague. She has lectured widely in the United Kingdom, Europe (Sorbonne, Pantheon), and the United States of America (Berkeley Law and New York University School of Law). In 2014 she was a Visiting Professor at the University of Kobe, Japan and has participated in many international conferences. She is the Editor in Chief of the *International Community Law Review*, and Editor-in-Chief of a book series 'Queen Mary Studies in International Law' (Martinus Nijhoff).

Mattia Fosci is a former visiting lecturer in International and EU law at the University of Leicester. He has recently founded, and is currently managing, an open-source think tank, and is also consulting on a number of government-sponsored projects on open access research. He has published in international environmental journals, as well as on legal and policy issues relating to research management and open access to academic research. Mattia Fosci holds a doctorate and LLM in International Law from the University of Nottingham, an MA in International Relations and a BA in Political Science from the University of Cagliari, Italy. His doctoral research focused on the multi-level governance of land and forests in the context of emerging international efforts to mitigate climate change. He has published on issues related to Reducing Emissions from Deforestation and Forest Degradation (REDD+), biodiversity law, and environmental governance in developing countries. In 2009–2010 he attended the UNFCCC negotiations on climate change as an accredited observer. He has worked with civil society organisations and the IUCN, and as a consultant in the areas of climate change law, policy and economics.

Duncan French is Head of the Law School and Professor of International Law at the University of Lincoln, UK. His research interests include international environmental law, international development law, law of the sea, and Antarctica. He was co-rapporteur of the International Law Association (ILA) Committee on International Law on Sustainable Development (2003–2012) and is presently Chair of the ILA Study Group on Due Diligence in International Law. He edited *Statehood and Self-Determination: Reconciling Tradition and Modernity in International Law* (CUP, 2013, p/back 2015).

Edward J. Goodwin is Assistant Professor of Law at the University of Nottingham, UK. He is a graduate of the University of Oxford (BA (Hons)) and the University of Nottingham (LL.M, PhD), and held a visiting position as a Grotius Research Scholar at the School of Law, University of Michigan in 2015. He is the Book Reviews Editor for the *International Journal of Marine and Coastal Law*. His research interests include international environmental law, the law of the sea, international heritage law, and land law. In addition to a range of journal articles and contributions to collected works in these areas, he is the author of *International Environmental Law and the Conservation of Coral Reefs* (Routledge, 2011).

Karen Hulme is Professor of Law in the School of Law at the University of Essex. She obtained her PhD in 2004 and has research interests in the laws of armed conflict and environmental rights. She has worked with the Essex Business and Human Rights Project (EBHR) on a number of reports and consultancies on the extractives industry, including legislation amendments and human rights impact monitoring, and, in particular, on issues of environmental law and environmental human rights. Consultancies include, on amendments to Afghanistan's new Mining Code and Regulations, for Global Witness, *A Shaky Foundation: Analysing Afghanistan's Draft Mining Law*, and for Amnesty International (Netherlands) to advise on necessary amendments to Senegal's (Gold) Mining Code in order to ensure compliance with their ECOWAS obligations. Karen Hulme also works with an NGO, the Toxic Remnants of War Project (part of ICBUW – International Coalition to Ban Uranium Weapons), as an adviser on humanitarian, environmental and human rights issues raised by toxic remnants of war. In 2009 she contributed, alongside the International Committee for the Red Cross, to the report for UNEP on 'Protecting the Environment during Armed Conflict: An Inventory and Analysis of International Law'. She is currently writing a report for the UN Independent Expert on the right to a healthy environment (on the right to a healthy environment in times of armed conflict) and she is a member of the IUCN, including their special expert group on Warfare and Environment. Karen Hulme's book entitled *War Torn Environment: Interpreting the Legal Threshold* won the American Society of International Law's Francis Lieber Prize for 2004 for 'outstanding scholarship in the field of the law of armed conflict'.

Elizabeth A. Kirk is Reader in Law at the School of Law, University of Dundee. She joined Dundee in 1995, having previously qualified as a solicitor in Scotland and worked as a Research Associate at the University of British Columbia. Elizabeth Kirk is the Western Europe representative on, and Deputy Chair of, the Governing Board of the IUCN Academy of Environmental Law. She also sits on the Managing Board of the European Environmental Law Forum. She is a member of the IUCN Commission on Environmental Law Specialist Group on Ocean Law and a member of the Council of the Society of Legal Scholars. Elizabeth's research focuses on adaptability within legal regimes, in particular the ability of regimes to respond to changing circumstances, scientific understanding, or actors. Her work, which has been supported by a number of research grants from, amongst others, the AHRC, British Academy, ESCR, European Commission, and Royal Society of Edinburgh, spans both the international and domestic law of marine governance (and marine resources).

Veit Koester is master of law; barrister and former external professor at Roskilde University in Denmark. As former Head of the Ecological Division of the Danish Nature Agency, he led work on national conservation issues and represented his country in international negotiations regarding, inter alia, the Ramsar Convention, the World Heritage Convention, the Migratory Species Convention, the CITES Convention, the CBD, the Cartagena Protocol on Biosafety, the Bern Convention on European Wildlife, and the Aarhus Convention. He served as chair at numerous of these negotiations and on several occasions represented the Danish EU Presidency. He has contributed to the work of the International Council of Environmental Law and the IUCN World Commission on Environmental Law. After his retirement he was the first chair of the Aarhus Convention Compliance Committee (2003–2011), and of the Compliance Committee of the Cartagena Protocol (2004–2009). Since 2010 he has been the chair of the Compliance Committee of the Protocol on Water and Health to the Convention on the Protection and Use of Transboundary Watercourses and International Lakes. He has published numerous articles and other publications in the field of national and international environmental law, including an extensive Commentary on the Danish Nature Protection Act (2009). His awards include, among others, the Elizabeth Haub Award for Environmental Diplomacy (2001) and the Environmental Law Prize of the Danish Society for Environmental Law (2010).

Nicole Mohammed is a PhD candidate at the University of Nottingham and a recipient of the University's Centre for Environmental Law scholarship for 2010. She is a graduate of the University of the West Indies (BA (Hons)) and the University of Nottingham (LLM). She is currently writing her thesis on environmental justice and public participation rights in the Commonwealth Caribbean. Her areas of interest include environmental governance, human rights and environmental law, and access to justice and participatory rights in environmental matters. Nicole Mohammed formerly worked as a barrister in Trinidad and Tobago with a focus on public law, maritime law and environmental law matters. She has also worked for the United Kingdom Environment Agency (Thames Region) in the Strategic Environmental Policy Unit.

Rosemary Rayfuse (LLB Queens, LLM Cambridge, PhD Utrecht, LLD h.c. Lund) is Professor of Public International Law in the Faculty of Law at UNSW Australia (the University of New South Wales). She holds a conjoint appointment at Lund University, a visiting professorship at the University of Gothenburg, and is an Associated Senior Fellow at the Fridtjof Nansen Institute and an Associated Researcher in the Centre for Water, Oceans and Sustainability Law at the University of Utrecht. Her main fields of interest are oceans governance, high seas fisheries, protection of the marine environment in areas beyond national jurisdiction, regulation of climate engineering and the normative effects of climate change on international law. In addition to her numerous book chapters and journal articles, she is the author of *Non-Flag State Enforcement in High Seas Fisheries* (Martinus Nijhoff, 2004) and editor of the *Research Handbook on International Marine Environmental Law* (Edward Elgar, 2015), *Protection of the Environment in Relation to Armed Conflict* (Martinus Nijhoff, 2014), *The Challenge of Food Security: International Policy and Regulatory Frameworks* (with N Weisfelt) (Edward Elgar, 2012), and *International Law in the Era of Climate Change* (with S Scott) (Edward Elgar, 2012). She is a member of the IUCN Commission on Environmental Law, Co-Chair of its Sub-Working Group on High Seas Governance, and the Chair's Nominee on the International Law Association's Committee on International Law and Sea Level Rise.

Karen N. Scott is Professor of Law at the University of Canterbury in New Zealand. Her research focuses on the areas of the law of the sea, the Polar regions and international environmental law. She has published over 50 journal articles and book chapters in these areas on issues such as ocean management, environmental governance, climate change and geoengineering. Recent publications include DR Rothwell, AG Oude Elferink, KN Scott and T Stephens (eds), *The Oxford Handbook of the Law of the Sea* (OUP, 2015). Karen Scott was the General Editor of the *New Zealand Yearbook of International Law* from 2009 to 2012 and remains a member of the Editorial Board. She is currently the Vice-President of the Australian and New Zealand Society of International Law.

Arie Trouwborst is Associate Professor of Environmental Law at Tilburg Law School in the Netherlands. He is a graduate of Utrecht University (BA, LL.M, PhD). His research mainly focuses on international and EU wildlife law, and he has published extensively on a range of topics within these fields. Recent topics include wilderness protection, invasive alien species, and large carnivore conservation. From 2014 to 2019 he is in charge of the project *Ius Carnivoris*, on law and large carnivores. He is a member of various international expert groups and regularly conducts research commissioned by (inter)national governmental bodies and NGOs.

Tom West is soon to complete his PhD thesis on Human and Nonhuman Rights at the School of Law, University of Nottingham. He has degrees in Mathematics (BSc) from the University of Warwick, and Law and Environmental Science (MSc) from the University of Nottingham. During 2013, he worked at the Research Institute for Humanity and Nature, a transdisciplinary research institution based in Kyoto. His research interests encompass a range of topics from law and philosophy including eco-phenomenology, analytical rights jurisprudence, and the emerging field of rights of nature.

Preface

Given the plethora of scientific findings and official reports to have emerged in recent years, there is no longer room for serious doubt regarding either the extremely parlous state of the world's biological diversity or the serious threats that this situation poses to human health, wealth and general welfare. Over and above these essentially prudential concerns, moreover, fundamental ethical questions continue to be raised, many of which embrace interests of other than a purely anthropocentric kind. In the light of such considerations, and of the practical intractability of the resulting problems, it is inevitable that the power of the law will be recruited in order to explore, and hopefully secure, the path to their alleviation or solution. Furthermore, the very nature of the issues involved is such that the elaboration of appropriate legal measures is likely to be needed not only at the national, but also at the regional and global levels.

To casual observers of international affairs – and even, indeed, to many seasoned international lawyers – it may come as something of a surprise to discover that governments first began to turn their attention to such matters in a serious way as early as the final quarter of the 19th century. Yet the measures ultimately adopted tended to be narrowly focused and essentially reactive in character, as well as being rather loosely drafted in legal terms and relatively ill-informed from an ecological perspective. It has only been over the last four decades, and more particularly over the final two, that anything resembling a coherent system and co-ordinated approach to such matters has begun to emerge. Even then, however, the emerging legal regime of biodiversity conservation has tended to be overshadowed to some extent by its environmental bedfellow, the international law of pollution. Part of the reason for this eclipse, arguably, has been the relative lack of attention paid by international lawyers to the wildlife dimension of environmental law, and it is therefore encouraging to note that there has been something of an acceleration of interest in this topic amongst academic lawyers in recent years.

With that in mind, it was extremely gratifying for us to receive an invitation from the publishers Edward Elgar to compile a volume in their highly regarded *Research Handbook* series, designed specifically to explore in greater breadth and depth the biodiversity issues that had already been briefly broached in the more widely focused handbook on international environmental law generally, published in 2010 under the joint editorship of Malgosia Fitzmaurice, David Ong and Panos Merkouris. Our intention in this work was not to offer an exhaustive account of biodiversity law, but rather to offer contributors a platform for their current research in this field whilst highlighting fruitful avenues for future research. We do, however, believe that taken collectively the chapters included engage with many of the core topics and most pressing challenges in this vibrant field.

The University of Nottingham Treaty Centre has a long and proud record of publication in the area of wildlife law, and on this occasion we were privileged to work with a wide array of scholars from around the world, at all stages of their respective careers and endowed with impressive expertise and experience of both an academic and practical kind. We are grateful to them all for their cooperation, erudition and endeavour – qualities which, we believe, are amply reflected within the covers of this work.

We would also like to thank Ben Booth and everyone at Edward Elgar for their very considerable professionalism, imagination, flexibility and forbearance, which have proved of such great assistance in the realisation of this endeavour.

Finally, Peter and Michael would like to express their particular appreciation to Edward for his exceptional contribution to the management of this project.

MJB/PGGD/EJG

Abbreviations

The majority of the acronyms and abbreviations used in this work are explained at the point at which they occur, while location references to periodical literature and other source materials have generally been given in full, unabbreviated form. For convenience, however, a small number of standard abbreviations have been employed, especially for the more familiar and commonly cited journals, law reports and treaty series. These include the following:

AJIL	American Journal of International Law
BFSP	British and Foreign State Papers
CETS	Council of Europe Treaty Series
COP	Conference of the Parties
CTS	Consolidated Treaty Series
CUP	Cambridge University Press
EJIL	European Journal of International Law
ETS	European Treaty Series (later to become the CETS)
FAO	Food and Agriculture Organization
ICJ	International Court of Justice
ICLQ	International and Comparative Law Quarterly
ILM	International Legal Materials
ILR	International Law Reports
IMO	International Maritime Organization
IUCN	International Union for the Conservation of Nature and Natural Resources (the World Conservation Union)
JEL	Journal of Environmental Law
JIWLP	Journal of International Wildlife Law and Policy
LNTS	League of Nations Treaty Series
MEA	Multilateral Environmental Agreement
NGO	Non-governmental organisation
OJEC	Official Journal of the European Communities
OJEU	Official Journal of the European Union
OUP	Oxford University Press
PCIJ	Permanent Court of International Justice (for reports of its decided cases)
RECIEL	Review of European, Comparative and International Environmental Law (formerly the Review of European Community and International Environmental Law)

RIAA	(UN) Reports of International Arbitral Awards
RSPB	Royal Society for the Protection of Birds (UK)
Stat	Statutes at Large (US)
TIAS	Treaties and Other International Acts Series (US)
UN	United Nations
UNEP	United Nations Environment Programme
UNESCO	United Nations Educational, Scientific and Cultural Organization
UNGA	United Nations General Assembly
UNJYB	United Nations Juridical Year Book
UNTS	United Nations Treaty Series
UP	University Press (as applicable)
UST	United States Treaties
USTS	United States Treaty Series
WTO	World Trade Organization
WWF	World Wide Fund for Nature (formerly World Wildlife Fund)
YIEL	Yearbook of International Environmental Law

PART I

VISIONS, VALUES AND VOICES

1. Law, legal scholarship and the conservation of biological diversity: 2020 vision and beyond

Michael Bowman

1. INTRODUCTION

It can hardly be doubted that the current state of the world's biological diversity provides cause for the gravest possible concern. As was candidly conceded by a wide-ranging survey prepared by the Convention on Biological Diversity (CBD) Secretariat for the purposes of the International Year which the UN had originally dedicated to the topic:[1]

> The target agreed by the world's Governments in 2002, 'to achieve by 2010 a significant reduction of the current rate of biodiversity loss at the global, regional and national level as a contribution to poverty alleviation and to the benefit of all life on Earth', has not been met.

The evidence of this failure was all around. Many of those species whose prospects of survival had been formally assessed were moving closer to extinction, with coral species experiencing the most rapid rate of deterioration and amphibians facing the greatest risk generally. Nearly a quarter of all plant species were judged to be threatened with extinction, while population surveys suggested that the overall abundance of vertebrate species had fallen by nearly one-third between 1970 and 2006, and was still falling, with the severest declines occurring in the tropics and amongst freshwater species. Despite some successes in slowing the process, natural habitats generally continued to decline in both extent and integrity. The services provided by forests, rivers and other natural ecosystems had progressively been compromised by fragmentation and degradation, while the genetic diversity of crops and livestock in agricultural systems remained in decline. The five principal drivers of biodiversity loss – habitat change, overexploitation, pollution, invasive alien species and climate change – were either undiminished or actually increasing in intensity. In sum, humanity's ecological footprint exceeded the biological capacity of the earth by an even wider margin than at the time the 2010 target was originally agreed.[2] This failure was acknowledged by the decision of the UN to follow the International Year of Biodiversity with the devotion of an entire decade to the global conservation project,[3]

---

[1] CBD Secretariat, *Global Biodiversity Outlook 3* (2010), Executive Summary, 9.
[2] Ibid.
[3] UNGA Resolution 65/161, specifying 2011–2020 for this purpose, in accordance with the current strategic plan of the 1992 Convention on Biological Diversity (CBD), (1992) 31 ILM 818.

in the hope of achieving significant progress, if not actually turning things around, by the year 2020.

We are now effectively half way through the decade in question, and there are still disturbingly few signs of the progress required. The latest major survey prepared by WWF, which announces itself to be 'not for the faint-hearted',[4] suggests that since 1970 (i.e., in less than two human generations) populations of vertebrate species globally have actually dropped by just over half.[5] Nevertheless, it retains a degree of confidence that we can 'close this destructive chapter in our history, and build a future where people can live and prosper in harmony with nature'.[6] The CBD's own most recent appraisal paints a broadly similar picture of the current state of biodiversity conservation as decidedly troubling, but not yet irredeemable.[7]

It is certainly not easy to be optimistic about the prospects for securing the transformation needed, however, since all these unmistakeable indicators of deterioration have become apparent during the very period in which we have also experienced an exponential increase in our knowledge and understanding of the centrality of biodiversity conservation to the preservation of the planet's life support systems, and hence of our ultimate absolute dependence upon the efforts that we make in that regard.[8] Thus, although we cannot, on any objectively plausible basis, any longer claim ignorance of either the nature or importance of the task, or of the most viable approaches to its performance, it would seem that the foibles of human nature are such that in practice we continue to experience the greatest of difficulty in bringing ourselves to address these issues effectively or even, in some quarters, to acknowledge their existence or seriousness at all. All endeavours to procure a committed and systematically rational response from humankind to challenges such as these have much in common with attempts to herd cats or to scoop up milk in the fingers: only a very small proportion of the total target, at best, seems likely to be secured.

This sorry saga of progressive ecological impoverishment certainly cannot, however, be attributed to any lack of formal legal regulation of the problem, since it has now been the subject of transnational regulatory attention, if only in extremely piecemeal fashion initially, for the best part of 150 years, and over the last few decades in particular an elaborate network of multilateral treaty regimes has been assembled to address the symptoms on a much more systematic basis. Over the course of that latter period, moreover, a significant body of legal scholarship devoted to the explication and analysis of the regimes in question has finally begun to emerge.[9] These developments,

---

[4] See <http://www.wwf.eu/media_centre/publications/living_planet_report/>. The report itself (WWF, *Living Planet Report 2014*) is downloadable from this site.

[5] The Index 'measures more than 10,000 representative populations of mammals, birds, reptiles, amphibians and fish'.

[6] *See* the WWF website (n 4).

[7] CBD Secretariat, *Global Biodiversity Outlook 4* (2014).

[8] In addition to the publications mentioned above, see also the 1980 *World Conservation Strategy*, devised by IUCN in collaboration with other interested organisations, and its 1991 revision, *Caring for the Earth*.

[9] The *Journal of International Wildlife Law and Policy*, published first by Kluwer International (1998–2002) and then by Taylor & Francis/Routledge has undoubtedly played a significant part in this process.

it might have been supposed, were very definitely to be regarded as entries on the credit side of the balance sheet on which the global conservation account is recorded, but there are those, it would seem, whose enthusiasm for them is decidedly muted. After all, if this substantial body of law has actually failed to counter the damaging trends towards biodiversity impoverishment which are mentioned above, then there must surely be grounds for questioning its efficacy.

Thus, in a recent review essay prompted by the publication of a new edition of *Lyster's International Wildlife Law* (hereafter *Lyster*),[10] of which the present writer was a co-author, the California-based political scientist Geoffrey Wandesforde-Smith expressed the view that the world had lost its appetite for 'making big splashes' with new treaty regimes, and might currently even be more interested in conducting its international conservation endeavours 'around, rather than through' those that already exist.[11] Accordingly, while he was fleetingly fulsome in his praise for the breadth, depth and quality of the analysis of the global conservation measures which the book contained,[12] his clear message was that this alone was not enough, and that the omission to address certain 'broader questions' – such as the demand by environmentally committed readers to know what they should do next in order to make a positive impact in this field – must be seen to reflect the lack of a strong and clear inspirational vision in the work as a whole.

2. THE VEXED QUESTION OF VISION

In the ordinary way, there is probably little to be gained from authors responding directly to reviews of their work, but in this particular case there are grounds for interpreting the essay as a calculated attempt to initiate a dialogue regarding the shape and direction that legal scholarship concerning biodiversity conservation should most profitably assume in the future.[13] This is a project that might justly be regarded as important both from the perspective of its contribution to legal education and research, and in the light of its potential practical impact upon the conservation process itself. Such questions are moreover, especially pertinent to such a work as the present, given that these are the very issues that it seeks to explore and illuminate. In these circumstances, the challenge is not an easy one to ignore, notwithstanding the fact that the gauntlet has been thrown down in terms that are likely to generate certain misgivings.

---

[10] MJ Bowman, PGG Davies and CJ Redgwell, *Lyster's International Wildlife Law* (2nd edn., CUP, 2010). The original volume was, of course, S Lyster, *International Wildlife Law* (Grotius, 1985).
[11] G Wandesforde-Smith, 'From Sleeping Treaties to the Giddy Insomnia of Global Governance: How International Wildlife Law Makes Headway' (2012) 15 JIWLP 80, 93. It is noticeable, however, that no real evidence is presented in support of this contention.
[12] *Ibid.*, 80, 90, 93–94.
[13] Thus, Professor Wandesforde-Smith not merely secured the publication of his review in the excellent journal of which he is currently associate editor, but took the additional step of sending copies to each of the authors of the new *Lyster* personally, as well as to its publishers, Cambridge University Press.

6 *Research handbook on biodiversity and law*

In particular, accusations of 'lack of vision' are nowadays all too likely to be interpreted merely as forlorn lamentations that others cannot be relied upon to share the speaker's own singular, and very possibly idiosyncratic, perspective upon the world.[14] A good part of the reason for that is doubtless that the word 'vision' itself became so devalued through systematic overuse in the self-serving rhetoric of political spin doctors and corporate PR operatives during that most wasteful, and wasted, of decades, the 1980s, that it is now less likely to serve as a source of inspiration than to provoke sighs of weary resignation or even cries of outright derision.[15] Yet, given the undoubted seriousness of the issues to which the reviewer's observations relate, it would surely be unforgiveable simply to give in to cynicism of this kind and thereby allow potentially important insights on his part to be marginalised or disregarded merely on account of the language in which his critique was couched. After all, the success of any political project would seem to require some reasonably clear sense of what it is trying to achieve and how it is going to achieve it, and in ostensible recognition of that point the institutional organs of the biodiversity treaties themselves continue to invoke the notion of 'vision' in their own policy pronouncements and planning arrangements.[16]

At the same time, if his observations are to be given due weight, it will be helpful as a preliminary to pay closer attention to what we might reasonably mean when deploying the terminology of 'vision'. It is, after all, an extremely wide-ranging and multi-faceted expression, applied with equal alacrity in everyday usage to both the *process* and the *product* of visual discernment, and in literal and metaphorical fashion alike: in this way it has come to embrace everything from general cognitive photo-sensitivity to sheer hallucination,[17] incorporating along the way such notions as dreams for the future, conceptual blueprints, mission statements, and general far-sightedness or perspicacity. By way of application of the last of these senses, the expression 'men of vision' is commonly deployed (in somewhat sexist fashion, no doubt) to denote those who display extreme or unusual perspicuity in some particular respect, rather than those with merely average or commonplace capacities, or – as a simple, strictly literal interpretation would surely require – all those who are not actually blind.

It is through such curious but commonplace linguistic processes that the word has come to be associated with the bestowal of an accolade, and the ensuing scramble to

---

[14] In this vein, note the recent assertion by the Turkish President that the principal obstacle to Turkey's full membership of the EU was not, in reality, any fault or shortcoming exhibited by Turkey itself but rather 'the EU's lack of vision', reported in Anon, 'Gül Accuses EU of Lacking Strategic Vision vis-à-vis Turkey' *Today's Zaman* (Istanbul, 27 September 2010), available at <http://www.todayszaman.com/>.

[15] See further on this point RE Ellsworth, *Leading with Purpose: The New Corporate Realities* (Stanford UP, 2002), 96–97: 'Unfortunately, … in recent years "vision" has taken the form of a management fad, and the term has become so widely used as to lose the essence of its meaning. Derisively, it has degenerated into "the vision thing".'

[16] GBO-4 (n 7), for example, speaks in terms of a 'vision' for biodiversity for the year 2050.

[17] As to the latter, *see*, for example, K Williams, 'My Psychedelic, Psychotic, Psychic and Spiritual Visions' (2014), available at <http://www.near-death.com/resources/editorials/my-psychedelic-psychotic-psychic-and-spiritual-visions.html>, especially Part 3.

claim 'visionary' status has undoubtedly proved a further impediment to the achievement of clarity in usage. As one commentator has put it, with regard to the corporate context:[18]

> For many CEOs, having a vision is a prerequisite to being considered an enlightened, modern leader. Consequently, vision statements have blossomed. Company after company has drafted statements to hang on walls and to place in public relations materials disseminated to employees, shareholders and the public. The terms vision, purpose, mission, shared values and strategy are often used indiscriminately and even interchangeably.

In an attempt to rescue the concept from this morass of misapplication, he continues:

> But what is vision? Vision is not something separate from purpose, mission, strategy, and shared values. It is the quality that is ingrained in each of these that defines a desired future state of the organization resulting from the fulfilment of the purposes and the strategy to get there. As John Young, Hewlett-Packard's former CEO, observes, 'Vision is simply mission and purpose made tangible in people's minds.' It 'refers to a vivid description of what it will be like when the mission is accomplished or the purpose fulfilled.' Vision combines an unusual discernment of the competitive future with foresight and wisdom as to how the company can make a valued social contribution in tomorrow's marketplace. This discernment and foresight are woven into the substance of the company's purpose, mission, strategy and values and are not separate from these concepts.

While this arguably represents a small step forward in terms of clarification, it still seems to leave a good deal to be desired, first of all because it risks overstating the specificity with which it is possible to make assessments about the needs and realities of 'tomorrow',[19] let alone all those days which are (hopefully at least) to follow. In addition, the 'marketplace' is hardly a forum where any meaningful evaluation of future reality can be guaranteed to occur.[20] Typically, a fundamental impediment to the development of a feasible and compelling agenda for the future is the failure to establish a sufficiently clear and accurate picture of the starting point for the process – that is, of the *current* condition and functional operations of the enterprise in question, in all their complexity. Ideally, moreover, this picture would be informed by a reasonably clear sense of how all these realities came about in historical terms. Indeed, until such a detailed and authentic appreciation of the status quo has been constructed, strategic planning for the future seems likely to prove largely futile.

Where the preservation of biological diversity is concerned, one key aspect of the planning problem is that it may in some cases take only days or months, but in others possibly years or even decades,[21] before the full impacts of conservation measures become discernible on the ground, rendering it extremely difficult for the complex

---

[18] Ellsworth (n 15).
[19] On the significance of unpredictability for decision making generally, *see* J Kay, *Obliquity: Why Our Goals Are Best Achieved Indirectly* (Profile Books, 2010).
[20] This is because within the traditional economic perspective future contingencies are heavily discounted: *see* A Gillespie, *International Environmental Law Policy and Ethics* (2nd edn., OUP, 2014), chapter 3.
[21] More realistically, some complex combination of the above interludes is likely to be involved.

relationships of cause and effect to be successfully unravelled. A currently deteriorating position in the vigour or abundance of protected wildlife might as easily be the result of imprudent policies that were implemented in the past (and perhaps even in a previous century) as of those that are presently in effect, rendering it extremely unwise to treat declining populations per se as a justification for launching into some dramatic programme of normative reform.[22] A further factor is that even if the substantive thrust of conservation law and policy is well judged, it may take a very long while before effective approaches to implementation and enforcement are fully operative, especially when the key mechanism to be employed is that of the criminal law.[23] All too often, the lack of due seriousness with which such matters are treated means that offenders are encouraged to believe that they may flout the law with relative impunity.

It is only very recently that indications have become apparent that the international community is finally beginning to wake up to the full seriousness of the problems involved.[24] A further element of the difficulties to be faced here is that the majority of the principal threats to biodiversity are actually attributable to factors and policies which originate entirely from outside what is typically regarded as the conservation sector, reinforcing the idea that the solution may lie not in the reconstruction or abandonment of existing conservation regimes but in the more effective integration and consolidation of their message across the governmental machine as a whole.

2.1 Vision, Visionaries and the Risks of Revisionism

Needless to say, it would make even less sense to contemplate some very drastic upheaval in existing conservation regimes where the weaknesses that had been attributed to them were actually the result of mere misunderstanding of their true substantive purport, past evolutionary development or current modes of operation. The point cannot be over-emphasised within the context of international wildlife law because, whereas there seems actually to be no shortage of extravagant and confidently expressed proposals of a strategic character, a striking deficiency has always been apparent with regard to the establishment of anything resembling a complete, cogent and coherent picture of the current state of affairs and operations, and of precisely what

---

[22] *See*, for example, D Tilman, and others, 'Habitat Destruction and the Extinction Debt' (1994) 371 Nature 65; A Helm, I Hanski and M Partel, 'Slow Response of Plant Species Richness to Habitat Loss and Fragmentation' (2006) 9 Ecology Letters 72; M Kuussaari and others, 'Extinction Debt: A Challenge for Biodiversity Conservation' (2009) 24 Trends in Ecology & Evolution 564; A Sang and others, 'Indirect Evidence for an Extinction Debt of Grassland Butterflies Half Century after Habitat Loss' (2010) 143 Biological Conservation 1405.

[23] For reports of an interesting array of promising recent developments in the prosecution of wildlife offences in the UK, see the RSPB's investigations newsletter, *Legal Eagle*, Issue No 75, March 2015.

[24] The declaration of the UN Decade for Biodiversity is one, along with the creation in 2012 of IPBES (the Intergovernmental Platform on Biodiversity and Ecosystem Services), the initiation of the UN's Harmony with Nature project (see n 103 below) and the recent recognition of wildlife crime as a threat both to sustainable development and to international peace and security, as to which *see* E McLellan and others, *Illicit Wildlife Trafficking: An Environmental, Economic and Social Issue* (Perspectives, Issue No.14, UNEP, May 2014).

the law is, and how it came to be so. Thus, a number of the earlier evaluations of the Ramsar Convention were gravely undermined by their failure to examine with sufficient precision or rigour the way in which the regime was actually operating and evolving in practice.[25]

To take another example, when evaluating the proposition that the wildlife trade Convention, CITES,[26] would operate much more efficiently and effectively if its basic procedures were turned on their head, in the sense that the listing process should be reconceived so as to specify those species in which trade was *permitted*,[27] it helps at least to be aware that just such a proposal was specifically examined and rejected as unworkable in the Convention's early years.[28] Otherwise, one might simply end up reinventing the wheel, and very possibly a square wheel at that! Similarly, when it is repeatedly asserted by certain cadres of commentator that the very adoption of a Convention regulating the global wildlife trade was born largely of a failure to appreciate the importance of *habitat loss* as the critical overall threat to wildlife species,[29] it becomes important to understand that the historical record lends no support whatsoever to this contention.[30] Rather, it seems that uncritical enthusiasm in certain quarters for trade as a vehicle of human development has given rise to the promulgation of a largely imagined historical narrative to support the ideological predispositions of its authors.[31] As a final illustration, it has been relatively rare to encounter appraisals of the 1946 Whaling Convention that go beyond the recitation of trite, and largely inaccurate, suppositions about its origins and objectives.[32]

The tendency towards glib revisionism where history is concerned is arguably no more strongly evident than in widely espoused (though generally under-evidenced) contemporary claims that the global conservation effort has traditionally been driven essentially by 'Romantic' thinking. It is troublesome enough to encounter such claims

---

[25] *See*, for example, D Farrier and L Tucker, 'Wise Use of Wetlands under the Ramsar Convention: A Challenge for Meaningful Implementation of International Law' (2000) 12 JEL 21, in response to which the Ramsar Bureau took the unusual step of issuing a public rebuttal. This comprised a letter to the editor of the journal in which the article was published and an internal paper commissioned by an Australian wetlands expert exposing the misapprehensions it was believed to contain.
[26] 1973 Convention on International Trade in Endangered Species of Wild Fauna and Flora, 993 UNTS 243.
[27] E Couzens, 'CITES at Forty: Never Too Late to Make Lifestyle Changes' (2013) 22 RECIEL 311.
[28] *See* CITES Conf. 3.21; Docs. 3.30, 3.30.1; ML Ditkof, 'International Trade in Endangered Species under CITES: Direct Listing vs Reverse Listing' (1982) 15 Cornell International Law Journal 107. Couzens, *ibid.*, acknowledges the point only in his conclusions, at 322, but suggests that circumstances have changed since then.
[29] *See*, for example, J Hutton and B Dickson (eds.), *Endangered Species, Threatened Convention: The Past, Present and Future of CITES* (Earthscan, 2000), xv, 47, 129.
[30] For discussion, see MJ Bowman, 'A Tale of Two CITES: Divergent Perspectives upon the Effectiveness of the Wildlife Trade Convention' (2013) 22 RECIEL 228, 233–235.
[31] *Ibid.*
[32] For an attempt to set the record straight, *see* MJ Bowman, '"Normalizing" the International Convention for the Regulation of Whaling' (2008) 29 Michigan Journal of International Law 293.

in writings in the literary field,[33] but more disturbing by far when they appear to be advanced by those whose work might impact more directly upon the actual application and development of conservation policy.[34] Romanticism was, after all, primarily a literary and artistic movement rather than one driven by philosophers[35] or politicians,[36] and its principal impact upon human attitudes to nature has been manifest more in the weekend musings of the leisured classes than in the workaday management of the natural landscape. The latter, in truth, has always been driven far more strongly by the cold, controlling, mechanistic and utilitarian thinking associated with the earlier age of so-called reason and enlightenment, to which Romanticism represented no more than a lively but limited backlash.[37] Thus, the traditional approach to 'sustainable' forest management has convincingly been traced back to 18th-century Prussia, where it was driven not by starry-eyed engrossment with all things silvestrian but by the urgent, ongoing demand for timber for the mining industry.[38] Accordingly, any glimmers of genuine holistic understanding that were revealed by the original architects of this brand of conservation policy were rather rapidly submerged in practice by more mundane and materialistic impulses,[39] leading ultimately to the establishment of vast

---

[33] Note in this vein the assertion that 'Romanticism forms the base from which Western society, especially American, views its natural surroundings': KA Mingey, *New Romanticism* (University of Montana Scholar Works, 2007), 2.

[34] *See*, for example, P Kareiva, M Marvier and R Lalasz, 'Conservation in the Anthropocene: Beyond Solitude and Fragility' (2011) <http://thebreakthrough.org/index.php/journal/past-issues/issue-2/conservation-in-the-anthropocene>; K Willis and C Fry, *Plants: From Roots to Riches* (John Murray, 2014), especially chapter 24. The first- and third-named authors of the former work are respectively Chief Scientist and Director of Science Communications at the Nature Conservancy, an influential US pressure group which pursues a global environmental programme, while the first-named of the latter is Director of Science at the Royal Botanic Gardens at Kew, in England, which fulfils a host of important functions in connection with the UK's national and international conservation commitments.

[35] Needless to say, it had its philosophical champions, amongst whom Rousseau, Goethe and (more controversially) Nietzsche would typically be numbered, though – as Richard Tarnas points out in his admirable and compendious study of Western thought – in Romanticism generally, and in Nietzsche in particular, the philosopher was effectively transformed into poet: R Tarnas, *The Passion of the Western Mind* (Pimlico, 1991), 371.

[36] No doubt it had its impact on certain politicians, most notably Teddy Roosevelt in the US. It also proved influential in Germany, where aspects of Nazi ideology drew heavily upon Romantic thought: *see*, for example, P Viereck, *Metapolitics: From Wagner and the German Romantics to Hitler* (expanded edn., Transaction Publishers, 2003).

[37] It has also been pointed out that over the course of time Romanticism came to embrace a powerful sense of *alienation* from nature: *see* Tarnas (n 35), 376. This was, moreover, largely attributable to the extent to which the mechanistic, enlightenment perception of the natural world had contrived to prevail.

[38] The term *Nachhaltigkeit* (i.e., sustainable forest management) was coined by Hans von Carlowitz, whose 1713 work *Sylvicultura Oeconomica* became the 'bible' in this field. Its author was not actually a biologist, however, but a mining administrator desperate to maintain the supply of timber for the mines: *see*, for example, JB Ball, 'Global Forest Resources: History and Dynamics' in J Evans (ed.), *The Forests Handbook, Volume 1: An Overview of Forest Science* (Blackwell Science, 2001).

[39] On the profound impact in other regions of the resulting theory and practice of forest management, *see*, for example, TJ Straka, 'Evolution of Sustainability in American Forest

arboreal monocultures that were supposedly far more 'efficient' than anything nature itself could produce, but have in reality all too often proved to be the graveyards both of biological diversity and of the more traditional forms of community benefit that can be derived from nature.[40] The engineers of this particular revolution, unfortunately, could only see the trees for their wood.

It would certainly present a major challenge to discern any trace of Romantic motivation behind the earliest legal measures regarding wildlife to be adopted at the international level, which were primarily concerned with conserving directly exploitable resources such as fish, fur seals and whales, protecting birds 'useful to agriculture', or countering the depredations of invasive alien species such as *Phylloxera vastatrix*, the aphid scourge of European vineyards.[41] Can it seriously be contended that those who, in the solitary preambular recital to the (now long superseded) Whaling Convention of 1937,[42] declared the conservation of whale stocks to be designed solely 'to secure the prosperity of the whaling industry' were really just a bunch of old romantics at heart? To reinforce the point, it might be noted that few creatures have seemed more powerfully evocative to poets, especially during the Romantic era, than birds of prey, which were celebrated as symbols not merely of freedom, but of speed, strength, visual acuity and closeness to heaven, albeit with more than a hint of menace.[43] Yet none of this aura availed them one jot in the early wildlife treaties, for the purposes of which they were typically not merely excluded from protection but positively earmarked for persecution.[44] Accordingly, when the Scottish naturalist Seton Gordon produced his 1927 paean to the Aquila *Days with the Golden Eagle*, it was

---

Resource Management Planning in the Context of the American Forest Management Textbook' (2009) 1 Sustainability 838; C Lang and O Pye, 'Blinded by Science: The Invention of Scientific Forestry and Its Influence in the Mekong Region' (2000–2001) 6 Watershed 25.

[40] Lang and Pye, *ibid.*; R Ziegler, 'Crooked Wood, Straight Timber – Kant, Development and Nature' (2010) 2 Public Reason 61.

[41] See generally *Lyster*, chapter 1.

[42] 1937 International Agreement for the Regulation of Whaling, 190 LNTS 79.

[43] The collection *Lyrical Ballads* (1798), which is generally regarded as having triggered the British Romantic movement, was inspired by the period spent by Coleridge and the Wordsworths living in Somerset, where the high heathland habitat of the Quantocks is noted for its birds of prey. Perhaps the best-known explicit allusion lies in Tennyson's short poem 'The Eagle' (1851–1854), though the albatross in Coleridge's own 'Rime of the Ancient Mariner' (1798) might be considered a marine equivalent (especially since it seems that part of the inspiration for this epic poem was an incident in which Coleridge himself witnessed the wanton killing of a hawk in the course of a voyage to Malta). More generally, references to raptors abound throughout the genre, in Keats, Hemans and others: for details, *see* D Wu, *Romanticism: An Anthology* (4th edn., Wiley-Blackwell, 2012), 339–357, 1329–1336, 1370, 1397, 1429, 1439, 1457. In North America, Longfellow is said to have glorified the 'unfettered self' with 'insistent images of birds of prey': A Kolodny, *In Search of First Contact* (Duke UP, 2012), 162.

[44] *See*, for example, the 1900 Convention for the Preservation of Wild Animals, Birds and Fish in Africa, 94 BFSP 715, Article II(13) and Table V; 1902 Convention for the Protection of Birds Useful to Agriculture, 102 BFSP 969, Article 9 and Schedule II. In similar fashion, the 1916 and 1936 bilateral migratory bird species treaties concluded by US with Canada (originally GB), 39 Stat 1702; USTS 628, and Mexico, 178 LNTS 309, respectively did not include any birds of prey in their lists of protected species, though some were added to the latter by virtue of the 1972 amendments, 23 UST 260, TIAS 7302.

considered 'a brave book to write', on account of 'the Victorian ethos that every bird with a hooked bill should be destroyed'.[45]

It would, of course, be foolish to deny the impact that Romanticism has exerted on the development of environmental ethics, especially in North America, or that the Romantic ideal of communion with wild nature has on occasion been translated into statutory language – thus, the Act establishing Yellowstone National Park declared it to be 'dedicated and set apart as a public park or pleasuring ground for the benefit and enjoyment of the people' and provided for the strict control of exploitation of natural resources and the removal of anyone who attempted to settle there.[46] Yet even the 80 million acres managed by the National Park Service represents only a tiny fragment of the total land area of the United States, and is in any event dwarfed by that controlled by the US Forest Service (193 million acres) and the Bureau of Land Management (248 million), both of whose mandates embrace sustainability, but whose focus is much more heavily upon resource exploitation.[47] It would therefore be gravely mistaken to assume that Romanticism had been the predominant motivation for conservation generally, even in the US. Within the international legal order, moreover, references to aesthetic and recreational considerations did not make a prominent entry until the 1940s, and even then only in the Americas and as an overlay upon more narrowly utilitarian considerations.[48] Such thinking was not really embraced by the international community as a whole until the adoption of the 1972 World Heritage Convention,[49] though that instrument, for all its undoubted importance, can hardly be regarded as the centrepiece of the global conservation effort.[50]

Furthermore, it should never be forgotten that such thinking is plainly no less *anthropocentric* in orientation than the purely materialistic reasoning that historically preceded – and comfortably survived – it, since both are essentially self-regarding and self-serving in character. It is simply that one approach attends to the needs of the body

---

[45] D Cobham, *A Sparrowhawk's Lament: How British Birds of Prey are Faring* (Princeton UP, 2014), 189.

[46] Yellowstone Act, 1872 (17 Stat. 32).

[47] *See* RW Gorte and others, *Federal Land Ownership: Overview and Data* (Congressional Research Report R42346, 8 February 2012).

[48] *See* the 1940 Convention on Nature Protection and Wildlife Preservation in the Western Hemisphere, 161 UNTS 193. Amongst the various types of protected area envisaged by Article 1 were not only national parks, but 'national reserves' – i.e., 'regions established for conservation and utilization of natural resources under government control, on which protection of animal and plant life will be afforded in so far as this may be consistent with the primary purposes of such reserves'. No explicit motivation was given in Article 5, which addressed conservation *outside* protected areas, while Article 7 called for the protection of migratory birds 'of economic or aesthetic value', or to prevent outright extinction. For further discussion, *see Lyster*, chapter 8.

[49] 1972 Convention for the Protection of the World Cultural and Natural Heritage (1972) UNJYB 89; 11 ILM 1358. The convention's adoption marked the centenary of the establishment of Yellowstone. For discussion, *see Lyster*, chapter 14.

[50] This assessment stems largely from the fact that, although its substantive scope (as defined by the concept of natural heritage) is potentially very wide, its primary focus has always been upon listed sites, which in the case of natural heritage extends to fewer than 200 globally, together with around 30 of mixed natural/cultural significance.

and the other to those of the spirit. The key point here, indeed, is to understand just how much the Romantic perspective actually shared with its supposedly contrasting Enlightenment predecessor:[51]

> Both tended to be 'humanist' in their high estimate of man's powers and their concern with man's perspective on the universe. Both looked to this world and nature as the setting of the human drama and the focus for human endeavour. Both were attentive to the phenomena of human consciousness and the nature of its hidden structures. Both found in classical culture a rich source of insight and values. Both were profoundly Promethean – in their rebellion against oppressive traditional structures, in their celebration of individual human genius, and in their relentless quest for human freedom, fulfillment [sic], and bold explanation of the new.

The notion that wildlife might merit protection *for its own sake*, entirely independently of human needs or interests, is a different matter altogether, being grounded in what has been described as nature's *intrinsic* (as opposed to instrumental or inherent) value.[52] This notion did not make an unambiguous appearance until much later, in the Council of Europe's 1979 nature conservation convention,[53] the focus of which is essentially restricted to *European* wildlife,[54] and the World Charter for Nature three years later,[55] which lacks legally binding effect. It was not, in fact, until the entry into force of the Biodiversity Convention a little over 20 years ago that the idea can be seen to have taken firm root in international law generally, and even then some eminent commentators were casting doubt upon its practical legal significance.[56]

Even though these doubts seem largely unjustified, it should be clear to everyone that the notion that nature has an independent, intrinsic value of its own represents merely the final, most recent strand in a complex skein of pluralistic justifications for conservation, the other components of which are much longer and more securely established. Indeed, whatever should be regarded as representing *in principle* the current philosophical underpinnings of global conservation measures, there are still undoubtedly some areas of human interaction with nature into which the concept of intrinsic value has yet to make any significant headway *in practice*: the most obvious instance concerns the realm of fisheries, and some may feel that it is unlikely to be a coincidence that it is here where the failure of conservation endeavours has been most

---

[51] Tarnas (n 35), 366. It will be noted that in their chapter in this present work, Fosci and West do not even consider it worthwhile to distinguish the two forms of anthropocentric justification.

[52] *Lyster*, chapter 3; Fosci and West, Chapter 2.

[53] 1979 Bern Convention on the Conservation of European Wildlife and Natural Habitats, ETS 104.

[54] While this much may seem evident from the title alone, the matter is in reality a little more complicated; for an explanation, *see Lyster*, 323–327.

[55] For the text of the instrument, together with its drafting history and commentary, *see* WE Burhenne and WA Irwin (eds.), *The World Charter for Nature* (2nd rev edn., Erich Schmidt Verlag, 1986).

[56] *See*, for example, PW Birnie, AE Boyle and CJ Redgwell, *International Law & the Environment* (3rd edn., OUP, 2009), 618.

14   *Research handbook on biodiversity and law*

profound, prolonged and pervasive.[57] Thus, any suggestion that it is the idea of 'protecting nature for its own sake' that 'has not worked',[58] and that we should accordingly turn to a more anthropocentric and utilitarian perspective, must count as one of the most muddle-headed and misconceived proposals of modern times.

Indeed, far too many of the more trenchant criticisms of existing treaties and strident calls for 'reform' – and especially those which tout the merits of so-called 'economic' solutions – seem ultimately to be grounded in the desire to advance the personal emotional and ideological preoccupations of their authors rather than conservation objectives as such, and tend for that reason to overlook or misrepresent crucial aspects of the contemporary legal regime and the way in which it came to be as it is.[59] While such solutions may undoubtedly have a place in the overall scheme of things, any attempt to present them as a universal panacea needs to be resisted and recognised for what it is – namely the product of a very singular perspective on human affairs, and one which has repeatedly failed to deliver the benefits it has so loudly and fervently promised.[60]

These failures, of course, extend far beyond the field of conservation itself to permeate every aspect of human affairs: thus, by the turn of the millennium, economists of the dominant neo-classical school had become so intoxicated by their own insistence that deregulated, market-driven policies would make the familiar boom-and-bust cycle of the global economy a thing of the past that none of them actually foresaw the all-too-obviously impending crash of 2008, the formal prediction of which was left to the few remaining mavericks within the discipline.[61] The susceptibility of the majority within this profession to becoming so easily blinded by the limitations of their own, still regrettably dismal, science[62] plainly offers an important lesson regarding the nature of vision in this context, which is that where it is only of the 'tunnel' variety it will all too readily become an encouragement towards detachment from reality, the engine for

---

[57] For recent assessments, *see Lyster*, chapter 5, part 6; R Barnes, 'Fisheries and Marine Biodiversity' in M Fitzmaurice, DM Ong and P Merkouris (eds.), *Research Handbook on International Environmental Law* (Edward Elgar Publishing, 2010); P Sands and J Peel, *Principles of International Environmental Law* (3rd edn., CUP, 2012), chapter 9, especially 447–448. McLellan and others (n 24), note that the scale of illegal fishing amounts to almost 20% of the total global catch, and is worth around 10 billion euros per year.

[58] Kareiva and others (n 34).

[59] For discussion of this problem in relation to CITES, for example, *see* Bowman (n 30).

[60] For a helpful recent exploration of these issues, *see* A Wiersema, 'Uncertainty and Markets for Endangered Species under CITES' (2013) 22 RECIEL 239–250.

[61] *See* S Keen, *Debunking Economics: The Naked Emperor Dethroned* (rev edn., Zed Books, 2011). For a variety of other valuable perspectives upon this sorry saga, *see* GA Akerlof and RJ Shiller, *Animal Spirits* (rev edn., Princeton UP, 2010); G Tett, *Fool's Gold* (Abacus, 2010); Ha-Joon Chang, *23 Things They Don't Tell You About Capitalism* (Penguin, 2011); R Layard, *Happiness* (2nd rev edn., Penguin, 2011).

[62] Chang, *ibid.*, has recently suggested in an interview that his discipline currently has more in keeping with science fiction than with genuine science: *see* M Reisz, 'Believing the Unbelievable: Where Adam Smith and Agent Smith Meet', *Times Higher Education* (London) 26 March 2015.

self-serving revisionism and the enemy of perspicacity of the genuine kind.[63] Accordingly, it is necessary to be alert to the risk that, while the adoption of a 'vision' may help to give purpose and direction to any particular project, this must always be balanced against the countervailing risk that it may merely constitute the formal enshrinement of particular prejudices and preconceptions, and on that account an active impediment to progress of a genuine kind, perhaps even precipitating the ultimate derailment of the entire enterprise.

Seemingly the latest in this long line of 'visionary' distractions entails the declaration of a putatively new geological era, the 'Anthropocene', in which we finally force ourselves to acknowledge the 'fast, deep and long-lasting effects that humans have on the planet' and indeed that we have now become 'the dominant force of change on Earth'. We must, it seems, recognise that 'humans are not an outside force perturbing an otherwise natural system but rather an integral and interacting part of the Earth system itself'. Accordingly, we 'need to start adjusting our lifestyles to nature – and then to turn our human systems into nature'.[64] Yet while there is not much of substance to quarrel with here, there is surely also decidedly little by way of novelty, since one could find an exposition of almost exactly the same set of ideas in the preamble to the Stockholm Declaration on the Human Environment,[65] concluded over 40 years ago!

Perhaps this propensity towards reinvention of the rhetorical wheel should simply be taken as confirmation of a strongly cyclical tendency within human affairs generally: it might even, indeed, represent something to celebrate, if only there were evidence that this latest upsurge of visionary zeal were to have resulted in the creation of a more effective rallying cry for contemporary conservation efforts. Unfortunately, however, the evidence is rather to the contrary. Indeed, the new slogan, as even its latest champion is forced to concede, suffers from the all too obvious weakness that it has already been misinterpreted as a reaffirmation of 'our right to rule the natural world for narrow human interests',[66] the very opposite of what was actually intended and what is now so clearly and urgently required. Yet surely it should not have taken much imagination to realise that any attempt to declare the onset of the 'Anthropocene' would be interpreted simply as a call for yet more anthropocentrism, especially when works published by way of advocacy of this idea are disseminated under the headline 'Time to Play God'.[67]

---

[63] For the latest example of recognition by a financial system insider of the damagingly 'blinkered' nature of modern economics, *see* 'Sackcloth and Ashes on Threadneedle Street', an interview with Bank of England research chief Andy Haldane, *New Scientist* (London) 28 March 2015, 28–29.

[64] C Schwägerl, 'Time to Play God', *The Independent* (London) 25 February 2015, heralding the impending publication of his book *The Anthropocene – The Human Era and How it Shapes the Planet* (Synergetic Press, 2015). For the origins of this idea, *see* PJ Crutzen, 'Geology of Mankind: the Anthropocene' (2002) 415 Nature 23; W Steffen, PJ Crutzen and JR McNeill, 'The Anthropocene: Are Humans Now Overwhelming the Great Forces of Nature?' (2007) 36 Ambio 614.

[65] UN Doc A/CONF/48/14/REV.1.

[66] Schwägerl (n 64).

[67] Though it should, perhaps, be acknowledged that the author himself may possibly have had no personal involvement in the choice of this title.

More worryingly still, these arguments display the all-too-familiar tendency to create a largely imagined historical and operational narrative to demonstrate (or inflate) the importance of the supposed new perspective. Thus, on the basis of a visit to a meeting of the CBD CoP in Nagoya, where the question of protected areas was discussed, the author of the piece in question interprets the existing global conservation regime as being based upon a desire to protect only nature which is 'pristine' and 'untouched', and believed on that account to be 'real'. Yet, however important the maintenance of nature reserves might be, he intones, this will not of itself be enough, as in the new era we will need to recognise, protect and even reconstruct nature everywhere if we are ultimately to flourish or even survive.[68] It seems somehow to have escaped his attention that the CBD from the outset established as its primary objective the *conservation of biological diversity* generally, whether wild or domesticated, and wherever it might be found; furthermore, as the substantive provisions explicitly confirm, this must necessarily include all the areas *beyond* nature reserves and even involve purely man-made facilities created specifically for the purpose.[69] Indeed, some 20 years before the CBD's adoption, the Ramsar Convention had introduced a system of protection for all wetland types, whether natural or man-made, and without regard to whether or not they had formally been accorded the status of a nature reserve in national law.[70] Accordingly, it enshrined as its guiding mantra the principle of 'wise use', deliberately eschewing any policy that might be described as 'hands-off'.[71] Several decades even before that, moreover, the Western Hemisphere Convention, while calling for the establishment of a graduated network of conservation areas – from those exhibiting strict wilderness values to those seen primarily as reserves set aside for resource exploitation – had in fact sought to protect and preserve wildlife both within and beyond them.[72]

2.2 Vision and Virtue in Legal Scholarship

In the light of the recurrent tendency amongst commentators to lose touch with historical reality in the course of these 'visionary' outpourings, it is surely to be regarded as one of the key virtues of Lyster's monograph when it originally appeared in the mid-1980s that it seemed so resolute in its determination to resist all temptations towards tub- or table-thumping,[73] and the drift into speculative, idiosyncratic or fanciful prescriptions for future action, to which this subject-matter so readily lends itself. Instead, the author's principal aim was to focus on providing a suitably detailed account of the workings of the international legal machinery of conservation that would simultaneously be lively, cogent and succinct, and yet at the same time scholarly, sober and restrained, and might through the exhibition of these qualities ultimately emerge as both engaging and authoritative. Through his success in capturing these virtues, he was

---

[68] Schwägerl (n 64).
[69] Note in particular Articles 8(c), (d), (e), (f), (h), (i), (j), (k), (l), 9.
[70] Note in particular the implications of Articles 3 and 4.
[71] The point was explicitly addressed in Lyster's own first edition, 206.
[72] 161 UNTS 229; *see* especially Articles 5, 7, 8, 9.
[73] It will doubtless have been of assistance in this regard that the work in question had its origins in a PhD thesis undertaken by the author within the Lauterpacht Centre for International Law at the University of Cambridge.

able to lay the intellectual foundations for the development of a viable new jurisprudence of international nature conservation, which had been sadly lacking up to that point. There had, admittedly, previously been isolated works which addressed selected aspects of the subject, but seemingly nothing which could be regarded as having addressed the full range of its subject-matter and drawn together the various strands into a coherent whole,[74] through the identification of the key legal principles, practices and procedures that permeated this fast expanding area of the law, as applied systematically in relation to each of the key modern treaties that populated the field. It was, indeed, really only from that point on that international wildlife law could lay credible claims to becoming a meaningful discipline at all, as commentators began to treat his work as a kind of intellectual base-camp for the serious and scholarly exploration of the many more specialised nooks and crannies of the subject-matter.[75] It was, moreover, not merely out of deference to Lyster personally, but precisely because his basic approach was thought still to have so very much to commend it, that it was so readily embraced by the authors of the second edition.[76] In that sense, therefore, the aim was very much to preserve the 'vision' of calm, coherent and conscientious analysis by which the original work was inspired.

Insofar as any attempt was made to modify this general approach in the new edition, it was only to the extent of seeking to contextualise the subject-matter a little more fully, through the insertion of entirely new sections at either end of the work. The introductory suite of chapters sought specifically to explain in greater depth and detail the historical development and philosophical underpinnings of international wildlife law, as well as its place in the wider global legal order. As to the former aspects, their critical importance to the development of a coherent picture of wildlife conservation has already been addressed in the discussion above, as indeed has the extent to which they have proved a fertile breeding ground for misapprehension and misrepresentation. As regards the latter, a key point is that many, and perhaps even the majority, of those who require to acquaint themselves with the operations of the major wildlife treaty regimes are likely to approach the subject from a position of relative unfamiliarity with the intricacies of the international legal system as such: their own background and expertise may lie, for example, in the realms of science or policy, or in the purely national dimension of conservation law. Accordingly, it was judged desirable to provide a brief overview of this broader legal backcloth, if only as a means of highlighting certain of its crucial aspects and directing readers to some of the more valuable sources of information about them. It may be, indeed, that it is not only the uninitiated who may benefit from a degree of exposure to such matters, since even those who can lay

---

[74] Perhaps the nearest attempt was SS Hayden, *The International Protection of Wildlife* (Columbia UP, 1942, repr 2007), though its orientation was largely North American and its scope, given the time of its publication, inevitably limited. R Boardman, *International Organization and the Conservation of Nature* (Macmillan,1981) had provided valuable insight into the political and institutional aspects of the subject, while MJ Bean's *The Evolution of National Wildlife Law* (first published 1977; 3rd edn., Praeger, 1997) had offered a useful domestic perspective.
[75] The extraordinary dearth of periodical literature available for citation in the original work, by comparison with the revised edition, is perhaps one of its most noticeable features.
[76] Only time will tell, of course, if the same level of success has been achieved.

claim to genuine expertise in this area may fall victim to perceptual shortfall on occasion. The point is particularly pertinent with regard to the review essay under discussion, for there are grounds for suggesting that the perspective of its author with regard to the existing realities of the international legal order seem a little blinkered or blurry here and there, even with regard to certain matters of a fairly fundamental kind.

3. PERCEPTIONS OF THE GLOBAL CONSTITUTIONAL ORDER

Any attempt to explore the workings of the international legal system must begin with the recognition that it is scarcely possible to conceive of a legal order without reference to the particular polity, or political community, to which it relates – legal norms must, after all, be created by someone, and with a view to regulation of the conduct of certain specified actors, as well as being motivated by certain needs or interests which they are intended to protect. In many cases, of course, they are designed to reconcile perceived conflicts of interest within the community in question, or to coax its members towards unified or at least harmonised approaches to significant problems. It is in that specific respect that certain of the comments made by Professor Wandesforde-Smith appear particularly problematic. Citing a passage from the original 1985 edition of *International Wildlife Law*,[77] he asserts that it

> is no longer useful … to speak of an international community as the chief source of pressure on States to protect wildlife and to see to it that there is 'somebody in government … doing something every working day' to make wildlife law and to make it effective. The idea of 'an international community' long ago fractured and multiplied into a profusion of international actors, the current number and variety of which far exceed anything Lyster had experienced when he put together his book.

He proposes instead that the expression be replaced by the term 'international environmental movement', which, he suggests, 'implicates a far more complex reality'.[78] It is, however, not at all easy to see how that could possibly be so, since it is surely more plausible to suppose that it is the entire political community within which international law is operative which represents the broader and more complex reality, of which 'the international environmental movement' (however that might be interpreted) is merely one element, functioning alongside the many other comparable movements devoted to alternative objectives, such as peace, disarmament, human rights, free trade, economic development and so on.

3.1 Reflections on the Nature of 'Community'

The passage in question also implies that it is somehow inconsistent with the essential nature of a 'community' for it to be 'fractured and multiplied into a profusion of actors', but it is again difficult to understand the basis for any such assumption. Assuredly, the expression 'community', like its Latin counterpart *communitas*, is often

---

[77] Wandesforde-Smith (n 11), 85, citing Lyster (n 10), 181.
[78] Wandesforde-Smith (n 11), 86.

used to convey some notion of shared purpose, conviction or 'agreement',[79] but there can be no justification for insistence upon that interpretation exclusively, since it is best seen as only a secondary application of the word, rather than reflecting the irreducible core of its meaning.[80] Indeed, the attribution to the word 'community' of a strong or reinforced sense of 'oneness' may possibly derive from the misconception that its Latin precursor represented a compound of the prepositional prefix *com-* (signifying 'with', or 'together') with the abstract noun *unitas* (unity). The true derivation of the word *communitas* (and hence the original thrust of the sense of sharing that it conveys) is, however, entirely different and altogether more subtle, since the word in reality represents a compound of *com-* with the plural noun form *munia*, alternatively rendered as *moenia*, a word conventionally deployed in the former spelling to signify public duties or functions and, in the latter, the walls or fortifications of a city.[81] Oddly, these two ostensibly very different meanings may well be mere semantic variants of the same basic idea, the verb *moenio* or *munio* meaning to build, prepare or make provision.[82]

While it therefore cannot be excluded that it was some sense of the need for, or the emergence of, public functions which originally underpinned the concept of 'community', the idea which tends more commonly to be emphasised in classical etymological references is that of co-location or confinement within defined boundaries.[83] Although the walls of a city represent a convenient paradigm here, the natural linguistic drift towards figurative usage inevitably results in the application of the same term regardless of whether the boundaries in question are actually man-made or even assume physical form at all. Accordingly, other commonly encountered English definitions include simply 'the public in general' and 'a body of persons in the same locality'.[84] If this group is of any size (and especially if it persists for any length of time) it seems almost inevitable that it will come to include numerous different sub-groups, factions and alignments within it.

In this vein, it is noteworthy that within the specialised lexicon of the life sciences, the term 'community' is used simply to indicate a group of wildlife populations that

---

[79] *See* C Schwarz and others (eds.), *Chambers English Dictionary* (1988); DP Simpson (ed.), *Cassell's New Latin Dictionary* (5th edn., 1968). Certainly the etymologically related word 'common' tends to be applied in that way, in international law as in more general usage: thus, the legal concepts of common property, common heritage and common concern all reflect that sense of universality of interest.

[80] For a singular (and seemingly extreme) view, note the assertion that '*communitas* is utterly incapable of producing effects of commonality, of association and of communion': R Esposito, *Communitas: The Origin and Destiny of Community* (Stanford UP, 2010), 140. For a justifiably sceptical review of this work, *see* M Cohen, (2012) 18 Common Knowledge 377.

[81] *See* Simpson (n 79). Also connected is *munus*, meaning some gift or service offered out of duty.

[82] *Ibid.* Thus, the two versions of the word can be taken respectively to connote the literal/physical and metaphorical/normative bulwarks and buttresses on which society depends. The compound verb *communio* meant to surround with fortifications.

[83] *Ibid.* It could, no doubt, be argued that, in the human case and in the long term at least, the one is a natural if not inevitable concomitant of the other.

[84] Schwarz and others (n 79).

inhabit the same area, and engage in interaction with each other.[85] Since these will inevitably include both predators and prey, any presumption of unity of purpose or perspective seems implausible to say the least. Instead, an ecological community must be viewed as a variety of complex system, an 'assemblage of organisms bound into a functional whole by their mutual relationships'.[86] Within this system, a kind of homeodynamic balance is preserved not by unity of purpose but by the richness and diversity of these interactions, which continuously recycle energy within the system and thereby forestall the drift into ultimate stillness, sterility and extinction of order that would otherwise represent the inevitable consequence of the operation of the second law of thermodynamics.[87]

That is not to suggest, of course, that the idea of community necessarily *excludes* the possibility of unity of purpose or perspective, since it will always remain possible, certainly in the human case, to create a physical, social or institutional enclave for the specific purpose of attracting a congregation of the like-minded, as for example in the case of a sports club, pressure group or religious retreat. It is also clear that the term has sometimes been explicitly employed in precisely that sense in international law, as in the case of the European Communities, ECOWAS or CARICOM.[88] Yet history suggests that even in cases such as these the task of maintaining the proclaimed state of unified purpose, let alone building effectively upon it, will prove challenging in the extreme. Nor will the deployment of some alternative label, such as 'union', to describe the organisation in question necessarily have the effect of enhancing the sense of motivational cohesion, for all its pretentions in that direction.[89] Furthermore, in what must surely be regarded as the more typical case of a political community, there will be no antecedent universal purpose at all, since the polity in question will only have come about in the first place as the result of mere historical happenstance rather than through an exercise of conscious planning on anyone's part. Undoubtedly, the chances of achieving a tolerable *modus vivendi* within the group as a whole are sure to be increased if all the various individuals and factions of which it is composed can somehow work their way towards achieving a shared sense of purpose and policy with regard to certain crucial questions; such concordance is, however, essentially a *political aspiration* of the community in question rather than a *constitutional precondition*.

---

[85] An ecosystem is commonly defined to comprise this biological community, taken together with the physical environment in which it is located: *see*, for example, A Allaby and M Allaby, *The Concise Oxford Dictionary of Earth Sciences* (OUP, 1990); C Park, *Oxford Dictionary of Environment and Conservation* (OUP, 2008).

[86] F Capra and PL Luisi, *The Systems View of Life: A Unifying Vision* (CUP, 2014), 67, with further elaboration at 341–351.

[87] On the second law of thermodynamics, which stipulates that the entropy of a closed system progressively increases with time, *see* J Daintith (ed.), *Oxford Dictionary of Physics* (6th edn., OUP, 2009), 546–547. 'Entropy' is a measure of the unavailability for work of the energy within a system.

[88] *See* respectively the 1973 Treaty establishing the Caribbean Community 947 UNTS 17 (later revised); 1975 Treaty of the Economic Community of West African States (1975) 14 ILM 1200 (later revised).

[89] For conspicuous examples, note the constituent treaties of the European Union, OJEU C 326/13, 26.10.2012, and African Union, text available from <http://www.au.int/en/content/constitutive-act-african-union>.

Where reference is made, moreover, to 'the international community' in a general political sense, it is clear that we are speaking of a polity of this latter kind, since it has come into being over the course of time simply as a casual by-product of the progressive intensification, diversification and globalisation of human affairs. It is also a highly complex, multi-faceted and pluri-dimensional entity, since the jumble of interlocking constituencies of which it is composed may validly be described and defined by reference to a multitude of highly diverse criteria, including those of a geo-political, ethnic, linguistic, socio-cultural, religious, ideological, occupational, recreational or even purely functional kind.[90] Whereas the multifarious activities associated with these different constituencies might once have been conducted predominantly within local or national boundaries, that has long since ceased to be the case, and the only encircling rampart or 'earthwork' by which they are now effectively contained is that represented by the planetary boundary itself.[91] That is certainly the reality in the case of the 'international environmental movement', and is no less true of the many other comparable constituencies mentioned above.

3.2 The International Community and the Environment

In the light of these considerations, the real challenge here, it might be suggested, is in the first instance to ensure that the voice of the environmental movement is not drowned out by the cacophony of claims that continuously emanate from these various other factions, as part of the ongoing process of competitive clamour that is played out amongst them. More ambitiously, however, and in the longer term, the aim must be to ensure that its message is actually absorbed into and effectively integrated with their own, in the sense that full and proper allowance is made for the basic principles of biodiversity conservation in the governance of operations that occur within and across the fields of economic development, international trade and finance, human rights, armed conflict and so on. In point of fact, this process is already under way, since formal recognition has long been accorded to conservation needs (if only, in some cases, in the form of lip service) in the formulation of the substantive legal norms by which these various areas of human endeavour are governed, above all through the principle of sustainable development.[92] What is now required is that this allowance be

---

[90] One notable example of a global community perceived from an exclusively *functional* perspective is the Universal Postal Union, founded originally in 1864 as the General Postal Union. The preamble to its current, 1964, constitution (as to which, see its website at <http://www.upu.int/>) defines the organisation as a 'union' of a 'universal' character designed solely to 'facilitate communication' through a system of 'inter-connected networks' of postal services potentially embracing all 'the inhabitants of the world'. Although this network is said to be designed with a view to 'contributing to the attainment of the noble aims of international collaboration in the cultural, social and economic fields', it would in fact persist and function even if the majority of the communications it permitted proved in practice to be fractious and divisive!

[91] Certain activities, of course, such as space travel, transcend even that limitation.

[92] Thus, the preamble to the 1994 Agreement establishing the World Trade Organisation, 1867 UNTS 3, reaffirms its goal of expanding trade and allowing for the optimal use of resources 'in accordance with the objective of sustainable development, seeking both to protect

transformed into full and effective implementation and endorsement at every level and on a systematic basis.

A significant spur to this process has very recently been provided by the explicit recognition of the need to reduce biodiversity loss as a key element of the 2015 Sustainable Development Goals,[93] following a much more muted reference in the Millennium Development Goals agreed by the UN in 2002.[94] It would seem that, where the environment is concerned, the practical dictates of living within a community in the original sense of the word may eventually give rise to the recognition of a community of interest in the extended, Wandesfordian, sense (i.e. *commonality*), for the simple reason that the earth's fundamental life support systems, although robust, are not endlessly resistant to unconstrained abuse, and collective action will therefore be necessary to preserve them. Accordingly, although the 'pressure on States to protect wildlife' may originally have emanated from what might be referred to as the 'international environmental movement', those particular cudgels would now seem to have been taken up, if only somewhat tentatively as yet, by the wider political community as a whole.

It is also important not to overlook the fact that this trend has been reflected in a series of highly significant individual initiatives from a wide variety of different sources. Wandesforde-Smith himself identifies 'the United States, the European Union, Canada, South Africa, Australia and New Zealand' as entities that have 'over recent decades … for one reason or another and in some policy areas more than another in each case, shown exceptional leadership' in the conservation field.[95] Yet it might be questioned whether he has been sufficiently alert here to avoid falling into the 'invidious distinctions' which he himself specifically cautions against,[96] resulting in a somewhat parochial assessment. For while the entities he highlights do indeed have certain significant achievements to their name, there are some vital policy areas – climate change for one – in which a number of them could as justifiably be assigned the label of laggard as that of leader. In addition, assessments of achievement in this field must surely depend to some extent upon what particular nations can feasibly afford, and it might on that account be questioned whether *all* of those entities mentioned could not, in the light of their affluence, reasonably have contributed significantly more than they actually have.

The Biodiversity Convention itself sees the transfer of resources from the richer nations as being critical to conservation success,[97] and Wandesforde-Smith interprets this as a movement 'from places where wildlife protection is highly valued to places

---

and preserve the environment and to enhance the means for doing so'. Note also the 2000 Cotonou Partnership Agreement between ACP and EC States, (2000) OJEC L317/3, especially Articles 1, 9, 32.

[93] Viewable at <http:www.un.org/sustainabledevelopment/sustainable-development-goals>; see especially Goals 14 and 15.
[94] Target 7.B, Goal 7, viewable at <http://www.un.org/millenniumgoals/>.
[95] Wandesforde-Smith (n 11), 89, footnote 31.
[96] *Ibid.*
[97] This concern is in fact pervasive throughout the Convention, but see especially Article 20(4).

where it is less so'.[98] Others, however, may see the matter as turning more on the relative ability of different states to fund the measures which they might ideally deem desirable, and on that account, for instance, to judge Kenya's decision in 1989 to destroy its ivory stockpile and forgo the considerable income it might have earned from it as indicative of very substantial conservation commitment.[99] It might equally be recalled, again simply by way of example, that it was the government of India that, through repeated promptings, secured formal recognition (albeit in rather equivocal terms) of the legitimate interests of other life-forms at the Stockholm Conference,[100] Zaire (as it then was) that spearheaded the project leading to the adoption of the World Charter for Nature 10 years later,[101] Costa Rica that has designated over 25 per cent of its land mass as protected areas[102] and Bolivia that triggered the current UN programme concerning elaboration of the idea of harmony with nature.[103] Conservation commitment of sorts can actually be found right across the international community, but is seldom as robust or consistent *anywhere* as might be judged ideal.

3.3 Legal Implications of Community Commitment

In addition, there are other, more subtle and specifically legal, senses in which it is appropriate to attribute the source of pressure for biodiversity conservation, however it is specifically manifest, to the international community as a whole rather than to some particular faction within it. This assignment of substantive sponsorship is in fact appropriate whenever a particular conservation programme becomes enshrined in *legal* form at the *transnational* level, since, once that process has been completed through the creation and assumption of appropriate legal undertakings, a new consortium of underwriters is automatically brought into the picture. This occurs for the simple reason that the more fundamental commitment to legal commitment as such – i.e., to the rule of law in international affairs – is not only dependent upon, but actually commands (in repeatedly reaffirmed fashion) the support of the *entire* community, and not just the environmental faction within it.[104] Indeed, within this broader polity, there would seem to be no politically significant entity, agency or constituency (however defined or

---

[98] Wandesforde-Smith (n 11), 91.
[99] For discussion of this incident, *see* R Leakey and V Morell, *Wildlife Wars* (St Martin's Press, 2001). Since that time, various other states have of course followed that example, most recently Ethiopia, as to which, *see* <http://www.bbc.uk/news/world-africa-31983727>.
[100] See LB Sohn, 'The Stockholm Declaration on the Human Environment' (1973) 14 Harvard International Law Journal 423, 459.
[101] *See* further Burhenne and Irwin (n 55).
[102] C Runyan, 'Forever Costa Rica' *Nature Conservancy Magazine* (Arlington, VA) June 2011.
[103] *See* the dedicated UN website at <http://www.harmonywithnatureun.org/>.
[104] The point is made, for example, both in the first recital of the preamble to the UN Charter, and in its first substantive provision. It is dealt with more substantively in UNGA Resolution 2625 (XXV), the 1970 Declaration on Principles of International Law concerning Friendly Relations and Co-operation among States in accordance with the Charter of the United Nations. *See* further UNGA Resolution A/RES/44/23, of 17 November 1989, on the United Nations Decade for International Law.

described) that seeks positively to deny, at least as a matter of abstract principle, the validity and importance of such commitment.

Accordingly, once any particular legal obligation is formally created in the environmental (or indeed in any other) field, it automatically attracts the support of the pre-existing and collectively endorsed constitutional infrastructure comprising the principles of state responsibility and, wherever the duty in question is couched in treaty form, the law of treaties. As a result, certain forensic opportunities for securing adherence to the commitment in question potentially become available. These possible sources of redress are, of course, overlaid upon whatever inducements to performance – whether in the form of carrots or sticks – may be available under specific MEA regimes themselves. Thus, in the recent *Antarctic Whaling* case,[105] the International Court of Justice (ICJ) made it clear that its decision should not be understood as an attempt to align itself with any of the sharply divergent perspectives on the environmental, scientific or ethical issues involved in the conduct of whaling, but only as an application of the general rules governing the implementation and interpretation of treaties,[106] which, as suggested above, can for most practical purposes be regarded as having received the endorsement of the international community as a whole.[107]

This process of collective underwriting is, moreover, significantly intensified whenever the substantive commitment in question is recognised as being grounded not merely in the particular interests of individual states, but in the shared interests of a certain group, or indeed of the community as a whole. In this vein, the 1946 Whaling Convention (ICRW), which was the treaty in issue in the *Whaling* case, explicitly recognised 'the interest of the nations of the world' in safeguarding whale stocks,[108] and accordingly permitted participation in the regime by *any state*, whether personally engaged in whaling or not.[109] In the modern context, the collective nature of the interest in the performance of conservation commitments might arguably be assumed even in the absence of express confirmation by the individual treaty in question, since it would seem that the conservation of biological diversity has been authoritatively declared to represent the 'common concern of humankind' as a matter of general principle.[110]

---

[105] *Whaling in the Antarctic* (Australia v Japan; New Zealand intervening), Judgment of 31 March 2014, ICJ.
[106] See in particular paragraph 69 of the Judgment.
[107] The substantive principles embodied in the 1969 Vienna Convention on the Law of Treaties, 1155 UNTS 331, are widely understood to represent contemporary customary international law (whether through codification or progressive development) and regularly applied as such, and largely regardless of whether the states in question have formally become parties to it.
[108] 1946 International Convention for the Regulation of Whaling, 161 UNTS 72, first preambular recital.
[109] Article X(2).
[110] CBD, preamble, third recital. For further consideration of this notion, and of its legal implications, see the chapter by Duncan French is this volume. Note also the affirmation, 20 years earlier, in the World Heritage Convention (n 49) that 'it is incumbent on the international community as a whole to participate in the protection of the … natural heritage of outstanding universal value': preamble, seventh recital.

This recognition has served only to bolster certain ideas of community interest that are already evident within the law of state responsibility,[111] and bear upon the ease with which it may be invoked against a delinquent state. Specifically, a breach of duty may be challenged (i) in a representative capacity, by any state to whom the duty is owed,[112] or (ii) in the more specific capacity of 'injured state', by any party within that group which is specially affected, or where the position of *all* group members with respect to further performance is radically transformed.[113] In either case, the state in question may demand cessation of the wrongful act and performance of the obligation of reparation.[114] The implications of these provisions are therefore such as to create considerable opportunities for any individual state with a sufficiently strong commitment to appoint itself as a guardian of conservation or other community interests on an ad hoc basis. Thus, in the light of their recent application by the ICJ to a treaty concerning accountability for torture in the case between Belgium and Senegal regarding the *Obligation to Prosecute or Extradite*,[115] Japan did not even consider it worthwhile challenging Australia's legal standing to complain of a breach of the ICRW by Japan in the *Whaling* case itself.

A further significant aspect of the case for present purposes concerned the decision by New Zealand to intervene in the proceedings in order to make representations regarding the proper interpretation of the Whaling Convention, as all contracting parties to a multilateral treaty which is in issue in proceedings before the ICJ are entitled to do.[116] In the course of a very fully elaborated opinion on the matter,[117] the Brazilian member of the Court, Judge Cançado Trindade, recognised this right as a vital means of diluting the heavy 'bilateralist bias that permeates dispute settlement under the procedure before this Court',[118] a transformation which was essential where the dispute related to 'domains of concern to the international community as a whole'.[119] There

---

[111] As to which, see the 2001 Articles on Responsibility of States for Internationally Wrongful Acts, formulated by the International Law Commission and available via its website at <http://www.un.org/law/ilc/> or from its Report of the 53rd Session, ILC (2001), GAOR 56th Session, Supp. 10.

[112] *Ibid.*, Article 48.

[113] *Ibid.*, Article 42.

[114] The key distinction is that only an injured state may demand reparation in *its own* interest; others may do so solely in the interest of the injured state or other beneficiaries of the obligation breached: Article 48(2)(b).

[115] (2012) ICJ Reports 422. The treaty in question was the 1984 UN Convention against Torture and Other Cruel, Inhuman or Degrading Treatment or Punishment, UN Doc A/Res/39/46; (1985) 23 ILM 1027, (1986) 24 ILM 535. Having determined Belgium's entitlement to institute proceedings simply by virtue of its participation in the treaty, the Court found it unnecessary to consider its further, more specific, claim to be an 'injured state': see especially Part III of its judgment, paras 64–70.

[116] See Article 63, Statute of the International Court of Justice. Note also Article 62, which provides that any state which considers that it has an interest of a legal nature which may be affected by a decision can *request to be permitted* to intervene.

[117] The opinion may be located via the 'Incidental Proceedings' link in the entry for this case on the Court's website.

[118] See para. 71 of his Separate Opinion on the Declaration of Intervention.

[119] *Ibid.*, para. 66.

was no doubt that the present case fell into that category, since 'The general policy objectives under the ICRW were ... – and remain – the protection of all whale species from overfishing, to the benefit of *future generations in all nations,* and the orderly development of the whaling industry was to abide by that'.[120]

He further emphasised the fact that the preambular reference to the whaling industry was expressed in terms of its *orderly* development rather than its development as such, indicating that its activities in the future were to be constrained by the collective interest in preserving the good order of the oceans through the adoption of proper mechanisms of conservation and the avoidance of conflicts or disputes, the 'mere profitability' of the industry having been discarded as an objective.[121] He duly concluded by hailing the 'resurrection' of the right of intervention in recent years as a most welcome development, 'propitiating the sound administration of justice ... , attentive to the needs not only of all states concerned but of the international community as a whole, in the conceptual universe of the *jus gentium* of our times'.[122]

In the light of these considerations, now would seem to be the least appropriate time imaginable to abandon commitment to the notion of the 'international community' as the underlying guarantor of conservation commitments and to reduce the project to the private fief of some ill-defined faction called the 'international environmental movement': however the matter may appear when viewed through the prism of political science, there can be no doubt of the continuing (and indeed growing) importance of community commitment to conservation from a strictly legal perspective. In this context, it is at last arguably appropriate to assert that we are all environmentalists now, at least in principle. Indeed, since the most fundamental objective of the international law of conservation is essentially to preserve the life support systems of the planet, it is difficult to see what rational grounds could exist for any particular constituency to seek to exempt itself from the project.[123]

At this juncture, however, it might well be objected that it is only in the very rare event of a dispute regarding the application or interpretation of a conservation treaty coming before an adjudicative body such as the ICJ that such a degree of attention is likely to be focused upon the precise nature and content of the legal commitments entailed, and that it would therefore be unwise to set too much store by any pronouncements that are generated in the process, especially since they will only produce binding effects for those states that were directly involved in the litigation

---

[120] *Ibid.,* para. 60 (emphasis added).

[121] *Ibid.,* para. 58. This should not, of course, be understood to imply that the prosperity of the industry was now to be regarded as entirely irrelevant: rather, it was effectively demoted from the status of an (indeed, *the sole*) objective, as it had been in the earlier 1937 Convention, to merely one factor to be taken into account in the fixing of quotas by the IWC: see Article V and especially para (2)(d) of the ICRW. These arguments had all been explored at much greater length in Bowman (n 32), which was cited by Judge Cançado Trindade in paragraph 71 of his further individual opinion delivered at the merits stage of the case.

[122] Para. 68, Individual Opinion regarding Intervention.

[123] Needless to say, this does not exclude the possibility that a host of imperfectly rational or blatantly irrational reasons may be raised in relation to particular projects or situations.

process itself.[124] There is no doubt some force in this point,[125] but there are also a number of countervailing considerations that should be borne in mind.

The first of these is that, whatever the formal limitations of the applicability of judicial decisions, in the modern context the echoes of any such judgment are sure to be heard far beyond the confines of the courtroom itself: it is difficult to believe, for example, that any International Whaling Commission (IWC) member that was minded to issue permits for scientific whaling would henceforth feel able to disregard altogether the considerations spelled out by the ICJ with regard to the interpretation in good faith of the relevant provision of the ICRW, Article 8.[126] Indeed, New Zealand has already secured the adoption of a resolution at the most recent meeting of the IWC, which was broadly designed to give effect to the key legal pronouncements of the Court.[127] Furthermore, the mere fact that issues concerning the conservation of biological diversity are now thought sufficiently important to generate litigation before the World Court serves of itself to heighten the profile of this area of policy concern in a more general way.

Secondly, and in reinforcement of the previous point, all the signs are that the incidence of such litigation is only likely to intensify in the future,[128] especially bearing in mind that the ICJ does not stand alone here, in the sense that the roll-call of judicial, quasi-judicial and arbitral tribunals available to process cases grounded upon the international responsibility of states in an environmental context has significantly increased of late.[129] Recent years have even witnessed the establishment by the Security Council of a wholly new institution designed to address on a purely ad hoc basis the legal consequences of the illegal Iraqi invasion of Kuwait in 1990, a

---

[124] That is to say, the actual parties to the litigation themselves, together with any state that opts to intervene in the case on the grounds that it is a party to a Convention which is in issue: ICJ Statute, Articles 59, 63.

[125] It seems, for example, that states other than Great Britain and the US that were involved in the exploitation of fur seals felt little compunction in ignoring the implications of the arbitral decision in the *Bering Sea Fur Seals* case (1898) 1 Moore's International Arbitration Awards 755: *see* generally S Barrett, *Environment and Statecraft: The Strategy of Environmental Treaty-Making* (OUP, 2005), chapter 2.

[126] Japan itself has already submitted a revised proposal which was intended to be compliant: *see* the Circular IWC.ALL.220 of 19 November 2014, 'Government of Japan: New Scientific Whale Research Program in the Antarctic Ocean', available on the IWC website at <http://iwc.int/circulars>.

[127] Resolution 2014-5, on Whaling under Special Permit. It must be conceded that this resolution was ultimately adopted only by 35 votes to 20, with five abstentions. It should also be recognised, however, that its failure to secure universal endorsement may well be attributable to the fact that, as observed by various states and accepted by NZ, it went beyond the strict terms of the judgment in certain respects.

[128] Note in this regard the consolidated cases currently pending before the ICJ between Costa Rica and Nicaragua, where each complains of a violation of the Ramsar Wetlands Convention.

[129] In addition to the forums noted in the text below, the International Tribunal for the Law of the Sea (ITLOS) represents a notable example.

development which has generated, inter alia, a mass of valuable precedents concerning the definition and valuation of environmental harm.[130]

In addition, the undeniable residual reluctance of states to institute international legal proceedings on their own behalf in respect of environmental harm may sometimes be circumvented by the recruitment or intervention of other agencies. In the case of the long-running saga concerning the Greek government's failure to fulfil its obligations under the Bern Convention with regard to the protection of turtle nesting sites at Laganas Bay, Zakynthos, for example, the nascent proposal by a small consortium of states to institute arbitration proceedings against Greece under the terms of Article 18 of that treaty was ultimately abandoned only in light of the willingness of the European Commission to institute proceedings for breach of the EU Habitats Directive – an option which also permitted access to the stronger remedies and enforcement mechanisms of EU Law.[131]

In other instances, the route to enforcement may be rather less direct and obvious. The government of Tanzania, for example, may possibly have anticipated the criticism to which it was subjected within the World Heritage Committee in 2010 on account of its proposal to develop an all-weather, bitumenised trunk road across the Serengeti National Park, since this was designed largely to stimulate economic growth in the surrounding regions and threatened significant harm to the park itself, a World Heritage site.[132] It may have been more surprised, however, to find itself subsequently instructed by the East African Court of Justice to abandon the project in its original form, as a result of proceedings instituted by an African animal welfare organisation.[133] Yet this decision was squarely based upon provisions of the East African Community Treaty which required partner states to co-ordinate their economic and other policies to the extent necessary to achieve Community objectives, and to promote the sustainable utilisation of their natural resources and the taking of measures designed to preserve, protect and enhance the quality of the natural environment.[134] It is therefore clear that,

---

[130] For details, see the website of the UNCC at <http://www.uncc.ch>.
[131] For discussion, see Lyster, 342–344.
[132] See World Heritage Committee Decision 35 COM 7B.5, which envisaged the possibility of irreversible damage to the property's outstanding universal value in the absence of a suitable environmental impact assessment. See further Decision 36 COM 7B.6 of 2011, in which the Committee welcomed Tanzania's decision in response to seek funding for a Strategic Environment and Social Assessment for the road.
[133] *African Network for Animal Welfare v A-G, Utd Republic of Tanzania*, Reference No 9 of 2010, Judgment of 20 June 2014, available via the website of the East African Community at <http://www.eac.int/>. Its right to undertake other programmes or policies which would not impact negatively on the environment and ecosystem of the Park was, of course, explicitly preserved. This was not seen to exclude some upgrading of the existing track for the purposes of tourism and park administration, a project towards which Tanzania itself now seemed to be inclining in any event.
[134] 1999 Treaty for the Establishment of the East African Community, text also available via its website, *ibid*. These obligations were generated by a combination of Articles 5(3)(c), 8(1)(c), 111(2) and 114(1). The fact that a later Protocol designed to give more concrete and detailed expression to these environmental commitments was not yet in force was not regarded as material. It is noteworthy that the court envisaged possible adverse effects not only on the Serengeti itself, but also on the neighbouring Masai Mara Park in Kenya.

even in the realm of an arrangement designed primarily for economic development, the commitment to conservation was not seen as mere window-dressing.

Yet, for all that, it would plainly be as much of an error to overstate the practical impact of these natural concomitants of international legal obligation as to overlook them altogether. Not only will the contribution of international courts and tribunals to the consolidation of conservation commitment remain relatively modest, it may also be further diluted by the possibility that disputes regarding the application of legal measures designed for conservation purposes may actually fall to be litigated within tribunals that perceive their most important policy priorities to lie elsewhere, and whose technical expertise is accordingly focused upon different areas of law entirely.[135] The vital need is therefore to prompt wide-ranging institutional action designed to counter the substantive fragmentation that is currently evident amongst these various bodies of norms, through the far more effective integration of the particular rules they contain.

As noted above, this integrative process is already under way in a formal sense. It is, of course, mandated in rudimentary fashion by the law of treaties itself, in the form of Article 31(3)(c) of the 1969 Vienna Convention, by which the process of treaty interpretation is to be informed by the principle of 'systemic integration' of legal norms generally.[136] In most cases, however, this minimum requirement is overlain by explicit provision in the key treaties themselves, which typically allows for (though without necessarily securing in practice) some more sophisticated mode of accommodation to occur.[137] The problem of fragmentation generally has, of course, been the specific subject of formal investigation by the International Law Commission,[138] though a case could certainly be made that this represents one of their least impressive and successful enterprises to date.

Accordingly, if there is one vital goal to which legal research and scholarship might most profitably be directed it would be the advancement and acceleration of this process of harmonisation and reconciliation between the general principles of biodiversity conservation and those governing other key areas of international endeavour, such as economic development, international trade, human rights and the protection of international peace and security.[139] Needless to say, this point has already been widely recognised in an abstract sense, being enshrined in the principle of *integration*, which

---

[135] A particularly notable example can be found in the dispute settlement arrangements of the WTO, though even here the bodies in question have moved gradually towards the more meaningful recognition of the sustainable development ideal (which is, after all, explicitly enshrined in its preamble). For an overview of the significance of litigation in the environmental field, *see* Birnie, Boyle and Redgwell (n 56), 250–262.

[136] For discussion, *see* C McLachlan, 'The Principle of Systemic Interpretation and Article 31(3)(c) of the Vienna Convention' (2005) 54 ICLQ 290; D French, 'Treaty Interpretation and the Incorporation of Extraneous Legal Rules' (2006) ICLQ 281–314.

[137] *See*, for example, CBD, Article 22; CITES, Article 14; CMS, Article 12.

[138] For details of the Commission's programme of work, *see* its website at <legal.un.org/ilc/guide/1_1.htm>.

[139] The importance of legal scholarship in the development of international conservation law was explicitly acknowledged by the Brazilian judge, Judge Cançado Trindade, in para. 71 of his individual opinion in the *Whaling* case (n 105).

finds reflection in the Rio Declaration and other key instruments.[140] Yet the undeniable reality is, regrettably, that this principle has itself not yet been fully integrated into the international legal order, or adequately reflected, respected and consolidated into the practice of international society, in either its governmental or non-governmental dimension. The principal focus of attention across the majority of areas of political activity is still almost invariably upon some very obvious and immediate – albeit commonly less essential – matter of human concern, whereas our absolute ultimate dependence upon the ecosystem services that other species and their interactions provide for us is something that requires a significantly more sophisticated, well-informed and reflective form of appreciation.[141]

To the extent, therefore, that there is currently an unmet requirement for the bringing to bear of a 'vision' of some kind, it would seem to relate to the means by which the imperative of biodiversity conservation can be woven more effectively into the fabric of global policy and practice. The key question accordingly becomes: what precise form should this vision ideally take? This is not, unfortunately, a question that can be answered with any certainty on the basis of Wandesforde-Smith's own account, for his pronouncements on the matter are decidedly Delphic.

4. THE ENVISIONING OF VISION ITSELF

One factor which contributes particularly to this ambiguity concerns the way in which his discussion appears to drift casually between the language of international law and international politics without any apparent need being felt for the establishment of any obvious lines of demarcation between the two.[142] This deficiency is a significant one, for, while the international *politics* of conservation plainly subsumes all the various elements of legal regulation that involve a supra-national dimension, it at the same time constitutes a much wider field of human endeavour, embracing many other forms of transnational activity. The World Conservation Strategy, for example, was a sophisticated attempt by a consortium of international organisations to elaborate a strategy for harmonising the goals of economic development and environmental protection, and setting out the means by which they might realistically be achieved. It does not of itself constitute *law* in any sense, however, since it is couched in descriptive and programmatic rather than normative terms.[143] The utilisation of legal measures is of course discussed, but only as one of the various means that might be employed to implement the strategy.[144] In similar fashion, the articulation of 'Proposed Legal Principles for

---

[140] See, for example, Rio Declaration, Principle 4; Stockholm Declaration, Principle 13; World Charter for Nature, Principle 7.
[141] Note in this context the potential contribution of IPBES (n 24).
[142] That being so, it is especially surprising to encounter within it (Wandesforde-Smith (n 11), 90) a complaint that 'it is striking in the new book [i.e., *Lyster*] ... how little discrimination there is among all the arenas or fronts in which the international politics of wildlife are played out', since that is exactly the charge that might justly be levelled against the reviewer himself.
[143] That said, it does of course form an important part of the policy backcloth onto which all subsequent legal developments should sensibly be projected.
[144] See especially chapter 15, which is merely one of 20 chapters in all.

Environmental Protection and Sustainable Development' appeared as only one element within the report of the Brundtland Commission.[145]

In international affairs, accordingly, law must be seen as only one of the many factors which serve to shape human conduct, much of which is scarcely affected by it at all, except perhaps in the negative sense that there is no explicit legal prohibition of the behaviour in question. Indeed, where ordinary private citizens are concerned, law of the public international variety is likely to appear especially peripheral and remote. Thus, in the particular field of conservation, it would be possible for a committed individual to engage in a lifetime of dedicated activity without overtly having to confront it at all. This commitment might take any one of a number of forms, including educating others in the practical importance of conservation, exploring and developing the scientific understanding through which it can be accomplished, participating in the maintenance or management of wildlife reserves, fund-raising for conservation activities or simply re-organising one's own personal lifestyle in such a way as to reduce the material demands it imposes upon the natural environment.

In the light of these considerations, it is surely bizarre to suggest, as Wandesforde-Smith appears to do,[146] that an environmental activist wishing to obtain guidance regarding what they should do next could sensibly expect to secure an answer from a monograph on international law! Such a work is intended neither as a clarion call to conservation action, nor as a practical manual for everyday activism – functions which are, after all, now amply served by numerous other publications.[147] Accordingly, the only reasonable response to someone who made the mistake of seeking enlightenment of this sort from a book like *Lyster* would be to echo the one helpfully offered by the apocryphal bystander, when asked by a would-be traveller for directions to the railway station: 'If I were you, I wouldn't start from here!'

4.1 Understanding the Roles of Actors in the International Legal Order

None of this should be taken to suggest, however, that the legal regime governing biodiversity conservation globally is of no importance or interest to individuals or other non-state entities, or even that they have no direct or significant role to play within it. To the contrary, Wandesforde-Smith is quite correct to note the substantial transformation that has occurred over the course of the last century with regard to the agencies through which the various activities that fall within the purview of international law are characteristically conducted. Although commerce and other forms of economic, social

---

[145] WCED, *Our Common Future* (OUP, 1987); *see* especially chapter 12 and annexe I.
[146] *See* Wandesforde-Smith (n 11), especially 92–93.
[147] For a brief sample of the numerous publications of this kind which are now available: *see* the Sierra Club's *Grassroots Organizing Training Manual* (Sierra Club, 1999); M Hawthorne, *Striking at the Roots: A Practical Guide to Animal Activism* (Changemakers Books, 2007); A Ricketts, *The Activists Handbook* (Zed Books, 2012); J Perlman, *Citizen's Primer for Conservation Activism* (University of Texas Press, 2014). Entire PhDs have in fact been written on the subject simply from the perspective of one particular country: for an example, *see* RB Wilcox, *The Ecology of Hope: Environmental Grassroots Activism in Japan* (Union Institute & University College of Graduate Studies, 2004), available at <http://www9.ocn.ne.jp/~asian/ecohope.pdf>.

32   *Research handbook on biodiversity and law*

and cultural co-operation between individuals, corporations and other private sector bodies have long been pursued on a transnational basis, and sometimes on a truly massive scale,[148] the vast bulk of international intercourse was traditionally conducted through the diplomatic organs of national government. During this period, the 'international community' was accordingly viewed as being essentially a community of *states* (and a rather small number of them at that), and the legal order by which it was served was therefore typically characterised as the *law of nations*.[149] In more recent times, however, the nature of this political community and legal order has been radically reconfigured through a series of significant developments, including (i) the vast proliferation of inter-governmental organisations that have come to assume technical or political functions on behalf of the community as a whole, (ii) the direct conferral of legal personality on individual human beings as a consequence of the recognition and development of the human rights regime and (iii) the emergence of NGOs as major players in many of the most important areas of international political activity.

In the wildlife field, NGOs have come to occupy a particularly crucial role, and to fulfil a vast array of functions which would once have been thought far beyond their remit or capacity. Thus, they have not only agitated for the adoption of treaty arrangements, but actually undertaken a major role in the drafting process itself; they have not simply monitored states parties in the performance of their conservation commitments but sometimes secured the opportunity to incorporate the results of this supervision within formal treaty compliance procedures; they have provided the initiative, expertise and resources that were required to make particular substantive conservation projects a reality; they have been allocated decision-making functions for the purpose of certain regimes and even provided secretariat services on occasion; in some cases, to conclude, they have been granted formal 'partnership' status that guarantees them, inter alia, a place at the table of every key committee that operates within the regime in question.[150] Accordingly, any account of the contemporary conservation regime is bound to make extensive reference to their involvement.

Yet for all that there have been certain features of the legal system which have remained essentially unchanged: most importantly, states themselves have never relinquished control of either the mechanisms of law-making or the processes through which legal personality is formally recognised, with the result that their traditional stranglehold over the legal order has been to a considerable extent retained.[151] Thus, when such organisations as IUCN, ICBP and the International Waterfowl Research

---

[148]   Perhaps the most obvious instances are provided by the para-statal activities of bodies such as the Dutch East India Company during the 17th and 18th centuries.

[149]   Thus JL Brierly's celebrated text, first published in 1928, has that as its very title (*see* now the 7th edition, by Andrew Clapham, published by OUP in 2012).

[150]   *See* generally Sands & Peel (n 57), 86–92 and the literature there cited. The Ramsar Convention – on which *see Lyster*, chapter 13 and GVT Matthews, *The Ramsar Convention on Wetlands: Its History and Development* (Ramsar Bureau, 1993) – provides a particularly instructive case study in all the respects mentioned.

[151]   Accordingly, Jutta Brunnée treats this fundamental truth as the very starting point of her discussion of the means by which international environmental law might ultimately be deployed to address 'the global concerns of humanity as a whole': *see* her 'Common Areas, Common

Bureau began to agitate during the 1960s for a reversal of the progressive loss of wetland habitat that was beginning to imperil the survival prospects of numerous species of birds, they immediately recognised that they would be unable to make significant substantive headway without sponsorship and support from states that were sympathetically disposed towards the project. Fortunately, this was in the event forthcoming from certain key quarters, including most notably the Dutch and Soviet governments, which worked with them on successive drafts of a treaty text, and that of Iran, who proved willing to host the diplomatic conference at which the evolving instrument was finalised and opened for signature.[152] No less importantly, some 23 governments actually turned up to attend this event, without which the entire project would have remained still-born in normative terms.[153]

Of course, given the vital role that these NGOs had played in the drafting and development of the agreement in question, it could hardly have been expected that they would thereafter simply stand back to watch passively how events unfolded. To the contrary, the intention was always that they would continue to participate directly in its ongoing implementation, and provision was therefore made for the convening of periodic conferences at which progress could be monitored by all the key stakeholders.[154] Yet the non-governmental bodies within this group would inevitably remain dependent upon the degree of latitude that governments themselves were prepared to grant them in that regard: full, front-line participation as of right in this Convention for wetland conservation would naturally only be open to statal entities,[155] since any other arrangement would inevitably have undermined the formal legal status of the entire agreement. Part of the reason for this, of course, is that it will be states themselves which, in formal terms, actually assume, and ultimately remain accountable for, the conservation responsibilities that such treaties impose.

The fact that, where law specifically is concerned, the formal primacy of *governmental* responsibility remains essentially unchanged seems largely to pass unnoticed in Wandesforde-Smith's analysis, and possibly stems from his failure to maintain a clear distinction between the political and the more narrowly legal aspects of international conservation endeavours. For example, he suggests at one point that

> [i]n retrospect, Lyster's insistence that the political will of governments and their investments in what today we would call implementation capacity are the critical factors for wildlife has about it a starchy whiff of the formal institutional stiffness that used to characterize international law and international lawyers before 1945. While people in government are certainly still part of the starburst constellation of actors seeking to shape international

---

Heritage, and Common Concern' in D Bodansky, J Brunnée and EC Hey (eds.), *The Oxford Handbook of International Environmental Law* (OUP, 2008) 550, 551.

[152] *See* generally the Final Act of the International Conference on the Conservation of Wetlands and Waterfowl (1971), available at <http://archive.ramsar.org/>, via the links 'Documents' and 'Conferences of the Parties'.
[153] Though five of these attended purely in the capacity of observers: *ibid*.
[154] Article 6, which was amended in 1987.
[155] On the question of participation, see Article 9.

wildlife law and its outcomes, the initiative and the balance of influence, if not power, now more often seems to lie with non-state actors.[156]

The observation seems open to question on a number of accounts. Read in context, Lyster's point was surely not so much that the political will and capacity of governments are the 'critical factors for wildlife', or even for its conservation, but rather for the much more specific question of *the capacity of international law as such to make a contribution in that regard*. As we have seen, international law is only one of the mechanisms and processes through which this objective might reasonably be pursued, and in many of these alternative modes of conservation activity the potential role of non-governmental agencies for the achievement of progress through their own unaided efforts may doubtless be considerably greater. Where international law itself is concerned, however, the role of governments will continue to be crucial, and would remain so even where (as may very well often be the case) they opt to place reliance upon the enthusiasm, initiative, expertise and resources of civil society in the course of discharging their responsibilities. Indeed, even a nation which goes so far as actually to delegate the performance of certain practical conservation functions to an NGO or other private sector agency will be obliged at the very least to equip itself with the powers needed to supervise the body in question, and, where necessary, to coax or coerce it into more effective action. For all their undoubted practical influence, non-state entities remain in the twilight zone of formal legal personhood on the international legal plane. In addition, the practical fulfilment of the conservation commitments imposed by treaty regimes will typically require the adoption of legal measures at the national level, and here again it is only states and governments that have ultimate capacity and authority in that regard. Consequently, Lyster's observations were not only perfectly apposite at the time of their publication, but remain so today.

As far as claims to have detected the 'whiff of starch' are concerned,[157] there are surely strong grounds for questioning whether starch itself should ever be seen as something merely to be sniffed at – it is, after all, the very fuel of life itself![158] And while it would plainly be excessive to make this claim of international law as such, its potential impact on human behaviour does at least resemble that of starch in its processed form, in terms of its specific capacity to stiffen the fabric of whatever conservationist apparel we have seen fit to assume at any given moment; it is, indeed, for that very reason that those in the 'international environmental movement' have so frequently sought to invoke its authority and support. The system undoubtedly has a great many weaknesses, but there has always been a very significant difference between, on the one hand, the charge that a government has failed to follow an

---

[156] Wandesforde-Smith (n 11), 86–87, reflecting on the observations of Lyster (n 10), 303–304.

[157] Starch in its pure form is actually odourless, it would seem: *see* the website of AAF, the European Starch Industry Association, at <http://www.aaf-eu.org/portfolio/>.

[158] Thus, it has recently been observed that starch, which is a carbohydrate produced by green plants as an energy store, 'remains at the very epicentre of the world's food and feed chains and has ... become one of the world's most important sources of biorenewable energy': PL Keeling and AM Myers, 'Biochemistry and Genetics of Starch Synthesis' (2010) 1 Annual Review of Food Science and Technology 271, Abstract.

appropriate policy in some particular respect and, on the other, the allegation that it has failed to comply with a specific legal obligation.

If, therefore, the authors of the new edition of *Lyster* should appear to have decided – as Wandesforde-Smith suggests – to 'stick to their legal last',[159] that will doubtless be because it is, after all, the law which constitutes the specific focus of their study. As it happens, moreover, the particular metaphor employed here seems more apposite than its creator may have realised, on the grounds that, where the fundamentals of the international legal system are concerned, we are indeed compelled for the time being to make do and mend with the cobbled-up normative order that we have inherited from our forebears.[160] Since this system has encountered considerable difficulty even in tackling its most loudly proclaimed concerns, such as the eradication of torture or of military aggression, it is only to be expected that it will struggle to cope effectively with the more subtle, complex and wide-ranging challenges posed by environmental degradation. Yet while international law undoubtedly remains a relatively primitive system, it is this very trait which suggests not only that it might retain an inherent capacity for evolutionary advance, but also that the serious, systematic and scholarly exploration of its current strengths and weaknesses might have a significant part to play in this process of ongoing development. And since the law governing conservation is predominantly treaty based, it stands to reason that treaty regimes must form the primary focus of scholarly attention.

4.2 Understanding the Role of Treaties in the International Legal Order

Accordingly, as Wandesforde-Smith himself observes:[161]

> Lyster put a focus on wildlife treaties. Now, twenty-five years later, there are more wildlife treaties than ever, and on top of that even more treaties addressing other global issues that have cross-cutting consequences for the future welfare of species and habitats. The new book comes to grips superbly with the details of this new legal reality in all of its complexity.

The authors will, I am sure, be extremely grateful for this generous endorsement of their efforts, especially since it suggests that the fundamental object and purpose of the book – namely to present 'a clear and authoritative analysis of the key treaties which regulate the conservation of wildlife and habitat protection, and of the mechanisms available to make them work' – has been amply achieved.

It seems, however, to be this very objective itself which the reviewer seeks to call into question, in suggesting that what the new edition

> does not seem to grasp and certainly does not venture to say anywhere is that the great age of wildlife and environmental treaty making, the age gathering a full head of steam as Lyster set pen to paper, is over. The political and economic conditions that brought it into existence are

---

[159] Wandesforde-Smith (n 11), 92.
[160] For the benefit of any readers unfamiliar with the cobbler's craft, it may be helpful to point out that a 'last' is a device comprising a number of variously sized, foot-shaped projections upon which shoes may be securely placed for the purposes of repair.
[161] Wandesforde-Smith (n 11), 93.

gone and may not soon, if ever, return. Doing more of the same and lots of it is not, therefore, a realistic option.[162]

Once again, this claim requires a good deal of unpicking. First, it can certainly be acknowledged that no statement of the kind alluded to will be found in the new edition, though the reasons why its inclusion would never have been contemplated have more to do with the doubts that must be entertained regarding its substantive pertinence and cogency than with any other consideration. To begin with the less important reservation that might be maintained, the statement is of questionable factual accuracy on a number of counts. For one, to the extent that it actually matters to identify the precise 'heyday' of conservation treaty making, it would surely be the 1970s, rather than the late 1980s, that had the strongest claim: indeed, it was very probably the spate of key treaties that emerged during that period that originally prompted Simon Lyster to consider it worthwhile 'setting pen to paper' in the first place.[163]

Equally, although the flow of international 'legislative' activity has arguably slowed a little since the original appearance of his work, the negotiation and adoption of wildlife treaties certainly has not halted:[164] most obviously, the Biodiversity Convention itself has been concluded, followed by three rather significant protocols and a separate treaty on plant genetic resources. This period has also witnessed the conclusion of important agreements on watercourses, mountains and forests and a revised version of the regional conservation regime for Africa, as well as a whole clutch of ancillary instruments adopted within the framework of the Migratory Species Convention. In addition, there has been a succession of crucial agreements addressing different aspects of fisheries conservation. A further noteworthy development, potentially heralding a new direction for wildlife protection, entailed the adoption of a treaty concerning individual animal welfare, conceived entirely independently of any threat to the survival of the species covered.[165] Furthermore, even had the procreative process regarding new legal instruments actually ground to a halt entirely, the most plausible explanation could surely have been found in the fact that the great majority of the principal substantive contexts where protection is currently required (barring, perhaps, additional measures to address the needs of particular categories of migratory species) are probably now covered.

Yet this element of factual shakiness is largely overshadowed by the fact that the reviewer's assertion is almost entirely beside the point, for there is surely no one who would suggest that the explication and analysis of international wildlife law can somehow be reduced to a mere inventory of instruments adopted. To the contrary, what

---

[162] Ibid.
[163] On this point, see Lyster's own 'Introduction', xxi.
[164] Indeed, this point is recognised by Wandesforde-Smith himself, as noted above. For a convenient snapshot of the ongoing programme of international treaty-making in the broad field of biodiversity conservation, together with text location references for the many instruments in question, including those mentioned in the text immediately following, see the 'Table of Treaties and Other International Instruments' in Sands and Peel (n 57), xlix–lix.
[165] 1997 Agreement on Humane Trapping Standards, OJEC 1998 L042, approved by Council Decision 98/142. For discussion, see Lyster, 686–688; SR Harrop, 'The International Agreement on Humane Trapping Standards' (1998) 1 JIWLP 387.

should form the proper focus of the attention of international wildlife lawyers is not 'the great age of wildlife and environmental *treaty making*' as such but rather the complex ongoing process of *application and development of treaty-based regulation*, which is a very much more wide-ranging phenomenon. Anyone who believes that this latter process has run out of steam has plainly not been paying sufficiently close attention. For the reality is that states continue to turn up in impressive numbers for the ongoing cycles of meetings of the Conferences of the Parties to these Conventions, to formulate and approve resolutions and recommendations regarding their interpretation and application, to devise and refine mechanisms and procedures for their more effective implementation and to support or participate in the many subsidiary bodies, committees and regional offshoot activities that the business of these regimes tends to generate. In addition, a great deal of time, effort and resources is being expended on the development of legal and institutional capacity to translate these international commitments into substantive progress on the ground at the national level.[166] Furthermore, it seems scarcely appropriate to suggest, as the reviewer does, that this amounts simply to 'more of the same',[167] because the tools, techniques, processes and procedures that are currently available for the environmental repair and restoration process are now of a decidedly superior order than those that were ever available to earlier generations.

This process effectively began, of course, with the incorporation into wildlife conservation treaties of institutional arrangements of the kind pioneered by Ramsar, transforming them into much more of an *organic* process, and thereby generating the potential for evolutionary change in the quality and quantity of normative responses to environmental problems. It is, however, only during the last few decades that these new institutional mechanisms have been established within the international community,[168] and given the very slow pace of change in international legal affairs it is certain to take time before their potential can be explored and exploited to maximum effect. It is only in the light of considerations such as these that the real evolutionary potential of international wildlife law can properly be assessed.

The conviction that a very great deal is to be gained from the meticulous application to international wildlife law of the traditional techniques of legal scholarship is, moreover, only strengthened by consideration of Wandesforde-Smith's personal prescription for progress in this area, which appears to involve dedication to the pursuit of what he describes as 'interstitial headway'.[169] Now it must first be observed that,

---

[166] *See*, for example, UNEP, 'Implementation of MEAs in National Law', a Powerpoint presentation viewable at <http://www.iucnael.org/en/documents/631-unit-9-implementation-of-meas-in-national-law/file>; GL Rose, 'Gaps in the Implementation of Environmental Law at the National, Regional and Global Level' (2011), available at <http://www.unep.org/delc/Portals/24151/FormatedGapsEL.pdf>; S Maljean-Dubois and L Rajamani, *Implementation of International Environmental Law* (Brill, 2011); A Chandra and A Idrisova, 'Convention on Biological Diversity: A Review of National Challenges and Opportunities for Implementation' (2011) 20 Biodiversity and Conservation 3295.
[167] Wandesforde-Smith (n 11), 93.
[168] Institutional arrangements have, admittedly, been a feature of fisheries treaty regimes for rather longer, but the precise nature of those arrangements has always left a great deal to be desired.
[169] Wandesforde-Smith (n 11), 94.

coming as it does from one who has specified as a key aim of legal scholarship the provision of a rallying cry to inject renewed impetus into environmental activism, this is scarcely the most inspirational of slogans. Part of the difficulty here, no doubt, lies in the obscurity of the concept itself, a problem that is compounded by the fact that its promulgator omits to offer any concrete illustrations to shed light on the matter. Only the following general explanation of the idea is presented:[170]

> This is the pursuit of initiatives that give the law new leverage and power by discerning and exploiting in the nooks and crannies of the law as it stands, and of the regimes it has already created, chances to put the pressure on, in Lyster's phrase, in fresh and unexpected and, therefore, potentially productive ways. It might come from framing problems in a new legal light or by putting before established institutions legal options to act in ways that they have not previously considered as within their remit.

It is at this point that one might reasonably conclude that what is being marketed here is essentially old wine in an old bottle, but under cover of a new label. For opportunities of the kind envisaged are exactly what might be expected to be unearthed by the systematic legal analysis of conservation treaty regimes, and in particular the arrangements for their implementation. This was, of course, precisely the reason why it was decided to incorporate an entirely new chapter concerning implementation and enforcement in the opening section of *Lyster*, and to greatly expand the attention devoted to such matters in the majority of the individual chapters. In each case, furthermore, due attention is given to the role of private sector actors in this context alongside that of states parties themselves.[171]

None of this, however, should be taken to suggest that the broad approach adopted in *Lyster* towards the structuring of its subject-matter – essentially a regime-by-regime analysis of international wildlife treaties, bolstered by the presentation of various cross-cutting issues and underlying contextual commonalities – represents the *only* legitimate way in which the topic can be addressed. To the contrary, there is a most vital and urgent need for the compilation of works which adopt alternative and complementary approaches, and in particular those which delve more deeply into the practical efficacy of particular treaty initiatives, identify and analyse more fully the recurrent themes and problems which pervade the regulatory landscape of conservation conventions generally, explore the many interfaces of international wildlife law with other specialist areas of legal concern, and evaluate its role and contribution within the international legal order as a whole. It is, indeed, precisely with these matters in mind that this present *Research Handbook* was conceived in the first place. Accordingly, it is hoped that it may serve as a modest means both of establishing and of advancing a regulatory vision and agenda for the future.

---

[170] *Ibid.*
[171] For a brief and random sample of references to opportunities for private sector actors to influence outcomes, *see* pp. 101–106, 111–112, 161, 183, 227, 371–372, 443, 476–477, 504, 565–566, 644, 677–678.

5. CONCLUSIONS: 2020 VISION AND BEYOND

In the light of all the above considerations, it should be possible to develop some sort of viable 'vision' for international wildlife law which can be drawn upon to guide the development both of the legal regime for the global conservation of biological diversity and the growing body of legal scholarship through which it is elucidated. In the first instance, this should entail the abandonment of any hopes for grand epiphanies, smart dodges or quick and easy solutions, and the continued dedication to work of a more grounded, potentially gruelling and unflaggingly meticulous character. More specifically, these endeavours should remain focused upon the deployment of traditional legal skills to best advantage in the thorough and thoughtful analysis and evaluation of the principles, practices and processes through which the conservation of nature is addressed and effectuated on the international plane. There is ample evidence that those charged with organisational responsibility for the implementation of the treaties in question will pay heed to serious scholarship which successfully exposes weaknesses and inconsistencies in or amongst the regimes in question.[172] There is, moreover, a long tradition of treaty bodies actively commissioning formal legal opinions on issues of current controversy and concern.[173] Elsewhere within the legal community, the members both of international courts and tribunals and of law reform agencies such as the International Law Commission have also shown themselves to be receptive to the lessons taught by legal scholarship.[174] The slowly growing impetus generated by such processes is now just beginning to enhance the profile of international environmental law, including that relating to the conservation of biological diversity.

As noted above, this enhanced sense of priority and understanding may ultimately prove to be of greatest significance in relation to the task of achieving a more effective accommodation of conservation interests with cognate areas of international law, as reflected above all in the regimes which govern such matters as international trade, the international economy, peace and security, and human rights. Although these have all traditionally been accorded a much higher priority than environmental protection, the realisation is very gradually beginning to dawn that none of the aspirations by which these regimes are motivated is remotely capable of achievement unless the life support

---

[172] The Ramsar Secretariat, for instance, began to publish examples of externally generated legal research relating to the Convention on its website in 2002, and the CoP definition of wise use was changed at its next meeting, partly with a view to achieving a better articulation between Article 3(1) and 3(2), a matter which had been the focus of considerable concern in the literature.

[173] The IWC, for example, has commissioned a number of opinions on legal questions, several of which are discussed in C Wold, 'Legal Opinion on the Competence of the IWC to Include Humane Treatment and Human and Cetacean Health Concerns in the RMS', 21 June 2000, available at <http://www.lclark.edu/live/files/183>.

[174] As regards the ICJ (which has been sparing in its explicit reference to doctrine), see Article 38(1)(d) of its Statute and M Peil, 'Scholarly Writings as a Source of Law: A Survey of the Use of Doctrine by the International Court of Justice' (2012) 1 Cambridge Journal of International & Comparative Law 136–161. As regards the ILC, a perusal of its reports reveals extensive reference to legal scholarship, which has proved influential in relation to the question of environmental liability, for example.

systems of the planet can be effectively assured. Thus, organisations such as the World Bank and the WTO have already taken some preliminary, tentative steps towards the more effective recognition of environmental protection as a cornerstone of sustainable development, while lawyers specialising in human rights and the use of force are very belatedly coming to realise the extent to which all their traditional efforts can effectively be set at naught by the ravages of environmental insecurity. As yet, however, the practitioners of these disciplines have barely begun to recognise the overriding importance of, still less to engage in any meaningful way with, the specific question of biodiversity conservation, or with the principles and processes through which it can best be secured.[175]

5.1 Discarding the Blinkers of the 'Enlightenment' Worldview

As suggested above, one of the principal challenges in this context undoubtedly derives from the extent to which the current global societal order, together with the academic disciplines by which it is ultimately underpinned (most notably politics, philosophy and economics) have remained trapped in the intellectual stranglehold imposed by a worldview that was originally developed by the likes of John Locke, Adam Smith and Immanuel Kant during the period of European history rather questionably known as the Enlightenment. It is a major irony of this era of intense disputation and debate that, although the one principal issue upon which every one of its major luminaries appeared to agree was the need for their theories to have a firm foundation in science, the only significant body of scientific expertise that was actually available to them comprised the Newtonian laws of motion and mechanics.[176] Needless to say, there are very severe limitations upon the level of understanding that can be attained with regard to the operations of human society and its relations with the natural world on the basis of analogies drawn from the circling of the planets and the collisions of billiard balls! While such a knowledge base, particularly when combined with the then nascent disciplines of chemistry and thermodynamics, proved an excellent springboard for launching the industrial and urban revolutions in human lifestyle, it had little or nothing of value to say about the preservation and cultivation of the natural resource base upon which all such developments must ultimately be posited. Indeed, such resources were widely asserted to be inexhaustible.[177] Accordingly, luminaries such as Kant and Hume explicitly advocated the clearance of forests and the draining of marshes as a means of advancing human welfare and convenience.[178] In Kant's mind, this was merely the

---

[175] See generally Sands and Peel (n 57), chapters 18–21.

[176] On the historical development of science, see, for example, G Johnson, *Fire in the Mind* (Penguin, 1995); E Mayr, *This Is Biology* (Belknap, Harvard UP, 1997), especially chapters 1–3; R Holmes, *The Age of Wonder* (Harper, 2008); Capra and Luisi (n 86). On the mechanistic worldview specifically, see, for example, DC Goodman and J Hedley Brooke (eds.), *Towards a Mechanistic Philosophy* (Open University Press, 1974).

[177] See, for example, CJ Glacken, *Traces on the Rhodian Shore* (1967; p/b edn., University of California Press, 1992), chapter 14, discussing the views of Buffon, Montesquieu and other Enlightenment luminaries.

[178] I Kant, *Critique of Judgment* (1790, 2nd edn., 1793; modern OUP edn., N Walker (ed.), 2007), part II, 1st Division, para 67; D Hume, 'Of the Populousness of Ancient Nations',

application of those powers of rational deliberation which had liberated humankind from the constraints of natural causation!

The crucial problem here was that, even by the very end of the period in question, the most basic principles of biology were only just beginning to be formulated, while those of genetics, psychology, ecology and complex systems still remained effectively in the realms of *terra incognita*, if not *terra improvisa*. Eventually, moreover, even the one sure source of scientific guidance at the time, Newtonian physics, was to be exposed by the emergence of quantum theory as providing no more than the most rough-and-ready depiction of underlying reality. Accordingly, for all the undoubted intellectual genius from which they were spun, and the nuggets of genuine insight that they contained, far too many of the pronouncements of these luminaries rested on little more than personal preconception, tendentious speculation and recycled superstition. Even at the very summit of their sophistication, Enlightenment conceptions of human nature and relationship to the rest of the living world were far too primitive and preposterous to merit any serious measure of continued credence. Yet much of our contemporary thinking is still inclined to treat them as if they were unchallengeable truth.[179] Indeed, it must surely be counted as the major intellectual tragedy of the 19th and 20th centuries that, as intellectual specialisation set in, disciplines such as philosophy and economics opted progressively to abandon the crucial Enlightenment injunction to ensure that all theorising remained continuously grounded upon the principles of natural science, while at the same time appearing to embrace wholeheartedly the crudely unenlightened 'junk' that was merely the product of its premature application.[180] As a result, we have allowed ourselves to be seduced by a worldview that is altogether too simplistic, atomistic, mechanistic, materialistic, individualistic, humanistic, hubristic and, in the final analysis, wholly unrealistic.[181]

Yet, as vital as it is for the flourishing of the human species that this worldview be brought back into line with reality, it cannot plausibly be supposed that international wildlife lawyers in their capacity as such can be at the forefront of the revolution in thinking that is so urgently required. As concerned citizens, they should undoubtedly endeavour to ensure that their voices remain prominent in the escalating chorus for

---

included in *Essays and Treatises on Several Subjects* (Vol. 1, new edn., T. Cadell, 1793), 370–441, 432. This was a common attitude of the time.

[179] For an extreme illustration of the corrosive effects of a traditional PPE education on the understanding of environmental issues, see N Lawson, *An Appeal to Reason: A Cool Look at Global Warming* (Duckworth Overlook, 2008).

[180] Foremost amongst these sources of misdirection were the writings of commentators whose primitive conceptions of evolutionary theory and natural selection were deployed in order to substantiate their own ideological perspectives upon issues of politics, sociology and economics: note especially in this context W Bagehot, *Physics and Politics* (Batoche Books, 1872) and the prolific output of Herbert Spencer, who produced multi-volume treatments on a wide range of subjects, including biology, psychology, philosophy and sociology.

[181] For a fascinating exposition of the psychological and neuro-scientific underpinnings of this imbalance, which is attributed to the unwelcome degree of supremacy that has emerged in our thinking on the part of the abstract and analytical processes of the left cerebral hemisphere over the more concrete, integrated and contextualising functions of the right, see I McGilchrist, *The Master and His Emissary: The Divided Brain and the Making of the Western World* (Yale UP, 2010).

reform, but it is to the practitioners of the disciplines that lie at the heart of the problem that the world must principally look for change.[182] With that in mind, it is encouraging to note that some of the most insistent demands for the transformation of economics are currently emanating from undergraduate students in that field, who not unnaturally feel entitled to expect that their studies should equip them with the kind of knowledge that might actually withstand scrutiny and application in the real world.[183] Perhaps a similar openness to radical reform within the discipline of philosophy might do something to counter the depressingly low levels of current student interest in the subject.

Nevertheless, it would be naive in the extreme to suppose that it will be easy to expunge the pervasive imprints of this intellectual legacy from popular consciousness, not least because it appears to accord so closely with the semi-detached, suburban, self-indulgent consumerist lifestyles that already represent the ongoing reality for many of us, and the ardent aspiration for countless others. It may, however, be feasible to ensure that such a mindset comes at least to be balanced by one that reminds us that our apparent disjunction from the natural world is a mere illusion of short-term perspective, and thereby serves to bring us back down to (mother) earth.[184] In particular, it will be salutary to be reminded that all the crucial ecological services and life support systems upon which our existence ultimately depends, including the very air that we breathe, do not represent an inevitable, natural outcome of the aggregation of matter into planetary form, but have been patiently assembled for us over time by the innumerable interactions of the diversity of life-forms that have chanced to emerge (perhaps uniquely in the universe) on one particular planet – ours. Furthermore, we disrupt the fundamental processes of this biosphere at our peril, since, however much we may flatter ourselves, the fleeting fads and fancies of *arriviste* decision-makers of our ilk may struggle to match the wisdom and rationality encapsulated in three-and-a-half billion years of small-scale, highly distributed and continuously sustained trial-and-error experiments in life support.

Nor would we seem well-advised to place our faith in those enduring but as yet inadequately substantiated Humanist assumptions regarding the unrelenting tendency towards progress in human affairs,[185] since the historical record (the absolute entirety of which is in any event currently far too short in evolutionary terms to permit any

---

[182] It is open to question whether even the more progressive elements within the discipline, who concede (p. 8) that nature 'has been ill-served by 20th century economics', can display the flexibility and imagination needed to transform their discipline: *see* D Helm and C Hepburn (eds.), *Nature in the Balance: The Economics of Biodiversity* (OUP, 2014).

[183] *See* Ha-Joon Chang and J Aldred, 'After the Crash, We Need A Revolution in the Way We Teach Economics' *The Observer* (London, 11 May 2014), which notes inter alia the launching of the International Student Initiative for Pluralism in Economics, now a collaboration of 65 student associations from 30 countries distributed across six continents, as to which *see* <http://www.isipe.net>. For a disappointingly (if predictably) muted account of these developments, *see* the *Economist* magazine (London, 7 February 2015), 27.

[184] This is part of the purpose of the UN's Harmony with Nature project, as to which, *see* note 103 above.

[185] For a sample of writings reflecting this way of thinking, *see* I Kramnick (ed.), *The Portable Enlightenment Reader* (Penguin, 1995), 351–395.

meaningful conclusions in this regard) might actually be interpreted to demonstrate that apparent steps forward in one area of activity tend to be offset or even occasionally outflanked by regression in others. So here it is evident that humans in pre-industrial societies will have maintained a much clearer grasp of our relations with the natural world, for the simple reason that their everyday lives more directly and immediately depended upon it. Subsequently, however, our intuitive sense of such matters has been progressively eroded by the very nature of urban living, occupational specialisation and participation in the exchange economy – the very phenomena that Enlightenment scholars trumpeted as the engines of progress. Today, we must recognise that, for all the benefits they may have brought us, they have also caused us to lose sight of certain fundamental truths. Accordingly, if we are to retain any realistic prospect of fulfilling our aspirations for sustainable development, we will need not only to establish a viable vision to govern biodiversity conservation for the year 2020 and beyond, but also, as a crucial part of that process, somehow to contrive to restore or even improve upon our original capacity for 20:20 vision with respect to the workings of the world around us.[186]

Hopefully, this process is already under way, since the accumulated wisdom of the chthonic communities is explicitly recognised in the CBD as a resource which states should protect and draw upon, wherever appropriate, in their formulation and application of conservation policy.[187] In that respect, it has the potential to serve variously as an inspiration, a supplement, a placeholder, a means of corroboration or even occasionally as a short-term corrective to the proceeds of more formal scientific endeavours, reference to which is also recurrent within the Convention.[188] Through deployment of these twin sources of elucidation in the form of an epistemological pincer movement,[189] it is hoped that the more ill-informed and environmentally destructive relics of Enlightenment thinking can progressively be expurgated, in the interests of the international community as a whole.

5.2 Sharpening the Focus upon 'Community'

This discussion should also serve incidentally as a reminder of the twin senses of the word 'community' itself, as highlighted in Section 3 above, and in particular of the virtues of the pluralism and diversity which is still to be found within our global society, and of the way in which it may be possible to draw upon it in circumstances

---

[186] '20/20 vision is a term used to express normal visual acuity (the clarity or sharpness of vision) measured at a distance of 20 feet. If you have 20/20 vision, you can see clearly at 20 feet what should normally be seen at that distance. If you have 20/100 vision, it means that you must be as close as 20 feet to see what a person with normal vision can see at 100 feet': American Optometric Association, available at <http://www.aoa.org/>.
[187] CBD, Article 8(j) and 12th preambular recital.
[188] Most obviously in the 7th preambular recital and then in Articles 12(a), 15–19 and 25, but reaffirmation of the relevance of science is actually pervasive throughout.
[189] By a curious coincidence, an interesting example of precisely this combinatorial process was reported on the very day that this passage was drafted: see E Sohn, 'Inuit Wisdom and Polar Science Are Teaming Up to Save the Walrus', available via <http://www.smithsonianmag.com/>, and derived from 'What Now Walrus?', *Hakai Magazine* (Victoria, BC, Canada, 22 April 2015).

where some broad uniformity of response must ultimately be adopted in response to the emergence of threats which are common to us all. Indeed, if there is anything to be identified as a guiding 'vision' for international wildlife lawyers, above and beyond the crucial need for them to 'stick resolutely to their legal last' in the analysis and evaluation of their subject-matter, and avoid the distractions generated by popular polemics, it might perhaps be found in the furtherance of understanding and practical realisation of the concept of *community* in all its various nuances. For a community in its most basic sense is, as suggested above, essentially no more than an aggregation of (potentially very disparate) entities interacting within a defined boundary. As a variety of complex living system, it is defined principally by simple reference to the fact that the exchange of energy within the encircling rampart is markedly more intense than that which occurs beyond it. The two things that will necessarily prove fatal to its capacity for self-maintenance are (i) if the barrier becomes *totally* impermeable to energy from without (for any closed system is doomed to the progressive dissipation and extinction of its internal order in accordance with the second law of thermodynamics) and (ii) that it should somehow lose the capacity to utilise this exogenous energy for the preservation of that internal order (in which case it will eventually disintegrate into and become indistinguishable from its surroundings in a functional sense). The biosphere itself represents a paradigm case here, because heterotrophs such as ourselves are entirely dependent upon the autotrophic life-forms to undertake the initial energy conversion (and starch creation) process which keeps us alive,[190] while the service of recycling energy through the system is performed on our behalf by innumerable unheralded organisms from bats, bees and beetles to the microbes that permeate the soil.[191] At the other end of the scale, apex predators and other keystone megafauna help to keep the whole system under overall operational control.[192] It is therefore to the preservation and protection of community in its biological sense, and hence to the very possibility of community *in any sense*, that the work of wildlife law is ultimately directed.

It follows from this that no conservation regime can hope to be remotely effective unless it pays adequate regard to the principles of conservation science, and especially its holistic focus, and it is precisely that consideration that has led to the gradual transformation of wildlife law from an essentially species-oriented endeavour to one that concentrates much more broadly on entire ecosystems and the biological communities of which they are composed. Accordingly, the ecosystem approach has become critical to the implementation of the CBD itself and of the other treaties that

---

[190] Autotrophs are organisms that can make their own food by synthesizing organic matter from inorganic substances, whereas heterotrophs can only derive their food from organic matter: Park (n 85), 36, 211.

[191] It is for such reasons that one very recent opinion piece highlights midges, soil mites, dung beetles, fruit bats, ants, parrotfish and the cyanobacterium *Prochlorococcus* as the life-forms deserving of our most urgent attention: P Barkham, 'Which Would You Save?' *BBC Wildlife* magazine (Bristol, February 2015), 28, 35.

[192] For an overview of the science, *see* C Fraser, 'The Crucial Role of Predators: A New Perspective on Ecology' *Yale Environment 360* (New Haven, CT, 15 September 2011), available via <https://e360.yale.edu/>.

operate within its broad sphere of influence.[193] For the lawyer, the crucial questions therefore concern the extent to which treaty regimes are currently giving adequate attention both to the major threats that have been identified to the functioning and integrity of biological communities, and to the optimal forms of response that have been specified to address them.

5.3 Maintaining Momentum: The Present Handbook

It will therefore come as no surprise that a key function of the present work is to address such questions. In that vein, Arie Trouwborst analyses the extent to which treaty regimes are successfully addressing the problem of habitat fragmentation, while Peter Davies investigates, through the prism of the EU regime, the ubiquitous problems posed by alien invasive species. The incidence of armed conflict represents another recurrent threat to wildlife, and this is examined in the chapter by Karen Hulme. The conservation of vulnerable marine ecosystems is considered by Edward Goodwin, while Rosemary Rayfuse also examines the marine environment, but specifically in the light of the hazards posed by climate change, not only in the form of global warming, but through the process of ocean acidification. With a particular emphasis on response strategies, Kees Bastmeijer considers the extent to which the managed restoration of ecosystems and their essential functioning processes represents a feasible solution or palliative to past failures to prevent harm.

In addition to illuminating the substantive conservation principles by which wildlife treaty regimes are informed, these chapters shed incidental light upon the institutions and procedures through which they are operationalised, and this issue of course represents a crucial focus of study in its own right. It should be noted that the discipline of international environmental law has made a very significant contribution to the development of the international legal system generally, through its exploration of the various ways in which the treaty instrument can be reconceived and elaborated in order to maximise its potential impact. In particular, the investment of treaty regimes with elaborate institutional arrangements of a dedicated, quasi-autonomous kind has helped to transform the 'sleeping treaties' – or perhaps, to adopt an alternative analogy, *abiotic mechanisms* – of the past into the familiar 'living instruments' of the modern era.[194] These biological metaphors are employed advisedly, for contemporary conservation regimes are organic in more than a purely figurative sense, by virtue of their creation of a specialised form of political community for the achievement of their objectives. Furthermore, it can be argued that the strength of any system of this kind is likely to depend considerably upon the extent to which it successfully replicates the essential features of a natural ecosystem.[195] Indeed, a political community, as noted above, is

---

[193] For an explanation, *see* <https://www.cbd.int/ecosystem/>.
[194] RR Churchill and G Ulfstein, 'Autonomous Institutional Arrangements in Multilateral Environmental Agreements: A Little-Noticed Phenomenon in International Law' (2000) 94 AJIL 623.
[195] *See* MJ Bowman, 'Beyond the "Keystone" CoPs: The Ecology of Institutional Governance in Conservation Treaty Regimes' (2013) 15 International Community Law Review 5–43

itself merely a variety of complex living system, and the plant geneticist Enrico Coen has recently argued that the same handful of basic principles in fact govern the functioning of all such entities, from individual living cells to entire human civilisations.[196]

Given the importance of diversity as a source of strength, of critical importance to the maximisation of the potential of treaties is the question of *participation*, and especially its extension beyond the sphere of statehood. This question must therefore be kept under continuous scrutiny, and for the purposes of this work Elizabeth Kirk examines specifically the role of non-state actors in certain marine biodiversity treaty regimes. Needless to say, the litmus test of the effectiveness of treaty institutions in this context will be seen by many as relating to the extent to which they prove successful in holding states to their commitments. Karen Scott's contribution is accordingly devoted to the investigation of the various procedures that have been adopted under wildlife treaties to date to address the crucial issue of non-compliance. One of the many interesting points to emerge from this study is the increasing occurrence of joint ventures by treaty bureaucracies in the implementation of such procedures, which serves as a reminder that such regimes operate not in splendid (or, perhaps one should say, wretched) isolation, but as part of a broader network or community of biodiversity-related Conventions. It would seem a reasonable aspiration that these regimes should not in any way compromise or undermine each other's work, and wherever possible seek synergies and opportunities for collaboration. The focus of Richard Caddell's chapter is therefore upon the extent to which such aspirations have so far been realised in practice, and the prospects for further instances of mutual reinforcement in the future.

This chapter also serves to confirm a point which is evident from certain other contributions already mentioned above, namely that many communities, whether of a normative or other character, commonly find themselves nested, after the fashion of Russian dolls, within broader systems or forms of polity. Thus, the network of biodiversity-related treaties to which Caddell refers can itself be located within a wider field of instruments devoted to environmental protection (as instanced by the interface between biodiversity and climate change, discussed by Rayfuse), and in turn within the still broader domain of international law generally (as exemplified in this work by the interaction of conservation norms with the regime governing armed conflict, explored by Hulme). As the perspective progressively widens in this way, the challenge of maintaining an appropriate degree of normative coherence tends to intensify, a point which is demonstrated very graphically in the chapter by Emilie Cloatre. This concerns the possibility of successfully reconciling – or at least reaching some form of workable accommodation between – the sharply contrasting regulatory demands of commerce and industrial property on the one hand and anthropology and human rights on the other in relation to the exploitation of biological resources. Since it is in the

---

and M Fitzmaurice and D French (eds.), *International Environmental Law and Governance* (Brill/Nijhoff, 2015).

[196] These are population variation, persistence, reinforcement, competition, cooperation, combinatorial richness and recurrence: for elaboration, *see* E Coen, *Cells to Civilizations: The Principles of Change that Shape Life* (Princeton UP, 2012).

Law, legal scholarship and conservation 47

conservation forum that these contrasting perspectives have chanced to meet head on, the question arises as to whether the CBD represents an appropriate crucible to achieve a viable amalgam, or whether some other venue or approach might be needed for this purpose. It should also be apparent that the tension here arises from the clash of two very different cultural communities and epistemic traditions; one is derived from the Enlightenment injunction to subjugate, control and assert ownership over the natural world, and the other from a long pre-existing and hopefully still active disposition to submit to the rhythms of its regenerative process and thereby be prepared, at least to some extent, to dance to its chosen tunes. On this view, *membership* rather than ownership, and *harmony* rather than hegemony, may have to be the vital watchwords for our relationship with nature in the future. A viable legal solution to the problems regarding intellectual property is only likely to emerge from practice, but it should come as no surprise that this practice might itself entail a very great deal of trial and error.

Certainly no single faction or interest group should feel entitled to assume that its own preferences or proclivities will automatically prevail in this regard, but the fact that conservation has been conclusively declared to represent the common concern of *all humankind* should at least give every relevant constituency a measure of standing in the matter. Even the hallowed, and strongly reaffirmed,[197] sovereignty of states over their living natural resources must now be exercised in conformity with the normative ethos of the CBD and other relevant instruments. Undeniably, the precise nature and implications of the common concern idea are not entirely pellucid, and Duncan French duly takes the opportunity in his chapter to offer further elaboration of the way in which the concept is currently evolving. Plainly, the attribution of this status stems from the acknowledged importance of biodiversity for the pursuit of sustainable development, which has become one of the defining objectives of the international community as a whole. Yet this idea too, for all the attention that it has commanded from commentators, remains open to ongoing interpretation and exegesis, and Veit Koester accordingly takes on the task of examining the extent to which it has been respected and applied in the practice of the Biodiversity Convention itself. What cannot be denied is that a commitment to equity, in both its inter- and intra-generational dimensions, represents a key component, and it is this issue which forms the focus of the chapter by Malgosia Fitzmaurice, as reflected in the ongoing controversies played out in the IWC, the constituent instrument of which was one of the first to contain explicit reference to such matters.

Each one of these fundamental principles of international environmental law would seem to place a premium on the question of *participation*, since it has become widely accepted that the dictates of good governance demand that all those who form part of a political community should have some role to play in the elaboration of the normative processes by which they are regulated. As noted above, the issue is tackled directly in relation to marine biodiversity treaty regimes in the chapter by Kirk, and also represents a pervasive theme of Cloatre's contribution insofar as it impacts upon indigenous communities and other traditional users of biological resources. The elucidation of this topic is enriched still further by Nicole Mohammed's instructive case

---

[197] See especially CBD, Article 3.

study of participatory resource management in the Caribbean, which suggests that there are very considerable problems to be overcome in order to make the aspiration of public participation a meaningful reality.

The significant practical problems associated with public participation produce the inevitable result, in many contexts, that it has to be realised in representative mode rather than personally.[198] That being so, it is relevant to ask whether there might be claims or interests that merit representation in appropriate forums and procedures regardless of the fact that the entities to which they relate could never be envisaged as direct participants in their own right and on their own behalf. To put it another way, many – indeed, the vast majority – of those who hold stakes in the ultimate success of our conservation endeavours are not human at all, raising the question of whether and how their interests are to be taken into account. At root, of course, this is a matter of values, and whether only those of an exclusively anthropocentric orientation are worthy and capable of accommodation. This key question is revisited in the chapter which immediately follows the present one, in which Mattia Fosci and Tom West conclude that it would be preferable to pursue policies that aim to be compatible with both intrinsic and instrumental modes of valuation than to continue to argue about their relative importance.

5.4 Final Reflections

This preference for pluralism seems very much in line with the conclusions that have begun to emerge from reflections on the financial crash and other policy disasters of recent years. As one leading financial expert puts it:[199]

> If you are clear about your high-level goals and knowledgeable enough about the systems their achievement depends on, then you can solve problems in a direct way. But goals are often vague, interactions unpredictable, complexity extensive, problem descriptions incomplete, the environment uncertain. That is where obliquity comes into play.

Where conservation is concerned, the ultimate truth is probably that even if we were disposed to be utterly selfish, we simply do not know enough to determine which elements of the natural world are critical to our own survival or success and which we can safely allow to go to the wall. The one thing of which we can be sure is that we are all part of the same global biotic community, and whereas a community of this kind does not necessarily involve much in the way of *commonality*, it does invariably involve highly complex networks of *interdependence*. It was considerations such as these that caused Aldo Leopold to conclude long ago that 'a thing is right when it tends to preserve the integrity, stability and beauty of the biotic community. It is wrong when it tends otherwise.'[200] In consequence, we shall probably arrive ultimately at the same

---

[198] To take an obvious example, the rules governing participation in CoP meetings – as to which *see* notes 194–195 and accompanying text – tend to focus upon 'bodies' or 'agencies' rather than individuals.

[199] J Kay, *Obliquity: Why Our Goals Are Best Achieved Indirectly* (Profile Books, 2010), 178.

[200] A Leopold, *A Sand County Almanac* (1949, Ballantine Books reprint 1986), 224–225.

broad policy conclusions regardless of whether we resolve initially to set off along the road of prudential self-interest or that of expansive ethical altruism: as the 1991 revised version of the *World Conservation Strategy* made clear, respect and care for the community of life represents the underlying ethic that provides the foundation for all the other principles on which sustainable living depends.[201]

Although we cannot easily calculate what level of investment in conservation is likely to be needed for this purpose, we should probably bank on it being considerably more than we have hitherto imagined we could possibly afford.[202] This realisation helps, indeed, to explain why it is difficult to understand how an unreformed discipline of economics can ever play more than an important supporting role in the determination of policy priorities: by systematically disenfranchising all those beings who do not participate in the exchange economy, it merely serves to perpetuate the manifold and intractable problems that have been caused by the commodification of nature in the first place. Perhaps, then, its only real hope for redemption would lie in the recognition of historic property rights over natural ecosystems on the part of all those species that have traditionally inhabited them, and asking them (or more plausibly some trustee acting on their behalf) how much they would be prepared to accept by way of compensation for any projected loss or despoliation of their habitat!

Since such an intellectual revolution might well take some considerable time to deliver, it would be preferable for the practitioners of other disciplines not simply to wait for it to happen, but rather to attempt in the meantime to make some more immediate progress of their own in the same general direction. For lawyers, this might entail exploring the various ways that might be found for giving a voice (if only through surrogates) to those countless stakeholders in the conservation process who, in the very nature of things, are constitutionally disabled from participation in verbal deliberation – that is to say, all those not of our own particular (sub-)species. If this should seem outlandish, it is to be remembered that the proposal already finds reflection in current arrangements to some extent through the admission to meetings of many treaty bodies of certain non-governmental entities 'technically qualified in protection, conservation or management of wild flora and fauna'.[203] Bodies with a proven record of positive contribution may in some cases see this recognised through the formal grant of 'partnership' status, which typically offers enhanced opportunities for participation. If there are treaties which deny or unduly hamper such engagement on the part of civil society, their practices might profitably be reconsidered.[204]

---

[201] IUCN/UNEP/WWF, *Caring for the Earth: A Strategy for Sustainable Living* (IUCN, 1991), *see* especially chapters 1 and 2. *See* further the chapter by Fosci and West in this volume.

[202] To set things in context, the total value of the ecosystem services provided for us by nature each year has been estimated to be worth around double global GDP: A Juniper, *What Has Nature Ever Done for Us? How Money Really Does Grow on Trees* (Profile Books, 2013).

[203] As per CITES, Article 11(7), for example.

[204] Fisheries treaties might seem to be a case in point: for discussion, *see* GM Wiser, 'Transparency in 21st Century Fisheries Management: Options for Public Participation to Enhance Conservation and Management of International Fish Stocks', a revised version of a WWF paper presented to a major fisheries conference in 2000, available via <http://www.ciel.org/Publications/trans21cenfisheries_options.pdf>; also D Symes, 'Fisheries Governance: A Coming of Age for Fisheries Social Science' (2006) 81 Fisheries Research 113–117.

Yet this approach does not of itself guarantee representivity of the kind envisaged, for it is not normally a requirement that such groups serve specifically as advocates for nature's interests: many doubtless do, but others may not at all, especially where their qualification for entry is derived, say, from the function of 'management' rather than 'protection'.[205] Consequently, there is scope for the exploration of other possibilities, especially since most of these are unlikely to be mutually exclusive. It should not be assumed, moreover, that states themselves have no part to play in this process, and it may indeed be worthwhile to explore further the process of designating particular states as 'champions' for certain species akin to that dabbled with by the Bonn Convention in connection with its 'concerted actions' procedure.[206] Perhaps other comparable avenues will come to light from a detailed exploration of practice under wildlife treaties generally – and if this is now to be described as making 'interstitial headway', then so be it.

A further point to note is that the route to more effective representation of nature's interests should certainly not be confined to the sphere of activities conducted under the aegis of existing conservation treaty institutions, as there may be scope for the utilisation also of judicial process for this purpose, whether at the national, regional or international level. Some examples have already been encountered earlier in this chapter, and it is for wildlife law researchers to ensure that there is a sufficient body of serious literature available to provide potential litigators with an adequate array of weapons for their forensic armoury. In addition to already familiar forms of legal action, the potential of new approaches should no doubt also be explored. There could hardly be a more auspicious moment than now, for while it was predictable that attempts to utilise the procedures of the European Convention on Human Rights for the benefit of hominids beyond our own species would be summarily rebuffed,[207] it cannot pass unremarked that various media sources have reported, though perhaps prematurely, a writ of *habeas corpus* being granted in national law for the benefit of just such an unconventional applicant – to wit, an orang-utang in the courts of Argentina, following an earlier case concerning a chimpanzee in Brazil, where the judge declined to rule out the possibility *in limine*.[208] While these cases obviously concern captive animals, seen purely as individuals, any such breaches in the barrier of legal personhood may conceivably carry implications in the longer term for their counterparts in the wild as well, and on a collective basis. Indeed, it is noteworthy that there is already one wildlife treaty, the 2007 Agreement for the Conservation of Gorillas and Their Habitats,[209] which appears to place unusually high emphasis upon the moral status and substantive interests of gorillas when set alongside that of their close relatives, our own

---

[205] Both activities are mentioned, alongside conservation, as relevant forms of expertise for the purposes of CITES, for example; see Article 11(7).

[206] For discussion, see *Lyster*, 574–576.

[207] ECHR admissibility decisions are no longer routinely published, but the cases in question are discussed by Judge Pinto de Albuquerque in his partly concurring/partly dissenting opinion in the recent case of *Herrmann v Germany*, Application No. 9300/07, Grand Chamber Judgment of 26 June 2012: see in particular footnote 22 and accompanying text.

[208] For information on such developments generally, *see* <http://www.nonhumanrights project.org>.

[209] Text available via <http://www.cms.int/gorilla/en/documents/agreement-text>.

species.²¹⁰ It does not go so far as to recognise formal legal rights on their part, however, and it remains uncertain whether any such development in international law could offer a viable route towards progress in the field of conservation. Nevertheless, the issue certainly seems worthy of further consideration and research.²¹¹

This suggests that the time may well have come when Christopher Stone's memorable question, 'Should Trees Have Standing?' must receive more active and systematic attention.²¹² Certainly, the 2012 Rio+20 Conference on Sustainable Development not only reaffirmed the need for promoting 'harmony' with nature but went on to acknowledge that some countries already recognise the *rights* of nature in this context.²¹³ This recognition may sometimes, perhaps, be essentially of a purely rhetorical character, but recent developments suggest that some states have been prepared to go further: thus, the website which the UN has dedicated to the Harmony with Nature project makes reference to what is presented as the first victory and explicit vindication of the rights of nature in a national court: a case in 2011 concerning the River Vilcabamba in Loja province, Ecuador.²¹⁴ There is, moreover, other evidence that such ideas are now beginning to acquire a degree of traction in the legal realm.²¹⁵

Finally, and back outside the courtroom, consideration could profitably be given to the creation of entirely new types of legal forum or institution, despite the alleged waning of enthusiasm amongst governments for initiatives of this kind. It is noteworthy that the coming to a close of the Decade for Biodiversity will coincide almost exactly with the centenary of the emergence onto the global stage of an institution that was entirely revolutionary for its time: the International Labour Organization.²¹⁶ At a time when international law was resolutely statist in orientation, and well before the formal consolidation of human rights into the system, the creation of a body in which a

---

²¹⁰ Note in particular Article III(2)(a), (j).

²¹¹ For an excellent foundation, though without any particular focus on international law, *see* CR Sunstein and MC Nussbaum (eds.), *Animal Rights: Current Debates and New Directions* (OUP, 2004).

²¹² His article of that name was sub-titled 'Toward Legal Rights for Natural Objects' and published in (1972) 45 Southern California Law Review 450. See also his book of the same title, published by Oxford University Press (3rd edn., 2010).

²¹³ See the outcome document *The Future We Want* (2012), and especially paragraph 39, available via <http://www.un.org/en/sustainablefuture/>.

²¹⁴ A report (in Spanish) is available via <http://www.harmonywithnatureun.org/rightsofnature.html>. For a summary in English, with accompanying commentary, *see* N Greene, 'The First Successful Case of the Rights of Nature Implementation in Ecuador', available at <http://therightsofnature.org/first-ron-case-ecuador>.

²¹⁵ Note in particular the 2012 Wanganui River Treaty, a framework agreement between the Crown and certain indigenous communities in New Zealand, which recognises the river as a living entity in its own right, incapable of being 'owned' in an absolute sense and enjoying legal standing of its own. The text may be viewed at http://www.harmonywithnatureun.org/content/documents/193WanganuiRiver Agreement–.pdf.

²¹⁶ The ILO was created pursuant to Part 13 of the 1919 Treaty of Peace with Germany, 225 CTS 188. For its website, *see* <http:www.ilo.org/>, and for discussion of its activities *see*, for example, S Hughes and N Haworth, *The International Labour Organization (ILO): Coming in from the Cold* (Routledge, 2011); JM Servais, *International Labour Organization (ILO)* (Kluwer Law International, 2011); F Maupain, *The Future of the International Labour Organization in the Global Economy* (Hart Publishing, 2013).

tripartite structure was created to give an equal voice to workers and employers alongside that of governments must surely be counted as one of the most imaginative of all developments within the global legal order. Perhaps there is now scope for a similar expansion of consideration to those who currently stand outside the pale of legal personhood.

Yet whatever the need for investigating such radical initiatives, it seems clear that wildlife lawyers should not allow themselves to become too carried away in their pursuit. The bulk of the research agenda, as reflected in the contributions to this work, should continue to be devoted to more mainstream issues, focusing on the tweaking of existing systems designed to secure measurable, if usually small-scale, improvements in the delivery of their fundamental objectives. Whatever the precise orientation of their efforts, moreover, researchers should prepare themselves for the dispiriting possibility that their labours may seem to have regrettably little discernible impact in practice, or even to go entirely unheeded. Hopefully, however, all such work will contribute in some way to the overall drip-feed of pressure by which governments and other key decision-makers are over the course of time imperceptibly affected. After all, the quantity of oxygen produced by a single cyanobacterium is doubtless infinitesimally small, but collectively they have created the essential conditions under which all the more complex life-forms, including ourselves, could develop.

So, too, in the sphere of international environmental governance, the chorus of pressure emerging from various disciplines has already produced some highly significant and promising developments during the Decade for Biodiversity. And if it should sometimes seem discouraging that more dramatic progress is so difficult to come by in this context, wildlife lawyers will simply have to find ways of stiffening their resolve, using whatever form of metaphorical starch might come to hand; perhaps it will help to be reminded that their collective efforts are devoted to securing the processes by which the literal version of this complex chemical is generated, and hence to preserving the very staff of life itself, not only for themselves, but for their children and their children's children.

If, on the other hand, the need is felt for some more tangible form of incentive and inspiration to motivate endeavours in this field, it always pays to remember that there are certainly plenty of individual conservation success stories to hold on to among the standard tales of doom and gloom.[217] Here is one that holds particular significance for me.

From my schooldays in the 1960s, I recall the graphic accounts in the media regarding the collapse of peregrine falcon numbers in the UK and elsewhere, as the calamitous effect of pesticides upon their breeding success served to exacerbate the more traditional threats posed to them by habitat encroachment and direct human persecution.[218] Indeed, their decline seemed almost as precipitate as the spectacular

---

[217] For a valuable selection of examples, see the FFI publications *Fauna & Flora,* Issue 18, December 2013 and its special supplement, 'Many Happy Returns from the Brink: Celebrating Fauna and Flora International's 110th Anniversary' and *Fauna and Flora International Update No. 27,* September 2015.

[218] JJ Hickey (ed.), *Peregrine Falcon Populations: Their Biology and Decline* (University of Wisconsin Press, 1969); DA Ratcliffe, *The Peregrine Falcon* (T and AD Poyser, 1980).

aerial 'stoops' by which they hurtle down on their unsuspecting prey. I knew little about such matters at the time, but, growing up in the London area, it was difficult to understand how such seemingly exotic creatures could possibly survive at all in a country that was so heavily urbanised and industrialised, whatever protective measures might be put in place.

Twenty years later, when I was established in a career as a university lecturer and contemplating the introduction of wildlife conservation into the Nottingham legal curriculum for the first time, I heard of the recolonisation by a pair of peregrines of one of their traditional breeding sites at Symonds Yat, a rocky outcrop along the River Wye, on the Gloucestershire/Herefordshire border. The Royal Society for the Protection of Birds had established a wardenship scheme and a public viewpoint across the river in the hope of garnering popular support for their attempts to translate formal protective legislation for this species into practical conservation success. Like thousands of others, I went to see these magnificent birds, though without really knowing whether their reappearance at this site signalled a genuine recovery in numbers, or was merely a last, defiant gesture in the death throes of the species in Britain. Whatever the truth of that particular saga, wildlife law duly secured a foothold in the curriculum, initially as part of a single, optional course in international environmental law designed to serve our new LLM degree in international law.

Now, a further 30 years on, the international law of biodiversity seems well entrenched here, with 50 undergraduate subscribers this session and a regular cohort of postgraduate students, some of whom devote their entire masters programme to the study of environmental law. And should you be more concerned (as I hope you will be) about the actual targets of all this academic attention, it may interest you to note that, as I left the building after the first seminar of the 2015–2016 session, a peregrine falcon could be seen hunting from the clock tower almost immediately above the room in which the class had taken place: so successful, indeed, has been the recovery programme for the species globally that their conservation status is now categorised as 'Least Concern' by IUCN. Iconic, once gravely imperilled, raptors silhouetted against the evening sky, apparently flourishing on the fringes of a major conurbation set in the heart of the English industrial midlands[219] – now that is a 'vision' worth holding on to!

SELECT BIBLIOGRAPHY

Bowman, MJ, PGG Davies and CJ Redgwell, *Lyster's International Wildlife Law* (2nd edn., CUP, 2010)
Capra, F and PL Luisi, *The Systems View of Life: A Unifying Vision* (CUP, 2014)
Cullinan, C, *Wild Law* (Siber Ink, 2002)
Fitzmaurice, M and D French (eds.), *International Environmental Law and Governance* (Brill/Nijhoff, 2015)
Gillespie, A, *Conservation, Biodiversity and International Law* (Edward Elgar Publishing, 2013)
Gillespie, A, *International Environmental Law, Policy and Ethics* (2nd edn., OUP, 2014)
Harmony with Nature, Report of the UN Secretary-General, UN Doc A/69/322, 18 August 2014
Helm, D and C Hepburn (eds.), *Nature in the Balance: The Economics of Biodiversity* (OUP, 2014)

---

[219] Indeed, peregrines have actually established nesting sites right in the very centre both of Nottingham and of neighbouring Derby, as can be observed over the internet through webcam arrangements serving each.

Kay, J, *Obliquity: Why Our Goals Are Best Achieved Indirectly* (Profile Books, 2010)
Mathews, F, *The Ecological Self* (Routledge, 1991)
McGilchrist, I, *The Master and His Emissary: The Divided Brain and the Making of the Western World* (Yale UP, 2012)
Mepham, B, *Bioethics: An Introduction for the Biosciences* (2nd edn., OUP, 2008)
Nussbaum, MC, *Frontiers of Justice* (Belknap, Harvard, 2006)
Wandesforde-Smith, G, 'From Sleeping Treaties to the Giddy Insomnia of Global Governance: How International Wildlife Law Makes Headway' (2012) 15 JIWLP 80

2. In whose interest? Instrumental and intrinsic value in biodiversity law

Mattia Fosci and Tom West

This chapter deals with the role of values in international biodiversity law. It begins by introducing the concepts of instrumental and intrinsic value (Section 1) and examining the extent to which they are reflected in international biodiversity law (Section 2). It then explores the two approaches in more detail, analysing their respective shortcomings and advantages (Sections 3 and 4). The conclusions look at reconciling these two theories of value within the same system of environmental protection.

1. DEFINING THE VALUE OF NATURE

The two main value-systems underpinning the development of biodiversity law are based on the seemingly competing concepts of instrumental and intrinsic value. Instrumental value typically describes the worth biodiversity derives from its human utility: for instance, the instrumental value of certain trees lies in timber production; animals can be used for food; plants and flowers for their medicinal properties and so forth. It relies on the assumption that a rational entity (the valuer) can attribute value to another entity (the valuee), and it endorses an anthropocentric view of the natural world founded on the conviction that humanity is superior to, and is thus permitted to exploit, the natural world. Instrumental valuation is a manifestation of instrumental rationality,[1] which also underpins rational choice as an economic, sociological and criminological theory.[2]

Bowman distinguishes instrumental value from inherent value, which he defines as the worth an entity possesses for its very existence rather than for its practical utility.[3] Instrumental value only considers the material and productive use of biodiversity, while inherent value also encompasses non-material and non-productive uses (such as those connected to aesthetic, religious, cultural and recreational considerations) as well as

---

[1] Instrumental rationality is a social phenomenon studied by psychologists, sociologists and philosophers; its key tenet is that rational beings seek the most efficient way to achieve a specific goal.

[2] *See* G Becker, *The Economic Approach to Human Behavior* (University of Chicago Press, 1976); D Blaikie Cornish and RVG Clarke, *The Reasoning Criminal: Rational Choice Perspectives on Offending* (Springer Verlag, 1986); G Homans, *Social Behaviour: Its Elementary Forms* (Routledge and Kegan Paul, 1961).

[3] MJ Bowman, 'Biodiversity, Intrinsic Value and the Definition and Valuation of Environmental Harm' in MJ Bowman and AE Boyle (eds.), *Environmental Damage in International and Comparative Law: Problems of Definition and Valuation* (OUP, 2002) 41–61.

non-use values (reflecting an individual's psychological preference).[4] Although important for understanding the historical evolution of the subject or for selecting appropriate methods of quantification, the separate recognition of inherent and instrumental value is not useful for the purpose of this discussion and will therefore be avoided. Instead, nature's utility for the external valuer is here included in the definition of (anthropocentric) instrumentalism regardless of whether it is productive, non-productive or psychological. The emphasis is placed on the difference between those values conferred by humans upon nature and the value possessed by nature without regard to human considerations: intrinsic value.

As to the latter, despite its widespread use in environmental ethics, intrinsic value has no agreed definition. It is thus worth providing some handles to understand the concept. O'Neill separates out three meanings of 'intrinsic value':[5] (i) value that is non-instrumental (value as an end in itself); (ii) value that is related to the intrinsic properties of entities (non-extrinsic); and (iii) value that exists independently from a human valuer (objective value). For the sake of clarity, all three meanings are combined here: intrinsic value is objective value that arises from an entity's intrinsic properties and renders a thing worthy of moral consideration.[6] Intrinsic value also conjures notions of being internally defined and self-referential: instead of there being a valuer and a valuee, the entity itself generates its own value. This notion implies that, although humans have constructed the concept of value, value exists outside of humanity: a world devoid of humans is not a world devoid of value.[7] Such a definition is not entirely uncontroversial: both the existence of value outside the psychological preferences of humans,[8] and the meaning of 'intrinsic properties' have room for discussion.[9]

---

[4] Non-use value, defined as the contentment people derive from knowing that a particular animal, species or ecosystem 'is preserved even if they will never directly use them', is more akin to an appreciation of intrinsic value yet this definition too retains, according to environmental philosophers, an element of utility.

[5] J O'Neill, 'The Varieties of Intrinsic Value' (1992) 75 The Monist 119. O'Neill discusses inherent value at 124–5. Inherent value is certainly compatible with the first meaning of intrinsic value, possibly with the second, but certainly not with the third.

[6] Because it is an end-in-itself; *see* W Fox, *Toward a Transpersonal Ecology* (SUNY Press, 1990), 193.

[7] One must remember that the human way of perceiving the world is only one such way and that looking for intrinsic value cannot seek out what is of value to humans in it: 'The value of a tree is not analogous, for example, to the greenness of its leaves. In "the world out there", there are only electromagnetic waves of 500 nanometres, not greenness. The human perceiver, however, sees the leaves as green. Thus, the greenness may be said to be a projection of the human perception onto the leaves themselves': K Lee, 'The Source and Locus of Intrinsic Value: A Re-examination' (1996) 18 Environmental Ethics 297, 298.

[8] This possibility is rejected by at least one well-known environmental philosopher, Callicott, who differentiates between the source and the locus of value: J Baird Callicott, 'On The Intrinsic Value of Nonhuman Species' in B Norton (ed), *The Preservation of Species* (Princeton University Press, 1986), 142. *Cf.* Lee (n 7).

[9] O'Neill's approximation (n 5) of 'non-relational' may be found wanting given the axiomatically relational nature of all living systems. In addition, it is unclear whether intrinsic properties are attached only to the very nature of an entity, or if they also include their state or condition. Excluding the latter means that endangered status cannot be used as an intrinsic value-based justification for protecting a species since rarity is, by definition, a relative property.

The above characterisations may seem rather simplistic. On the one hand is a selfish, egocentric human who takes decisions based only on personal gain, not accounting for the needs of any other organism. Similarly to Hobbes' self-interested man,[10] they are only able to instinctively appreciate their own individual value and not that of other humans, let alone other species. On the other hand lies an abstract conception of value existing in an intellectual realm. Intrinsic value requires looking outside our own immediate experience for guidance to actions and using empathy, sympathy and compassion[11] to adopt a more all-encompassing moral outlook. This depiction unfairly excludes the fact that both selfish and selfless impulses can and do cohabit in the human mind as well as in the law. However, the distinction holds for explicative purposes: selfishness and selflessness represent two extremes in a range of ethical positions that symbolise two approaches to the legal protection of biodiversity.

2. LOCATING VALUE IN INTERNATIONAL BIODIVERSITY LAW

The type and level of protection accorded to biodiversity in legal instruments is connected to the value-systems that underpin them. Instrumental value has historically been the dominant consideration in international environmental law. Early biodiversity treaties were concerned with the conservation of wildlife and fauna as exploitable 'natural resources'. Their focus resided in, inter alia: protecting whales as a source of food and oil;[12] managing fisheries to prevent their depletion;[13] protecting selected bird species useful to fight pests in agriculture;[14] and preserving wildlife as a hunting resource.[15] The main concern of these Conventions was to establish a management regime to promote the fair and equal access of all states parties to the exploitation of resources sited beyond national jurisdiction (i.e. through the imposition of quotas, particularly for fishing and whaling),[16] or to protect useful migratory animals from

---

[10] Thomas Hobbes, *Leviathan* (1651).
[11] See E Aaltola, 'Empathy, Intersubjectivity, and Animal Philosophy' (2013) 10 Environmental Philosophy 75.
[12] See, for example, the 1931 Convention for the Regulation of Whaling, 155 LNTS 349, which was replaced by the 1937 International Agreement for the Regulation of Whaling, 190 LNTS 79 and ultimately by the 1946 International Convention for the Regulation of Whaling, 161 UNTS 72.
[13] See, for example, the 1875 Convention Establishing Uniform Regulations concerning Fishing in the Rhine and its Tributaries, including Lake Constance between Baden and Switzerland, 149 CTS 139; 1911 Treaty for the Preservation and Protection of Fur Seals, 104 BSFP 175; 1958 Convention on Fishing and the Conservation of Living Resources of the High Seas, 516 UNTS 205.
[14] The 1902 Convention for the Protection of Birds Useful to Agriculture, 4 IPE 1615, replaced by the 1950 International Convention for the Protection of Birds, 638 UNTS 186.
[15] The 1900 Convention for the Preservation of Wild Animals, Birds and Fish in Africa, 4 IPE 1607, replaced by a succession of later treaties; the 1940 Convention on Nature Protection and Wildlife Preservation in the Western Hemisphere, 161 UNTS 193.
[16] In fisheries treaties, this idea was commonly expressed by the concept of 'maximum sustainable yield', a scientific assessment of fish stock depletion that underpins the assignment of quotas to countries.

threats occurring outside their jurisdiction. International jurisprudence confirmed this supremacy of instrumental considerations: the oft-cited *1898 Bering Sea Fur Seals Arbitration* between the US and the UK is an enlightening example of the early approach and contributed to establishing a common property regime over resources (including animals) located in areas beyond national jurisdiction.[17] Reference by the US in the arbitration itself to the need to minimise cruelty alongside the regulation of the methods and means of seal hunting suggests that at least some limited consideration was given to the intrinsic value of animals, but these were arguably ancillaries of a markedly instrumental approach.[18]

Later multilateral environmental agreements (MEAs) have expanded the interpretation of instrumental value while retaining a broadly anthropocentric approach. The 1971 Ramsar Convention on the Protection of Wetlands considers 'the fundamental ecological functions of wetlands as regulators of water regimes and as habitats supporting a characteristic flora and fauna, especially waterfowl' and is 'convinced that wetlands constitute a resource of great economic, cultural, scientific, and recreational value, the loss of which would be irreparable'.[19] The 1979 Convention on Migratory Species (CMS) acknowledges the value of wild animals from 'environmental, ecological, genetic, scientific, aesthetic, recreational, cultural, educational, social and economic points of view',[20] and strives to conserve them for the benefit of future generations.

Anthropocentrism also guides two soft-law documents that can be considered the international community's environmental manifestos. The 1972 Stockholm Declaration on the Human Environment proclaims that 'man is both creature and moulder of his environment, which gives him physical sustenance and affords him the opportunity for intellectual, moral, social and spiritual growth'[21] and affirms that 'natural resources … must be safeguarded for the benefit of present and future generations'.[22] The 1992 Rio Declaration on Environment and Development goes even further, pointing out that

---

[17] In this case, the US claimed a right of protection or property over the fur seals when they were outside the ordinary three-mile limit of coastal territory. The arbitrators decided that every nation had the right to accede and exploit the high seas, but proposed regulations for the protection and preservation of fur seals outside jurisdictional limits. They also recommended both governments to prohibit any killing of fur seals within their jurisdictions for a period of two years: neither government accepted the recommendation. The regulations were followed by treaties in 1911, 1942 and 1957, and shaped the content of subsequent fisheries and whaling agreements. In the 1950s a series of multilateral Conventions established detailed obligations, but still recognising the right of all states to engage in fishing on the high seas.

[18] See generally S Barrett, *Environment and Statecraft: The Strategy of Environmental Treaty-Making* (OUP, 2003), 19–48.

[19] 1971 Convention on Wetlands of International Importance especially as Waterfowl Habitat, 996 UNTS 245, preamble. The 1973 Convention on International Trade in Endangered Species of Wild Fauna and Flora (CITES), 993 UNTS 243, attributes these same values, as well as aesthetic considerations, to 'wild fauna and flora'.

[20] (1980) 19 ILM 15, preamble.

[21] 1972 Declaration of the United Nations Conference on the Human Environment, UN Doc A/CONF.48/14/ Rev.1, preamble.

[22] *Ibid.*, Principle 2.

'human beings are at the centre of concerns for sustainable development'[23] and that states have 'the sovereign right to exploit their own resources pursuant to their own environmental and developmental policies'.[24] The idea of 'rational use'[25] mutated into the 'sustainable use' of natural resources, including flora, fauna and genetic resources, which became the dominant paradigm of international environmental law. This does not mean that intrinsic value is absent from the Rio Declaration, but simply that it does not carry the same weight as instrumental value. In fact, the idea stated in Principle 1, that human beings are entitled to a life 'in harmony with nature' opened the door to an ethic that is not exclusively anthropocentric. The 'harmony with nature' theme has in fact recently been revisited by the UN, demonstrating its permanent relevance to environmental governance.[26]

This development is consistent with the acknowledgement, made in several international instruments, that nature also possesses intrinsic value. The 1979 Convention on the Conservation of European Wildlife and Natural Habitats (Bern Convention) recognises that 'wild flora and fauna constitute a natural heritage of aesthetic, scientific, cultural, recreational, economic and intrinsic value that needs to be preserved and handed on to future generations'.[27] The Bern Convention establishes a system for the protection of all wild flora and fauna, not only endangered and vulnerable migratory species.[28] Such an approach is certainly more in tune with intrinsic value, which applies to organisms regardless of their relative rarity or usefulness. In addition, the requirement to maintain populations at a level which corresponds to, inter alia, ecological requirements[29] could be read in line with ecosystemic intrinsic value.[30]

The 1982 World Charter for Nature (WCN) attempts to codify intrinsic value by setting out general principles to guide the formulation of more substantive international law. In the preamble to the resolution, the General Assembly of the United Nations declared 'that every form of life is unique, warranting respect regardless of its worth to man, and to accord other organisms such recognition, man must be guided by a moral code of action'.[31] However, given its non-binding character and paucity of substantive

---

[23] 1992 Rio Declaration on Environment and Development, UN Doc A/CONF.151/26/Rev.1, Principle 1. Arguably, this is not quite the same as stating that humans are also at the centre of concern for *environmental protection*: sustainable development in fact results from a compromise between environmental interests (which may or may not have an eminently human dimension) and social and economic ones (which are clearly anthropocentric).

[24] *Ibid.*, Principle 2.

[25] The term was used as early as 1980: see the 1980 Convention for the Conservation of Antarctic Marine Living Resources (CCAMLR), 1329 UNTS 4, Article 2.

[26] See generally the UN website at <http://www.harmonywithnatureun.org/> and in particular UNGA Res 68/216 (12 February 2014); UNGA Report of the Secretary-General, 'Harmony with Nature' (2013) UN Doc A/68/325.

[27] 1979 Convention on the Conservation of European Wildlife and Natural Habitats, ETS 104, preamble.

[28] *Ibid.*, Article 1.

[29] *Ibid.*, Article 2.

[30] This interpretation holds if one considers the ecological requirements to be those of the ecosystem itself: see Section 4.1.2 below.

[31] UNGA, World Charter for Nature (adopted 29 October 1982) UN Doc A/RES.737/7, preamble.

obligations, the impact of the WCN is better appreciated by looking at whether its principles have been reflected in subsequent biodiversity treaties.

More recent environmental agreements have in fact begun to contain references to intrinsic value. The 1991 Protocol on Environmental Protection to the Antarctic Treaty notes that 'the intrinsic value of Antarctica, including its wilderness and aesthetic values ... shall be fundamental considerations in the planning and conduct of all activities in the Antarctic Treaty area'.[32] The meaning of intrinsic value is left open to interpretation, since, as elsewhere in international law, no formal definition is given. It appears that intrinsic value is here used as an umbrella term that includes some of the instrumental values of the Antarctic environment. However, the inclusion of intrinsic value suggests that legal protection is also justifiable on non-instrumental grounds. And the fact that both are included may even suggest that neither is enough on its own to warrant protection.[33]

Intrinsic value also features prominently in what is likely the most eminent biodiversity treaty in force, the Convention on Biological Diversity (CBD).[34] The document begins by stating that the 193 Parties are 'conscious of the intrinsic value of biological diversity', and then goes on to list a vast array of instrumental values, including 'the ecological, genetic, social, economic, scientific, educational, cultural, recreational and aesthetic values of biological diversity and its components'. Although the latter values shape the objectives of the CBD – which are not dissimilar to those of early wildlife treaties: to conserve and sustainably use biodiversity and ensure the fair and equitable sharing of the benefits arising[35] – intrinsic value does have some bearing on the substantive provisions. Indeed, the distinction drawn in the Convention between 'biological diversity' and 'biological resources' confirms that the object of protection is not only those biotic components and ecosystems 'with actual or potential use or value for humanity' but also the 'variability among living organisms from all sources' in itself. Whereas provisions on sustainable use of biological *resources* are clearly and exclusively instrumental, biological *diversity* as such is to be conserved for reasons arguably ascribable to both instrumental and intrinsic value.

Finally, the recently revised African Nature Conservation Convention[36] makes some interesting references to intrinsic value. In particular, Article IV places a 'fundamental obligation' on parties to enhance environmental protection 'with due regard to ethical and traditional values', and Annex 2 advocates 'maintaining respect for the ecological, geomorphological, sacred or aesthetic attributes' of landscapes. It amends its predecessor by building upon an approach that had already 'moved away from a concept of natural resources conservation solely centred on utilitarian purposes'.[37] Although the African Convention is deeply rooted in post-Rio Declaration anthropocentrism, it will

---

[32] (1991) 30 ILM 1461, Article 3.
[33] See Section 5 below.
[34] 1992 United Nations Convention on Biological Diversity, 1760 UNTS 79.
[35] Article 2.
[36] The text may be found in IUCN, *An Introduction to the African Convention on the Conservation of Nature and Natural Resources* (IUCN, 2004); for the original, see the 1968 African Convention on the Conservation of Nature and Natural Resources, 1001 UNTS 3.
[37] IUCN (n 36), 4.

be seen whether specific obligations to protect nature will develop based on ethical considerations once the treaty enters into force.

In sum, despite a number of references to the intrinsic value of the natural world, international law has not developed mechanisms or standards specifically directed to protect this value. Although the notion of value has expanded over time, instrumentalism has not been abandoned. Indeed, instrumental considerations are still dominant and intrinsic value seems more ornamental than influential to the substantive law. Intrinsic value within the law is broadly focused at a relatively high level of abstraction: it is attached to ecosystems, diversity or species, whereas the intrinsic value of individual organisms has received little direct attention.[38] The challenges in reconciling these two possible manifestations of intrinsic value will emerge later.

3. IN HUMAN INTEREST? PROBLEMS IN APPLYING INSTRUMENTAL VALUE

Before that, however, it must be acknowledged that, despite its apparent predominance, the application of the instrumental approach has not been entirely free of problems.

3.1 The Rationale of Instrumental Value

Supporters of anthropocentric instrumentalism readily accept that there are times when environmental protection may not be feasible. The interest in protecting the environment must be measured against other legitimate objectives: protecting wildlife is good, but so is producing food, harvesting timber, building settlements and so forth. A key contribution of instrumental value is that it allows individuals and governments to make decisions based on a case-by-case evaluation of the human benefits of a certain course of action.[39] Complicated moral decisions become management decisions. For instance, a farmer may decide to preserve a woodlot within a plot of land because it harbours bird species that help control pests.

Yet this approach has not been particularly effective in protecting biodiversity. On the one hand, the formal protection provided by international biodiversity law has increased. This is evident in the conclusion of a great number of treaties at the global and regional levels;[40] in the recognition of non-productive and non-use instrumental

---

[38] Although this is not to say it has been entirely disregarded by the international community, see, e.g., the various agreements adopted within the Council of Europe for the protection of individual animals, such as the 1976 European Convention for the Protection of Animals Kept for Farming Purposes, ETS 87, and the 1986 European Convention for the Protection of Vertebrate Animals used for Experimental and Other Scientific Purposes, ETS 123.

[39] By contrast, intrinsic value is unsuitable to make trade-offs between competing objectives because it is not measurable: see F Mathews, *The Ecological Self* (Routledge, 1991), 88. Note, however, that being immeasurable is not a decisively negative characteristic.

[40] See Section 2 above; for a more detailed review of major biodiversity instruments, *see* MJ Bowman, PGG Davies and CJ Redgwell, *Lyster's International Wildlife Law* (2nd edn., CUP, 2010).

values made in such treaties;[41] in the expansion of protected areas, which cover 12.7 per cent of land surface and 1.6 per cent of the oceans;[42] in the creation of a list of endangered species and the maintenance of adequate monitoring capacity;[43] and, more generally, in the universal affirmation of biodiversity protection as a legitimate policy objective. On the other hand, species are disappearing at a rate unseen since the extinction of dinosaurs,[44] biodiversity protection is generally underfunded[45] and the international community has failed to meet its global biodiversity goals.[46] This is because other legitimate human interests have prevailed. As the global population has grown fourfold and the global economy has increased tenfold in the last 60 years,[47] the human footprint on the global biota has increased exponentially.[48]

Does this mean that high rates of biodiversity loss are a necessary evil in the trade-offs of (instrumental) decisions? Or does it demonstrate that humans acting in their apparent best interests do not necessarily reach sound conclusions? The latter reasoning seems more plausible. The next section will show that, when it comes to evaluating biodiversity in instrumental terms, humans may not be able to act in their true self-interest because of the inadequacies of the valuation process.

3.2 The Limits of Instrumental Value

Instrumentalism is still a distorted approach to fully understand the value of nature. Its first limitation is methodological. So far, the instrumental value of nature has been measured as the monetary value of direct, material and productive uses: the revenues of timber sale, the price of agricultural commodities, the lost profit caused by fisheries collapse and so forth. But economic valuation is incapable of reflecting the whole (instrumental) value of the environment. For instance, the public good character of biodiversity means that no individual would necessarily gain a direct material benefit from its protection. This makes it difficult to assess the overall cost for the public of its depletion.

---

[41] *See* Bowman and others, *ibid.*, especially chapters 1 and 3.

[42] B Bertzky and others, *Protected Planet Report 2012: Tracking Progress towards Global Targets for Protected Areas* (IUCN and UNEP-WCMC, 2012), iv.

[43] *See* IUCN, *The IUCN Red List of Threatened Species* (International Union for the Conservation of Nature, 2013), available at <http://www.iucnredlist.org>.

[44] *See* the Millennium Ecosystem Assessment, available at <http://www.unep.org/maweb/en/index.aspx>.

[45] CBD, 'State of financing for biodiversity: draft global monitoring report 2012 on the Strategy for Resource Mobilization under the Convention', a Note by the Executive Secretary (2012), UN Doc UNEP/CBD/ COP/11/INF/16.

[46] UN, 'World Population Prospects: The 2012 Revision, Volume I: Comprehensive Tables' (2013), UN Doc ST/ESA/SER.A/336.

[47] On this point, see CIA, *The World Factbook* (Central Intelligence Agency, 2014), available at <https://www.cia.gov/library/publications/resources/the-world-factbook/index.html>.

[48] The environmental impact of human activity is such that some have argued that we are now in the 'Anthropocene', a new geological era in which mankind is the dominant force of environmental change: E Ehlers and T Craft (eds.), *Earth System Science in the Anthropocene* (Springer, 2006).

Given the current dominance of capitalist economic thinking in public policy, there is a clear need to refine our capacity to measure environmental values in these terms.[49] The question with which the various instruments concerned with environmental protection have struggled is how to measure instrumental value that is not comprised in the financial worth of the resources used in productive activities. This question is of the utmost importance: *ex ante*, development decisions should be based on a cost-benefit analysis of projects and programmes, which includes an accurate appraisal of environmental costs (i.e. the environmental values that are depleted if the project or programme is executed); *ex post*, the valuation of environmental damage should reflect the *entire* value of the damaged resources, particularly in cases where the cost of such damage is borne by the public and thus does not fit into the traditional categories of private property.

Some steps have been made to address these issues. International law was historically concerned with the environmental damage caused by one state to the territory of another.[50] Case law has long established that states have an obligation to prevent serious transboundary harm or damage to the environment of other states and that, should they fail to do so, they have a duty to provide reparation or compensation under the law of state responsibility.[51] Courts initially adopted a narrow view, deciding that compensation should be limited to property damage, personal injuries and economic loss,[52] and so did early civil liability Conventions.[53] Consistent with the broadening concept of instrumental value in MEAs, this view was later supplanted in modern civil liability Conventions by a broader definition of environmental damage, which now contemplates non-productive use values of the environment.[54] Recent developments in environmental economics have begun to encompass Bowman's inherent value (which in

---

[49] For an overview of the problems of 'economism' in this area, *see* A Gillespie, *International Environmental Law, Policy and Ethics* (2nd edn., OUP, 2014), chapter 3.

[50] The principle of 'good neighbourliness', sometimes expressed using the maxim *sic utere tuo ut alienum non laedas*, has found expression in many international and bilateral instruments and has been upheld by international courts in a number of cases; the origin of the rule can be traced to the *Trail Smelter Arbitration* (United States v Canada, awards of 16 April 1938 and 11 March 1941) 3 RIAA 1907.

[51] International Law Commission, Responsibility of States for Internationally Wrongful Acts, Yearbook of the International Law Commission 2001, Volume II, Part Two, Report of the Commission to the General Assembly on the work of its fifty-third session (UN 2007) UN Doc A/CN.4/SER.A/2001/Add.1 (Part 2).

[52] This was the approach taken, for example, in the *Trail Smelter* case itself (n 50), which concerned the emission of pollutants from a smelter in Canada, which caused damage to people and property in the United States.

[53] *See*, for example, the 1963 Vienna Convention on Civil Liability for Nuclear Damage, 1063 UNTS 265; 1969 Convention on Civil Liability for Oil Pollution Damage, 973 UNTS 3.

[54] 1992 Convention on Civil Liability for Oil Pollution Damage, IMO LEG/CONF 9.15; 1993 ECE Convention on Civil Liability for Damage Resulting from Activities Dangerous to the Environment (1993) 32 ILM 1228; 1996 Convention on Civil Liability and Compensation for Damage in Connection with the Carriage of Hazardous and Noxious Substances by Sea (1996) 35 ILM 1406; 1997 Protocol to amend the Vienna Convention on Civil Liability for Nuclear Damage (1997) 36 ILM 1454; 1999 Protocol on Liability and Compensation for Damage Resulting from the Transboundary Movements of Hazardous Wastes and their Disposal, UNEP/CHW.5/29, Annex III.

this chapter is considered a subcategory of instrumental value). A growing approach is the concept of 'ecosystem services', which are defined as 'the benefits people obtain from ecosystems'[55] or, more specifically, 'the ecological aspects of ecosystems utilised (actively or passively) to produce human well-being'.[56] Ecosystem services encompass use values that are not linked to production, such as cultural values, recreational values and ecological benefits that are non-excludable and non-rivalrous and therefore cannot be economically evaluated through standard mechanisms.[57]

Some countries have already introduced large-scale schemes to provide incentives for the maintenance of ecosystem services,[58] and the concept has also gained ground in a number of international environmental regimes. The CBD encourages the advancement of 'studies on approaches to develop markets and payment schemes for ecosystem services at local, national and international levels'.[59] By way of example, 'The Economics of Ecosystems and Biodiversity' is an international initiative managed by the UN Environment Programme that aims to develop appropriate valuation techniques to encompass the whole range of instrumental use values provided by nature. A mechanism to reduce carbon emissions from deforestation under the United Nations Framework Convention on Climate Change (UNFCCC) is also set to provide incentives for a host of ecosystem services.[60]

Methodologies have also been developed to measure the non-use value of the environment, which has been recognised as relevant to the quantification of environmental damage.[61] 'Maximum willingness to pay' and 'minimum willingness to accept compensation' are two methodologies used to attribute a monetary value to nature, and can be applied to measure the worth of ecosystems, particular ecological status (such as the absence of pollution) and inanimate natural features, as well as species and

---

[55] Millennium Ecosystem Assessment, Ecosystems and Human Well-being: Biodiversity Synthesis (World Resources Institute, Washington, DC, 2005).

[56] B Fisher, RK Turner and P Morling, 'Defining and Classifying Ecosystem Services for Decision Making' (2009) 68 Ecological Economics 645. For other definitions, *see* J Boyd and S Banzhaf, 'What are Ecosystem Services? The Need for Standardized Environmental Accounting Units' (2007) 63 Ecological Economics 616; KJ Wallace, 'Classification of Ecosystem Services: Problems and Solutions' (2007) 139 Biological Conservation 235; RS de Groot, MA Wilson and RMJ Boumans, 'A Typology for the Classification, Description and Valuation of Ecosystem Functions, Goods and Services' (2002) 41 Ecological Economics 393.

[57] In other words, ecological benefits have the characteristics of public goods, which are difficult to evaluate using economic tools developed for the private property regime. Ecosystem services are affected by the same problems of public goods, in particular the problem of free-riding (the ability to benefit from a service without paying for it or otherwise contributing to its maintenance or regeneration). *See* also G Hardin, 'The Tragedy of the Commons' (1968) 162 Science 1243.

[58] Payments for Ecosystem Services (PES) schemes have been developed in the US, Mexico, Costa Rica, the Dominican Republic and Japan, for example. See generally OECD, *Paying for Biodiversity: Enhancing the Cost-Effectiveness of Payments for Ecosystem Services (PES)* (OECD, 2010).

[59] CBD (n 34), COP Decision IX/6.

[60] The programme, called REDD-plus, is mostly concerned with carbon sequestration and storage services, but also aims to preserve biodiversity and the productive uses of forests in a sustainable manner.

[61] *State of Ohio v US Department of the Interior* 880 F. 2d. 432 (DC Cir., 1989).

biodiversity.[62] Yet these valuation techniques, which attempt to reflect how much people are willing to sacrifice in order to protect nature, seldom provide objective and accurate measurements of value and have in fact scarcely been used.[63] What they do represent is an attempt to integrate instrumental environmental value within the economic subsystem and they can be further improved to make more rational instrumental choices about what is in our best interest.

The second limitation of instrumental value is cognitive and has to do with our limited capacity to anticipate the environmental consequences of certain activities. First, there is a limited understanding of the cumulative environmental impact of individual actions. A fishing vessel would have an interest in catching as much fish as possible, but if all fishing boats maximise their catch this would lead to the depletion of fish stocks and to future shortages. The clash between individual and collective interests in the management of non-excludable resources is known as the 'tragedy of the commons'.[64] Second, there is a lack of understanding of the cumulative impact of activities over time. The tendency is to privilege decisions that yield an immediate gain and to ignore or underestimate future negative consequences, particularly when these are little known as in the case of biodiversity loss.[65] The algal blooms triggered by slow releases of fertilisers over time are a clear example of this.

It must therefore be asked whether we are truly able to take environmental management decisions based solely on human utility. There are reasons for optimism. Scientific inquiry on the many ways in which nature is useful to humans is a driving force behind the development of international environmental law. Looking forward, instrumental valuation is likely to motivate a greater level of legal protection for nature as our understanding of the natural world advances. This generally requires three steps. The first step is to increase the understanding of natural processes and their dependence on the complex relations among biota and landscape. The second step is the translation of scientific knowledge into public consciousness: when information is made widely available, individuals recognise that nature protection is in their best interest. The third step is the articulation of these interests into legal protection through a political process. The progression of scientific understanding of the environment, and hence the

---

[62] *See* N Hanley, 'The Economic Value of Environmental Damage' in MJ Bowman and AE Boyle (eds), *Environmental Damage in International and Comparative Law: Problems of Definition and Valuation* (OUP, 2002); RK Turner, 'The Place of Economic Values in Environmental Valuation' in IJ Bateman (ed.), *Valuing Environmental Preferences: Theory and Practice of the Contingent Valuation Method in the US, EU, and Developing Countries* (OUP, 2001).

[63] For instance, one problem is that people value potential losses more highly than equivalent gains. Another problem is that willingness to pay or accept compensation depends on subjective perceptions that may not account for the whole range of services provided by nature: Hanley (n 62).

[64] Following the influential essay by Garrett Hardin (n 57). For a relevant criticism of the concept, however, see T Morton, *The Ecological Thought* (Harvard University Press, 2010) 122–3.

[65] Limited foresight entails a poor understanding of the consequences of environmental overexploitation. It must not be confused with concerns for inter-generational equity, where there is at least some awareness of negative environmental impacts. For interesting discussion on this theme, *see* J Diamond, *Collapse* (Penguin, 2006).

instrumental rationale for its protection, is not the only instigator of legal development (consider religious beliefs or other ethical principles), but it is certainly a key one within environmental law.

Scientific progress has so far been unidirectional, piling up evidence that the health of species and ecosystems is crucial to the human economy,[66] society[67] and psychology.[68] We are gaining a new awareness of living systems. Such findings give credibility to broad applications of ecological principles in a way that emphasises the relational nature of all living beings and how each individual organism can only thrive through connections with other organisms and the physical environment that sustains them. Just because humans do not appreciate the instrumental utility of an organism, it does not mean that such an organism does not have any utility. In fact, as all organisms depend on other organisms, it is hard to set arbitrary boundaries to what is useful and what is not. If instrumental utility is ascribed to all interrelated organisms, then according to the precautionary principle, they should all be worthy of at least some legal protection, as well as the network itself.

Such an approach has informed some international legal instruments. For instance, the 1982 Law of the Sea Convention, the 1985 Ozone Convention and the 1992 Climate Change Convention all establish a general obligation to protect ecosystems and species from the harmful effects of pollution without reference to the human usage of these resources.[69] In order to do this, they use an extremely broad interpretation of instrumental value which 'focuses on the interdependence of human activity and nature'.[70]

A similar rationale was used, and further expanded, by the UN Compensation Commission (UNCC). The UNCC was set up to pay compensation for loss and damage arising from Iraq's invasion and occupation of Kuwait, including for damage to the environment under category 'F4' claims.[71] 'Environmental damage' included the value of nature beyond its practical utility and, to this end, a Working Group looked at

---

[66] See P Sukhdev and others, *The Economics of Ecosystems and Biodiversity – Mainstreaming the Economics of Nature: A Synthesis of the Approach, Conclusions and Recommendations of TEEB* (TEEB, 2010).

[67] See T Goeschl and T Swanson, 'The Social Value of Biodiversity for R&D' (2002) 22(4) Environmental and Resource Economics 477; DA Posey, *Cultural and Spiritual Values of Biodiversity* (UNEP, 1999), particularly chapter 6.

[68] See J Sempik, R Hine and D Wilcok (eds), *Green Care: A Conceptual Framework. A Report of the Working Group on the Health Benefits of Green Care, COST Action 866, Green Care in Agriculture* (Loughborough: Centre for Child and Family Research, 2010); R Louv, *Last Child in the Woods: Saving Our Children From Nature-Deficit Disorder* (Algonquin Books, 2008); I Alcock and others, 'Longitudinal Effects on Mental Health of Moving to Greener and Less Green Urban Areas' (2014) 48 Environmental Science & Technology 1247.

[69] 1982 United Nations Convention on the Law of the Sea, 1833 UNTS 3, Articles 192–196; 1985 Convention for the Protection of the Ozone Layer, 1513 UNTS 293, preamble and Articles 1(2) and 2(1); 1992 United Nations Framework Convention on Climate Change, 1771 UNTS 107, Articles 1(1), 2.

[70] PW Birnie, AE Boyle and CJ Redgwell, *International Law & the Environment* (3rd edn, OUP, 2009), 183.

[71] UNSC Resolution 687 (8 April 1991), UNSC Resolution 692 (20 May 1991).

hedonic pricing, replacement cost and contingent valuation.[72] In total, over US$5bn was awarded for environmental damages.[73] Although the majority of this sum was for instrumental valuations of the environment, there are examples of compensation paid for damage to the environment per se, for example for damage to Kuwaiti vegetative cover, which 'provides an essential mechanism for desert surface stabilization [and] also helps to regulate the distribution of rainfall and provides sustenance for wildlife'.[74] Such awards, however, were in the minority. A major difficulty in establishing claims was the issue of causation,[75] a frequent thorn in the side for international environmental law. Moreover, the imposition of a deadline for the submission of claims ruled out possible longer-term environmental problems being accounted for.[76] Despite these problems, the UNCC methodology for doing this presents a potential pathway to expand legal protection for nature.

A final challenge concerns the logical coherence of instrumental value. As O'Neill has observed, 'it is a well-rehearsed point that, under pain of an infinite regress, not everything can have only instrumental value'.[77] That is, any instrumental value approach must itself be premised on some theory of ultimate value: A has value because of reason B, which in turn is valued because of C, and so on. Eventually, we must stumble upon something which is of value simply because it is. At the bottom of the instrumental pile lies something which is considered to be of value in itself (i.e. intrinsically valuable) underpinning it all. It is uncontroversial to state that humans possess such value.[78] However, wary of the charge of speciesism,[79] it is worthwhile considering whether such value can be located elsewhere too.

---

[72] UNEP, *Report of the Working Group of Experts on Liability and Compensation for Environmental Damage Arising from Military Activities*, UNEP/Env.Law/3/Inf.1 (1996), cited in R Juni, 'The United Nations Compensation Commission as a Model for an International Environmental Court' (2000) 7 The Environmental Lawyer 53, 69.

[73] See <http://www.uncc.ch/sub-category-f4-claims-environmental-damage>.

[74] Paras 133–148 of S/AC.26/2003/31, available from the UNCC website, *ibid*. See also S/AC.26/2001/16; S/AC.26/2002/26; S/AC.26/2004/16 (especially paras 139–147); S/AC.26/2004/17; S/AC.26/2005/10 (especially paras 353–366) for details of the claims.

[75] UNCC Decision No. 15, Compensation for Business Losses Resulting From Iraq's Unlawful Invasion and Occupation of Kuwait Where the Trade Embargo and Related Measures Were Also a Cause, UN Doc. S/AC.26/1992/15 (1992). See further T Lee, 'Environmental Liability Provisions under the UN Compensation Commission' (1999) 11 Georgetown International Environmental Law Review 209, 219. See also, however, UNCC Decision No. 7, Criteria for Additional Categories of Claims, UN Doc S/AC.26/1991f7/ Rev. 1 (1992) for the Commission's listing of how this is to be interpreted for environmental claims.

[76] Lee (n 75), 221.

[77] O'Neill (n 5), 119.

[78] *See*, for example, the Universal Declaration of Human Rights (1948) GA Res 217 A (III), preamble.

[79] The term was coined by Richard Ryder, but made most familiar by Peter Singer, *Animal Liberation* (HarperCollins, 1975).

4. THE INCOMPLETENESS OF INTRINSIC VALUE

Intrinsic value also has problems in building a workable ethic out of its principles. Do all organisms have intrinsic value? If so, do we have a moral obligation to respect and protect the intrinsic value of all of them, and what does this entail? When interests are in conflict, how can we decide which one prevails? Building an ethic that coherently guides human action requires finding an answer to these questions. This section introduces some of the key ideas underlying intrinsic value and then asks what the consequences of taking a maximalist stance might be.

4.1 Where Is Intrinsic Value?

Defining intrinsic value (asking *what* intrinsic value is) is by no means the same as locating it (asking *where* intrinsic value is). Having defined what intrinsic value is in Section 1, this section considers some of the properties that have been proposed by environmental philosophers as demonstrating the existence of intrinsic value. The general idea is to find activities, behaviours or tendencies that suggest that an entity values itself. In other words, the search is for a 'locus of valuational activity'.[80]

It is important to note that intrinsic value need not be tied to one particular intrinsic property but could arise from a number of different properties. Here, common themes are identified to loosely categorise candidate properties and allow clearer understanding of the underlying justifications. These categories can then be grouped further into two broad approaches: those that focus mostly on the value of individual organisms (biocentric) and those that focus on the value of the whole as something independent from the sum of its parts (ecocentric).

4.1.1 Biocentric approaches
The underlying principle behind biocentric approaches to intrinsic value is to demonstrate that organisms somehow have a 'point of view'[81] of the world that allows them to interact with it in such a way as to generate value in themselves. Each and every organism has its own story and perspective and this makes them 'loci of valuational activity' since no particular way of perceiving and interacting with the world ought to be given precedence a priori: any given point can be considered the centre of the universe.[82]

---

[80] *See* B Morito, 'Intrinsic Value: A Modern Albatross for the Ecological Approach' (2003) 12 Environmental Values 317, who uses this term in opposition to 'repository of intrinsic value' because it is 'thoroughly dialectical' (332) and 'each locus, in this scheme, whether human or non-human, contributes to the ways in which values are constituted, because each transmutes the network of valuations according to its perspective' (328).

[81] This 'point of view' is not meant to be taken in a literal, anthropomorphic sense. Instead it stands for the manner in which an organism experiences and interacts with the world. See Jakob von Uexküll's 'Umwelt' in *A Foray into the Worlds of Animals and Humans* (Joseph O'Neill trans, first published 1934, University of Minnesota Press, 2010) and Tom Regan's 'subject-of-a-life' in *The Case for Animal Rights* (Routledge, 1984).

[82] To be more precise, at every point of the universe, the rest of the universe is moving away. There is no centre of our universe.

The intrinsic value approach is not about moral extensionism, but about acknowledging the numerous ways in which life, and so value, pervades. What is important is that all forms of life exhibit 'their own existence, their own character and potentialities, their own forms of excellence, their own integrity, their own grandeur'.[83] This idea may be viewed from a variety of perspectives.

(a) Needs: Biologically speaking, all living beings have certain 'needs'[84] and the first route to locating intrinsic value focuses on the observation that states of affairs are not neutral for living organisms: things can go well, or badly, for each and every one. The existence of 'needs' demonstrates that an organism has a 'good-of-its-own':[85] a set of conditions that together are beneficial to it. This line of reasoning echoes Aristotle's idea of *eudaimonia*, or flourishing, which can be defined as: 'success in life: the good composed of all goods; an ability which suffices for living well; perfection in respect of virtue; resources sufficient for a living creature'.[86] At least some of these concepts are certainly applicable beyond the human species.

(b) Agency and Strategy: Instead of focusing on needs, other theories consider the mechanisms by which such needs are met in pursuit of an organism's or a species' survival. This concept has been expressed using terms such as desire[87] or striving.[88] These risk anthropomorphising the nonhuman as they evoke a psychological process that is not present in the majority of organisms. Other terms have been used in the literature to express similar ideas, which may avoid this problem. Spinoza's concept of 'conatus', interpreted by Freya Mathews as 'the impulse for self-preservation or self-maintenance'[89] can perhaps be extended more seamlessly to nonhumans. Another term in use for a similar idea is that of 'teleological centres of life',[90] which are 'unified, coherently ordered systems of goal-orientated activities that have a constant tendency to protect and maintain that organism's existence'.[91] Overall, the idea is that organisms' behaviour is directional and interactive: their manifestly purposeful re-arrangement of their surroundings is evidence for the intrinsic value of organisms because it demonstrates that things *matter* to them. It may not be relevant whether this process is a psychological one or not, but what is relevant is that organisms have strategies which

---

[83] J Rodman, 'The Liberation of Nature?' (1977) 20 Inquiry 83, 94.

[84] Alternative handles for similar ideas include: 'interests' (Regan (n 81), 87); 'benefits' or 'tasks' (K Goodpaster, 'On Being Morally Considerable' (1978) 75 The Journal of Philosophy 308, 319).

[85] P Taylor, *Respect for Nature* (Princeton University Press, 1986), 60. For a similar concept, see John Muir's discussion on rattlesnakes: J Muir, *Our National Parks* (Scholarly Press, 1901), 57–8.

[86] 'Definitions' in J Cooper (ed.), *The Complete Works of Plato* (Hackett Publishing Company, 1997), 1680. See also O'Neill (n 5).

[87] G Varner, *In Nature's Interests?* (OUP, 1998), 26–30, 51–4. The exact interplay between interests and desire is complicated: part of it lies in the difference between preference satisfaction and what is actually good for an entity.

[88] Lee (n 7) talks about 'striving to maintain functional integrity' especially at 158.

[89] Mathews (n 39), 109.

[90] Taylor (n 85), 122. Or simply 'teleology': V Plumwood, *Feminism and the Mastery of Nature* (Routledge, 1993) 10.

[91] Taylor (n 85) 122.

they set out to achieve. They have agency in the universe, rather than being simply inert with respect to it.

(c) Autopoiesis: The biological term 'autopoiesis' has entered the argot of environmental philosophers as a way to underpin intrinsic value. Originating as a way to define life, the key features of an autopoietic entity are that it is bounded, self-organising and self-regenerating.[92] It thus introduces the idea of 'selfhood' into intrinsic value considerations. The 'self' is a handle to understand both why an entity may be generating its own value and what it seeks to achieve in doing so: the maintenance of a 'self' of some description. It is this existence of a notion of self (delimited by its boundaries), together with behaviour that seeks overall to organise and regenerate that self, which implies that autopoietic entities have intrinsic value. Things matter to them, which is witnessed by their tendencies to maintain some continuing identity through time.[93]

4.1.2 Ecocentric theories of intrinsic value

Given the ready acceptance of the intrinsic value of ecosystems by some conservationists,[94] and its tentative acceptance by the law,[95] it is necessary to consider theories that support such assertions. Some of the properties considered above can in fact explain the intrinsic value of ecosystems too. Thus, it may be intelligible to talk of ecosystems 'striving' in some metaphorical or allusive sense;[96] it is meaningful to talk about an

---

[92] F Varela, H Maturana and R Uribe, 'Autopoiesis: The Organization of Living Systems, Its Characterization and a Model' (1974) 5 Biosystems 187; H Maturana and F Varela, *Autopoiesis and Cognition* (Reidel, 1980); J Mingers, *Self-Producing Systems: Implications and Applications of Autopoiesis* (Plenum Press, 1995); PL Luisi, 'Autopoiesis: a Review and a Reappraisal' (2003) 90 Naturwissenschaften 49.

[93] The question of how a continued self exists through time is a long-standing one: the thought experiment known as the Ship of Theseus (or Trigger's broom) explores some of the complexities of this. *See also* A Gallois, 'Identity Over Time' in E Zalta (ed), *The Stanford Encyclopedia of Philosophy* (Summer 2012), available at <http://plato.stanford.edu/archives/sum2012/entries/identity-time/>.

[94] *See* WF Butler and TG Acott, 'An Inquiry Concerning the Acceptance of Intrinsic Value Theories of Nature' (2007) 16 Environmental Values 149. Interestingly, there are a number of conservationists-turned-philosophers: Aldo Leopold, James Lovelock and Edward O Wilson are among the most famous.

[95] See Section 2 above.

[96] Ecosystems may be thought of as 'striving' through complex processes of emergent phenomena (that is, phenomena that cannot be explained through a reduction to lower-level processes and activities). In the words of Ernst Mayr, 'systems almost always have the peculiarity that the characteristics of the whole cannot (not even in theory) be deduced from the most complete knowledge of the components, taken separately or in other partial combinations. This appearance of new characteristics in wholes has been designated as emergence': E Mayr, *The Growth of Biological Thought: Diversity, Evolution, and Inheritance* (first published 1982, 12th printing, Harvard University Press, 2003) 63; see also J-F Ponge, 'Emergent Properties from Organisms to Ecosystems: Towards a Realistic Approach' (2005) 80 Biological Reviews, Cambridge Philosophical Society 403. One example of an emergent phenomenon of ecosystems is in their response to catastrophes such as forest fires or hurricanes. In these cases, ecosystems will often gradually move back towards a state similar to the pre-catastrophic one, unless the disturbance is so great as to force the ecosystem to move towards another state. Although this

ecosystem flourishing;[97] and it is possible to view ecosystems as part of organisms' selfhood through the idea of 'intrinsic relations'[98] and even as a 'Self' in themselves, since they do exhibit some autopoietic qualities or tendencies.[99] However, it may be inappropriate to adopt reasoning for organisms to ecosystems since they are different categories of living entity.

Other philosophers seek to defend the intrinsic value of ecosystems in other ways, which tend to elevate the 'whole' above the 'parts' by focusing on the value in a functioning ecosystem.[100] Such a view can be criticised as being 'eco-fascist' owing to its reduction of the individual to the whole,[101] potentially overlooking the fact that each organism has its own intrinsic value too.

An alternative argument in favour of the intrinsic value of ecosystems is put forward by Rolston's 'systemic value'[102] and Mathews' 'background value' theories.[103] Ecosystems (or indeed the whole biosphere) possess their own brand of intrinsic value as the *generators* of intrinsically valuable organisms: they provide the conditions for life to emerge and develop to its fullest potential, hence their value is derived from their ability to produce life in itself rather than borrowed from the value of individual organisms. This 'systemic value' theory has, however, been criticised as creating intrinsic value from instrumental value.[104]

Perhaps ecocentric and biocentric value work best in tandem: the fact that it is equally defensible to see organisms generating ecosystems as vice versa can help to understand this. Organism and ecosystem are both mutually interdependent and defined in terms of one another: their intrinsic value depends on that of the other, since their

---

process is not a perfectly predictable, neatly linear one – for a lucid description of how ecosystems can settle in more than one steady state, see M Scheffer and others, 'Catastrophic Shifts in Ecosystems' (2001) 413 Nature 591 – it does exhibit a certain directionality, as to which *see* FA Bazzaz, *Plants in Changing Environments* (Cambridge University Press, 1996).

[97] Although this flourishing may be by human-imposed standards, rather than the standards of the ecosystem itself.

[98] A term coined by Arne Naess as follows: 'An intrinsic relation between two things A and B is such that the relation belongs to the definition or basic constitutions of A and B, so that without the relation, A and B are no longer the same things': A Naess, 'The Shallow and the Deep, Long-Range Ecology Movement: A Summary' (1973) 16 Inquiry 95, 95.

[99] Most famously in J Lovelock, *Gaia: A New Look at Life on Earth* (OUP, 1982). See also W Devall and G Sessions, *Deep Ecology* (Gibbs Smith, 1985); Fox (n 6); Mathews (n 39).

[100] A Leopold, *A Sand County Almanac* (first published 1949, OUP edn., 1968); J. Baird Callicott, *In Defense of the Land Ethic* (SUNY Press, 1989); A Naess, 'The Deep Ecological Movement: Some Philosophical Aspects' (1986) 8 Philosophical Inquiry 10.

[101] *See* Regan (n 81); T Morton, 'Architecture without Nature' in *Not Nature*, Tarp Architecture Manual (Spring 2012) 20, available at <http://issuu.com/tarp/docs/notnature_finaldraft_041012>.

[102] See H Rolston III, *Environmental Ethics – Duties to and Values in the Natural World* (Temple University, 1988), 187–8, 220–30, 257–8 and 'Value in Nature and the Nature of Value' (1994) <http://ebooks.cambridge.org/ebook.jsf?bid=CBO9780511524097>.

[103] Mathews (n 39); note especially that at 119 Mathews claims background value to be 'objective and absolute, in the sense that it inheres in the things which possess it and is not relativized to the needs and desires, or interests, of external observers or agents'.

[104] R Elliot, 'Instrumental Value in Nature as a Basis for the Intrinsic Value of Nature as a Whole' (2005) 27 Environmental Ethics 43.

intrinsic properties are themselves mutually dependent. The relationship between organisms and their ecosystem (or, alternatively, ecosystems and their organisms) is not composed entirely of instrumentality, but is in fact an 'intrinsic relation' between them. This does not deny the possibility of finding instrumental dependence in either direction but rather moves towards a realisation that there is a tight interplay between intrinsic and instrumental valuation.

To conclude, although the approaches examined above are grouped under the biocentric and the ecocentric, there is actually no decisive divide between these categories: a subtle interplay between biocentrism and ecocentrism begins to emerge. Somewhere among the ideas of agency and autopoiesis, there is a realisation that neither organisms nor ecosystems are totally isolatable and that perhaps they ought best to be considered as 'organisms-in-environment'.[105] It is also notable that, although the law tends to focus practical measures at the species level, and on the intrinsic value of highly abstract entities such as 'biological diversity',[106] philosophical investigations into intrinsic value are more advanced with respect to individual organisms or ecosystems.[107] In accordance with the caricatured positions adopted here, let it be assumed that all individual organisms, ecosystems and the biosphere are, in their own way, intrinsically valuable so that the practical implications of this may be investigated.

4.2 The Consequences of Intrinsic Value

Although international environmental law has encompassed intrinsic value to some extent, it is by no means clear precisely what the legal consequences of intrinsic value are. Neither extant law nor philosophical theories make this immediately apparent,[108] and it is therefore worthwhile probing whether intrinsic value would bolster or antagonise protection based on instrumental value. The key question is how the law ought to reflect the fact that natural organisms and ecosystems have value for themselves and in themselves. In theory, this would cause a radical shift in legal

---

[105] Mathews (n 39), 106, citing G Bateson, *Steps Towards an Ecology of Mind* (Paladin, 1973). Similar ideas can be found in the work of Richard Dawkins, *The Extended Phenotype* (OUP, 1999); John Dewey, as to whom, *see* S Odin, *The Social Self in Zen and American Pragmatism* (SUNY, 1992); H. Macdonald, *John Dewey and Environmental Philosophy* (SUNY, 2003); Martin Heidegger, as to whom, *see* M Wrathall (ed), *The Cambridge Companion to Heidegger's Being and Time* (CUP, 2013), 6; H Dreyfus, *Being-in-the-World* (first published 1991, 8th printing, Massachusetts Institute of Technology, 1999), 43; and H Gordon, *Heidegger-Buber Controversy: The Status of the I-Thou* (Greenwood, 2001), 87.

[106] See in particular the CBD. 'Diversity' itself is an even more abstract concept than an 'ecosystem' or an 'organism'. If it becomes possible to unpick the idea of 'diversity' itself having value, this may be compatible with both instrumental and intrinsic value.

[107] Although the intrinsic value of species is not entirely absent from the philosophical literature: *see* H Rolston III, *Philosophy Gone Wild: Essays in Environmental Ethics* (Prometheus, 1986); Callicott (n 100); Bowman and others (n 40), chapter 3.

[108] General rules have of course been formed by philosophers, but these are often too vague or too exacting to be of immediate practical legal significance: *see*, for example, Taylor (n 85); Devall and Sessions (n 99); Mathews (n 39) 118. Note also Leopold's famous maxim that 'A thing is right when it tends to preserve the integrity, stability, and beauty of the biotic community. It is wrong when it tends otherwise': Leopold (n 100), 189.

thought: the world becomes a place no longer composed exclusively of resources, but one which is imbued with living legal subjects.[109] The most obvious and immediate consequence of this is the need for acknowledgement of moral and legal obligations to nonhumans. This would lead to greater levels of environmental protection and could be accomplished by establishing rights,[110] or by adopting a duty-based[111] or a care-based[112] ethic.

The actual detail of these obligations is currently unclear. However, it is not the case that the tide of intrinsic value would necessarily render human life impossible by legislating against any use, disturbance or killing of other living beings. This would respect neither the intrinsic value of humans nor that of ecosystems; it is in the nature of all organisms to rely on the consumption of other ones.[113] New general principles could emerge based on working with, rather than against, the processes of life. These might include: the prevention of wanton and indiscriminate destruction; incorporation of the long-term needs of all life forms into management schemes; acceptance of the aesthetic and non-aesthetic value of wildness – not only in national parks but in cities and airports too; and stricter welfare requirements for domesticated species. Such principles could find legal meaning through instruments that acknowledge and apply them in various contexts such as agriculture or urban planning.

For example, agricultural policies such as factory farming and mono-cultural practices do not align with the intrinsic value of either individuals or ecosystems. In contrast, practices such as organic farming and permaculture[114] seek to work alongside other species and natural processes and could be promoted through legal instruments. An example of a codified approach to incorporating intrinsic value in a particular domain is the 2002 Melbourne Principles for Sustainable Cities.[115] These present a vision for urban planning which 'recognises the intrinsic value of biodiversity and

---

[109] See G Francione, *Animals, Property and the Law* (Temple University Press, 1995); H Rolston III, 'Is There an Ecological Ethic?' (1975) 18 Ethics 93; C Cullinan, *Wild Law* (Green Books, 2003).

[110] See, for example, Regan (n 81); P Cavalieri and P Singer (eds), *The Great Ape Project* (St. Martin's Press, 1993); The Universal Declaration of Rights of Mother Earth (World People's Conference on Climate Change and the Rights of Mother Earth 2010) available at <http://therightsofnature.org/universal-declaration/>; Cullinan (n 109).

[111] Taylor (n 85).

[112] Devall and Sessions (n 99) and Fox (n 6) both propose a spontaneously emerging ethic of care. See further J Howarth, 'Neither Use nor Ornament: A Consumers' Guide to Care' in J Benson (ed.), *Environmental Ethics: An Introduction With Readings* (Routledge, 2000).

[113] Rather, the trick is in working out the relevance of the various characteristics exhibited by different organisms. As Warwick Fox puts it (n 6, 166) regarding sentience: 'This view does not deny the moral relevance of sentience; it simply denies that sentience is an appropriate criterion of moral considerability.'

[114] That is, permanent agriculture, as to which see P Whitefield, *The Living Landscape: How To Read and Understand It* (Permanent Publications, 2009); Aranya, *Permaculture Design: A Step-by-step Guide* (Permanent Publications, 2012); <http:www.permaculture.org.uk>.

[115] For the text, see UNEP, Division of Technology, Industry & Economics, Integrative Management Series No. 1, available at <http://www.unep.or.jp/ietc/focus/melbourneprinciples/english.pdf>.

natural ecosystems'.[116] This principle can be transformed into policy objectives supported by legal requirements, such as creating urban habitat corridors and 'rebuilding real connections between city dwellers and the living world so as to foster attitudes of respect and care'.[117] In addition, more biting legal tools could develop, as witnessed by the movement to create environmental crimes, specifically the crime of ecocide as the fifth crime against peace. This has been proposed as a crime to be heard by either the International Criminal Court or a new International Court for the Environment.[118] Such a mechanism would give teeth to the understanding that human behaviour towards nonhumans has moral, and consequently legal, implications. At the very least it would open the doors to the valuation of environmental damage in cases where no direct consequence for humans can be established.

However, there are some conundrums encountered by attempting to create legal schemata based on intrinsic value. Three problems arise in this regard, which are avoided rather than countered by the above examples. First, if all living beings have intrinsic value, and this is not measurable, countable or comparable,[119] then how do we know what to protect and what not? Why protect the panda over the common garden slug? Why feed a chimpanzee in a zoo but not one in the wild? The inability to prioritise protection could, in practice, result in a lack of real protection. The obvious route out of this is to state that it is the nature of protection that should differ, but this still does not inform us about prioritisation. Although purely instrumental approaches may still face a similar issue due to the difficulty of putting accurate economic values on ecosystem services, the difficulty here is somewhat more fundamental since there is not even a metric by which to measure. The problem of prioritisation is a matter common to many areas of law (for example human rights or prison management), and so may not be a product of intrinsic value analysis as such, but is certainly compounded by the intricacies of the underlying ethical justifications and the complexity of the subject matter itself.

Secondly, protecting the intrinsic value of individual organisms may do harm when certain individuals are damaging the wider ecological community. The promotion of the needs of one particular life form, or collection of life forms,[120] over the needs of the

---

[116] Principle 3.

[117] P Newman and I Jennings, *Cities as Sustainable Ecosystems: Principles and Practices* (Island Press, 2008), 69.

[118] Ecocide has been defined as the 'extensive damage to, destruction of or loss of ecosystem(s) of a given territory, whether by human agency or by other causes, to such an extent that peaceful enjoyment by the inhabitants of that territory has been or will be severely diminished': P Higgins, *Eradicating Ecocide* (Shepheard-Walwyn, 2010), 63; <http://eradicatingecocide.com/> (Note that it is already a crime to cause widespread, long-term and severe damage to the non-human environment during wartime if it is disproportionate to the military advantage anticipated: 1998 Rome Statute of the International Criminal Court, 2187 UNTS 90, as rectified, Article 8.2.b (iv).) Although ecocide is not specifically referential of intrinsic value, it is certainly inspired by it.

[119] Examples of attempts to perform this sort of prioritisation can be found in Taylor (n 85); N Agar, *Life's Intrinsic Value* (Columbia University Press, 2001); and Mathews (n 39) in her discussion of blue whales and krill, 123ff.

[120] Usually a species, such as *Homo sapiens* (which is listed by the IUCN as one of the worst 100 invasive species).

whole community could prove short-sighted. Take the example of alien species: should a cull of the American grey squirrel take place in the UK? This requires balancing the (biocentric) intrinsic value of individual grey squirrels with that of whole ecosystems: again, there is no metric by which to measure or compare these. Practical solutions that meet both biocentric and ecocentric intrinsic value requirements may require long-term, systemic thinking. For example, re-introduction of the European pine marten may help suppress grey squirrel numbers while supporting the integrity of the ecosystem, given that a recent study has shown that 'pine marten abundance may be a critical factor in the American grey squirrel's success or failure as invasive species'.[121] Perhaps what is needed is a more ready acceptance of death and suffering (when these take place as natural processes): the intrinsic value of the organism does not mean that it must never die, and the continuation of the ecosystem is in fact reliant on the death of its constituents. Ecocentric intrinsic value also avoids our having to understand what the intrinsic value of each slug, beetle, flowering shrub and fungus demands of us individually. And although these aspects of their intrinsic value are not to be ignored, they are perhaps able to be respected alongside an ecosystemic approach.

Neither biocentrism nor ecocentrism can provide simple answers to the third problem: that intrinsic valuation struggles to provide a moral justification to protect species. Species have, in themselves, no intrinsic value according to many philosophical definitions. Compounding this, as noted above, intrinsic value does not depend on the relative state or condition of an entity but only on its own intrinsic properties: rarity or endangered status cannot be intrinsic in this sense.[122] Given that species protection has been the main thrust of biodiversity law, a problem emerges. More nuanced arguments are required to allow the focus to glide onto keystone species, endangered species and holistic management practices.

Presented with several '-centrisms', which conflict and disagree at times, environmental valuation (when applied across many levels simultaneously) requires a balance between different approaches. Such a balance needs tools from instrumentalism to assist with prioritisation as well as justifications from ethics. Considered approaches to intrinsic value also indicate that the ecosystem (and all its components) is of instrumental value to *all* its inhabitants, not just the human ones. This reimagines instrumental value as now including all organisms and ecosystems as bearers of intrinsic value (after pain of a finite regress) and considers how species, rivers,

---

[121] E Sheehy and C Lawton, 'Population crash in an invasive species following the recovery of a native predator: the case of the American grey squirrel and the European pine marten in Ireland' (2014) 23 Biodiversity and Conservation 753, 753.

[122] A potential route around this is to consider that part of an entity's intrinsic value is as a representation of a particular life form, strategy for survival, or exemplar of the diversity of life. The relatively rarer individuals of endangered species carry a larger share of this value and thus ought to receive additional protection. Interestingly, this can then be extended to justify further protection of even more phylogenetically isolated life forms (i.e. ones of rare genus, family, order or other taxonomic rank). *See* further MJ Bowman, 'Biodiversity, Intrinsic Value and the Definition and Valuation of Environmental Harm' in MJ Bowman and AE Boyle, *Environmental Damage in International and Comparative Law: Problems of Definition and Valuation* (OUP, 2002) 56–7. *See* also T Vernimmen, 'Sorry, Tiger: Why We Should Save Weird Species First' (2014) 2978 New Scientist 38.

decomposing microorganisms and so forth are all of instrumental value to each other.[123] Not only does intrinsic value pervade, but so too does instrumental. Perhaps neither the tenets of biocentrism nor ecocentrism nor egocentrism nor anthropocentrism should be rejected, but rather the '-centrism' aspects of them all.

5. CONCLUSIONS: AN AGENDA FOR RECONCILIATION?

This chapter has shown that neither instrumental nor intrinsic values provide, on their own, an entirely convincing ethic and strategy for conservation. On the one hand, instrumental value, particularly when it is yoked to evaluating all with a common metric (viz. currency), does not have the capacity to reflect all value which exists; on the other hand, intrinsic value presents unresolved theoretical contradictions which do not yield practical legal solutions. It seems to be the case that entities are frequently of *both* instrumental *and* intrinsic value simultaneously: life may well be valuable in itself, but it also requires the consumption or use of other forms of life in order to be perpetuated. At the same time, the sustainability of instrumental values, and of renewable resources, depends on the preservation of the autopoietic processes (and associated intrinsic values) that generate life forms, and their utility, in the first place.[124] However, it is not the intention here to convince the reader that a just ethic (and a just law) must explicitly combine both value-systems. Instead, it is sufficient for adherents to one value-system to acknowledge the reasonableness of the other. As Turner states, 'the term "valuing the environment" means different things to different people depending on which of the world-views they find acceptable'.[125]

The discussion in this chapter has tried to encompass a wide range of world-views by describing two rather extreme positions. Yet, what emerged is that the different interpretations of 'value' are bound to converge. At one end of the value spectrum, instrumental value can become 'informed instrumentalism', which expands valuation to previously non-marketable ecosystem services, non-material uses and even non-use values. This expansion is driven by science and disciplined by economics, and it is likely to justify a decisive increase in environmental protection. Looking at the well-established principle of interconnectedness among organisms, ecosystems and the physical environment, the whole biosphere may be considered instrumentally valuable and so worthy of legal protection. In addition, a clearer understanding of intrinsic value allows the realisation that it is not only humans that value things instrumentally, but all forms of life since they are all 'loci of valuational activity'. In this sense, intrinsic value's scope is not one of absolute responsibility, but of a contextualised responsibility to protect life forms considered as 'organisms-in-environment' and the environment itself. Recognition that all organisms have intrinsic value gives rise to a form of

---

[123] See Section 3.2 above.
[124] This suggests a shift in focus for environmental protection: *see* KE Limburg and others, 'Complex Systems and Valuation' (2002) 41 Ecological Economics 409.
[125] Turner (n 62), 20.

'non-anthropocentric instrumentalism based on the promotion of nonhuman nature'.[126] Refined intrinsic value thus is able to build guidance for international biodiversity law.

The above is in line with Norton's 'convergence hypothesis', the belief that 'policies serving the interests of the human species as a whole, and in the long run, will serve also the "interests" of nature, and vice versa'.[127] This can lead to growing awareness that our self-preservation as individuals, as species and as ecosystems cannot be performed piece by piece but rather that it must be done by understanding the inter-relations between individuals, species, and ecosystems. Built on the ecological interdependence of all living and non-living things, and thus mirroring both environmental philosophy and its movement to 'enlarge the boundaries of the community to include soils, waters, plants and animals'[128] and modern ecology, which acknowledges that 'complex systems must be studied at every level, because each level has properties not shown at lower levels',[129] a new law can emerge where the differences between the protection demanded of biodiversity law by instrumental value and intrinsic value dissipate.

For this to be a good omen for the future development of biodiversity law, awareness of the value(s) of nature must be spread across society and materialised into clear political demands. The role of the environmental movement is crucial: environmentalists should therefore avoid conflict between supporters of intrinsic valuation and instrumental valuation and must instead develop compatible theories and manifestos.

SELECT BIBLIOGRAPHY

Bateman, IJ, *Valuing Environmental Preferences: Theory and Practice of the Contingent Valuation Method in the US, EU, and Developing Countries* (OUP, 2001)
Bowman, MJ and AE Boyle (eds.), *Environmental Damage in International and Comparative Law: Problems of Definition and Valuation* (OUP, 2002)
Ehlers, E and T Craft, *Earth System Science in the Anthropocene* (Springer, 2006)
Gillespie, A, *International Environmental Law, Policy and Ethics* (2nd edn., OUP, 2014)
Lee, K, 'The Source and Locus of Intrinsic Value: A Re-examination' (1996) 18 Environmental Ethics 297
Mathews, F, *The Ecological Self* (Routledge, 1991)
Norton, BG, *Toward Unity Among Environmentalists* (OUP, 1991)
O'Neill, J, 'The Varieties of Intrinsic Value' (1992) 75 The Monist 119
Rolston III, H, 'Value in Nature and the Nature of Value' in R Attfield and A Belsey (eds.), *Philosophy and the Natural Environment* (1994) <http://ebooks.cambridge.org/ebook.jsf?bid=CBO9780511524097>
Sukhdev, P and others, *The Economics of Ecosystems and Biodiversity – Mainstreaming the Economics of Nature: A Synthesis of the Approach, Conclusions and Recommendations of TEEB* (TEEB, 2010)
Varner, G, *In Nature's Interests?* (OUP, 1998)

---

[126] E Katz and L Oechsli, 'Moving beyond Anthropocentrism: Environmental Ethics, Development, and the Amazon' (1993) 15 Environmental Ethics 50.
[127] BG Norton, *Toward Unity Among Environmentalists* (OUP, 1991), 240.
[128] Leopold (n 100), 39. Consider also Naess's notion of 'intrinsic relations' (n 98), and Morton (n 64) 28.
[129] Mayr (n 96), 64.

3. Participatory resource management: a Caribbean case study

Nicole Mohammed

1. INTRODUCTION

Public participation is a difficult concept to define. Pring and Noe note that while the term is used frequently, 'it is rarely defined other than by implication'.[1] One of the broadest definitions is that adopted in the 2000 Organisation of American States Inter American Strategy for the Promotion of Public Participation in Decision Making for Sustainable Development.[2] The Strategy defines public participation as 'all interaction between government and civil society ... including the process by which government and civil society open dialogue, establish partnerships, share information and otherwise interact to design, implement and evaluate development policies, projects and programmes'. Public participation provisions began to appear in the planning and environmental regulations of some states during the 1960s and 1970s.[3] Richardson and Razzaque note that in developing countries the trend often manifested itself in the form of greater community involvement in development planning and poverty alleviation projects.[4] On an international level, participation was addressed in 1972 at the Stockholm Conference on the Human Environment.[5] It was also addressed in 1982 by

---

[1] GR Pring and SY Noe, 'The Emerging International Law of Public Participation Affecting Global Mining, Energy and Resource Development' in A Zillman, A Lucas and GR Pring (eds.), *Human Rights in Natural Resource Development: Public Participation in the Sustainable Development of Mining and Energy Resources* (OUP, 2002), 14. The authors also note that there are a myriad of corollary terms including 'citizen rights', citizen action', 'citizen participation', 'stakeholder participation' and 'community involvement'; *ibid.*, 15.

[2] Organisation of American States Inter-American Council for Integral Development (OAS CIDI), 'Inter-American Strategy for the Promotion of Public Participation in Decision Making for Sustainable Development', CIDI/RES.98 (V-O/00), OEA/Ser.W/II.5, (20 April 2000), available at <http://www.oas.org/dsd/PDF_files/ispenglish.pdf>. *See* also E Dannenmaier, 'Democracy in Development: Toward a Legal Framework for the Americas' (1997) 11 Tulane Environmental Law Journal 1.

[3] *See* B Richardson and J Razzaque, 'Public Participation in Environmental Decision-Making' (2006) Environmental Law for Sustainability 165.

[4] *Ibid.*, 168.

[5] 1972 Action Plan for the Human Environment, Recommendation 7(a) states: '[i]t is recognised that Governments and the Secretary General provide equal possibilities for everybody, both by training and ensuring access to relevant means and information, to influence their own environment by themselves'. Principle 19 of the 1972 Stockholm Declaration and Action Programme also recognises the need for improving education and public awareness in the field of the environment; see Report of the United Nations Conference on the Human Environment, UN Doc. A/CONF/48/14/Rev.1 (1972) 11 ILM 1416.

the UN General Assembly in the World Charter for Nature, although it did not become a significant topic in international law until the 1990s and in particular, in 1992, at the Rio Conference on Environment and Development.[6]

Principle 10 of the 1992 Rio Declaration emphasises the role of participatory rights at a national level in the achievement of the goal of sustainable development. The three pillars of procedural participation are defined as:

(a) appropriate access to information concerning the environment that is held by public authorities;
(b) an opportunity to participate in decision making;
(c) effective access to judicial and administrative proceedings, including redress and remedy as regards environmental issues.

These three pillars form the backbone of the 1998 Convention on Access to Information, Public Participation in Decision-Making and Access to Justice,[7] which entered into force in 2001 and was concluded as part of the United Nations Economic Commission for Europe's 'Environment for Europe Process'.

Caribbean nations were among the more than 170 states that adopted the Rio Declaration in 1992. In the last two decades, the region has continued to demonstrate support for the goal of sustainable development and the participatory concepts contained in Principle 10. In 1994, Caribbean states were among the nations that adopted the Declaration of Barbados and the Programme of Action for Small Island Developing States ('Barbados Programme Of Action') at the UN Global Conference on Sustainable Development of Small Island Developing States. The Barbados Programme of Action reaffirms the principles set out in the Rio Declaration.[8] In 2005, the principles endorsed by the Rio Declaration and the Barbados Programme of Action were again reaffirmed in the Mauritius Declaration.[9] Most recently, 20 Latin American and Caribbean countries (including Jamaica, Trinidad and Tobago, and St Vincent and the Grenadines) have signed the 2012 Declaration on the application of Principle 10 of the Rio Declaration on Environment and Development in Latin America and the

---

[6] 1992 United Nations Declaration on Environment and Development, 13 June 1992, UN Doc. A/CONF.151/26/Rev.1 (1992) 31 ILM 876. See particularly Principles 10 and 22.

[7] Readers may refer to <www.unece.org/env/pp/welcome.html> and <europa.eu.int/comm/environment/aarhus/> for more information on the Convention.

[8] The Barbados Programme of Action sets out 14 priority areas and makes repeated references to the importance of participation in the shaping of policies and decisions and the special role of NGOs, women, indigenous communities and youth. The final section of the programme recognises that 'genuine involvement of all social groups' will be critical to the implementation of the programme of action and Agenda 21, and notes that 'broad public participation in decision-making' is a fundamental prerequisite to the achievement of sustainable development.

[9] United Nations, *International Meeting to Review the Implementation of the Programme of Action for the Sustainable Development of Small Island Developing States* (2005) (document A/CONF.207/L.6).

Caribbean ('the LAC Declaration on Principle 10').[10] The Declaration outlines an agreement to support the development of 'a regional instrument, ranging from guidelines, workshops and best practices to a regional convention' to strengthen access to information, encourage public participation, and facilitate access to justice in sustainable development decision making.

2. PARTICIPATORY APPROACHES TO RESOURCE MANAGEMENT

Participatory approaches to forest management have grown in popularity over the past three decades. A growing recognition that top-down approaches have failed to resolve inevitable conflicts between economic goals and environment protection has led to more decentralised forms of environmental governance based on community participation.[11] The Caribbean region is no exception to this trend.[12] Renard and Geoghegan note that: '[t]he concept of co-management, although not always well understood, has become increasingly popular in the Caribbean where it is seen as a mechanism for improving management by supplementing the limited resources available to most of the region's governments with those of the community, private sector and NGOs.'[13]

One of the primary rationales for community participation in forest management is that it leads to better decision making.[14] Steele suggests that the increasing prevalence of participation in environmental law can be explained in terms of a 'new understanding' of the public's potential to contribute to environmental decisions.[15] Decision makers are increasingly accepting the validity and importance of local knowledge and local voices. Agencies are often short-staffed and under-resourced.[16] Members of the public and NGOs can bring expertise to bear on the issues being decided. Further, persons with local knowledge may also be able to see potential solutions overlooked by those who are less familiar with local conditions. These participatory processes therefore aim to give a voice to members of the community who would otherwise be

---

[10] Available at <http://www.cepal.org/rio20/noticias/paginas/8/48588/Declaracion-eng-N1244043.pdf>.

[11] B Akbulut and C Soylu, 'An Inquiry into Power and Participatory Natural Resource Management' (2012) 36 Cambridge Journal of Economics 1143, 1144.

[12] T Geoghegan, *Participatory Forest Management in the Insular Caribbean: Current Status and Progress to Date* (CANARI Technical Report No. 310) (Caribbean Natural Resources Institute ['CANARI'] 2002), 1, available at <www.ema.co.tt/new/images/policies/particip_forest.pdf>.

[13] T Geoghegan and Y Renard, 'Beyond Community Involvement: Lessons Learned from the Insular Caribbean', (2002) 12(2) PARKS 16, 21.

[14] See N O'Reilly, 'Come One Come All? Public Participation in Environmental Law in Ireland: A Legal, Theoretical and Sociological Analysis' (2005) 8 Trinity College Law Review 72.

[15] J Steele, 'Participation and Deliberation in Environmental Law: Exploring a Problem Solving Approach' (2001) 21 Oxford Journal of Legal Studies 415, 418.

[16] See Geoghegan (n 12), 7, who notes that shrinking budgets for forestry management in the Caribbean region have been a major driver towards participatory approaches to forest management.

ignored or by-passed by the government officials and bureaucrats who hold decision-making powers.

As the discussion below also demonstrates, the benefits of these types of participatory arrangements are not confined to the resource they aim to protect, but can also extend to the local participants engaged in the decision-making process. This is because when participatory processes work well, these local voices become more confident, empowered and even more protective of the resource which they are helping to manage. In the Jamaican example, which is explored in detail below, participatory processes have not only improved forest management, but have also contributed to capacity building, improved leadership, sustainable livelihood projects and even environmental activism among participants.

In the context of forest management, participatory arrangements may take a variety of forms:

- contract-based relationships where objectives and outputs are defined by the contracting parties;
- loose collaboration where objectives are generally defined by the initiating party and the project is open to others based on interest;
- formal collaboration where objectives are defined jointly by the parties to agreements – roles, responsibilities, rights and benefits are clearly spelled out and are to some extent binding;
- multi-stakeholder management or advisory committees where objectives are defined and managed by multiple stakeholders.[17]

Some authors have classified these various forms of participatory arrangements using categories essentially based on the extent of distribution of power between the members of the public and the relevant government agencies. Arnstein's 'ladder of participation' proposes a typology of eight levels of participation.[18] At the bottom rungs of the ladder are (1) manipulation and (2) therapy. The real objective of this kind of engagement is to enable power-holders to educate or persuade participants. She regards this as contrived 'non-participation'. The next three rungs involve 'tokenism'. This entails information (3), consultation (4) and placation (5). At this level, citizens lack the power to ensure their views will be heeded by the powerful. At the very top of the ladder is what Arnstein refers to as 'citizen power' (partnership (6), delegated power (7), citizen control (8)). At this level the public has full decision-making power. The ladder metaphor assumes that higher rungs, i.e. active engagement, self-mobilisation or citizen control, are preferable to lower rungs of the ladder. Arnstein warns that without a genuine redistribution of power, citizens are simply undergoing an 'empty ritual of participation', which results in a decision-making process that is not reflective of the public's needs or priorities. It is arguable, however, that a participation process characterised at either end of the 'ladder' would be undesirable.

---

[17] CANARI, *Participatory Forest Management in the Caribbean: Impacts and Potential* (CANARI Policy Brief No. 1) (2002).
[18] S Arnstein, 'A Ladder of Citizen Participation' (1969) 35(4) Journal of the American Institute of Planners 216.

The participatory scheme which is the subject of this discussion arguably aims to achieve some kind of power sharing or co-management between bureaucrats and the local community. In theory, it should fall somewhere between the middle and upper rungs of Arnstein's ladder of participation; i.e., with some degree of genuine collaboration but also some degree of centralised decision making. In practice, however, an approach such as this can lead to local participants being treated like hired hands rather than a true local voice.[19]

Goodwin comments on this tension between the 'hired hand' versus the 'local voice' approach to collaboration in his discussion of community participation in industrial countries:

> [P]articipation is not an uncontested concept ... Although ostensibly, it suggests a renegotiation of power between 'expert outsiders' and 'ordinary people', there is a great deal of difference between local initiatives that are seen as contributing to, but not conflicting with centrally determined objectives, and those in which participants are given the power to participate in the production, as well as implementation of central control.[20]

The question arises, however, as to whether the success of a participatory scheme should be judged simply by the degree of power sharing which is achieved between bureaucrats and members of the public. While much of the literature makes the assumption that participation that results in power sharing is ideal,[21] some authors recognise that different forms of engagement are appropriate, depending on the objectives of the participation exercise and the capacity of the individuals to make a contribution.[22] Lawrence, in fact, suggests that 'transformative participation' may be an alternative highest rung of the ladder of participation.[23] In such cases communities are transformed through education and empowerment even where central control is still exercised outside the community. She argues that power can take many forms and can be equated with knowledge, social inclusion and social capital. Transformative types of participation may therefore result in greater commitment to the environment and its conservation, through learning among different stakeholders and even change in values. When viewed this way, ranking different levels of power sharing may be of limited

---

[19] P Goodwin, '"Hired Hands" or "Local Voice": Understandings and Experience of Local Participation in Conservation' (1998) 23 Transactions of the Institute of British Geographers 481.

[20] *Ibid.*, 483.

[21] *See* MS Reed, 'Stakeholder Participation for Environmental Management: A Literature Review' (2008) 141 Biological Conservation 2417.

[22] C Richards, KL Blackstock and CE Carter, *Practical Approaches to Participation*, SERG Policy Brief No. 1 (Macauley Land Use Research Institute, 2004) available at <www.macaulay.ac.uk/socioeconomics/research/SERPpb1.pdf>; J Tippett, JF Handley, J Ravetz, 'Meeting the Challenges of Sustainable Development – A Conceptual Appraisal of a New Methodology for Participatory Ecological Planning' (2007) 67(1) Progress in Planning 9. Davidson has also proposed an alternative metaphor described as 'the wheel of participation', which illustrates the legitimacy of different degrees of engagement; *see* S Davidson, 'Spinning the Wheel of Empowerment' (1998) 3(4) Planning 14.

[23] A Lawrence, '"No Personal Motive?" Volunteers, Biodiversity and the False Dichotomies of Participation' (2006) 9(3) Ethics, Place and Environment 279.

utility. The extent to which power has been shared in the collaborative process is just one aspect of that process and, in some cases, may not be the key factor which allows for sustainable use and long-term community buy-in. Arguably education, empowerment, sustainable livelihoods and a management system which makes the local communities part of the solution, rather than the problem, are all important indicators of a successful participatory process, regardless of whether or not some decision making remains centralised.

Goodwin also notes that in contrast to conservation professionals, who tend to measure the success of participatory initiatives in qualitative terms, local participants see participation as a much more open-ended, transformative process. Hence local participants in conservation projects in the county of Kent, spoke of the success of the programmes in terms of 'a sense of belonging', the 'development of community spirit' or of a 'better quality environment'.[24] Conservation projects with participatory mechanisms were therefore not just an end in themselves, but had the potential to lead to a greater consciousness among locals and even spark new ideas and new areas of interest. Goodwin argues that in this way, participation can be perceived as a more self-regulating process, capable of generating its own targets and realising its own goals.[25]

The discussion below will focus on an example of participatory resource management in the Caribbean island of Jamaica, which has its origins in the Jamaican forestry legislation. The aim is to examine the challenges faced by the local forest management committees over the period of a decade and to assess how the existing mechanism has evolved to meet those challenges. In so doing, the author will also look at the extent to which this scheme has been able to bring about genuine power sharing as well as the transformative types of participation referred to above. It will be argued that it is the flexibility of the scheme and its ability to adapt to the challenges faced during its implementation, plus these transformative forms of participation, rather than the extent of power sharing achieved, which represent the true successes of this Jamaican model.

3. LOCAL FOREST MANAGEMENT COMMITTEES IN JAMAICA

During the 1970s and 1980s government policies and shrinking budget allocations created a reduction in the capacity of the Jamaican Forestry Department to adequately manage forest reserves. The result was an increase in illegal timber extraction and destructive agricultural practices. By the 1990s, decades of poor management had led to increasing soil erosion, landslides, flooding and declining water quality. However, the Forestry Department was revitalised in the 1990s with the support of the United Nations Development Programme and 'Trees for Tomorrow', a project funded by the Canadian International Development Agency.[26] This project emphasised participatory

---

[24] Goodwin (n 19), 490.
[25] *Ibid.*, 489–490.
[26] T Geoghegan and N Bennett, *Risking Change: Experimenting with Local Forest Management Committees in Jamaica* (CANARI Technical Report No. 308) (Caribbean Natural Resources Institute, 2003).

approaches to forest and watershed management and resulted in the creation of new forestry legislation and policy.

Section 12 of the Forest Act of 1996[27] provides that the Minister[28] may 'after consultation with the Conservator, appoint a forest management committee for the whole or any part of a forest reserve, forest management area or protected area'.[29] It also expressly provides for the participation of the local community on such committees: 'Whenever possible, each forest management committee shall include at least two members having local knowledge of the area ...'[30]

Section 13 of the Forest Act sets out the functions of these local forest management committees (LFMCs) as follows:

- monitoring of the condition of natural resources in the relevant forest reserve;
- holding of discussions, public meetings and like activities relating to such natural resources;
- advising the Conservator on matters relating to the development of the forest management plan;
- proposing incentives for conservation practices in the area in which the relevant forest reserve is located;
- assisting in the design and execution of conservation projects;
- any other functions provided for under the Forest Act.

The role of LFMCs as contemplated under section 12 is therefore primarily functional and advisory. The management structure is a centralised one, which allows for input from the community.[31] The only real hint of co-management and a move towards less centralised decision making lies in the provision for 'assisting in the design' of conservation projects by the LFMCs.

Section 2.1 of the Forest Policy 2001 also emphasises the importance of community participation and makes reference to 'direct participation' in forest management:

> Sustainable use, management and protection of the Nation's forest resources require the participation and co-operation of local communities, particularly those living on the fringes of the forest ...

> The Forest Act provides for the formation of Local Forest Management Committees for forest reserves, forest management areas and protected areas. These committees will be the institutional bodies for enabling the direct participation of communities in forest management ...

---

[27] Forest Act, Act No. 17 of 1996.
[28] 'Minister' in the Act refers to the Minister of Environment.
[29] Forest Act s. 12(1).
[30] Forest Act s. 12(2).
[31] N Brown and N Bennett, *Consolidating Change: Lessons from a Decade of Experience in Mainstreaming Local Forest Management in Jamaica* (CANARI Technical Report No. 390) (Caribbean Natural Resources Institute, 2010), available at <http://canari.org/canaribackup/documents/FINALLFMCCaseStudyJan2011.pdf>.

The recently updated Draft Forest Policy[32] places a great deal of emphasis on the importance of public participation, as well as the management role of LFMCs. It states as follows:

> Jamaica's Strategic Forest Management Plan explicitly recognizes the importance of stakeholder participation in the sustainable management and conservation of Jamaica's forests. Participatory management approaches will be adopted to facilitate access to and sharing of some of the benefits that can be derived from forests by these rural communities, in order for example to reduce rural poverty levels. One such initiative is the opportunity for persons who reside on the fringes of State-owned forests, with an interest in the sustainable management of these areas, becoming part of a local forest management committee.[33]

The new draft policy also places greater emphasis on the development of co-management plans and capacity building for LFMCs.[34] The current Strategic Forest Management Plan (SFMP) 2010–2014 further identifies 'community participation and public awareness' as a strategic objective and states that it is 'essential' to forest management.[35]

The first two LFMCs were created in 2000 in the Buff Bay and Pencar watershed. Those areas were among the poorest in Jamaica, with farming as the residents' main occupation and much of the forest on the Buff Bay side converted to coffee plantations.[36] Funding for the launch was provided by the Trees for Tomorrow project. At the time the LFMC model represented a significant departure from traditional forest management in Jamaica. However, at the time of writing, seven LFMCs had been established with a target for the development of a further six by 2015. The new Draft Forest Policy also states that the Forestry Division has plans to create an LFMC for every Forest Reserve in Jamaica.[37]

Over the last decade the implementation of the LFMC model has been studied and documented by the Caribbean Natural Resource Institute as well as the Forestry Department and other participating NGOs.[38] These reports provide considerable insight into some of the strengths and weaknesses of the LFMCs as a mechanism for effective participation in environmental decision making and management. They also serve to

---

[32] Jamaica Draft Forest Policy Green Paper 2/15 available at <http://www.forestry.gov.jm/sites/default/files/Resources/forest_policy-green_paper_0.pdf>.
[33] *Ibid.*, 31.
[34] *Ibid.*, 18, 34.
[35] Jamaica Forestry Department, Strategic Forest Management Plan – Final Draft, Objectives 2 and 3, available at <http://www.forestry.gov.jm/?q=jamaicas-strategic-forest-management-plan-final-draft>.
[36] Geoghegan and Bennett (n 26), 3.
[37] Jamaica Forestry Department (n 35), para. 3.1.
[38] Geoghegan and Bennett (n 26); Brown and Bennett (n 31); M Headley, *Participatory Forest Management – the Jamaica Forestry Department Experience* (Jamaica Forestry Department, 2003); RN Oliphant, 'Local Forest Management Committees: the Jamaican Perspective' (2010) 18th Commonwealth Forestry Conference. See also A Hayman, *National Report on Management Effectiveness Assessment and Capacity Development Plan for Jamaica's System of Protected Areas* (Capacity Development Working Group, 2007).

document the way in which this participatory model has evolved in order to address barriers and obstacles to effective participation.

3.1 Barriers to Participation under the LFMC Model

3.1.1 Lack of equitable stakeholder representation
In 2002 a study undertaken of the first LFMCs in the Buff Bay and Pencar watersheds criticised the existing committees for their failure to truly represent a broad spectrum of the local community.[39] While in theory, LFMC membership was open to all stakeholders, in reality only formal organisations and legal entities were targeted and invited to join. The result was that the poorest and marginalised segments of the community, who did not tend to be involved in citizens' associations and other groups, had no representation at all. Further, those directly affected by the management of the watershed, such as farmers and local residents, were represented by weak and undemocratic organisations with ineffective feedback systems for their members. In contrast, government, national conservation NGOs, landowners and large agricultural business interests were well informed and strongly represented on the committees.[40] This focus on organisational membership was recognised as a weakness and a broader spectrum of participants was later targeted by the Forestry Division. A focus on individuals as well as organisational membership has resulted in much more equitable stakeholder representation. Brown and Bennett note that 10 years on, the LFMCs have successfully facilitated the coming together of groups and individuals who would otherwise not have done so. The Buff Bay LFMC now works in as many as 18 communities in the valley and has as many as 143 people on its roll.[41] Similarly the Northern Rio Minho LFMC contains representatives from over 23 communities located in the parish of Clarendon.[42] The legitimacy of the LFMC membership has also been one of the factors which has attracted international funding for development work in LFMC areas from partners such as the United States Agency for International Development.[43]

3.1.2 Lack of power sharing and a management role for LFMCs
In 2002 Geoghegan and Bennett observed that while the policy rhetoric in Jamaica seemed supportive of the decentralisation of decision making and the devolution of management responsibility to local communities, the institutional context, and in particular the Forest Act, still centralises decision making within the Forestry Division.[44] When one examines the function of LFMCs as defined by the Act it is clear that the role envisaged for LFMCs is more of a functional and advisory one, with decision-making power remaining in the hands of the Forestry Department.

---

[39] Geoghegan and Bennett (n 26), 12–13.
[40] *Ibid.*, 23.
[41] Brown and Bennett (n 31), 26.
[42] Oliphant (n 38), 7.
[43] Brown and Bennett state that the inclusive and broad-based membership of the Cockpit Country LFMC was a factor in the USAID/PARE project's decision to partner with The Nature Conservancy on work in the area, Brown and Bennett (n 31), 15.
[44] Geoghegan and Bennett (n 26), 13.

Section 12 of the Forest Act provides that:

> The Conservator shall, from time to time, make available to any forest management committee such technical advice and assistance as may be necessary to assist the committee in its functions.[45]

The Act therefore contemplates a lack of technical knowledge on the part of committee members at times. The provision suggests that the committees are expected to provide local knowledge and community feedback, rather than technical direction in the management of the forest resources.

However, early documents such as the 2001 Forest Plan and reports coming out of the Forestry Division make reference to the idea of co-management, with the Forestry Division playing a supporting 'secretariat' role.[46] These diverging objectives and viewpoints about the role of LFMCs, and the extent and type of participation expected of them, were also reflected among government officials. Geoghegan and Bennett observed that while some officials saw their role as advisory, others expected that they would undertake management responsibilities for specific sites.[47]

Over a decade after the establishment of the first LFMCs, the issue of decentralisation remains a live one. In their 2010 study, Brown and Bennett reported that for the most part the policy objective of co-management has not been achieved.[48] Although recent documents like the Draft Forest Policy[49] suggest a greater role for co-management, management decision making has been retained by the Forestry Division, albeit with input from the LFMCs.[50]

The question arises as to whether this can be seen as a failure of the LFMC system or simply a reflection of the fact that the type of participation which currently takes place is appropriate to the capacity of the local communities. As noted in the introduction, there are several different meanings attached to the concept of participation. During the history of its development and in the different contexts in which it has been applied, participation has become loaded with ideological, socio-political and methodological meaning, giving rise to a wide range of interpretations.[51]

As noted earlier, Arnstein's ladder of participation seeks to explain the different approaches and methodologies of participation as a continuum of increasing stakeholder involvement moving from passive dissemination of information to active engagement. Pretty et al employ a similar approach, using seven different types of participation ranging from 'passive participation' to 'participation in information

---

[45] Forest Act s. 12(3).
[46] Headley (n 38) and 'Formation of a Local Forest Management Committee for the Buff Bay/Pencar areas' (November 2000) Position Paper No. 6, 46.
[47] Geoghegan and Bennett (n 26), 9.
[48] Brown and Bennett (n 31), 5.
[49] *Ibid.*
[50] Oliphant (n 38), 2.
[51] Lawrence (n 23), 298.

giving' to 'functional participation' to 'self-mobilisation'.[52] Brown and Bennett use Pretty's typology in their 2010 study of LFMCs in Jamaica and conclude that the LFMCs largely engage in 'functional participation', with participants primarily executing predetermined objectives.[53] They note that while LFMCs often establish their priorities for local action and develop local plans of action, these always fall within the parameters set by the Forestry Department. In their view, therefore, participation through LFMCs cannot be categorised in the highest rungs of Arnstein's ladder or Pretty's scale of participation.

However, the fact that the LFMC model has not achieved participation at the highest rung of the ladder does not mean that it has not succeeded in creating an effective participatory environment. As noted earlier, participatory approaches may bring about transformative benefits. Increased community activism, empowerment and leadership among those who previously had no voice, as well as increased environmental awareness, are all benefits which cannot be measured in terms of power sharing.

These transformative characteristics are very much present in the Jamaican LFMC model. As early as 2002, Geoghegan and Bennett reported that one of the greatest successes of the LFMC model was that it had greatly enhanced the local understanding of the value of forests and the requirements for effective watershed management.[54] This was largely due to the extensive outreach programme undertaken by the Forestry Department and the sensitisation and awareness campaign designed by the Trees for Tomorrow project. In their 2010 case study Brown and Bennett also comment that LFMCs have been most successful in the role of advocate for the forest and its resources.[55] Most now run environmental education and outreach programmes and are active in area schools, churches and community groups. Anecdotal evidence[56] from Forestry Department officials also indicates that LFMCs have successfully helped to reduce problems such as illegal logging. The LFMCs have also lobbied on issues which go beyond strict forest management decisions. The Cockpit Country LFMCs have played a role in lobbying against bauxite mining in Cockpit Country, lending their support to a broader coalition of groups. The Buff Bay group has also lobbied the National Solid Waste Management Agency to begin regular garbage collection in the area in order to address illegal dumping of garbage in the river. Brown and Bennett have also noted that there has been an increase in community confidence as well as an improvement in conflict resolution and mediation skills since most LFMCs are used as a forum to resolve disputes and mediate with state agencies. They have also identified

---

[52] JN Pretty, S Bass and B Dalal-Clayton, 'Participation in Strategies for Sustainable Development' (1995) 7 Environmental Planning Issues 32, available at <http://pubs.iied.org/pdfs/7754IIED.pdf>.

[53] The term is defined as follows: 'People participate by forming groups to meet predetermined objectives related to the project which can involve the development or promotion of externally initiated social organisation. Such involvement tends not to be at early stages of project cycles or planning, but rather after the major decisions have already been made. These institutions tend to be dependent on external initiators and facilitators, but may become self-dependent'; *ibid.*, 32.

[54] Geoghegan and Bennett (n 26), 12.

[55] Brown and Bennett (n 31), 25.

[56] *Ibid.*, 29.

empowerment as one of the key outcomes of the LFMC process with strong leaders emerging from the communities.[57]

Brown and Bennett also note that, while the LFMCs engage in a type of participation that can largely be described as functional, there have been examples of self-mobilisation and interactive participation. At a project implementation level many of the community groups are drivers determining what they want to do and at what pace. In Buff Bay, for example, the LFMC now gives the Forestry Division advice on what to plan in certain areas. It also determines what maintenance activities need to take place, rather than waiting for direction from the Forestry Division.[58] Many of the LFMCs have also on their own initiative used the groups as a platform to launch sustainable livelihood projects.

The Forest Act is currently under revision and it is expected that amendments to the Act will reflect an expansion of the advisory and monitoring roles of LFMCs to one which allows them to function as full management partners.[59] The drafting instructions for the amendment of the Act propose that the new provisions should expressly include 'sustainable co-management' as one of the LFMCs' functions.[60] Other functions that will be introduced by the amendment include:

- development of a management plan for the area; and
- seeking funding for the management of forest reserves.[61]

The Strategic Forest Management Plan also demonstrates an intention to strengthen the co-management role of LFMCs. It states that in meeting the objective of 'increased community participation' the Forestry Department will:

- Expand the LFMC initiative to create fully functioning committees representing the variety of community interests for all the main areas identified as having priority for forest conservation
- Conduct community research and mobilisation leading to LFMC formation and memoranda of agreement with Forestry Department
- Train and provide guidance to LFMCs in proposal writing, management, and capacity building for forest management, conservation and sustainable livelihood projects.[62]

---

[57] *Ibid.*, 30.
[58] Some of the LFMCs (notably the Buff Bay and Northern Rio Minho group) have been clamouring for a greater management role and have even requested that a portion of the forest reserve be turned over to them for management. However, in the absence of an instrument that would allow such devolution, the LFMCs have had to use Memoranda of Agreement.
[59] Oliphant (n 38), 10. *See* also newspaper reports, e.g. 'Jamaica's Forestry Law to be Revised', *The Gleaner* (23 February 2014) available at <www.jamaica-gleaner.com/latest/article.php?id=51275>.
[60] Oliphant (n 38), 12.
[61] *Ibid.*, 10.
[62] Strategic Forest Management Plan (n 35), 16.

The LFMCs may therefore be well on their way to participation at the highest rung of the participation ladder, provided that the proper steps are taken to ensure that they build the capacity to take on that role.

3.1.3 Difficulty in maintaining interest in LFMCs among the local community

One of the initial problems faced when implementing LFMCs was the difficulty in attracting sustained interest from participants in the local community. In 2000, when the first two LFMCs were launched, a considerable amount of resources were spent on a sensitisation and awareness campaign to recruit participants. Within a four-month period, over 88 visits to communities, organisations and group meetings were held to promote the idea of LFMCs. The initial groundwork for the committees was also laid in the preceding two-year period, with community engagement taking place through the Trees for Tomorrow project, which undertook a forest inventory and important socio-economic and agroforestry studies. Funds were also allocated to ensure that LFMC members were compensated for their travel costs and lunches and refreshments were provided for special events.

Despite these measures, two years after the outreach campaigns Geoghegan and Bennett reported dwindling interest from committee members.[63] There was a significant decline in attendance and participation, with one study estimating that after the first year, 14 of the original 19 members were dormant or very weak participants.[64] They concluded that incentives and longer-term benefits were key to keeping stakeholders involved. Lack of monetary compensation and constraints due to poverty were identified as key factors in the growing lack of interest experienced by local stakeholders.

The importance of incentives and rewards is also expressly recognised in the legislation which establishes the committees. Section 13 of the Forest Act lists the proposing of 'incentives for conservation practices' in the relevant forest area as one of the functions of the LFMCs. Over the years the Forestry Department has given preferential access to the forest for use in a sustainable manner as an incentive to encourage participation. In addition LFMCs have sought to become more involved in the management of forest reserves by seeking grant funding to implement projects related to reforestation, ecotourism and the development of cottage industries. By 2012 LFMCs were able to access over US$600,000 in funding from agencies such as the Forest Conservation Fund, the Environmental Foundation of Jamaica and the International Institute for Environment and Development.[65]

Some LFMCs have also used the committees as a platform for sustainable livelihood projects. The Cockpit Country LFMC, with the assistance of international partners and the Forestry Department, has set up a historical tour, a medicinal plant nursery and a gift shop with local artisan crafts at a recently opened visitor centre.[66] These projects have generated occasional project-specific income for participants, though no projects

---

[63] Geoghegan and Bennett (n 26), 26.
[64] W Mills, 'Characterisation of At-risk Populations in the Pencar Watershed and Existing and Potential Benefits from Forest Resource Use' (2001) CANARI Technical report No. 306, 15.
[65] Brown and Bennett (n 31), 28.
[66] Oliphant (n 38), 6.

Participatory resource management 91

have yet been able to generate long-term sustainable income. The result is that, while in its initial stages there was no local demand for LFMCs,[67] 10 years on local groups are now approaching the Forestry Department to form LFMCs in areas where none currently exist. This is because many rural border communities are now beginning to recognise a direct correlation between the protection/restoration of the forest and the generation of wealth via sustainable livelihood projects.[68] Brown and Bennett also note that while the income benefits have not been long term, the capacity building and opportunity for empowerment and growth are benefits that are recognised and valued by local communities.

The importance of this element of the LFMC model has now been recognised in the revised Strategic Forest Management Plan (2010–2014). The 2001 Forest Plan recognised income-earning activities as an incentive for sustainable forest management. However, the revised plan now includes the 'number of LFMCs implementing sustainable livelihood, community development and/or conservation projects supporting sustainable forest management' as a performance measure for the implementation of its objective of increased community participation.[69] The drafting instructions for the amended Forest Act also include 'seeking funding for the management of forest reserves' as one of the functions of LFMCs.[70]

3.1.4 Lack of capacity for effective participation among stakeholders

One of the problems identified in the earlier studies of the LFMC process was the lack of capacity among the local communities to manage themselves as committees and to make meaningful contributions to forest management planning. The Forest Act in fact contemplates lack of technical expertise on the part of local communities. As previously noted, Section 12(3) provides that:

> The Conservator shall, from time to time, make available to any forest management committee such technical advice and assistance as may be necessary to assist the committee in its functions.

However, even with the initial funding provided by the Trees for Tomorrow project, the Forestry Department has struggled to provide the kind of 'on the ground support' that is needed.[71] The lack of capacity among the committees has also made it difficult for them to participate in the drafting of management plans for the forest reserves in which they work. These draft plans are written by trained foresters and often contain complex data, which lay-persons may have difficulty understanding. This has had an impact on what Geoghegan and Bennett refer to as the 'equity' of the participation process. They argue that effective participatory management requires all parties to come to the table with a power base of knowledge, skills and resources. Hence they argue that equity can

---

[67] Geoghegan and Bennett (n 26), 16.
[68] Oliphant (n 38), 2.
[69] Jamaica Strategic Forest Management Plan (2010–2015) (n 35), para 3.1.
[70] This is due to a recognition that the sustainability of these LFMCs is highly dependent on the receipt of grant funding; *see* Oliphant (n 38), 10.
[71] Geoghegan and Bennett (n 26), 24.

only be achieved if the position of the stakeholders representing the local communities is strengthened.

Since the establishment of the first LFMCs, a significant amount of resources has gone into providing training for this purpose. The Forestry Department has organised on-going training in basic forest management, customer service, product development, entrepreneurship, and tour guiding. LFMC members have also been invited to attend workshops for foresters in such areas as tree identification and forest fire prevention. However LFMCs still have capacity constraints in terms of technical knowledge, group management and development, organisational management and financial management.

Now that steps are being taken to amend the Forest Act to give LFMCs greater management autonomy and a greater role in the design of forest management plans, the need to address these constraints is even greater. Jamaica's experience of parks and protected area management has shown that groups without the correct mix of skills for management are doomed to failure.[72] Hence the move towards genuine co-management, while a step up the metaphorical participation ladder, will have to be accompanied by the training necessary to ensure its success.

4. SOME CONCLUSIONS

The LFMC model has evolved to tackle many of the barriers to participation identified above. However, the benefits of this model go beyond better decision making or even the improved management of the resource it was designed to help protect. One of its greatest successes is that it has helped to empower the local communities, creating livelihood opportunities, building capacity and giving strength to voices of sections of the community that could otherwise have easily been excluded from the environmental governance process. The model serves as an excellent example of the types of 'transformative participation' referred to by authors such as Lawrence[73] and Goodwin.[74] The LFMC process has brought with it benefits among participating communities in terms of awareness building, education and empowerment opportunities, and strengthened avenues for environmental advocacy. As Goodwin notes there is a distinction to be made between a person's local knowledge and their perception of their own competence to speak about that knowledge. What the LFMCs have achieved among locals is a growing recognition of both their capacity and entitlement to speak.[75] The LFMCs have not only given them a platform to express their views, but have also helped to engender a confidence in the validity and importance of that viewpoint. These kinds of benefits have taken place despite the fact that the process has not fully

---

[72] Hayman (n 38), 58.
[73] Lawrence (n 23), 279.
[74] Goodwin (n 19), 490.
[75] '[T]he capacity to have a viewpoint to answer questions is seen as being understood in terms of the individual's perception of their competence to speak. Competence in this sense is both technical, in terms of a person's skills and knowledge, and social through a recognition of their capacity and entitlement to speak and have a viewpoint'; *ibid.*, 487.

decentralised the management decisions related to the forested areas with which the LFMCs are associated.

The model has also been refined and altered to take into account feedback and problems encountered in its implementation. Through the LFMCs, local communities have been able to have a real influence on decision making, with their input influencing the design of management plans and projects within the forest reserves in which they operate.[76] The fact that the LFMCs have become a platform for livelihood projects also means that there is now scope for sustainability, provided that they are able to continue attracting funding for conservation and development projects.

The impact on the Forestry Department is another major success of the model. Brown and Bennett note that there has been a significant change in attitudes among Forestry Department staff over the last 10 years.[77] Resources have been allocated to train forestry staff in communicating and working with communities. A rural sociologist has been hired by the Department to assist in building and managing these relationships. LFMCs are now regarded as the 'eyes and ears' of the Department and community participation in decision making is now regarded as the norm, rather than the exception. Further, LFMCs are now considered active co-stewards by foresters and are valued for their contributions to forest activities.[78]

The LFMC model also demonstrates that there is no easy solution for the creation of institutional arrangements that seek to distribute rights and responsibilities more efficiently and more equitably. The degree to which management authority is devolved to local communities depends on the conditions of each case. Even where capacity exists for increased responsibility among local people, the role of the state as a facilitator and catalyst for implementation remains vital.[79] State agencies must strengthen their capacity for facilitating policy processes, while recognising the need to create a fair participatory environment. Civil society organisations do not always reflect the views of the poor and the powerless. Hence the challenge of creating effective multi-stakeholder partnerships is a difficult one. However, a flexible approach like that adopted in Jamaica, which allows for feedback and modification, can go a long way towards meeting these challenges and creating the space and the opportunity for true community participation and better environmental decision making.

---

[76] Contributions have, however, been mainly on a field level as LFMCs lack the capacity to make technical contributions to management plans. Recently efforts have been made to make management plans shorter and simpler in order to facilitate a greater level of participation; *See* Brown and Bennett (n 31), 24.
[77] Brown and Bennett (n 31), 24.
[78] *Ibid.*, 6.
[79] Y Renard and V Krishnarayan, 'Participatory Approaches to Natural Resource Management and Sustainable Development: Some Implications for Research and Policy' 5 (paper presented at the Regional Conference, 'Managing Space for Sustainable Living in Small Island Developing States, Port of Spain, Trinidad and Tobago', 16–17 October 2000).

SELECT BIBLIOGRAPHY

Akbulut, B and C Soylu, 'An Inquiry into Power and Participatory Natural Resource Management' (2012) 36 Cambridge Journal of Economics 1143

Arnstein, S, 'A Ladder of Citizen Participation' (1969) 35(4) Journal of the American Institute of Planners 216

Barton, B, 'Underlying Concepts and Theoretical Issues in Public Participation in Resource Development' in D Zillman, A Lucas and GR Pring (eds.), *Human Rights in Natural Resource Development: Public Participation in the Sustainable Development of Mining and Energy Resources* (OUP, 2002)

Brown, N and N Bennett, Consolidating Change: Lessons from a Decade of Experience in Mainstreaming Local Forest Management in Jamaica (CANARI Technical Report No. 390) (Caribbean Natural Resources Institute, 2010)

CANARI, *Participatory Forest Management in the Caribbean: Impacts and Potential* (CANARI Policy Brief No. 1) (2002)

Dannenmaier, E, 'Democracy in Development: Toward A Legal Framework for the Americas' (1997) 11 Tulane Environmental Law Journal 1

Ebbesson, J, 'The Notion of Public Participation in International Environmental Law' (1997) 8 Y.B. Int'l. Envtl. L 51

Geoghegan, T, *Participatory Forest Management in the Insular Caribbean: Current Status and Progress to Date* (CANARI Technical Report No. 310) (Caribbean Natural Resources Institute, 2002)

Geoghegan, T and N Bennett, *Risking Change: Experimenting with Local Forest Management Committees in Jamaica* (CANARI Technical Report No. 308) (Caribbean Natural Resources Institute, 2003)

Geoghegan, T and Y Renard, 'Beyond Community Involvement: Lessons Learned from the Insular Caribbean' (2002) 12(2) PARKS 16

Goodwin, P, '"Hired Hands" or "Local Voice": Understandings and Experience of Local Participation in Conservation' (1998) 23 Transactions of the Institute of British Geographers 481

Krishnarayan, V, *Participatory Approaches to Natural Resource Management and Sustainable Development: Some Implications for Research and Policy* (2000)

Lawrence, A, '"No Personal Motive?" Volunteers, Biodiversity and the False Dichotomies of Participation' (2006) 9(3) Ethics, Place and Environment 279

O'Reilly, N, 'Come One Come All? Public Participation in Environmental Law in Ireland: A Legal, Theoretical and Sociological Analysis' (2005) 8 Trinity College Law Review 72

Pretty, JN, S Bass and B Dalal-Clayton, 'Participation in Strategies for Sustainable Development' (1995) 7 Environmental Planning Issues 32

Pring, GR and SY Noe, 'International Law of Public Participation' in A Zillman, A Lucas and GR Pring (eds.), *Human Rights in Natural Resource Development: Public Participation in the Sustainable Development of Mining and Energy Resources* (OUP, 2002)

Reed, MS, 'Stakeholder Participation for Environmental Management: A Literature Review' (2008) 141 Biological Conservation 2417

Richards, C, KL Blackstock and CE Carter, *Practical Approaches to Participation*, SERG Policy Brief No. 1 (Macauley Land Use Research Institute, 2004)

Richardson, B and J Razzaque, 'Public Participation in Environmental Decision-Making' (2006) Environmental Law for Sustainability 165

4. The role of non-state actors in treaty regimes for the protection of marine biodiversity
*Elizabeth A. Kirk**

1. INTRODUCTION

Recent discoveries of the richness of life around deep ocean hydrothermal vents vividly highlight the relative lack of understanding of the resources of the oceans and of the impacts of human activity on them. To this must be coupled the rapid rate of expansion of activities and consequent threats to marine biodiversity in general. These discoveries raise hard questions about how best to manage and conserve biodiversity in the oceans. As these questions arise, so do questions about who has authority to make decisions about the future of the myriad of species that combine to make up that biodiversity. Will legal regimes be effective if established by States alone, or is there a need for other actors to be engaged in decision-making or in the development of the regimes? If there is a need, then what should the nature of that involvement be?[1]

Most regimes aimed at protecting marine biodiversity demonstrate a willingness to engage with non-State actors in the development or implementation of the regimes. Indeed in international law more generally, non-State actors have been involved in the development and implementation of the law for a considerable length of time.[2] Their role has gone through various stages with their influence ebbing and flowing across the decades. More recently the perception has been that the influence of both non-State actors, and NGOs in particular, has increased and that their role ought to be enhanced. This chapter examines the provision for non-State actor involvement in the legal regimes concerned with the conservation of marine biodiversity. Non-State actors do, of course, engage with decision-making processes in other less formal ways, lobbying States being one example. While the importance of such activities is acknowledged, this chapter focuses solely on formal engagement with non-State actors.

The chapter begins with a brief review of some of the theoretical underpinnings of the involvement of non-State actors, before moving on to review the provision made for their engagement in the regimes addressing the protection or management of marine biodiversity. Consideration is given to the justifications given for having participatory processes, the form that participation takes (focusing on participation within treaty and relevant soft law regimes), and the fit between form and any stated objectives of

---

  * With thanks to the Secretariats of the regional seas organisations for their help in tracking down various rules of procedure and to the editors for their helpful comments on earlier drafts. All errors remain the author's alone.
  [1] This chapter focuses on participation at the international (global or regional) level.
  [2] S Charnovitz, 'Two Centuries of Participation: NGOs and International Governance' (1997) 18 Michigan Journal of International Law 183.

participation. It also considers the impact that the chosen form may have on the levels and types of participation, and ultimately the impact participation may have on the quality or legitimacy of the decisions taken. The objective is to provide a taxonomy of participation within the regimes that are focused upon the protection of marine biodiversity.

2. ENGAGING WITH NON-STATE ACTORS: JUSTIFICATIONS AND LIMITATIONS

There are a variety of reasons for engaging non-State actors in decision-making. Although it may not be possible to point to a binding norm in international law requiring participatory rights for non-State actors,[3] it is possible that such a right is emerging or will emerge in the future.[4] In the meantime there are other practical reasons for providing such rights. For example, there is a substantial body of literature predicated upon the argument that the effectiveness of international regimes rests upon their legitimacy.[5] While there are a number of possible routes to legitimacy, such as the constructivist route expounded by Brunnée and Toope,[6] one route that has gained considerable support is to ensure that regimes are based upon deliberative democratic processes that engage with non-State actors.[7] The latter route is followed in this

---

[3] A Boyle and C Chinkin, *The Making of International Law* (OUP, 2007), 57; A Peters, 'Membership in the Global Constitutional Community' in J Klabbers, A Peters and G Ulfstein (eds), *The Constitutionalization of International Law* (OUP, 2009), 22; A Tanzi, 'Controversial Developments in the Field of Public Participation in the International Environmental Law Process' in P-M Dupuy and L Vierucci (eds), *NGOs in International Law: Efficiency in Flexibility?* (Edward Elgar Publishing, 2008), 135–152.

[4] Peters (n 3); S Charnovitz, 'The Illegitimacy of Preventing NGO Participation, The Symposium: Governing Civil Society: NGO Accountability, Legitimacy and Influence' (2010–2011) 36 Brooklyn Journal of International Law 891, 898.

[5] *See*, for a summary, D Bodansky, 'Legitimacy' in D Bodansky, J Brunnée and E Hey (eds), *The Oxford Handbook of International Environmental Law* (OUP, 2007), chapter 30. *See* also S Bernstein, 'Legitimacy in Global Environmental Governance' (2004–2005) 1 Journal of International Law & International Relations 139; J Brunnée and SJ Toope, 'International Law and Constructivism: Elements of An Interactional Theory of International Law' (2000) 39 Columbia Journal of Transnational Law 1; J Brunnée and SJ Toope, *Legitimacy and Legality in International Law* (CUP, 2010); TM Franck, *The Power of Legitimacy Among Nations* (OUP, 1990).

[6] Brunnée and Toope, 'International Law and Constructivism' and *Legitimacy and Legality in International Law* (n 5).

[7] *See*, for example, A Buchanan and RO Keohane, 'The Legitimacy of Global Governance Institutions' (2006) 20 Ethics and International Affairs 405; J Ebbesson, 'Public Participation' in D Bodansky, J Brunnée and E Hey (eds), *The Oxford Handbook of International Environmental Law* (OUP, 2007), chapter 29. *See* also, TM Franck, 'Remarks' in R Hofmann and N Geissler (eds), *Non-State Actors As New Subjects of International Law: International Law – From the Traditional State Order towards the Law of the Global Community: Proceedings of an International Symposium of the Kiel Walther-Schücking-Institute of International Law, March 25 to 28, 1998* (Duncker & Humblot, 1999) 151, 152; *Report of the Panel of Eminent Persons on United Nations-Civil Society Relations*, 37, 46, UN Doc. A/58/817, (11 June 2004).

chapter. The next few paragraphs provide a brief introduction to the relationship between engagement with non-State actors and legitimacy, and that relationship is returned to throughout the chapter.

Where once theory and practice may have led one to conclude that the requirement for deliberative democracy may be satisfied by ensuring that any and all interested States have an opportunity to participate, that no longer appears to be sufficient, and practice across many areas of international law now demonstrates engagement with non-State actors.[8] Such practice has led to a positive demonstration that engagement with NGOs (and other non-State actors) can make a substantial contribution to improving the legitimacy of decision-making within international regimes.[9] This may be as a result of simple engagement with non-State actors, or because engagement allows those actors to highlight poor decisions,[10] thus leading to better quality decision-making. They may also improve the quality of decision-making through their relative freedom to champion certain developments, which States may lack the freedom to do.[11]

Equally, the failure to include non-State actors directly in the decision-making process does not preclude their influence on decisions through lobbying activities,[12] but the perception of undue influence that may arise as a result of such action could undermine the perceived legitimacy of an organisation or regime. Thus there may be a need to provide input to decision-making by non-State actors through formal transparent participatory processes, which may be viewed as more legitimate.

The substantial body of general literature that explains why participatory decision-making is beneficial also supports the international law literature promoting the use of participatory processes.[13] The arguments presented in the general literature include that

---

[8] *See*, for example, P-M Dupuy and L Vierucci (eds), *NGOs in International Law: Efficiency in Flexibility?* (Edward Elgar Publishing, 2008); Charnovitz (n 2); S Charnovitz, 'Nongovernmental Organizations and International Law' (2006) 100 American Journal of International Law 348.

[9] *See*, for example, R Moloo, 'The Quest for Legitimacy in the United Nations: A Role for NGOs?' (2011) 16 UCLA Journal of International Law and Foreign Affairs 1, who provides an excellent critique of some of the literature on this topic.

[10] K Raustiala, 'The "Participatory Revolution" in International Environmental Law' (1997) 21 Harvard Environmental Law Review 537, 553.

[11] Charnovitz (n 8).

[12] *See*, for example, EA Kirk, 'Marine Governance, Adaptation and Legitimacy' (2011) 22 Yearbook of International Environmental Law 110.

[13] *See*, for example, M Appelstrand, 'Participation and Societal Values: The Challenge for Lawmakers and Policy Practitioners' (2002) 4 Forest Policy and Economics 281; J Black, 'Proceduralizing Regulation: Part I' (2000) 20 Oxford Journal of Legal Studies 597; J Dryzek, *The Politics of the Earth: Environmental Discourses* (OUP, 2005); DJ Fiorino, 'Citizen Participation and Environmental Risk – A Survey of Institutional Mechanisms' (1990) 15 Science Technology & Human Values 226; K Getliffe, 'Proceduralisation and the Aarhus Convention: Does Increased Participation in the Decision-making Process Lead to More Effective EU Environmental Law?' (2002) 4 Environmental Law Review 101; M Lee and C Abbot, 'The Usual Suspects? Public Participation under the Aarhus Convention' (2003) 66 Modern Law Review 80; J Steele, 'Participation and Deliberation in Environmental Law: Exploring a Problem-Solving Approach' (2001) 21 Oxford Journal of Legal Studies 415.

participatory processes enhance problem-solving abilities and provide access to additional information and perspectives not otherwise available to decision-makers. A further argument is based on the acknowledgement that science and the technocracy can never have the capacity to provide absolute certainty of result, which leads to the conclusion that uncertainty is a normal part of scientific understanding. Thus, if society is to make decisions on the basis of what is acknowledged to be uncertain information, then participatory processes are necessary to ensure that the resulting decisions maintain legitimacy. Without deliberative democracy the decisions risk (in the context of uncertainty) becoming obviously based upon the preferences or prejudices of the bureaucrats or political elite that make them.

There are, however, certain counter-arguments to the involvement of non-State actors in decision-making processes. Central to these is the positivist argument that the only recognised subjects of international law with the capacity to make international law are States.[14] The argument is that non-State actors, and NGOs in particular, are not recognised as full subjects of international law. They therefore do not have capacity to engage in law-making activities as such activities are, under the positivist rule of recognition, reserved for subjects with full international legal capacity i.e. States. This argument, of course, ignores the fact that international law does now recognise the various forms or degrees of personality granted to non-State actors.[15]

Further arguments have been developed with regard to the inclusion of NGOs, in particular, in international decision-making. One is that NGOs representing a particular narrow interest group or section of global society may change the power balance in an international organisation. If, for example, the result of their involvement is to marginalise developing States, the efficacy of engaging with NGOs may be questioned.[16] In addition, some have argued that allowing non-State actors, and NGOs in particular, a seat at the decision-making table, allows certain groups to have 'two bites at the cherry'. One bite occurs at the national level through lobbying of national governments and then a second at the international level. This might be particularly problematic if only certain sections of society have representation at the international level.[17] A further argument for the exclusion of NGOs in particular is that they may

---

[14] P-M Dupuy, 'Conclusions: Return on the Legal Status of NGOs and on the Methodological Problems Which Arise For Legal Scholarship' in P-M Dupuy and L Vierucci (eds), *NGOs in International Law: Efficiency in Flexibility?* (Edward Elgar Publishing, 2008) 204–215.

[15] Thus, for example, private individuals are granted rights under human rights regimes, including procedural rights that enable them to hold States to account for breaches of certain obligations. Individuals may also be prosecuted for international crimes. Inter-governmental organisations have likewise been recognised as having such personality as they require to carry out their functions. While these developments in the law may leave the question of the rights and obligations of NGOs under international law unanswered, they suggest that the argument that only States should be involved in decision-making, as only they have personality in international law, is now rather dated.

[16] S Ripinsky and P Van Den Bossche, *NGO Involvement in International Organizations: A Legal Analysis* (British Institute of International and Comparative Law, 2007).

[17] See Charnovitz (n 8); JR Bolton, 'Should We Take Global Governance Seriously?' (2000) 1 Chicago Journal of International Law 205, 217.

lack 'legitimacy, meaning they are neither accountable to an electorate nor representative in a general way'.[18]

These arguments give cause to consider carefully the nature of participatory rights and the reasons for granting any such rights at the international level. If the object of participation is to ensure that any decisions best reflect society's preferences, then the criticisms weighed against NGO involvement are significant. To ensure that the decision-making processes maintain legitimacy in this context it would be necessary to design participatory rights in a manner that facilitated representation by a broad cross-section of society. If, by contrast, the objective is to improve decision-making through improving the quality of information available within a particular regime, then questions of representativeness or of marginalisation become less significant. What would be significant in this context is whether those involved in deliberations have access to, or bring with them, key information necessary for an informed decision.

These arguments give rise to further questions: what do we mean by participation and should participatory rights take a particular form?

3. THE MEANING OF PARTICIPATION

In both theory and practice, there are a number of possible answers to the question of what participation involves. Arnstein first categorised the possibilities in her ladder of participation, and the possibilities have been expanded upon since then. Arnstein's categories are based upon the degree to which decision-making involves the public.[19] At the bottom of the ladder are processes that inform the public that the decision has been made. Above these, in terms of opportunity to participate, sit processes involving consultation (where the public respond to questions or information given by the decision-maker, but have no role in making the final decision), to co-decision-making (partnership) and, finally, to fully delegated decision-making. Each level may, in practice, take a number of forms, such as: hearings, focus groups, (internet) forums, roundtables, citizen forums, multiple stakeholder conferences, consensus-oriented meetings, and multiple discussion circles.[20]

One of the challenges in participatory decision-making is in determining the appropriate form and level of participation and appropriate limitations on engagement. Thus consideration has to be given to questions such as which actors should be given a right to participate, how they should be informed of such rights, and what assistance if any should be provided to enable them to participate. These questions are in addition to the initial question of what level of participation should be granted. It could be argued

---

[18] Ripinsky and Van Den Bossche (n 16), 12.

[19] SR Arnstein, 'A Ladder of Citizenship Participation' (1969) 26 Journal of American Institute of Planners 216. *See* also, for example, Organisation for Economic Co-operation and Development, *Stakeholder Involvement Techniques: Short Guide and Annotated Bibliography, A Report from Nuclear Energy Agency Forum for Stakeholder Confidence* (2004) 5, available at <http://www.oecd-nea.org/rwm/reports/2004/nea5418-stakeholder.pdf>.

[20] O Renn and P-J Schweizer, 'Inclusive Risk Governance: Concepts and Application to Environmental Policy Making' (2009) 19 Environmental Policy and Governance 174.

that any limitation on participatory rights is inappropriate; that deliberative democracy demands that all should be free to participate fully in decision-making. While it is possible to construct endless arguments as to the theoretical appropriateness of the breadth or depth of participation, pragmatism requires that consideration be given to both the workability of any scheme and State responses to participatory rights. A completely open forum would make international decision-making unworkable and thus participation must be limited in some way.

Participation may also need to be limited to help ensure the continuing participation of States within a regime. There is a danger that too broad a forum may undermine State perceptions of the legitimacy of a regime, particularly if they view some of the participants as having more influence than their interest in an issue is thought to warrant. A relatively simple example of this is provided by the debates surrounding the legitimacy of the International Whaling Commission (IWC). While the rights of non-State actors to participate directly within the regime are quite limited,[21] their role in ensuring change within the IWC has led some States to question the legitimacy of the regime.[22] For example, from the 1970s onwards NGOs lobbied States inside and outside of the IWC,[23] and in other forums,[24] presenting what some have identified as propaganda[25] in the (ultimately successful) pursuit of a whaling moratorium. As a result of the adoption of the moratorium certain States left the IWC, objecting to the decision and decision-making process (though some of those returned at a later date).

It is clear, then, that there is no simple or singular approach to defining participation or the appropriate level of participation for international regimes, and that any answer constructed on a theoretical basis may be of little utility in guiding the practice of international law. Participatory processes, if they are to enhance the effectiveness or legitimacy of regimes, will require to be designed to fit the needs of the particular regime. In the following sections, consideration is given to the stated objectives for providing participatory processes within regimes addressing marine biodiversity, with a view to establishing what the needs of these regimes actually are. Attention is then

---

[21] *See* IWC Rules of Procedure and Financial Regulations as amended by the Commission at the 64th Annual Meeting, July 2012, and in particular Rule C and E, available at <https://iwc.int/home>.

[22] Kirk (n 12); P Birnie, 'The Role of Developing Countries in Nudging the International Whaling Commission from Regulating Whaling to Encouraging Nonconsumptive Uses of Whales' (1984) 12 Ecology Law Quarterly 937; K Mulvaney, 'The International Whaling Commission and the Role of Non-Governmental Organizations' (1996–1997) 9 Georgetown International Environmental Law Review 347; PJ Stoett, 'Of Whales and People: Normative Theory, Symbolism and the IWC' (2005) 8 Journal of International Wildlife Law and Policy 151; D Wagner, 'Competing Cultural Interests in the Whaling Debate: An Exception to the Universality of the Right to Culture Symposium: Whither Goes Cuba: Prospects for Economic & Social Development: Part II of II: Student Notes' (2004) 14 Transnational Law & Contemporary Problems 831.

[23] Mulvaney (n 22); Birnie (n 22), 954–955.

[24] *See*, for example, S Andresen and T Skodvin, 'Non-State Influence in the International Whaling Commission, 1970 to 2006' in MM Betsill and E Corell (eds.), *NGO Diplomacy: The Influence of Nongovernmental Organizations in International Environmental Negotiations* (The MIT Press, 2008), 127.

[25] Stoett (n 22), 160–165.

turned to the form in which participatory rights are granted. The analysis that follows thereafter draws out the impact of form on the ability of the participatory processes to meet the regimes' objectives vis-à-vis participation and their impact on the legitimacy of the regimes.

4. OBJECTIVES FOR PARTICIPATORY PROCESSES

None of the treaties examined here contain an express statement as to the purpose of participatory processes. A few do, however, imply that the objective is to improve the implementation of the treaty in question. Within the Convention on Biological Diversity (CBD)[26] regime, for example, the primary function of participatory processes appears to be to ensure the effective implementation of its obligations. This aim is evident in the preamble to the CBD where the Parties '[stress] the importance of' NGOs in conservation. In addition the preamble also contains reference to the role of indigenous peoples and women in preserving biodiversity. The relative importance of indigenous peoples and their knowledge in biodiversity protection is highlighted again by Article 8, which, in addressing *in situ* conservation, refers to the need to respect, preserve *etc.* indigenous and local knowledge and practices. Similarly, Article 10 refers to the protection of customary uses and local knowledge in ensuring sustainable use of the components of biodiversity.

While neither of these articles actually provide for civil society involvement in decision-making, what they do is highlight the reasons such participation may be needed. They highlight that in particular it is participation from indigenous groups within society that is thought to be particularly important, with participation from NGOs more generally following behind this. The motivation for such involvement may be either to improve the quality of information available in the decision-making process, or to improve the perceived legitimacy of the regime amongst certain sections of society with a view to enhancing their compliance with the regime or their involvement in decision-making relating to implementation. The former at first sight seems more credible since the lack of attention to other groups, such as industry, and the focus on particular sections of society, indicates that achieving legitimacy through deliberative democracy was not a concern of the Parties to the regime. It is, however, worth considering the fact that industry may be represented through NGOs and so, while not mentioned by name, its representation is not precluded by these provisions. Secondly, it may also be that a decision was taken to prioritise rights for indigenous peoples and NGOs generally to redress the fact that business interests are more likely to be taken into account in decision-making because of the historically greater access to governments that business has had compared to other sections of society. Thus, while it appears more likely that the focus of the biodiversity regime is on improved information rather than delivering deliberative democracy, a clear conclusion on this point is not entirely possible.

Other regimes appear to aim to improve the performance of the regime through drawing on non-State actors to increase the capacity and expertise of those tasked with

---

[26] 1992 Convention on Biological Diversity (1993) 1760 UNTS 79.

102   *Research handbook on biodiversity and law*

implementation of the regime. The Convention on International Trade in Endangered Species of Wild Fauna and Flora (CITES)[27] provides that intergovernmental and non-governmental agencies may be called upon to assist the Executive Director of UNEP in providing a Secretariat for CITES (Article XII). Something similar exists under Ramsar[28] where the International Union for the Conservation of Nature (IUCN) provides the Bureau or Secretariat to the Convention under Article 8 and also under the Convention on the Conservation of Antarctic Marine Living Resources (CCAMLR).[29] The CCAMLR points to the involvement of non-State actors in the development of the regime as well as in its implementation. Under Article XXIII the Commission and Scientific Committee are to 'seek to develop co-operative working relationships, as appropriate, with inter-governmental and nongovernmental organisations which could contribute to their work'. And where cooperative arrangements are entered into the Commission may invite the organisations to send observers to its meetings. Similarly the 2003 African Convention on the Conservation of Nature and Natural Resources provides that the Secretariat may seek the cooperation of various bodies, including non-governmental organisations, to aid with implementation.[30] While this focus on improving the quality of decision-making may in turn improve the legitimacy of the regime, legitimacy derived from deliberative democracy again does not appear to be a concern of the Parties.

In the context of marine biodiversity conservation, the CBD, CITES, CCAMLR and the Ramsar Convention are rather unusual in their provisions regarding non-State actors. The other relevant regimes in this context are the regional seas regimes and few of these make the role of non-State actors so explicit in their founding treaties. Even those that do address the role of non-State actors tend to focus on the provision of information to them and education of them. For example, the Convention for the Protection, Management and Development of the Marine and Coastal Environment of the Eastern African Region[31] (Nairobi Convention) encourages the Parties to take part in activities designed to provide information to the public and public education (Article 15). Similarly the Framework Convention for the Protection of the Marine Environment of the Caspian Sea[32] provides in Article 20(2) that:

---

[27]   1973 Convention on International Trade in Endangered Species of Wild Fauna and Flora (1973) 993 UNTS 243.
[28]   1971 Convention on Wetlands of International Importance especially as Waterfowl Habitat (1971) 996 UNTS 245.
[29]   1980 Convention on the Conservation of Antarctic Marine Living Resources (1982) 1329 UNTS 47.
[30]   2003 African Convention on the Conservation of Nature and Natural Resources, Article 28, text available at <http://faolex.fao.org/docs/pdf/mul45449.pdf>.
[31]   Convention for the Protection, Management and Development of the Marine and Coastal Environment of the Eastern African Region, Nairobi, 1985 (Nairobi Convention), text available at <http://www.unep.org/NairobiConvention/The_Convention/Nairobi_Convention_Text/index.asp>.
[32]   Framework Convention for the Protection of the Marine Environment of the Caspian Sea, Tehran, 2003 (Tehran Convention), text available at <http://www.tehranconvention.org/spip.php?article4>.

> The Contracting Parties shall endeavour to ensure public access to environmental conditions of the Caspian Sea, measures taken or planned to be taken to prevent, control and reduce pollution of the Caspian Sea ... taking into account provisions of existing international agreements concerning public access to environmental information.

Similar provisions are found elsewhere, such as in Article 17 of the 1992 Convention on the Protection of the Marine Environment of the Baltic Sea Area.[33] Other Conventions and Protocols that appear at first sight to instigate a proactive approach are, on closer inspection, found to contain rather weak provisions. For example, the Protocol Concerning Protected Areas and Wild Fauna and Flora in the Eastern African Region[34] to the Nairobi Convention makes mention in Article 12 of the need to take account of traditional activities in areas that are to become protected areas. However, there is no reference to involvement of the indigenous peoples in the establishment or management of the areas. It is, therefore, quite conceivable that information could be gathered on traditional activities, and that information could be factored into decision-making without any involvement of those living within the area.

This focus on the provision of information suggests that the majority of regimes addressing marine biodiversity aim to provide participatory rights at the lower end of Arnstein's ladder. In addition, the regional seas regimes appear to mirror the biodiversity regime in making scant, if any, direct reference to business. Their focus is, instead, on engaging members of the public in general and, in some regions, indigenous peoples. Only in the Mediterranean is there an apparent reference to business, where it calls upon NGOs to engage with the private sector, which one may assume includes business.

It also appears that some of the regimes, such as the Caspian and Baltic regime, have engaged in the provision of information to improve the transparency of decision-making and so enhance the legitimacy of the regime in that way. Others, such as the Eastern African regime, may aim to improve the quality of decision-making in relation to implementation, but the commitment to doing so is rather limited.

In general, then, the conclusion to be drawn on the objectives of participatory rights, as stated or implied from the founding documents of these regimes, is that they are included to improve the quality of decision-making and the quality of implementation of decisions within regimes. While this may ultimately help improve the legitimacy of each of the regimes, there appears to be little desire amongst the Parties to use participatory processes as a means to improve the regimes' democratic legitimacy. In this context, the competing arguments of writers such as Anderson[35] and Charnovitz[36] as to the appropriateness or otherwise of giving non-State actors full participatory

---

[33] Convention on the protection of the marine environment of the Baltic Sea, Helsinki 1992, (2002) 2099 UNTS 195 (Helsinki Convention).

[34] Protocol Concerning Protected Areas and Wild Fauna and Flora in the Eastern African Region, Nairobi, 1985, text available at <http://www.unep.org/NairobiConvention/The_Convention/Protocols/Protocol_Protected_Areas.asp>.

[35] K Anderson, 'Accountability as Legitimacy: Global Governance, Global Civil Society and the United Nations' (2011) 36 Brooklyn Journal of International Law 841, 844.

[36] Charnovitz (n 8).

rights, become less important. And, while the apparently limited approach to participatory rights may serve to disappoint some sections of society, it appears to accord with the general approach in international law as expressed in the Almaty Guidelines on Promoting the Application of the Principles of the Aarhus Convention in International Forums. Although these guidelines call for participation to be as broad as possible[37] they also recognise, in paragraph 31, the need for participation to be restricted at times. Practice elsewhere also shows a similarly limited approach to participatory rights. Bettin,[38] for example, notes the limited role generally granted to NGOs in relation to development policy. In the field of marine biodiversity, there is, however, one notable exception to the general rule that participation is limited: the Protocol Concerning Specially Protected Areas and Wildlife to the Convention for the Protection and Development of the Marine Environment of the Wider Caribbean Region,[39] which requires Parties to develop public awareness programmes and to involve the public 'in the planning and management of protected areas' (Article 6). As a counter to this exception it is also worth noting that there are some biodiversity regimes that make no express provision for participation by non-State actors, such as the Bonn Convention Memorandum of Understanding Between the Argentine Republic and the Republic of Chile on the Conservation of the Ruddy-Headed Goose[40] and the Bonn Convention Wadden Sea Seal Agreement.[41]

To base conclusions on the participatory nature of the regional seas regimes on the treaty provisions alone would, however, be misleading. More progressive approaches are frequently found in the soft law attached to these regimes. Many of the regional seas regimes have provisions on participation within programmes or plans of action or within recommendations or decisions and some address the participation of non-State Parties through rules of procedure. Nevertheless, although more progressive than the treaty provisions, they could not be described as providing leading examples of participatory processes. The provisions instead tend to focus on capacity-building and improving implementation rather than improving the quality of normative decisions.

The most progressive of the regimes include partnership agreements between the States Parties and civil society organisations. Generally these agreements provide for identified projects to be carried out either by non-State actors alone or in consortia of States and non-State actors. For example, the Parties to the Nairobi Convention have entered into memoranda of understanding with regional and global NGOs such as WWF, the IUCN, the Western Indian Ocean Marine Science Association (WIOMSA), BirdLife International and the Wildlife Conservation Society (WCS). The Parties to the

---

[37] Almaty Guidelines, Aarhus Convention MOP 2 (2005) DECISION II/4, Promoting the Application of the Principles of the Aarhus Convention in International Forums, paragraph 30.
[38] V Bettin, 'NGOs and the Development Policy of the European Union' in P-M Dupuy and L Vierucci (eds), *NGOs in International Law: Efficiency in Flexibility?* (Edward Elgar Publishing, 2008), 116–134.
[39] Protocol Concerning Specially Protected Areas and Wildlife to the Convention for the Protection and Development of the Marine Environment of the Wider Caribbean Region, Kingston, 1990, (2004) 2180 UNTS 101 (SPAW Protocol).
[40] <http://www.cms.int/ruddy-headed-goose/en/documents/agreement-text>.
[41] For further information *see* <http://www.waddensea-secretariat.org/management/seal-management>.

Barcelona Convention have similarly made provision for the establishment of a partnership agreement within the Mediterranean Action Plan. The Plan draws a range of intergovernmental organisations and NGOs into the implementation process.

There are also examples of fuller cooperation between States and non-State actors. The East Asia Seas Action Plan, for example, is administered by the Coordinating Body for the Seas of East Asia (COBSEA) which had as its original objective to bring together interested States and to create

> an environment, at the regional level, in which collaboration and partnership (between all stakeholders and at all levels) in addressing the environmental problems of the South China Sea is fostered and encouraged and to enhance the capacity of the participating governments to integrate environmental considerations into national development planning.[42]

The South Asia Co-operative Environmental Programme (SASCEP) goes further still in that not only does its Action Plan provide for cooperation with NGOs, but it is reliant on funding from external actors including NGOs. There are, in addition, some organisations that set out to engage more fully in multi-stakeholder governance. For example, the Coral Triangle Initiative was established through the work of a consortium of States and NGOs and in its Action Plan repeatedly refers to working groups being established which encompass NGO representatives as well as other actors.[43]

In respect of all of these regimes, the aim of involving non-State actors appears to be focused entirely on enhancing the quality of implementation. Few of the arrangements are designed to create equal partnerships between States and non-State actors. In some instances it might be more accurate to describe the non-State actors as carrying out tenders for projects, rather than as partners. Nevertheless these arrangements are likely to improve implementation and better implementation may ultimately improve the legitimacy of the regime, but that does not appear to be the primary objective of participatory decision-making in these regimes.

The same may be said of other regions, which, while not containing such progressive features, do note the importance of engaging with non-State actors to ensure that the provisions of their treaties and programmes of action are implemented effectively.[44] For some, such as the Baltic, this acknowledgement leads to calls to the States Parties to ensure engagement with civil society during implementation.

While the objectives of participatory processes may set the tone for engagement with non-State actors, the nature of the rights granted is frequently dictated by rules on participation. There are broadly two aspects to the granting of participatory rights in regimes addressing marine biodiversity: the adoption of threshold requirements for entities seeking participatory rights and the level or degree of participation granted to such entities. These aspects are now addressed in turn.

---

[42] New Strategic Direction for COBSEA 2008–2012, 3.
[43] Coral Triangle Initiative 'Regional Plan of Action' 2009.
[44] *See*, for example, HELCOM Recommendation 28E/9 'Development of Broad-Scale Marine Spatial Planning Principles in the Baltic Sea Area'.

5. RULES ON PARTICIPATION

5.1 Threshold Requirements for the Granting of Participatory Rights

Many of the regimes considered in this chapter take a 'restrictionist'[45] approach to the engagement of non-State actors. For example, under Article 23(5) of the Biodiversity Convention and Article XI of CITES, any governmental or non-governmental body can apply for observer status provided that they meet certain requirements for qualification. In the Biodiversity regime the requirement is that they be 'qualified in fields relating to conservation and sustainable use of biological diversity'.[46] In CITES it is that they be 'qualified in protection, conservation or management of wild fauna and flora'.[47] No definition or explanation of the word 'qualified' is given in either regime. Further detail is provided within the Rules of Procedure for both conventions, though in neither case does this address what it means to be qualified in the particular field. What they do provide is a rule (CBD Rule 7 and CITES Rule 2) that observer status will be granted to applicants unless at least one-third of the Parties object. By contrast the Whaling Convention Rules of Procedure simply provide that non-party States and inter-governmental organisations may become observers and that '[a]ny non-governmental organisation which *expresses an interest* in matters covered by the Convention, may be accredited as an observer' (emphasis added).[48] While proving an interest in a matter should be easier than proving that an entity is qualified in a particular area, both types of provision leave considerable discretion to the Parties. Such discretion may appear unproblematic where the objective of participatory processes is to improve the quality of information feeding into decision-making, but it relies on the Parties being fully aware of gaps in the information they have. Yet research has shown that, on occasion, regulators unconsciously place more emphasis on some types of information than on others.[49] They may equally be likely to invite certain types of organisations rather than others.[50] Moreover, if the objective of the participatory processes were to help improve the legitimacy of decision-making, then the existence of such wide discretionary powers would be problematic, as it may be used to block participation by otherwise qualified participants.

---

[45] Ripinsky and Van Den Bossche (n 16), 15.
[46] CBD, Article 23(5).
[47] CITES, Article XI(7).
[48] IWC Rules of Procedure and Financial Regulations as amended by the Commission at the 64th Annual Meeting, July 2012, Rule C(1)(b).
[49] *See*, for example, H Valve and J Kauppila, 'Enhancing Closure in the Environmental Control of Genetically Modified Organisms' (2008) 20 Journal of Environmental Law 39; M Kritikos, 'Traditional Risk Analysis and Releases of GMOs into the European Union: Space for Non-Scientific Factors?' (2009) 34 European Law Review 40; B Hedelin and M Lindh, 'Implementing the EU Water Framework Directive – Prospects for Sustainable Water Planning in Sweden' (2008) 18 European Environment 327.
[50] *See*, in a different context, K Sherlock, EA Kirk and AD Reeves, 'Just the Usual Suspects? Policy Networks and Environmental Regulation' (2004) 22 Environment and Planning C: Government and Policy, 651–666.

Despite these potential problems, the same types of provisions as found in the CBD apply under other Conventions, such as under Article VII of the Bonn Convention,[51] and the Barcelona Convention for the Mediterranean,[52] though the latter requires that two-thirds of the Parties positively approve the invitation, rather than simply providing that unless one-third object the invitation will stand. The Bonn Convention also has a series of subsidiary agreements and memoranda of understanding, which contain some interesting variations on these participatory rights. While most contain similar provisions – providing for participatory rights for suitably qualified organisations – some, such as the Agreement on the Conservation of Cetaceans of the Black Sea, Mediterranean Sea and Contiguous Atlantic Area (ACCOBAMS)[53] and the Agreement on the Conservation of Small Cetaceans of the Baltic, North East Atlantic, Irish and North Seas (ASCOBANS),[54] also introduce deadlines by which objections must be received.[55] This highlights a potential problem with the provisions under the Bonn Convention and the CBD, as, theoretically, under the provisions found in these Conventions the last objection necessary to prevent an organisation obtaining observer status could be received at any time up until the start of the meeting. This potential threat, whether likely or not, creates uncertainty within the system of participatory rights, as potentially does the lack of time limits on the application process under these Conventions. ASCOBANS, by contrast, also has agreed time limits across the process for applying for observer status, from reception of the application to the last date on which objections may be received. Although none of these regimes contain ideal provisions in terms of encouraging participation, they do at least contain some form of criteria for the granting of observer status. A number of regimes, such as the Eastern Africa regional seas regime,[56] simply contain provisions requiring that the Executive Director, or Chair of the Conference of Parties or equivalent body, invite non-State actors or non-Party States to become observers to meetings.

These approaches to participation can be contrasted with the approaches seen in some of the non-binding mechanisms established to conserve marine biodiversity. In these mechanisms non-State actors often play a significant role. For example, some are involved in gathering data through tagging and monitoring various species so helping to determine the success or otherwise of particular programmes and the need for new

---

[51] Convention on the Conservation of Migratory Species of Wild Animals, 1979 (1991) 1651 UNTS 333 (Bonn Convention).

[52] Barcelona Convention Rules of Procedure for Meetings and Conferences of the Contracting Parties to the Convention for the Protection of the Mediterranean Sea Against Pollution and its Related Protocols UNEP/WG.83/3, Annex II, Rule 8.

[53] Agreement on the Conservation of Cetaceans of the Black Sea, Mediterranean Sea and Contiguous Atlantic Area (ACCOBAMS), Monaco, 1996 (2001) 2183 UNTS 303.

[54] Agreement on the Conservation of Small Cetaceans of the Baltic, North East Atlantic, Irish and North Seas (ASCOBANS) 1992 (1994) 1772 UNTS 217.

[55] Article III(4) of ACCOBAMS and Article 6.2.2 of ASCOBANS.

[56] Rules of Procedure for the Meetings and Conferences of the Contracting Parties to the Convention for the Protection, Management and Development of the Marine and Coastal Environment of the Eastern African Region UNEP (DEPI)/EAF/CP.7/Inf2/en, Rule 47.

ones.[57] The Memorandum of Understanding for the Conservation of Cetaceans and Their Habitats in the Pacific Islands Region seems to go further. When considering assessment of implementation of the Memorandum, it appears to place non-State actors on an equal footing with States by including them as full partners in the future monitoring of implementation of the Memorandum of Understanding (MoU) (paragraph 6).[58] More significantly, the MoU was signed by States and non-State actors, as equal partners.

In addition to the restrictions discussed above, many of the regimes place stricter limitations on some non-State actors than on others, depending on the type of right to be granted. Thus there are variations in the criteria applied as between those seeking observer and those seeking partner status, and there are variations in relation to the types of non-State actors engaged with. In the Mediterranean, for example, international and regional NGOs are distinguished from national and local NGOs. In both cases, the NGOs must satisfy a long list of criteria contained in the annex, which basically require the NGOs to have expertise relevant to the Barcelona Convention and to be operative within the relevant region and able to contribute to the regimes activities through participating in projects and distributing information etc. National NGOs must also be able to show a genuine interest in marine issues. Procedures are set out for approval of partnership status, and accreditation is valid for six years. Similar criteria are laid down by a number of other regional seas regimes, including the Baltic,[59] and the North-East Atlantic.[60]

Despite some variations between and within regimes in respect of different types of participation, the criteria used in determining whether or not to grant observer, or partner, status is relatively uniform across the regimes dealing with marine biodiversity. The focus tends to be on the existence of relevant expertise or proof of sufficient interest in the issue, rather than on the ability of the non-State actors to enhance the legitimacy of the regime through broadening representation.[61] Although this accords with a widespread practice in marine resources regimes of limiting participation to those with a real interest in the issue being regulated – a practice that includes a limitation on State membership – and while arguably necessary to ensure that the Convention regimes can operate effectively (and while recognised as a legitimate act

---

[57] *See* generally <http://www.cms.int/en/about/partnerships> and <http://www.cms.int/en/projects>.

[58] See also M Prideaux, *The Natural Affiliation: Developing The Role Of NGOs in the CMS Family. Part One Summary of The Review: Defining the Relationship Between the NGO Community and CMS For the 40th Meeting of the CMS Standing Committee* (Wild Migration, Australia, 2013).

[59] HELCOM Guidelines on Granting Observer Status to Intergovernmental Organizations and Non-Governmental International Organizations to the Helsinki Commission 1999 as amended in 2001, 2002 and 2008 <http://www.helcom.fi/about-us/internal-rules/rules-of-procedure>

[60] Rules of Procedure of the OSPAR Commission (Reference Number: 2005-17) as revised at OSPAR 2001 (Annex 29), OSPAR 2002 (Annex 10), OSPAR 2005 (Annex 25). Editorial amendments made at OSPAR 2012 (see OSPAR 12/22/1, §§12.5–12.6).

[61] Anderson (n 35).

under the Almaty Guidelines for that reason),[62] nevertheless this approach has the effect of excluding a large section of society from direct participation. It can also be contrasted with other areas of law where criteria range across such matters as: expertise; the nature of the organisation (for example, is the organisation a non-profit organisation); the character of the organisation (for example, NGOs may be scrutinised to establish whether or not they are truly non-governmental in nature); the international structure or scope of the organisation; its membership base; geographical coverage; aims and objectives; formal institutional structures; representative nature; and the length of time that it has been in existence.[63] It is not possible to determine at this time whether the focus of these criteria on expertise, and the relative uniformity of approach in marine biodiversity regimes, have arisen by design as a result of careful consideration of the needs of the regime, or are simply the outcome of chance or habit. It is, however, possible to conclude that this relative uniformity has the advantage of making the navigation of participatory rights somewhat easier for non-State actors. At the same time, however, it points to some potential concerns.

If the justification for ensuring participation by non-State actors is to improve decision-making by ensuring that a range of views and types of information can be fed into the process, or to maintain the legitimacy of the decision-making process by ensuring that an appropriate cross-section of society is represented in the process, then these criteria fall short of what is required. If either of these scenarios applies then it may be appropriate to consider the true pedigree of certain non-State actors. For example, some NGOs, such as the Marine Stewardship Council, have been established as a partnership between environmental and industry groups. Other bodies present themselves as independent NGOs but are in fact wholly financed by industry. Should these bodies apply for observer status it may be appropriate to consider whose interests they represent. In such cases, the availability of clear criteria to assess the nature of the applicants would be useful in ensuring that their involvement is appropriate to the aims of the participatory processes concerned. Equally, a closeness between States and non-State actors may reduce the possibility of different perspectives being brought to bear on an issue and so reduce a regime's capacity for effective protection of biodiversity. Again, then, if the objective is to draw in a range of perspectives, clear criteria might need to be adopted to ensure the pedigree of the non-State actors joining the regime.

The discussion so far has revealed certain limitations upon participatory rights. What it does not tell us is the degree to which participation is limited to meetings or conferences of the Parties and the degree to which, once granted, observer or partner status allows non-State actors or non-member States to attend and participate in other types of meetings. For example, many of the regimes reviewed here have provision for a scientific committee, but may not make explicit mention of the role of non-State actors in these committees. In addition, as Le Prestre notes (in relation to the CBD)

---

[62] Almaty Guidelines, paragraph 31.
[63] Ripinsky and Van Den Bossche (n 16), 217–218.

'[p]articipation need not be defined in the same way or take a similar form in every case.'[64] The next sub-section considers the modes of participation granted to non-State actors.

5.2 Modes of Participation

Under all of the agreements reviewed, non-State actors, when granted observer or partner status,[65] are described as being able to participate in meetings but not vote. In most they are entitled to make oral or written submissions to meetings and in some they may make proposals that will be voted on if supported by a State Party. For example, the Mediterranean has adopted a code of conduct for its 'partner' NGOs, which addresses their rights and duties.[66] These provide that NGOs are to be afforded the opportunity to make oral or written contributions to meetings, that their comments are to be reflected in the report of the meeting and they have the right to information. It is expressly noted that they do not, however, have the right to vote.

These types of participatory provisions accord with the recommendations of the Almaty Guidelines on public participation in international forums.[67] Under some agreements, however, certain further limitations on participation apply. In the CBD, actors granted observer status may participate in meetings only if invited by the President of the meeting to do so and so long as the Parties have not vetoed their participation. Under the CCAMLR, observers may be present at both public and private meetings unless Members of the Commission request a restricted meeting, in which case either only certain observers (which excludes civil society, NGOs etc.) may be present, or no observers at all may be present.[68] In other words, the participatory rights granted to non-State actors might be described as equating to consultation and the provision of information, which are rather low in the hierarchy of participatory rights discussed in the literature. Thus, the rights enable them to be informed and to provide information to underpin decisions, but do not guarantee that the information they provide will actually be taken into account in any decision-making. Indeed as Holder notes, a right to submit oral or written comments is not quite the same as a right to influence decisions[69] in so far as a right to submit comments does not guarantee that those comments will in any way influence the decision-makers. One might not, therefore, put much store by the granting of these rights in terms of improving citizen

---

[64] PG Le Prestre, 'Studying the Effectiveness of the CBD' in PG Le Prestre (ed), *Governing Global Biodiversity: The Evolution and Implementation of the Convention on Biodiversity* (Ashgate, 2002), 83.

[65] While the term observer has a fairly uniform meaning across regimes there are some variations as to what is meant by partner. In some regimes partner organisations will have the right to attend all meetings, in other organisations their rights appear to be limited to what is necessary to enable participation in implementation projects.

[66] Barcelona Convention Decision IG.19/6 MAP/Civil Society Cooperation and Partnership.

[67] *See*, in particular, Almaty Guidelines, paragraph 34.

[68] Commission for the Conservation of Antarctic Marine Living Resources Rules of Procedure, Basic Documents December 2012, Rule 33.

[69] J Holder, *Environmental Assessment: The Regulation of Decision Making* (OUP, 2004), 205.

participation. Indeed there are other provisions that, while not directly related to the granting of participatory rights, also have a potentially negative impact upon their exercise. For example, CBD Rules 10 and 13 provide that meeting agendas need only be distributed to Parties. Moreover, the provisions in the CBD can be compared and contrasted with OSPAR's[70] provisions on the distribution of agendas. Rule 56 of OSPAR's Rules of Procedure provides that all documents and decisions are to be made available to anyone who requests them (on payment of a reasonable fee if appropriate) unless there is a reason as set out in Rule 57 (e.g. commercial confidentiality) that has led to a request that the documents not be released. Seen in this context the CBD provisions appear particularly restrictive and not designed to induce effective public participation. That said, the OSPAR regime does also differentiate between actors in the provision of access to documents. Parties, observer States and observer international organisations automatically have access to all documents, under Rule 58. By contrast NGOs have automatic rights only to some documents and may have to request other documents under Rule 56 and pay a fee to obtain them.

Further distinctions are applied in other regimes. In CITES the Rules of Procedure provide that during plenary sessions and in committees and working groups observers are to sit in specially designated seating areas and may only enter the delegates' area if invited by a delegate to do so (Rule 11) and that the media are to be similarly segregated (Rule 13). While observers and the media may variously be invited or permitted to enter other areas, this segregation points to the limitation of rights and opportunities for non-State actors to engage with States during the decision-making processes.[71] The Mediterranean code of conduct referred to earlier also provides an example of differentiation in the level of participation afforded to different types of non-State actors. A distinction is drawn between international and regional NGOs on the one hand and national and local NGOs on the other. The former are automatically entitled to participate in meetings, the latter must request special permission to attend a meeting or conference of direct concern to them and such requests are described as 'exceptional'.

Similar distinctions between different types of non-State actors are drawn in other regimes. For example, the Wider Caribbean Region draws a distinction between, on the one hand, Non-Party States and the UN and its subsidiary bodies, all of which are automatically entitled to attend meetings once granted observer status,[72] and, on the other hand, NGOs and other IGOs which are, under Rule 54, allowed to participate 'in matters of direct concern to them'. OSPAR contains similar provisions and also

---

[70] Convention for the Protection of the Marine Environment of the North-East Atlantic 1992 (2006) 2354 UNTS 67.

[71] Though it might be argued that much of the important decision-making takes place outside of the plenary hall.

[72] Rules of Procedure of the Caribbean Environment Programme, Report of the Fourteenth Intergovernmental Meeting on the Action Plan for the Caribbean Environment Programme and Eleventh Meeting of the Contracting Parties to the Convention for the Protection and Development of the Marine Environment of the Wider Caribbean Region, October 2010, UNEP (DEPI)/CAR IG.30/6 Annex VIII Rules 52 and 53. *See* also Rules of Procedure of the Commission on the Protection of the Black Sea Against Pollution available at <http://www.blacksea-commission.org/_od-commission-rulesofproc.asp>.

distinguishes between general and specialist observers, limiting the number of seats available at meetings for specialist observers; in other words those NGOs dealing with a very specific area of OSPAR's work and which will likely have an interest in attending only a particular meeting or part of a meeting.[73] The Black Sea regime places more extreme limitations on the numbers of observers or representatives of observers that may attend meetings. Here NGOs are entitled to attend a forum from which they may elect two representatives to attend meetings of the Commission.[74] This type of arrangement is found in other international organisations where, as Ronit and Schneider note, the organisations often go further in recognising 'only one organization for each interest category'.[75] By contrast, although the IWC also limits the number of seats available to observers, in that case the limitation applies per observer, not to the number of observer organisations.

The drawing of distinctions between different groups of observers or partners is found in areas beyond the conservation of marine biodiversity. Kamminga, for example, discusses it in relation to NGO activities in the UN and other bodies.[76] But there are apparent differences in the nature of rights granted in marine biodiversity regimes and participatory rights under some other regimes. Whereas, as noted earlier, participatory rights in the marine biodiversity regimes are largely limited to rights to receive or provide information, in some other areas such as global health, education and agriculture non-State actors, and NGOs in particular, are engaged in full decision-making.[77]

Any criticism of the marine biodiversity regimes should, however, be tempered by the fact that an examination of the rules of procedure (or their equivalent) does not tell the full story on participation any more than an examination of treaty provisions alone would. Parties to the various agreements have repeatedly recognised the importance of participation by different sectors of society in the decisions they adopt. For example, in their calls for participation by stakeholders and civil society generally, the Parties to the CBD have recognised that the varying interests of stakeholders and others require the Parties to engage with them in different ways. Thus, they have over time adopted decisions relating to the provision of information to the public,[78] called upon business to be engaged in the process of implementation of the Convention and its Strategic

---

[73] OSPAR, Rules of Procedure of the OSPAR Commission (Reference Number: 2005-17).
[74] Rules of Procedure of the Commission on the Protection of the Black Sea Against Pollution, Rule 1(6).
[75] K Ronit and V Schneider, 'Private Organisations and Their Contribution to Problem-Solving in the Global Arena' in K Ronit and V Schneider (eds), *Private Organisations in Global Politics* (Routledge, 2000), 15.
[76] MT Kamminga, 'What Makes an NGO "Legitimate" in the Eyes of States?' in Anton Vedder (ed), *NGO Involvement in International Governance and Policy* (Brill Academic Publishers, 2007), 175.
[77] D Gartner, 'Beyond the Monopoly of States' (2010–2011) 32 University of Pennsylvania Journal of International Law 595, 621.
[78] *See*, for example COP 9 Decision IX/32 Communication, education and public awareness (CEPA); MOP 2 Decision BS-II/13 Public awareness and participation; MOP 4 Decision BS-IV/17 Public awareness, education and participation.

Plan,[79] and agreed to work on capacity-building for indigenous peoples including through the provision of a knowledge-sharing web page.[80] All of these actions point to, at least, a desire on the part of the Member States for greater participation by non-State actors in the implementation processes. This desire is mirrored in calls for greater participation of non-State actors in the CBD's implementation process. An illustration of this can be found in the recommendations of the Working Group on the Review of Implementation (WGRI). This Group has made a number of calls that appear designed to increase the involvement of certain groups of non-State actors in the implementation of the Convention. Recommendation 1/5,[81] for example, calls for consideration to be given to involving indigenous and local communities in the plan of implementation and notes the need for the plan to address '[t]he full range of potential audiences, including key stakeholders, the general public and donors'. This suggests that the members of the WGRI were aware that the participation of non-State actors has been too limited. Similarly ASCOBANS has recognised the importance of providing information to the public to help them to become involved in monitoring programmes.[82]

The provision of participatory rights through decisions or recommendations points to the possibility that Parties to regimes addressing marine biodiversity are growing increasingly aware of the benefits that participatory decision-making can bring. It is, however, also potentially problematic in that these provisions are easily revised. While this may bring the advantage that States can refine the participatory rights as practice points to issues arising with them, it also means that such rights can be withdrawn more easily than if they were enshrined in a treaty document. It means also that the only conclusion that can be drawn at this point in time is that participatory rights tend to be afforded at the lower level of Arnstein's ladder of participation, though there may be a move towards granting greater rights.

6. A TAXONOMY OF PARTICIPATORY RIGHTS IN BIODIVERSITY REGIMES

As indicated earlier in the chapter, the rights granted in relation to decision-making in binding regimes tend to be rights of access to information or to provide information. In some non-binding regimes, such as the Coral Triangle Initiative, non-State actors are accorded rights to partnership, or co-decision-making powers. This categorisation of rights can be refined further by consideration of the limitations placed on the exercise of participatory rights by non-State actors, with some regimes operating more restrictive regimes for the granting of rights than others. Table 4.1 illustrates the categories of rights that exist.

---

[79] *See*, for example, COP 10 Decision X/2; Decision XI/7 Business and biodiversity.
[80] *See*, for example, Cop 10 Decision X/40 Mechanisms to promote the effective participation of indigenous and local communities in the work of the Convention.
[81] WGRI 1 Recommendation 1/5 Mechanisms for Implementation: Review of the Global Initiative on Communication, Education and Public Awareness.
[82] *See* ASCOBANS, Annex para. 5.

Table 4.1 Categories of participatory rights

Fullest participatory rigths →

Full partnership			
Implementation partnership			
Observer status, attend all meetings, criteria based on interest and time limited process for granting		Observers and partners have full access to paperwork	Observers mix with Parties and may present oral and written information
Observer criteria, qualified, time limited process		Observers may request access to paperwork	Observers may be invited to sit with Parties but may present oral and written information
Observer Criteria Interest, no process	Observer time limited process, Parties decide, no criteria	Observers have access to limited paperwork	Observers sit separately from Parties but may present oral and written information
Observer criteria, qualified	Observer time limited process, Parties decide, no criteria		Observers sit separately from Parties and may be invited to present oral and written information
Observer – process Executive or Secretariat decide	Observer process parties decide	Observers have no access to paperwork	Limited numbers of observers, limited invitations to present information

Least participatory rights

The fullest rights are those towards the top and left of the table, with the most restrictive towards the bottom and right. It is possible at this stage to note that Parties appear to be reluctant to grant full partnership rights in relation to binding decision-making processes. Such rights tend only to be granted in relation to soft law arrangements or in the implementation of binding agreements. Instead, the rights in binding decision-making processes tend to be quite restricted, with limited rights to receive or present information being given in many instances, and in some only limited rights to be present granted to (certain) non-State actors. This distinction throws up an interesting geographical distinction too. Amongst the regional seas regimes it is also possible to distinguish between those providing greater or lesser participatory rights on a geographical basis. The soft law regimes placing stronger reliance on the roles of non-State actors, both in terms of partnership agreements and cooperation generally and in terms of funding, tend to be situated in Africa and Asia. This focus on non-State actors may not be surprising when one considers that these regions are more likely to need assistance in providing the capacity to implement the regime, than is, for example, Europe, but it does raise the question of whether or not such partnerships are beneficial and, if they are, why regimes in other parts of the world are less willing to engage with non-State actors.

Implementation procedures are more likely than decision-making procedures to involve those rights towards the top and left of the table. Of course regimes and their rules are developed through implementation. Implementation actions show how decisions are interpreted and action that is deemed to be non-compliant may prompt development of new provisions, or revision of existing ones.[83] In so far as such development may happen, the distinction between direct engagement in the generation of new rules and amendment of existing ones on the one hand, and engagement through the implementation processes on the other, may seem rather artificial and perhaps a little dated. Though it accords with the tradition in international law of recognising only States as having authority to make international law, it ignores the reality of the influence that non-State actors wield in the development of the law of the sea.[84]

Reviewing the provisions on the granting of observer or partner status throws up an additional categorisation issue. While in some regimes the provisions on the granting of observer (or other status) are found in treaty texts, none of the regional seas treaties contain such provisions. Instead the granting of observer status is addressed, if at all, in rules of procedure. While the net effect in the different regimes might very well be the same in practice – observer status is granted to non-State actors – the fact that some regimes address the question of such status in treaties, while others address it in soft law rules of procedure points to better protection of observer status in the former than in the latter. It is, after all, much easier to amend rules of procedure than it is to amend a treaty, not only procedurally, but also in that there tends to be less scrutiny by the media and the public of amendments to what may be termed regulatory instruments, than of legislative instruments such as treaties. These distinctions are, as noted earlier, magnified where the rights to participate are contained in decisions or recommendations of the regimes. Thus we can add additional categories to our types of participatory rights – permanent and fixed versus reversible and malleable. Given that in the majority of regimes, participatory processes appear to be designed to meet similar objectives, the fact that different categories of rights are granted may be presumed to impact on the legitimacy of regimes in different ways. The next section examines these issues.

7. PARTICIPATORY RIGHTS AND LEGITIMACY

In general, the degree of participation afforded to non-State actors in regimes addressing marine biodiversity could be described as fitting with the objectives of the regimes in respect of participatory processes, in that improving the flow of information between States and non-State actors should improve implementation of the regimes. Equally, however, none of the organisations discussed here could be described as providing cutting edge participatory rights. While some organisations do provide for partnership status, those that do tend to be soft law organisations, or to provide for partnership status in implementation. They could not then be described as being of an equivalent status to, for example, the rights provided by the Parties to the Aarhus

---

[83] EA Kirk, 'Non-Compliance and the Development of Regimes Addressing Marine Pollution from Land-based Activities' (2008) 39(3) Ocean Development and International Law 1.
[84] Kirk (n 12).

Convention, where NGO representatives sit as members of compliance committees[85] holding States to account for breaching binding obligations. This may not be terribly surprising given that, in the context of marine biodiversity the reason for engaging non-State actors appears to be to enhance the capacity to take better quality decisions and ensure more effective implementation of agreed provisions, rather than to improve the democratic legitimacy of the regime. Yet the review of the levels of participation granted suggests that the rights may not be sufficient to ensure better-quality outputs. The lack of guaranteed access to information, combined with the lack of guaranteed opportunities to present information to decision-makers in almost all contexts, are significant stumbling blocks to improved decision-making. The provisions on participation generally make no mention of a requirement to inform non-State actors of their rights, let alone of an obligation to seek out non-State actors to admit to the decision-making procedures. This approach to participation means that States remain in control of decision-making. If a regime's legitimacy is assumed to be dependent upon the use of participatory decision-making, the fact that States remain in control is unlikely to be perceived as improving the legitimacy of the regime.[86] In addition, the restriction of participation to organised interests only may undermine the overall legitimacy of the regimes. First it may reduce non-State actors' confidence in the regimes and willingness to participate in them. Whether a reduction in confidence and willingness to engage arises or not, the restrictions on participation by non-State actors may mean that the regime outputs are weaker than could be hoped for as they may rest on a less than ideal range of information. Deliberation may also be diminished if there is no opportunity, or only a limited opportunity, to take account of alternative views.[87]

These conclusions then point to two areas for further research: establishing the nature of the outputs of decision-making in the marine biodiversity regimes and establishing the impact on the willingness of States and non-State actors to become, or remain, actively involved in the regime. At this point in time we might hypothesise that certain types of rights will lead to better quality decisions and to improved legitimacy as a result. Those offering the fullest participatory rights (as illustrated in Table 4.1) appear more likely to achieve this than those offering only limited rights to a select group of observers.

The findings with regard to levels of participation point to a second legitimacy issue. Whether differentiation between different groups of non-State actors in terms of the level of participatory rights, or the ease with which they may be exercised, is appropriate to ensuring legitimacy or otherwise achieving the aims of the regimes is again debatable. While it is recognised that some limitation on the numbers of bodies participating may be necessary, the blanket criteria that are applied may not lead to the best possible outcomes. It creates a hierarchy amongst non-State actors and may lead

---

[85] First Meeting of the Parties to the Aarhus Convention, Decision 1/7, UN Doc. ECE/MP.PP/21Add.8, annex, para. 4.

[86] M Suškevičs, 'Legitimacy Analysis of Multi-Level Governance of Biodiversity: Evidence from 11 Case Studies Across the EU' (2012) 22 Environmental Policy and Governance 217. *See also* Charnovitz (n 4).

[87] DC Esty, 'Good Governance at the Supranational Scale: Globalizing Administrative Law' (2006) 115 Yale Law Journal 1490.

some actors to see the privileging of rights of others as reducing the legitimacy of the regime. Where that happens it may make successful implementation of the regime more difficult. Even without a reduction in perceived legitimacy, the creation of a hierarchy of rights may diminish opportunities for implementation. If, for example, there is a question regarding protection of biodiversity in the coastal zone, it is possible that local community groups may be best placed to provide information on the uses made of particular areas by them. Yet, some of the regional seas agreements make it very difficult for such groups to participate in decision-making and instead give preference to international groups, or to specialist non-State actors. The drawing of such distinctions is not, however, unique to marine biodiversity. Kamminga, for example, discusses it in relation to NGO activities in the UN and other bodies.[88] ECOSOC, for example, distinguishes between the observer rights of non-Party States and the consultative rights of non-State actors. The latter are further split into general, special and roster status, with different criteria applied to the granting of each status.[89] Again, though, this potential issue with legitimacy in marine biodiversity regimes points to an area in which further research should be undertaken.

In addition, the fact that the rights to participation are generally contained in rules of procedure, and in decisions and recommendations, means that such rights may easily be revised or removed. The potential uncertainty created by the nature of the provisions could again be felt to undermine the legitimacy of the regimes in the eyes of non-State actors. It does, as noted earlier, also provide greater opportunity for revision and improvement of such rights compared to enshrining them in treaty. This then points to a further issue for future research: it might be fruitful to establish both the nature and degree of change in participatory rights within marine biodiversity regimes, and the impact of any such change on the regimes' perceived legitimacy.

8. CONCLUSIONS AND RECOMMENDATIONS FOR FURTHER RESEARCH

This chapter has mapped out the types of participatory rights granted in regimes addressing marine biodiversity. The broad conclusion to draw from this mapping process is that guaranteed rights for participation by non-State actors are rather limited, though their rights to participate in soft law regimes, or through soft law provisions attached to binding regimes, are rather fuller. In addition, it appears that certain groups tend to be privileged in the granting of participatory rights, in that international NGOs have greater rights than national or local organisations, and that indigenous or local communities are more likely to be referred to than the business community in the documents providing for participatory processes. In many respects the rights provided for non-State actors in the regimes discussed here are similar to rights in other areas of international law, though they do not accord with best practice. The mapping exercise is, however, just the beginning of the necessary research. The next stage is to establish

---

[88] Kamminga (n 76).
[89] ECOSOC Res. 1996/31, 49th Plenary Meeting 1996, 'Consultative Relationship Between the United Nations and Non-Governmental Organizations'.

how closely practice mirrors the provisions discussed in this chapter. Do, for example, Parties to regimes strictly apply the threshold criteria contained in rules of procedure? Have they interpreted these criteria in set ways? And in this context, the impact that threshold criteria have on the willingness or desire of non-State actors to engage with regimes is also critical. It may be that other factors play a more significant role in influencing how non-State actors engage with regimes. For example, some NGOs have indicated that they prioritised working with certain international organisations (such as the climate change regime and CITES) over others (such as the Bonn Convention) as they saw more chances of their work having a positive impact.[90]

The discussion has also shown a variety of routes being adopted to ensure flexibility in the criteria for the granting of observer or partner status. In common with other areas of international law,[91] criteria tend to be contained in rules of procedure. The advantage of this practice is that these are flexible documents, which may be amended more readily than treaties. It also makes it possible to adopt different rules for different aspects of the work of the organisation. Some of the organisations provide for even greater flexibility by giving discretion to the Executive Director, or Chair, to invite entities to become observers. The question that then arises is which of these routes provides for the more objective approach to the granting of observer status, i.e. which is the least subject to political influence. Kamminga suggests that the granting of observer status by the Secretariat alone, without the need for State approval,[92] has proved more objective in practice in other areas of law. Whether or not marine biodiversity follows these areas is a possible topic for further research. In addition, in the many marine biodiversity regimes where little guidance is given as to the criteria to apply when considering the granting of observer or partner status, it would be useful to establish whether or not criteria have been developed in practice. It would also be useful to establish whether or not any such criteria, or the lack of such criteria, impact upon the perceived legitimacy of the regimes.

A further set of research questions centre on the impact of participation on the regime. Broadly these questions are: what impact do the different forms of participatory rights outlined here have on the legitimacy or effectiveness of the regimes; has participation influenced the type of decisions made within the regime, or influenced the ways in which the regime has been implemented?

Finally, there is further work to be done in establishing the impact of the various participatory processes on the legitimacy of the regimes. In addition to the question of whether in fact the decision-making outputs are of a better quality as a result of participatory processes, further research is needed to establish whether or not this improves the legitimacy of the regime. In addition it might be fruitful to consider the comparative impact of participatory processes and lobbying on the impact of the perceived legitimacy or effectiveness of marine biodiversity regimes.

---

[90] Prideaux (n 58), 5.
[91] *Ibid.*, 209.
[92] Kamminga (n 76).

SELECT BIBLIOGRAPHY

Abbott, KW and D Gartner, 'Reimagining Participation in International Institutions' (2012) 8 Journal of International Law and International Relations 1

Arnstein, SR, 'A Ladder of Citizenship Participation' (1969) 26 Journal of American Institute of Planners 216

Bernstein, S, 'Legitimacy in Global Environmental Governance' (2004–2005) 1 Journal of International Law & International Relations 139

Black, J, 'Proceduralizing Regulation: Part I' (2000) 20 Oxford Journal of Legal Studies 597

Boyle, A and C Chinkin, *The Making of International Law* (OUP, 2007)

Brunnée, J and SJ Toope, *Legitimacy and Legality in International Law* (CUP, 2010)

Charnovitz, S, 'Two Centuries of Participation: NGOs and International Governance' (1997) 18 Michigan Journal of International Law 183

Charnovitz, S, 'Nongovernmental Organizations and International Law' (2006) 100 American Journal of International Law 348

Charnovitz, S, 'The Illegitimacy of Preventing NGO Participation, The Symposium: Governing Civil Society: NGO Accountability, Legitimacy and Influence' (2010–2011) 36 Brooklyn Journal of International Law 891

Dupuy, P-M and L Vierucci (eds), *NGOs in International Law: Efficiency in Flexibility?* (Edward Elgar Publishing, 2008)

Ebbesson, J, 'Public Participation' in D Bodansky, J Brunnée and E Hey (eds), *The Oxford Handbook of International Environmental Law* (OUP, 2007)

Franck, TM, *The Power of Legitimacy Among Nations* (OUP, 1990)

Gartner, D, 'Beyond the Monopoly of States' (2010–2011) 32 University of Pennsylvania Journal of International Law 595

Kamminga, MT, 'What Makes an NGO "Legitimate" in the Eyes of States?' in A Vedder (ed), *NGO Involvement in International Governance and Policy* (Brill Academic Publishers, 2007)

Le Prestre, PG, 'Studying the Effectiveness of the CBD' in PG Le Prestre (ed), *Governing Global Biodiversity: The Evolution and Implementation of the Convention on Biodiversity* (Ashgate, 2002)

Peters, A, 'Membership in the Global Constitutional Community' in J Klabbers, A Peters and G Ulfstein (eds), *The Constitutionalization of International Law* (OUP, 2009)

Raustiala, K, 'The "Participatory Revolution" in International Environmental Law' (1997) 21 Harvard Environmental Law Review 537

Ripinsky, S and P Van Den Bossche, *NGO Involvement in International Organizations: A Legal Analysis* (British Institute of International and Comparative Law, 2007)

Ronit, K and V Schneider, 'Private Organisations and Their Contribution to Problem-Solving in the Global Arena' in K Ronit and V Schneider (eds), *Private Organisations in Global Politics* (Routledge, 2000)

PART II

SIGNIFICANT THREATS TO BIODIVERSITY

5. Climate change, marine biodiversity and international law
Rosemary Rayfuse

1. INTRODUCTION

Covering more than 70 per cent of the earth's surface, the oceans are rich in biodiversity. Described by the Census of Marine Life (COML) as containing 'an unanticipated riot of species',[1] the oceans are estimated to contain nearly 250,000 known species with at least one million 'kinds of marine life that earn the rank of species' estimated to be found.[2] These species, more often than not rare, include everything from the largest whales to the smallest microbes, bacteria and archaea. While endemism is high, marine biodiversity is found in all ocean areas, from the water column to the deep seabed, from coral reefs to the hostile and extreme environments surrounding hydrothermal vents.[3]

Unfortunately, the Census also found corroborating evidence of the unmistakeable detrimental impacts of human activities on marine biodiversity caused, inter alia, by overfishing, habitat destruction and marine pollution.[4] It is a matter of public record that almost one-third of the world's fish stocks are over-exploited and a further 60 per cent are exploited at absolute maximum levels.[5] It has been estimated that if current catch levels continue, global fish stocks will be commercially extinct by 2100.[6] Moreover, record levels of pollution from land-based sources are contributing to coastal degradation[7] and dead zones in the oceans,[8] while pollution from shipping, dumping and marine litter, and activities such as seabed mining, construction of artificial islands,

---

[1] JH Ausubel, DT Crist and PE Waggoner (eds), *First Census of Marine Life 2010: Highlights of a Decade of Discovery* (Census of Marine Life, 2010) 3. See further <http://www.coml.org>.

[2] *Ibid.*

[3] *Ibid.*

[4] *Ibid*, 28–29, 31.

[5] FAO, *The State of World Fisheries and Aquaculture: 2012* (FAO, 2012), 37 and 41.

[6] B Worm and others, 'Impacts of Biodiversity Loss on Ocean Ecosystem Services' (2006) 314 Science 787.

[7] Millennium Ecosystem Assessment, *Ecosystems and Human Well Being: Current State and Trends. Findings of the Condition and Trends Working Group* (Island Press, 2005) Vol 1, 514, 516.

[8] C Nelleman, S Hain and J Adler (eds), *Dead Water – Merging of Climate Change with Pollution, Overharvest and Infestation in the World's Fishing Grounds* (UNEP, 2008); RJ Diaz and R Rosenberg, 'Spreading Dead Zones and Consequences for Marine Ecosystems (2008) 321 Science 926.

oil and gas exploration, and bioprospecting, have all been identified as posing individual, cumulative and synergistic threats to marine biodiversity.[9]

These 'classical' threats to marine biodiversity are now compounded by climate change. In its Fifth Assessment Report (AR5), released in March 2014, the Intergovernmental Panel on Climate Change (IPCC) unequivocally confirmed that the oceans are warming[10] and becoming more acidic.[11] Heat absorption is leading not only to thermal expansion and sea level rise but also to ocean stratification and declining oxygen concentrations, causing deep-ocean anoxia and hypoxia, which have already been implicated in mass mortalities among some deep water benthic communities.[12] Absorption of anthropogenic carbon dioxide (CO_2) is acidifying the surface layers of the oceans, threatening coral reefs and other calcareous organisms including krill, one of the keystone species in the global food chain.[13] These changes are leading to increasingly rapid biological responses and ecological shifts including species depletion, migration and range shifts and the increasing prevalence of disease-causing agents, exotic and potentially invasive alien species and other threats to marine biodiversity.[14]

The transformation of ocean ecosystems as a result of climate change poses significant challenges for existing international regimes for the conservation and management of marine biodiversity.[15] These regimes, designed for more biologically stable conditions, are generally considered to be deficient in the essential capacities for adaptive, integrated governance and management required to effectively support the resilience of marine ecosystems in an increasingly dynamic, climate change-challenged environment.[16] Within areas under national jurisdiction (AUNJ), where coastal states enjoy jurisdictional competence in respect of marine biodiversity, marine governance

---

[9] Joint Group of Experts on the Scientific Aspects of Marine Environmental Protection ('GESAMP'), *Protecting the Oceans from Land-Based Activities* (UNEP, 2001); UNEP, *Marine Litter: A Global Challenge* (UNEP 2009).

[10] IPCC, 'Summary for Policymakers' in *Climate Change 2013: The Physical Science Basis. Contribution of Working Group I to the Fifth Assessment Report of the Intergovernmental Panel on Climate Change* (CUP, 2013) 24–25.

[11] IPCC, 'Ocean Systems', in *Climate Change 2014: Impacts, Adaptation, and Vulnerability. Part A: Global and Sectoral Aspects. Contribution of Working Group II to the Fifth Assessment Report of the Intergovernmental Panel on Climate Change*, 418 (IPCC, WG II, Ocean Systems).

[12] *Ibid.*

[13] MT Burrows and others, 'The Pace of Shifting Climate in Marine and Terrestrial Ecosystems' (2011) 334 Science 652–655.

[14] IPCC, 'Ocean Systems' (n 11), 432-443; IPCC, 'The Ocean', in *Climate Change 2014: Impacts, Adaptation, and Vulnerability. Part B: Regional Aspects. Contribution of Working Group II to the Fifth Assessment Report of the Intergovernmental Panel on Climate Change*, 1655 ('IPCC, WGII, The Ocean').

[15] R Rayfuse 'Climate Change and the Law of the Sea' in R Rayfuse and S Scott (eds), *International Law in the Era of Climate Change* (Edward Elgar Publishing, 2012) 147.

[16] C Folke, 'Social-ecological Systems and Adaptive Governance of the Commons' (2007) 22 Ecological Research 14; M Lockwood and others, 'Marine Biodiversity Conservation Governance and Management: Regime Requirements for Global Environmental Change' (2012) 69 Ocean and Coastal Management 160.

regimes are largely characterised by high levels of sector-specific, uncoordinated institutional fragmentation.[17] The situation is even more fraught in areas beyond national jurisdiction (ABNJ) where sectoral fragmentation is compounded by substantive inadequacy and regulatory ineffectiveness. While a complex array of treaty regimes exists, governance, regulatory and substantive gaps hinder the ability of these regimes to adequately address both existing and emerging threats to marine biodiversity, including from the impacts of climate change.[18] General consensus on the need to promote international cooperation and coordination to achieve long-term conservation of marine biodiversity in ABNJ has not yet resulted in agreement on the legal and institutional mechanisms required to meet this objective.[19] Moreover, even if a new agreement is forthcoming this will not render redundant the need for both improved implementation of and coordination and cooperation between existing regimes.[20] More importantly, however, a new agreement will not negate the challenges for existing marine biodiversity-related agreements posed by climate change-induced changes in species composition, distribution and productivity. In addition, responding to some climate change-related threats, such as ocean acidification, will require a level of cooperation and coordination between various global legal regimes that currently does not exist.

This chapter examines the implications of climate change for international legal regimes relating to the protection of marine biodiversity with specific focus on two major climate change-related effects: thermal-induced changes in species abundance and distribution, and ocean acidification. The chapter focuses both on how these climate change-related stressors are currently addressed in international law and on the effects of those stressors on the law itself. It begins with a brief summary of the physical effects of climate change on marine biodiversity followed by an overview of the international legal framework for the protection of marine biodiversity, before focusing on the high seas fisheries regime as a case study of the legal challenges posed by biological range shifts. It then turns to an examination of the legal framework relating to the regulation of ocean acidification. It will be demonstrated that the effects of climate change on the oceans raise questions not only of ecosystem resilience, but also of institutional resilience and that the achievement of neither is simple or guaranteed.

---

[17] RK Craig, 'Marine Biodiversity, Climate Change, and Governance of the Oceans' (2012) 4 Diversity 224, 231; LB Crowder and others, 'Resolving Mismatches in U.S. Ocean Governance' (2006) 313 Science (5787) 617; B Cicin-Sain and S Belfiore, 'Linking Marine Protected Areas to Integrated Coastal and Ocean Management: A Review of Theory and Practice' (2005) 48 Ocean and Coastal Management 847; Lockwood and others (n 16).
[18] K Gjerde and others, 'An Analysis of the Regulatory and Governance Gaps in the International Regime for the Conservation and Sustainable Use of Marine Biodiversity in Areas beyond National Jurisdiction' *IUCN Marine Law and Policy Paper No 1* (IUCN, 2008).
[19] E Druel and KM Gjerde, 'Sustaining Marine Life beyond Boundaries: Options for an Implementing Agreement for Marine Biodiversity beyond National Jurisdiction under the United Nations Convention on the Law of the Sea' (2014) 49 Marine Policy 90.
[20] JA Ardron and others, 'The Sustainable Use and Conservation of Biodiversity in ABNJ: What Can Be Achieved Using Existing International Agreements?' (2014) 49 Marine Policy 98.

This chapter does not address the effects of sea level rise on marine biodiversity. Admittedly, the IPCC has warned that coastal ecosystems, particularly low-lying areas of high biodiversity such as barrier islands, mangrove forests and near-shore coral reefs, will suffer increasingly adverse impacts due to submergence, coastal flooding and coastal erosion.[21] Some plants and animals may drown while others will be affected by changes in parameters such as available light, salinity and temperature.[22] These impacts will add to the existing sources of coastal degradation – predominantly land, but also marine-based pollution – documented in the 2005 Millennium Ecosystem Assessment.[23] However, the impacts of sea level rise are predominantly pertinent to impacts on human systems and coastal communities.[24] While sea level rise may affect the capacity of animals, such as corals, and plants, such as mangroves, to keep up with the vertical rise of the sea,[25] it is ocean temperature change and ocean acidity that are the two key drivers of change for all marine biodiversity.[26]

2. THE EFFECTS OF CLIMATE CHANGE ON MARINE BIODIVERSITY

Since the publication of the IPCC's Fourth Assessment Report (AR4) in 2007, in which the oceans received modest attention, a rapid increase in studies focusing on climate change impacts on marine species has provided the evidence needed to assess more

---

[21] IPCC, 'Summary for Policymakers' in *Climate Change 2014: Impacts, Adaptation, and Vulnerability. Part A: Global and Sectoral Aspects. Contribution of Working Group II to the Fifth Assessment Report of the Intergovernmental Panel on Climate Change* (CUP, 2014) (IPCC WG II Summary for Policymakers), 17.

[22] IPCC, 'Coastal Systems and Low-Lying Areas' in *Climate Change 2014: Impacts, Adaptation, and Vulnerability. Part A: Global and Sectoral Aspects. Contribution of Working Group II to the Fifth Assessment Report of the Intergovernmental Panel on Climate Change* (CUP, 2014) ('IPCC WG II Coastal Systems and Low-Lying Areas'), 374.

[23] Millennium Ecosystem Assessment (n 7), 516.

[24] IPCC WG II Coastal Systems and Low-Lying Areas (n 22), 364.

[25] Studies indicate that a number of fast-growing corals may keep up with sea level rise, even at the maximum projected rate of 15.1 mm/yr by the end of the century. However, regional variation in vertical accretion rates during the last deglaciation period show that not all corals have been able to keep up with sea level rise. This trend is expected to continue and worsen, particularly with increases in turbidity due to coastal erosion. See, for example, NE Chadwick-Furman, 'Reef Coral Diversity and Global Change' (1996) 2 Global Change Biology 559, 566; GF Camion and others, 'Reef Response to Sea-level and Environmental Changes During the Last Deglaciation: Integrated Ocean Drilling Program Expedition 310, Tahiti Sea Level' (2012) 40(7) Geology 643; WC Dullo, 'Coral Growth and Reef Growth: A Brief Review' (2005) 51(1–4) Facies 33–48; CD Storlazzi and others, 'Numerical Modelling of the Impact of Sea Level Rise on Fringing Coral Reef Hydrodynamics and Sediment Transport' (2011) 30 Coral Reefs 83; IPCC WG II Coastal Systems and Low-Lying Areas (n 22), 369.

[26] IPCC WG II Ocean Systems (n 11).

extensively, and more conclusively, the effects of climate change on marine biodiversity.[27] The IPCC's AR5 provides significantly more focus on the oceans than previous reports – and the results are sobering.

According to the IPCC, climate change has already had widespread impacts on species distribution, abundance, phenology (timing of biological events), and therefore on species richness and community composition across a broad range of taxonomic groups (plankton to top predators).[28] The timing of many biological events has changed, resulting in shifts in biological events such as peak abundance of phytoplankton and zooplankton and the reproduction and migration of invertebrates, fish and seabirds. Distributions of benthic, pelagic and demersal species and communities have shifted 10s to 1000s of kilometres. While these range shifts have not been uniform across taxonomic groups or ocean regions, the rates of migration of marine species have been much higher than even the 'potential maximum rates reported for terrestrial species, despite slower warming of the Ocean than land surface'.[29] Poleward distribution shifts of many fishes and invertebrates, the natural result of the inherent thermal sensitivity of most organisms, have resulted in increased species richness in mid- to high-latitude regions and changing community structure. Warmer water species are increasingly dominating in sub-tropical and higher-latitude regions, while the abundance of sub-tropical species in equatorial waters is declining. Most vulnerable are polar and tropical species due to their narrow temperature ranges and the fact that they are already living close to their thermal limits.[30]

These shifts in abundance and distribution are all related to changes in ocean temperature and conditions. While ecosystems are naturally subject to climate-related variability, which drives fluctuations in productivity, many of these shifts, particularly those in phyto- and zooplankton (the very basis of the marine food chain) have exceeded natural 'climate velocities'. In other words, climate change-related changes in water temperature are amplifying natural variations. Further complicating the picture, these biogeographic shifts are also influenced by other factors such as nutrient and stratification changes, species' interactions, habitat availability and fishing. Rates and pattern of shifts may also be affected by local dynamics such as coastal upwelling or topographic features such as islands, channels and lagoons or may be wholly constrained by geographical barriers that make it impossible for endemic species to migrate. According to the IPCC, the capacity of present-day fauna and flora to compensate for or keep up with the rate of ongoing thermal change is limited.[31] In other words, although many uncertainties still exist, it appears that the survivability of marine species and of marine biodiversity is at risk.

The IPCC concludes that 'the warming-induced shifts in the abundance, geographic distribution, migration patterns, and timing of seasonal activities of species have been and will be paralleled by a reduction in their maximum body size. This has resulted and will further result in changing interactions between species, including competition and

---

[27] *Ibid.*, 417.
[28] *Ibid.*, 463.
[29] *Ibid.*, 414, 461.
[30] *Ibid.*, 414, 456-460.
[31] *Ibid.*, 414.

predator-prey dynamics'.[32] The Report predicts ocean-wide changes in ecosystem properties to continue (under a warming trend of 1°C by mid-century) and that these large irreversible shifts in the spatial distribution of species and seasonal timing of their activities (feeding, growth development, behaviours, and productivity) will have implications for species composition, and ecosystem goods and services.[33] By the mid-21st century, species richness will increase at mid to high latitudes, but decrease at tropical latitudes, with high local extinction rates anticipated in the tropics and semi-enclosed seas. In particular, by 2055, assuming a 2°C warming above pre-industrial levels, fisheries yields are expected to increase by 30–70 per cent in some high-latitude regions (relative to 2005). There will be a redistribution of existing yields at mid-latitude levels, but in tropical areas and the Antarctic yields are expected to drop by 40–60 per cent. Expansion of oxygen-deprived anoxic 'dead zones' is expected to further constrain fish habitat while projected falls in open ocean net primary production will further decrease the overall global fisheries catch potential. As the IPCC notes, this global redistribution of catch potential for fish and invertebrates has profound implications for fisheries and for global food security.[34]

3. THE FURTHER EFFECTS OF CO_2 EMISSIONS ON MARINE BIODIVERSITY – THE PROBLEM OF OCEAN ACIDIFICATION

However, ocean warming is not the only problem. Increasing concentrations of atmospheric CO_2, the primary cause of global warming, have led to increasing ocean uptake of CO_2, which has resulted in decreased ocean pH, fundamentally changing the ocean carbonate chemistry. This process, known as ocean acidification, has been observed in all ocean areas but is particularly noted at high latitudes.[35] While evidence of its biological effects is largely still limited to historical observations, an increasing number of studies show that it will increasingly affect marine biota and interfere with ecological and biogeochemical processes in the oceans.[36] Critically, the current rate of ocean acidification is unprecedented within the last 65 to 300 million years[37] and it will continue to affect marine ecosystems for centuries if CO_2 emissions continue.[38]

The adverse effects of ocean acidification are increasingly being documented.[39] Ocean acidification poses substantial risks to both organism physiology and behaviour

---

[32] *Ibid.*, 414.
[33] *Ibid.*, 415.
[34] *Ibid.*, 415.
[35] IPCC WG II The Ocean (n 14), 1673.
[36] IPCC WG II Ocean Systems (n 11), 436.
[37] IPCC WG II The Ocean (n 14), 1675. B Hörnisch and others, 'The Geological Record of Ocean Acidification' (2012) 335 Science 1058, 1062.
[38] IPCC WG II Ocean Systems (n 11), 1675.
[39] *Oceans and Law of the Sea: Report of the Secretary General*, UN Doc A/68/71 (2013); *See* also, K Kroecker and others, 'Impacts of Ocean Acidification on Marine Organisms: Quantifying Sensitivities and Interaction with Warming' (2013) 19 Global Change Biology 1884; C Turley and others, 'Future Biological Impacts of Ocean Acidification and Their

as well as to population dynamics in everything from phytoplankton to animals. It may lead to more frequent harmful algal blooms and, by decreasing the rate of calcification, it is predicted to have potentially catastrophic consequences for corals and shelled marine life such as molluscs and echinoderms.[40] While its direct effect on fish is currently considered to be minimal, its indirect effect, through destruction of prey and habitat, is predicted to have detrimental impacts on fisheries, broader ecosystem services and livelihoods. When combined with ocean warming the effects of ocean acidification are amplified.[41] When combined with other stressors, the IPCC suggests that evolutionary rates may not be fast enough for sensitive animals and plants to adapt to the projected rate of future change.[42] The collectivity of these known and projected 'interactive, complex and amplified impacts'[43] poses significant threats for marine species and ecosystems.

4. THE INTERNATIONAL LEGAL FRAMEWORK FOR THE CONSERVATION AND MANAGEMENT OF MARINE BIODIVERSITY

It will be immediately apparent that the physical changes to spatial distribution, abundance, phenology and species interactions being wrought by climate change, particularly when compounded by ocean acidification, have profound implications for marine conservation and management regimes at both the national and international level. Interestingly, the IPCC notes the critical importance of international frameworks for collaboration and decision-making in providing 'opportunities for global cooperation and the development of international, regional and national policy responses to the challenges posed by the changing ocean',[44] and calls for the 'international community to progress rapidly to a "whole of ocean" strategy for responding to the risks and challenges posed by anthropogenic ocean warming and acidification'.[45] However, the current international legal framework allocates jurisdictional responsibility for the protection of marine biodiversity either to coastal states or to a complex, and not necessarily coherent, array of international treaty regimes. These

---

Socioeconomic-policy Implications' (2012) 4 *Current Opinion in Environmental Sustainability* 278; J-P Gattuso and L Hanson (eds), *Ocean Acidification* (OUP, 2011).
 [40] IPCC WG II Ocean Systems (n 11), 443.
 [41] IPCC WG II Ocean Systems (n 11), 1675, 1708.
 [42] *Ibid.*, 415.
 [43] IPCC, 'Summary for Policymakers' in *Climate Change 2014: Impacts, Adaptation, and Vulnerability. Part A: Global and Sectoral Aspects. Contribution of Working Group II to the Fifth Assessment Report of the Intergovernmental Panel on Climate Change* (CUP, 2014) (IPCC WG II Summary for Policy Makers), 17.
 [44] IPCC WG II The Ocean (n 14), 1711.
 [45] *Ibid.*, 1661.

130 *Research handbook on biodiversity and law*

agreements, some global and some regional in scope, include a number of sector-specific agreements for the management of marine resource exploitation (predominantly fish) and maritime activities, as well as conservation agreements related to species, habitats or biodiversity in general.

Of prime importance is the United Nations Convention on the Law of the Sea[46] (LOSC), which provides the overarching global legal framework for the oceans. The LOSC requires states to cooperate in the conservation of the living resources of the high seas[47] and to preserve and protect the marine environment.[48] In particular, states are to prevent, reduce and control pollution of the marine environment, and to protect and preserve rare or fragile ecosystems as well as the habitat of depleted, threatened or endangered species and other forms of marine life.[49] They are not to transfer, either directly or indirectly, damage or hazards from one area to another or transform one type of pollution into another,[50] and they are to ensure that technologies under their jurisdiction or control do not pollute the marine environment or result in the introduction of harmful alien species.[51] These general obligations articulated in the LOSC are buttressed by the more specific obligations relating to flora and fauna of the deep seabed found in Part XI of the LOSC and the 1994 Implementing Agreement,[52] and the obligations found in various sectoral agreements, such as those establishing Regional Fisheries Management Organisations (RFMOs), the International Whaling Convention (IWC)[53] and agreements relating to marine pollution such as the MARPOL Convention,[54] and the London Dumping Convention (LC)[55] and its London Protocol (LP).[56]

Also important to the conservation of marine biodiversity are a number of global and regional conservation agreements, the most important of which is the Convention on Biological Diversity (CBD),[57] which addresses the protection and sustainable use of the components of biological diversity, including marine biodiversity. The CBD defines 'biodiversity' as referring to the 'variability among living organisms from all sources

---

[46] 1982 United Nations Convention on the Law of the Sea 1833 UNTS 397 ('LOSC').
[47] LOSC, Articles 117–120.
[48] LOSC, Article 192.
[49] LOSC, Article 194.
[50] LOSC, Article 195.
[51] LOSC, Article 196.
[52] 1994 Agreement relating to the Implementation of Part XI of the United Nations Convention on the Law of the Sea of 10 December 1982 1836 UNTS 3.
[53] 1946 International Convention for the Regulation of Whaling 161 UNTS 72.
[54] 1973 International Convention for the Prevention of Pollution from Ships 1340 UNTS 184 as amended by the Protocol of 1978 relating to the International Convention for the prevention of pollution from ships 1340 UNTS 61.
[55] 1972 Convention on the Prevention of Marine Pollution by Dumping of Wastes and Other Matter 1046 UNTS 138.
[56] 1996 Protocol to the Convention on the Prevention of Marine Pollution by Dumping of Wastes and Other Matter (1997) 36 International Legal Materials 1.
[57] 1992 Convention on Biological Diversity 1760 UNTS 79 (CBD).

including, inter alia, terrestrial, marine and other aquatic ecosystems and the ecological complexes of which they are part; this includes diversity within species, between species and of ecosystems'.[58] Thus the conservation of biodiversity requires protection of both species and their habitats, and their variability.[59] In other words, what is required is a holistic ecosystem approach to the protection of biodiversity, including marine biodiversity. Unfortunately, particularly in the marine biodiversity context, achievement of this goal is made difficult by the division of jurisdictional competences in respect of areas under and areas beyond national jurisdiction. In terms of its jurisdictional scope, the CBD covers marine biodiversity within the limits of national jurisdiction as well as processes and activities carried out under the jurisdiction or control of the member states.[60] In ABNJ the CBD applies only to processes and activities carried out under the jurisdiction or control of the member states.[61] It does not, per se, protect the processes and components of biodiversity that reside in ABNJ. Protection of marine biodiversity in ABNJ is therefore left to be governed by the overarching LOSC framework together with the existing ad hoc range of international agreements. These include the agreements establishing RFMOs and the IWC referred to above, as well as other species- or habitat-specific global conservation agreements such as the Convention on the Conservation of Migratory Species of Wild Animals[62] and its various relevant subsidiary agreements, and regional agreements such as the Barcelona Convention,[63] the OSPAR Convention,[64] the Noumea Convention[65] and the Convention on the Conservation of Antarctic Marine Living Resources,[66] and the Antarctic Treaty.[67] In addition, a number of other global and regional agreements are relevant to marine biodiversity in AUNJ. These include the IWC, the Ramsar Convention on International Wetlands,[68] the World Heritage Convention,[69] the regional agreements adopted under the United Nations Environment Programme (UNEP)

---

[58] CBD, Article 2.
[59] R Rayfuse, 'Biological Resources' in D Bodansky, J Brunnée and E Hey (eds), *Oxford Handbook of International Environmental Law* (OUP, 2007) 362, 366.
[60] CBD, Articles 1 and 4.
[61] CBD, Articles 1 and 4.
[62] 1979 Convention on the Conservation of Migratory Species of Wild Animals 1651 UNTS 333.
[63] 1976 Convention for the Protection of the Mediterranean Sea Against Pollution 1102 UNTS 44.
[64] 1992 Convention for the Protection of the Marine Environment of the North-East Atlantic (the 'OSPAR Convention') 2354 UNTS 67.
[65] 1986 Convention for the Protection of the Natural Resources and Environment of the South Pacific Region available at <http://www.sprep.org/legal/the-convention>.
[66] 1980 Convention on the Conservation of Antarctic Marine Living Resources 1329 UNTS 47 (CCAMLR Convention).
[67] 1959 Antarctic Treaty 402 UNTS 71.
[68] 1971 Convention on Wetlands of International Importance Especially as Waterfowl Habitat 996 UNTS 241.
[69] 1972 Convention for the Protection of the World Cultural and Natural Heritage 1037 UNTS 151.

Regional Seas Programme,[70] the more general regional European,[71] African,[72] and Western Hemisphere[73] wildlife treaties and the Polar Bears Agreement.[74]

Assessments of this fragmented legal framework show that it is singularly ill-equipped to respond to the challenges of spatial shifts and changes in phenology, species abundance and species interactions caused by climate change.[75] These challenges can be clearly demonstrated by reference to the regime for the conservation and management of high seas fisheries.

5. CLIMATE CHANGE AND THE CONSERVATION AND MANAGEMENT OF HIGH SEAS FISHERIES

Given their role in ensuring global food security, fish arguably constitute the element of marine biodiversity of most immediate interest and importance to humans. As noted above, the LOSC requires states to cooperate in the conservation and management of the marine living resources of the high seas, generally taken to mean fish. The recognised *modus operandi* for this cooperation is through the establishment of subregional or regional fisheries organisations and arrangements. The institutionalisation of such arrangements is further recognised, at least in respect of straddling fish stocks (SFS) and highly migratory fish stocks (HMFS) in the 1995 United Nations Fish Stocks Agreement (FSA),[76] which requires all states having a 'real interest' in the fisheries concerned to join, or where none exists, to establish and participate in a relevant organisation or other arrangement.[77] While some regional fisheries organisations have an advisory function only, the obligation to cooperate has predominantly manifested itself in the establishment of organisations having a management function. These Regional Fisheries Management Organisations (RFMOs) seek to regulate exploitation of either particular ocean spaces or particular species through the adoption of regulatory measures relating, inter alia, to acquisition of fishery-related data and

---

[70] See further, <http://www.unep.org/regionalseas/>.
[71] 1979 Convention on the Conservation of European Wildlife and Natural Habitats (Bern Convention) 1284 UNTS 199.
[72] 1968 African Convention on the Conservation of Nature and Natural Resources 1001 UNTS 3.
[73] 1940 Convention on Nature Protection and Wild Life Preservation in the Western Hemisphere 161 UNTS 193.
[74] 1973 Agreement on the Conservation of Polar Bears UNTS Registration I-50540-0800000280363c19.pdf.
[75] Lockwood and others (n 16); Rayfuse (n 15); A Proelss and M Krivickaite, 'Marine Biodiversity and Climate Change' (2009) 4 Carbon and Climate Law Review 437.
[76] 1995 Agreement for the Implementation of the Provisions of the UN Convention on the Law of the Sea of 10 December 1982 relating to the Conservation and Management of Straddling Fish Stocks and Highly Migratory Fish Stocks 2167 UNTS 3 ('FSA').
[77] FSA, Article 8. See further, R Rayfuse, 'The Interrelationship between the Global Instruments of International Fisheries Law' in E Hey (ed), *Developments in International Fisheries Law* (Kluwer Publishing International, 1999) 107.

stock assessment, management of fishing effort, allocation of fishing opportunities, compliance and enforcement and, increasingly, protection of the wider marine environment.

A significant challenge for RFMOs relates to their ability to manage under conditions of uncertainty. The LOSC requires states to adopt measures on the basis of the best scientific evidence available, to consider effects on associated and dependent species, and to maintain or restore populations of harvested species at levels that ensure maximum sustainable yield.[78] The precautionary and ecosystem approaches are specifically called for in the FSA,[79] the adoption of which led to the establishment of a number of new RFMOs[80] and spurred many existing RFMOs to amend their Conventions or adopt new management plans to reflect these modern management approaches.[81] However, as evidenced by practice within the Commission on the Conservation of Antarctic Marine Living Resources (CCAMLR), the application of these approaches is a complicated and difficult task. By way of example, despite the fact that CCAMLR is specifically mandated to apply an ecosystem approach[82] and that its parties have specifically agreed to establish marine protected areas (MPAs), the Commission has thus far been unable to reach agreement on any.[83] UNGA resolutions on bottom fisheries[84] have led a number of RFMOs to adopt measures for the protection of vulnerable marine ecosystems (VMEs) from the adverse effects of bottom trawling, including closed areas and 'move-on' rules which, require vessels to stop fishing and move off a certain predetermined distance if they bring up more than prescribed quantities of listed indicator species such as deep water corals. However, both the implementation and the efficacy of these measures have been seriously questioned by NGOs, who point to the extreme complexity of the regulations, the need

---

[78] LOSC, Articles 117–119.

[79] FSA Articles 5(c), (d), 6 and Annex II.

[80] Including the South East Atlantic Fisheries Organisation (SEAFO), the Western and Central Pacific Fisheries Commission (WCPFC) and the South Pacific Regional Fisheries Management Organisation (SPRFMO).

[81] For example, in 1997 the General Fisheries Commission for the Mediterranean (GFCM) adopted a revised Convention, which entered into force on 29 April 2004. On 28 September 2007, the Northwest Atlantic Fisheries Organisation (NAFO) adopted a document entitled 'Amendment to the Convention on Future Multilateral Cooperation in the Northwest Atlantic Fisheries', constituting the first formal step towards a reformed Convention for NAFO. As of 2015 six parties have ratified the amended Convention, which has yet to enter into force. In 2007 the North East Atlantic Fisheries Commission (NEAFC) adopted a revised Convention, which the parties have agreed to apply on a provisional basis pending its ratification. In 2003 the Inter-American Tropical Tuna Commission (IATTC) adopted a new Convention for the Strengthening of the Inter-American Tropical Tuna Commission (the Antigua Convention), which came into force on 27 August 2010.

[82] CCAMLR Convention, Article II.

[83] See CCAMLR, Report of the Thirty-Second Meeting of the Commission, Hobart 23 October–1 November 2013, available at <http://www.ccamlr.org/en/system/files/e-cc-xxxii.pdf>, para 7.32.

[84] UNGA Res 61/105 (2007) and UNGA Res 64/72 (2009) on Sustainable Fisheries and relevant paragraphs on protection of deep-sea species and ecosystems beyond national jurisdiction from the harmful impacts of bottom fishing.

for enhanced mapping of VMEs, the need for new requirements for exploratory fisheries, the need to ensure that any closures triggered by the 'move-on' rules in existing fishing areas are applicable to all vessels fishing in that area and not just the vessel reporting the encounter, and the need to ensure that mechanisms are in place to monitor closed areas to ensure no fishing takes place there.[85]

The challenges of grafting climate change impacts such as changes in species distribution and productivity onto the range of factors already to be considered in implementing the precautionary and ecosystem approaches are significant. Stock or species management boundaries often do not align with ecological boundaries. In addition, there is a tendency in many RFMOs to ignore both precaution and the best scientific evidence available and to adopt measures, instead, on the basis of political, rather than biological, considerations.

Moreover, despite the increasing recognition of the requirements of precautionary and ecosystem management, RFMOs and arrangements generally manage stocks either on a species-specific[86] or geographic basis.[87] In geographical terms, global RFMO coverage is not comprehensive. For example, no RFMO exists for non-tuna species in the central and south Atlantic or Indian Oceans and no RFMO exists at all in the Central Arctic Ocean.[88] Thus, many areas are unregulated and many stocks and species remain unmanaged. The problems that climate change-induced species migration pose here are threefold. First, the efficacy of RFMO management measures will be significantly impacted by declines in species abundance. Second, where species or stocks migrate outside the area of competence of the RFMO, they may move into an area regulated by another, leading to conflict between the two RFMOs as to the proper locus of managerial competence. The issue has already arisen in the context of Southern Bluefin Tuna (SBT) migrating southwards into the CCAMLR area with CCAMLR claiming managerial and enforcement competence over vessels fishing for SBT within the CCAMLR area and the CCSBT contesting this competence. Third, the species or stock may migrate into a wholly unregulated area of ocean, rendering the existing management regime obsolete and leaving the species or stock vulnerable to unregulated and unstoppable over-exploitation. It is precisely this concern that is

---

[85] M Gianni and others, *Unfinished Business: A Review of the Implementation of the Provisions of United Nations General Assembly Resolutions 61/105 and 64/72, Related to the Management of Bottom Fisheries in Areas Beyond National Jurisdiction*, (Deep Sea Conservation Coalition, 2011) available at <http://www.savethehighseas.org/publicdocs/DSCC_review 11.pdf>.

[86] The Indian Ocean Tuna Commission (IOTC), the Western and Central Pacific Fisheries Commission (WCPFC), the Inter-American Tropical Tuna Commission (IATTC), the International Commission for the Conservation of Atlantic Tunas (ICCAT) and the Commission on the Conservation of Southern Bluefin Tuna (CCSBT) manage highly migratory species (mostly tuna).

[87] The Northwest Atlantic Fisheries Organization (NAFO), the North-East Atlantic Fisheries Commission (NEAFC), the South East Atlantic Fisheries Organisation (SEAFO), and the Commission on the Conservation of Antarctic Marine Living Resources (CCAMLR) manage non-tuna stocks within their geographical areas of coverage.

[88] *See* further, R Rayfuse, 'Melting Moments: The Future of Polar Oceans Governance in a Warming World' (2007) 16(2) RECIEL 196 and R Rayfuse, 'Protecting Marine Biodiversity in Polar Areas Beyond National Jurisdiction' (2008) 17(1) RECIEL 3.

driving calls for a moratorium on fishing in the Arctic Ocean pending the establishment of appropriate fisheries regimes in both AUNJ and ABNJ.[89]

Even where regulation exists, the difficulties of managing migratory species and stocks which cross biologically arbitrary geo-political and legal jurisdictional lines will only increase as species distributions change. In the case of transboundary or shared stocks (those shared between two states), climate-induced stock migration may affect the share of a stock in each state. As both states seek to maintain their share of the catch, this will have adverse implications for stock management and for stock status. If the state with the diminishing percentage of the stock fails to reduce its catch this will undermine conservatory efforts and catch limits in the other country. In a worst case scenario, continued take by the state losing the stocks coupled with increased take by the state acquiring more of the stock will lead to the stock being fished possibly to extinction.

Similarly, in high seas fisheries, Article 7 of the FSA requires coastal states and RFMOs to adopt 'compatible' conservation and management measures in respect of SFS and HMFS. A range shift away from the coastal state will weaken its conservation incentives and aggravate management as between that state and the relevant RFMO. A range shift to a coastal state will similarly aggravate management and conservation status if it leads to increased fishing pressure within the areas under national jurisdiction and no corresponding reduction in the high seas, RFMO, area. Dramatic shifts in migration will be particularly problematic in the case of highly migratory species, such as tuna, in areas where pockets of high seas are interspersed with areas under national jurisdiction such as in the Western and Central Pacific.[90] According to Axelrod, states most vulnerable to these anticipated changes are more likely to push for greater control over the resources which straddle their Exclusive Economic Zones through more stringent RFMO measures including reduction of quota for distant water fishing fleets. The latter, however, are more likely to insist on maintaining their quota and less restrictive measures while shifting the organisational goals away from catch limits to enable them to maintain their presence in the fishery even in the face of increasing catches in AUNJ.[91] The robustness and institutional resilience of these fisheries regimes will therefore be threatened.

Lockwood and others have identified a number of regime requirements for the governance and management of marine biodiversity conservation in a climate change-challenged world.[92] Key requirements identified include: (a) institutional flexibility based on formal and informal networks, redundancy and modularity, and polycentricity;

---

[89] *See* further, the Declaration Concerning the Prevention of Unregulated High Seas Fishing in the Central Arctic Ocean adopted on 16 July 2015 by the five Arctic coastal states, Canada, Denmark, Norway, Russia and the United States, available at <https://www.regjeringen.no/globalassets/departementene/ud/vedlegg/folkerett/declaration-on-arctic-fisheries-16-july-2015.pdf>.

[90] IPCC WG II The Ocean (n 14), 1701–1704.

[91] M Axelrod, 'Climate Change and Global Fisheries Management: Linking Issues to Protect Ecosystems or to Save Political Interests?' (2011) 11(3) Global Environmental Politics 64.

[92] Lockwood (n 16).

(b) effective and transformational leadership concerned with enabling learning, innovation and adaptive capacity; (c) inclusive, effective and meaningful participation and decision-making; (d) strategic connectivity, coordination and cohesion through spatial and temporal alignment of purpose, strategy and action; and (e) quality governance based on legitimacy, transparency and accountability. Performance reviews of RFMOs indicate that some progress has been made towards better achieving these requirements, although continuing declines of global fish stocks suggest the need to resist complacency.[93] Fundamentally, there is a need for RFMOs to take seriously the need to directly address climate change in their activities and decisions. Since 1992, of the 15 existing high seas RFMOs, Axelrod identifies only six that have taken 'action' related to climate change either by 'deciding to undertake climate adaptation or mitigation activities or to allocate funds towards climate change research'.[94] Ultimately, there remains a need to ensure that decision-making processes within RFMOs are strengthened to enable them to respond expeditiously and effectively to the challenges posed by climate change to the stocks they manage. Mechanisms for cooperation between existing RFMOs will need to be strengthened and wholly new regulatory regimes may need to be developed to ensure the sustainable conservation and use of fisheries resources in a climate-changed ocean.

Similar challenges will be faced by the other existing species, habitat and regional conservation agreements. At the extreme end, the Polar Bears Agreement is in danger of losing its very raison d'être. The plight of the charismatic marine mega-fauna, the polar bear, has received considerable attention in the popular press. The parties to the Agreement have made clear that continued and increasing loss and fragmentation of sea ice brought about by climate change is 'the most important threat to polar bear conservation'.[95] Studies presented to the Convention on Migratory Species show the high vulnerability of a number of species listed in its appendices including marine mammals and fish.[96] Parties to the World Heritage Convention (WHC) have explicitly recognised climate change as a 'real danger' to some World Heritage properties, including the iconic Belize Barrier Reef and the Australian Great Barrier Reef.[97] The

---

[93] In 2006 the FAO identified 75% of the world's fish stocks as either fully exploited or over-exploited or depleted and recovering from depletion. In 2010 the number had risen to 85% representing 53% of global fish stocks as fully exploited and 32% as either over-exploited (28%), depleted (3%) or recovering from depletion (1%). In 2012 the FAO reported 87% of global fish stocks were either over-exploited (30%) or fully exploited (57%). See FAO, *The State of World Fisheries and Aquaculture: 2006* (FAO, 2006); FAO, *The State of World Fisheries and Aquaculture: 2008* (FAO, 2008); FAO, *The State of World Fisheries and Aquaculture: 2010* (FAO, 2010); and FAO, *The State of World Fisheries and Aquaculture: 2012* (FAO, 2012).

[94] Axelrod (n 91). The six are CCAMLR, IOTC, NAFO, NASCO, NPAFC, WCPFC. Axelrod also includes the International Pacific Halibut Commission (IPHC), which is a bilateral arrangement between Canada and the United States, and the IWC as having taken similar 'action'.

[95] Meeting of the Parties to the 1973 Agreement on the Conservation of Polar Bears, Tromsø, Norway, 17–19 March 2009, 'Outcome Document' available at <http://pbsg.npolar.no/export/sites/pbsg/en/docs/Outcome_MOP2009.pdf>.

[96] Convention on Migratory Species Secretariat, *Migratory Species and Climate Change: The Impacts of a Changing Environment on Wild Animals* (CMS, 2006).

[97] WHC Decision 32 COM 7A. 32 (2008).

CBD has similarly recognised the threats to marine and coastal biodiversity[98] and has called on its parties to implement plans to improve ecosystem resilience and to restore degraded ecosystems as a means of contributing to both the conservation of biodiversity and to climate change mitigation and adaptation.[99] Other specifically marine focused regimes such as the IWC, CCAMLR and the FAO Committee on Fisheries have also taken action to study the effects of climate change on species and the UNGA has called for specific studies on the topic of climate change and the marine environment. However, as the parties to the WHC[100] and the CBD[101] have noted, it is difficult to separate climate change impacts from those of existing stressors. Thus, while biological regime shifts will necessitate more than ever the functioning of governance structures 'across multiple scales and levels, from local to sub-national, national and global' and their alignment 'between and across jurisdictional, sectoral and geographic boundaries',[102] the solution to the problem of saving marine biodiversity from the threat of climate change lies not in these regimes, but rather in the international climate regime. The challenge of working between and across the various marine biodiversity-related regimes is a daunting one. However, this challenge pales in significance when the problem of ocean acidification and the need to work with the climate regime are grafted into the equation.

6. MARINE BIODIVERSITY AND OCEAN ACIDIFICATION

As noted above, ocean acidification refers to the changes in ocean chemistry that are caused by absorption of CO_2 from the atmosphere. It is important to note that ocean acidification is not caused by climate change or global warming. Rather, it is a parallel impact resulting from higher atmospheric CO_2 concentrations. Ocean acidification is thus closely linked with climate change because CO_2, while not the only greenhouse gas, is the main one by volume.[103] Approximately one-third of anthropogenic CO_2 emissions have been absorbed by the oceans since the industrial revolution.[104] As emissions have increased, so, too, has ocean absorption of CO_2, leading to changes in the ocean's pH levels. Also affecting ocean pH, at least on local or regional scales, is coastal acidification due to nitrogen and phosphate run-off from agricultural, industrial, urban and domestic sources, and the as yet theoretical although increasingly possible

---

[98] CBD Decision X/29 Marine and Coastal Biodiversity (COP 10, 2010).
[99] CBD Decision IX/16 Biodiversity and Climate Change (COP 9, 2008); Decision X/33 Biodiversity and Climate Change (COP 10 2010) and Decision X/2 Strategic Plan for Biodiversity (COP 10 2010), Target 15.
[100] WHC Decision 31 COM. 7.1 (2007). *See* also 'Impacts of Climate Change on World Heritage Properties' WHC-07/31.COM/7.1 (2007).
[101] *See* CBD Decision X/29 (n 98).
[102] Lockwood (n 16), 167.
[103] IPCC, *Radiative Forcing Report; Climate change 1995, The science of climate change, contribution of working group 1 to the second assessment of the intergovernmental panel on climate change* (UNEP and WMO, CUP, 1996).
[104] D Archer, *The Global Carbon Cycle* (Princeton University Press, 2010), 116.

release of methane from hydrates on the seafloor as a result of thermal warming of the deep ocean.[105] However, CO_2 absorption is the main driver of ocean acidification.

While the consequences of ocean acidification will differ across species and communities, the IPCC is clear that it threatens many marine species and ecosystems.[106] According to the geological record, a period of mass extinctions of marine organisms occurred the last time ocean acidification reached the levels predicted by the end of 2100.[107] The current rate of acidification is 10 times faster than during that period.[108] If acidification continues at current rates it will have profound effects on marine biodiversity and marine ecosystems globally, and on the human sectors dependent on them, such as fishing, aquaculture and tourism.[109] This will only compound the challenges to marine biodiversity and to regime resilience discussed above.

Responding to ocean acidification represents both a policy and a legal challenge of increasingly pressing importance and complexity.[110] Indeed, according to Rau and others, 'short of stabilizing if not reducing atmospheric CO_2 there may ultimately be no perfect or even satisfactory conservation option for the oceans, either globally or regionally'.[111] Yet, ocean acidification is not defined or specifically referred to in any binding international legal instrument.[112] Rather, as with marine biodiversity, a number of international legal regimes are relevant to ocean acidification, depending on whether the regulatory focus is on its causes, its effects or both.

Given that the main driver of ocean acidification is CO_2 absorption, the only way to address ocean acidification is to reduce CO_2 emissions thereby reducing the amount of CO_2 being absorbed by the oceans. Reduction of emissions is the concern of the climate regime established by the UN Framework Convention on Climate Change (UNFCCC)[113] and the Kyoto Protocol.[114] However, that regime is concerned with global warming and the atmospheric effect of emissions, not their effects on the

---

[105] IDDRI, 'Ocean acidification – what can we do?' Policy Brief No 17/12 (IDDRI, Paris, 2012), 2.
[106] IPCC WG II Ocean Systems (n 11), 432–443 and The Ocean (n 14), 1673–1677 and 1714.
[107] Q Schiermeier, 'Earth's Acid Test' (2011) 471 Nature, 154.
[108] IGBP, IOC, SCOR, *Ocean Acidification Summary for Policymakers – Third Symposium on the Ocean in a High-CO_2 World* (International Geosphere-Biosphere Programme, 2013).
[109] IPCC WG II Ocean Systems (n 11), 464; IPCC WG II The Ocean (n 14), 1708, 1714.
[110] C Turley and others, 'The Ocean Acidification Challenges Facing Science and Society' in Gattuso and Hanson (eds), *Ocean Acidification* (n 39), 249; N Hilmi and others, 'Towards Improved Socio-economic Assessments of Ocean Acidification's Impacts' (2013) 160 Marine Biology 1773. See also *Oceans and Law of the Sea: Report of the Secretary General* (2013) (n 39).
[111] GH Rau, EL McLeod and O Hoegh-Guldberg, 'The Need for New Ocean Conservation Strategies in a High-Carbon Dioxide World' (2012) 2 Nature Climate Change 720, 723.
[112] UN Secretary General, Oceans and the Law of the Sea: Report of the Secretary General to the General Assembly, UN Doc A/68/71 (2013).
[113] 1992 United Nations Framework Convention on Climate 1771 UNTS 107.
[114] 1997 Kyoto Protocol to the United Nations Framework Convention on Climate Change 2303 UNTS 148.

oceans.[115] Article 2 of the UNFCCC articulates the objective of the climate regime as being the stabilisation of greenhouse gas concentrations in the atmosphere at a level that would prevent dangerous anthropogenic interference with the climate system, which is defined as the totality of the atmosphere, hydrosphere, biosphere and geosphere and their interactions. 'Climate change' is defined in Article 1(2) as the change of climate attributed to human activity that alters the composition of the global atmosphere, while Article 1(1) states that the adverse effects of climate change are those involving alterations in the physical environment or biota resulting from climate change which have significant deleterious effects on composition, resilience or productivity of natural and managed ecosystems. Of course this can be read as including effects on marine biota and marine ecosystems as well. However, the atmospheric focus is reinforced by reference to the Kyoto Protocol which calls not for the reduction of CO_2 emissions but rather of 'CO_2 equivalent' emissions of listed greenhouse gases.

Moreover, the UNFCCC contains provisions that may actually exacerbate ocean acidification. Pursuant to Article 4(1)(d) of the UNFCCC states are to promote and cooperate in the conservation and enhancement of 'sinks' and 'reservoirs' of all greenhouse gases, including the oceans. This has been read as meaning not only that parties must act to enhance the passive absorption of anthropogenic CO_2 into the oceans but that they may also act to encourage the active sequestration of CO_2 into the oceans.[116] This, in turn, has been taken by some as an invitation to promote and engage in controversial marine geoengineering schemes such as ocean fertilisation and deep-ocean injection aimed at further increasing ocean absorption of CO_2. While, as discussed below, regulation of this active ocean sequestration poses a number of challenges, adequate regulation of 'passive' ocean acidification is even more challenging, particularly given the climate regime's focus on the oceans as part of the solution to climate change.[117] Although great strides have been made in recent years in bringing the oceans into the climate negotiations, they have yet to be given serious consideration in the development of international climate policy.[118] Thus it is necessary to consider the other regimes that may potentially be applicable. As with the case of marine biodiversity, the picture that emerges is a complicated one of inadequate, overlapping, sometimes inconsistent and almost always uncoordinated treaty and customary rules, the applicability of which depends on how CO_2 emissions are characterised.[119]

Characterising CO_2 emissions as pollution opens up a range of possible avenues for addressing, at least in part, their effect on the oceans. Pollution of the marine environment is defined in the LOSC as the direct or indirect introduction, by humans, of substances or energy into the marine environment, which results or is likely to result

---

[115] EJ Goodwin, *International Environmental Law and the Conservation of Coral Reefs* (Routledge, 2011), 259.

[116] Scientific Committee on Ocean Research and International Oceanographic Committee, 'The Ocean in a High CO_2 World' (2004) 17 Oceanography 72.

[117] D Freestone, 'Climate Change and the Oceans' (2009) 4 Climate and Carbon Law Review 383.

[118] G Galland, E Harrould-Kolieb and D Herr, 'The Ocean and Climate Change Policy' (2012) 12 Climate Policy 764, 765.

[119] R Baird, M Simons and T Stephens, 'Ocean Acidification: A Litmus Test for International Law' (2009) 4 Climate and Carbon Law Review 459.

140 *Research handbook on biodiversity and law*

in such deleterious effects as harm to living resources and marine life, hazards to human health, hindrance to marine activities, including fishing and other legitimate uses of the seas, impairment of quality for use of sea water and reduction of amenities.[120] The inherent quality of the substance being introduced is not what matters. Rather, it is the effect that the introduction may have on the marine environment that is relevant. Thus it is possible to characterise the introduction, either directly or indirectly, of CO_2, the main driver of acidification, into the oceans as pollution.

All states have the obligation to protect and preserve the marine environment[121] and to take, individually or jointly, all measures necessary to prevent, reduce and control pollution of the marine environment from any source.[122] The framework provisions of the LOSC that establish the parameters of these obligations are further detailed in a number of treaties dealing with marine pollution from vessel, terrestrial and atmospheric sources. The regime that has been the focus of most attention to date has been that established by the 1972 LC and its 1996 LP, which aim to prohibit dumping at sea unless it can be shown that the substance dumped poses no threat to the marine environment.

Dumping, which is defined in identical terms in the LOSC and the LC/LP, consists of 'any deliberate disposal of wastes or other matter from vessels, aircraft, platforms or other man-made structures at sea'.[123] The LOSC requires all states to adopt national laws to prevent and regulate dumping, which must be no less effective than the internationally agreed upon rules and standards,[124] which, for their parties, are considered to be those found in the LC and LP respectively. The LC adopts a listing approach, prohibiting outright the dumping of certain listed wastes and making the dumping of others subject to a special prior permit. Dumping of non-listed wastes requires a general permit.[125] The LP takes the reverse approach, prohibiting outright the dumping of any wastes other than the few materials specifically listed in its Annex I.[126]

When the LC was negotiated no mention was made of the seabed or sub-seabed. However, when the LP was negotiated, the definition of dumping was specifically expanded to prohibit 'any storage of wastes or other matter in the seabed and the subsoil thereof from vessels, aircraft, platforms or other man-made structures at sea'.[127] Thus, while sub-seabed CO_2 storage might be permitted under the LC it would not have been permitted under the LP. However, with some states increasingly promoting sub-seabed carbon capture and storage as a means of alleviating atmospheric CO_2 levels, the LP was amended at the First Meeting of its Contracting Parties immediately after it came into force in 2006. These amendments now permit the dumping of 'carbon

---

[120] LOSC, Article 1(4).
[121] LOSC, Article 192.
[122] LOSC, Article 194(1).
[123] LOSC, Article 1(5); LC, Article I; LP Article 2.
[124] LOSC, Articles 194 and 210.
[125] LC, Articles III and Annex III.
[126] LP, Articles 2 and 4.
[127] LP, Article 1(4)(3) (1996).

dioxide streams from carbon dioxide capture processes for sequestration' into sub-seabed geological formations[128] subject to a range of precautions set out in Annex I to the LP and in the Specific Guidelines for the Assessment of Carbon Dioxide Disposal into Sub-Seabed Geological Formations adopted by the International Maritime Organisation (IMO) in 2012.[129] The Guidelines are considered by the parties to the LC/LP to be sufficient to minimise the risks associated with sub-seabed sequestration, including the risk of CO_2 leakage into the water column.

While sub-seabed sequestration may not contribute to ocean acidification, the same cannot be said for sequestration of CO_2 directly into the water column. Particularly in the case of ocean fertilisation, the threat of uncontrolled large-scale commercial fertilisation activities and the risks of possible adverse effects on the marine environment have been considered serious enough to warrant binding regulation. Ocean fertilisation refers to the deliberate addition of fertilising agents such as iron, phosphorous or nitrogen, or the control of natural fertilising processes through, for example, the artificial enhancement of deep-ocean mixing, for the purposes of stimulating the primary productivity in the oceans to increase CO_2 absorption from the atmosphere. From a legal perspective, the main issue that has been addressed regarding ocean fertilisation is whether it falls under the exception in the LOSC and the LC/LP, which exempts from the dumping regime the 'placement of matter for a purpose other than the mere disposal thereof, provided that such placement is not contrary to the aims of' the LOSC or the LC/LP.[130]

In 2008, rising concerns as to the potential negative impacts of ocean fertilisation on the marine environment[131] led to the adoption by the parties to the LC/LP of a moratorium on all ocean fertilisation activities except those being conducted for legitimate scientific research.[132] For the purposes of the moratorium, ocean fertilisation was defined as 'any activity undertaken by humans with the principle intention of stimulating primary productivity in the oceans, not including conventional aquaculture or mariculture, or the creation of artificial reefs'. In 2010 the parties to the LP adopted an Assessment Framework for ocean fertilisation activities, which requires proof of 'proper scientific attributes' and a comprehensive environmental impact assessment to

---

[128] LP, Annex 1 Article 1, as amended.
[129] IMO Doc. LC 34/15, Annex 8, 2 November 2012.
[130] LOSC, Article 1(5)(b)(ii); LC, Article I; LP, Article 2. *See* further, R Rayfuse, MG Lawrence and K Gjerde, 'Ocean Fertilisation and Climate Change: The Need to Regulate Emerging High Seas Uses' (2008) 23(2) International Journal of Marine and Coastal Law 297; R Rayfuse and D Freestone, 'Ocean Iron Fertilization and International Law' (2008) 364 Marine Ecology Progress Series 227.
[131] In July 2007 a 'Statement of Concern' was adopted by the Scientific Groups of the LC and LP followed by formal agreement by the parties at their meeting in November 2007 that knowledge about ocean fertilisation was at that time insufficient to justify large-scale projects and that they would further study the issue with a view to its regulation. *See Report of the Twenty-Ninth Consultative Meeting of the Contracting Parties to the Convention on the Prevention of Marine Pollution by Dumping of Wastes and Other Matter and Second Meeting of the Contracting Parties to the 1996 Protocol Thereto*, IMO Doc. LC 29/17 (2007).
[132] Resolution LC-LP.1, 31 October 2008. *Report of the Thirtieth Consultative Meeting of Contracting Parties to the London Convention and Third Meeting of Contracting Parties to the London Protocol*, IMO Doc. L 30/16, paras 4.1–4.18, Annexes 2, 5.

ensure that the proposed activity constitutes legitimate scientific research that is not contrary to the aims of the LC/LP and should thus be permitted to proceed.[133] Recognising that other marine geoengineering schemes beyond ocean fertilisation may also adversely impact the oceans, the LP was amended in 2013 to require its parties to prohibit the placement of matter into the sea from vessels, aircraft, platforms or other man-made structures at sea for any marine geoengineering purposes unless the activity is authorised under a permit.[134] At the same time, the Assessment Framework was made mandatory and extended to all marine geoengineering processes listed in a new Annex to the Convention.[135] Currently ocean fertilisation is the only process listed.

For present purposes the main issue here is the extent to which ocean fertilisation contributes to ocean acidification. The answer to this remains unknown for some fertilisation processes.[136] However, since cold water absorbs more CO_2 than warm water, processes that involve artificially pumping nutrient-rich water from the ocean depths up to the surface to stimulate phytoplankton blooms (or for any other purpose) would contribute significantly to ocean acidification.[137] It is not altogether clear that the LP, even as amended, applies to all such processes. While the LP clearly applies to processes that involve the placement of pipes or pumps into the water column from vessels, aircraft, platforms or other man-made structures at sea, it does not apply to processes involving the use of shore-based fertilisation technologies, arguably even where these involve the use of pipelines extending into the sea. Admittedly, the LOSC obliges states to prevent, reduce and control marine pollution from land-based sources,[138] however, both the development of the specific content of the obligation and its enforcement are the responsibility of the individual state on whose territory the land-based source is located. The Global Programme of Action for the Protection of the Marine Environment from Land-Based Activities (GPA)[139] provides guidance on the development and implementation of national and regional pollution reduction plans for certain pollution source categories but these categories do not capture CO_2 emissions. It is thus doubtful that the GPA has any applicability to the ocean

---

[133] Assessment Framework for Scientific Research Involving Ocean Fertilisation, Resolution LC-LP.2 (2010), *Report of the Thirty-Second Consultative Meeting of Contracting Parties to the London Convention and Fifth Meeting of Contracting Parties to the London Protocol*, IMO Doc. 32/15, Annex 5.

[134] Article 6 *bis* (2013).

[135] Resolution LP.4(8) 2013.

[136] P Williamson and C Turley, 'Ocean Acidification in a Geoengineering Context' (2012) 370 Philosophical Transactions of the Royal Society A 1974; L Cao and K Caldeira, 'Can Ocean Iron Fertilisation Mitigate Ocean Acidification?' (2010) 99(1–2) Climate Change 303.

[137] IPCC WG II Ocean Systems (n 11), 444–455; RA Feely and others, 'Evidence for Upwelling of Corrosive "Acidified" Water onto the Continental Shelf' (2008) 320 (5882) Science 1490. Upwelling can occur naturally. However, schemes have been suggested that would involve the placing of pipes into the water column to artificially enhance this upwelling. See, Williamson and Turley (n 136).

[138] LOSC, Article 196.

[139] Global Programme of Action for the Protection of the Marine Environment from Land-Based Activities, UNEP(OCA)/LBA/IG.2/7, 1995. *See* further, <http//www.gpa.unep.org/>.

acidification context.[140] Nevertheless, given that the sole purpose of ocean fertilisation – and other marine geoengineering schemes – is the deliberate placement into the oceans of excess atmospheric CO_2 for the purpose of disposing of that CO_2, any such activities may well fall foul of the prohibition in Article 195 of the LOSC on the transfer of damage or hazards from one area (the atmosphere) to another (the ocean) or the transformation of one type of pollution (atmospheric) into another (marine).

More certain contributions to ocean acidification come from shipping, which currently accounts for 2.7 per cent of global CO_2 emissions. In light of projected increases of up to 6 per cent by 2020, and even greater increases beyond,[141] the IMO has amended Annex VI of the MARPOL Convention, which deals with the 'Prevention of Air Pollution from Ships'.[142] Originally aimed at reducing emissions of sulphur and nitrous oxides, the Annex was amended in 2011 with the addition of a new Chapter 4 entitled 'Regulations in energy efficiency for ships', which includes mandatory technical and operational energy efficiency measures aimed at reducing CO_2 and other greenhouse gas emissions from ships.[143]

The problem of ocean acidification has also been acknowledged in a number of other regional oceans-related regimes, including the OSPAR Commission for the Protection of the Marine Environment of the North East Atlantic,[144] the CCAMLR,[145] and the Arctic Council.[146] However, short of noting their concerns, calling on their member states to reduce their CO_2 emissions, and adopting revised conservation and management measures that take into account the effect of ocean acidification on regulated stocks and species, there is little these regimes can do to provide for a concerted, coherent response to the problem. The same is true for the CBD. Like the LC/LP, the CBD has imposed a moratorium on ocean fertilisation and other marine geoengineering activities.[147] Its parties have also committed to ensuring, by 2015, the minimisation of 'anthropogenic pressures on coral reefs and other vulnerable ecosystems impacted by climate change or ocean acidification … so as to maintain their integrity and

---

[140] D VanderZwaag and A Powers, 'The Protection of the Marine Environment from Land-Based Pollution Activities: Gauging the Tides of Global and Regional Governance' (2008) 23 International Journal of Marine and Coastal Law 423, 439.

[141] IMO, Second IMO GHG Study 2009 (IMO, 2009); See also *Oceans and Law of the Sea: Report of the Secretary General*, UN Doc A/64/66/Add.1 (2009), para 349.

[142] Adopted in 1997 and entered into force 19 May 2005. Annex VI was revised to provide for even stricter emission limits in October 2008. The revised Annex entered into force on 1 July 2010.

[143] Entered into force on 1 January 2013.

[144] ICES, *Report of the Joint OSPAR/ICES Ocean Acidification Study Group* (SGOA), 7–10 October 2013, Copenhagen, Denmark (ICES CM 2013/ACOM:31 2013).

[145] Report of the Thirty-First Meeting of the Scientific Committee for the Conservation of Antarctic Marine Living Resources (2012) 182–183; See also *Oceans and the Law of the Sea: Report of the Secretary General* (n 141).

[146] Arctic Monitoring and Assessment Programme, *Arctic Ocean Acidification Assessment: Summary for Policy Makers* (AMAP, 2013).

[147] CBD, COP-9 Decision IX/16, Biodiversity and Climate Change (2008), sec. C.

functioning'.[148] Nevertheless, the precise mechanics of how this target is to be met appear to be left to the climate regime.

Neither the law of the sea nor any other international treaty regime acknowledges ocean acidification as a regime focus. This is perhaps not surprising given its causes and wide-ranging effects. Proposals have therefore been made for the adoption of a protocol to the LOSC to deal specifically with the issue of ocean acidification.[149] However, given that the primary cause of ocean acidification is anthropogenic CO_2 emissions, the reduction of which is a primary focus of the climate negotiations, it may be that the most appropriate place to deal with the causes of ocean acidification, if not its effects, is within the climate regime. Nevertheless, given the intractability of those negotiations and the existential threat posed by ocean acidification to the marine environment, its effects must also continue to be addressed, even if they cannot be fully ameliorated, within the law of the sea and other conservation regimes.

7. CONCLUSION

Climate change is now recognised as 'the defining issue of our era',[150] 'a serious and long-term challenge that has the potential to affect every part of the globe',[151] and 'one of the greatest challenges of our time'.[152] With their ability to absorb CO_2, the oceans have long been considered part of the solution to the climate change problem.[153] However, given current and projected thermally induced changes in species abundance and distribution, and the compounding effects of ocean acidification caused by CO_2 absorption, it is increasingly clear that the oceans, and the marine biodiversity contained within them, are as much, if not more, a victim of climate change.

There can no longer be any doubt that climate change and ocean acidification pose major challenges to marine ecosystem resilience. They also pose major challenges to the institutional resilience of national, regional and international regimes for the conservation and management of marine biodiversity. Improved national and global frameworks for policy development and decision-making are already needed to ensure the successful management of existing stressors. Further improvement and coordination between, among and across jurisdictions and regimes, will be critical to achieving the

---

[148] CBD, COP-10, Decision X/2, The Strategic Plan for Biodiversity 2011–2020 (2010); CBD, COP-11, Aichi Biodiversity Targets (UNEP/CBD/COP/DEC/X/2, 29 October 2012), Target 10.

[149] *See*, for example, RE Kim, 'Is a New Multilateral Environmental Agreement on Ocean Acidification Necessary?' (2012) 12 RECIEL 258; V Gonzalez, 'An Alternative Approach for Addressing CO_2-Driven Ocean Acidification' (2010) 12 Sustainable Development Law and Policy 45.

[150] UN News Centre, '"Climate Change defining issue of our era" says Ban Ki-moon, hailing G8 action' 8 June 2007, available at <http://www.un.org/apps/news/story.asp?NewsID=22836#.Vbts5mfbKUl>.

[151] World Summit Outcome 2005, UN Doc A/Res/60/1 (24 October 2005), para 51.

[152] 2009 Copenhagen Accord, para 1, available at: <http://unfccc.int/documentation/decisions/items/3597.php?such=j&volltext=/CP.15#beg>.

[153] Freestone (n 117).

'whole of ocean' approach called for by the IPCC and the effective governance and management of marine biodiversity in a climate-change challenged world.

SELECT BIBLIOGRAPHY

Axelrod, M, 'Climate Change and Global Fisheries Management: Linking Issues to Protect Ecosystems or to Save Political Interests?' (2011) 11(3) Global Environmental Politics 64

Baird, R, M Simons and T Stephens, 'Ocean Acidification: A Litmus Test for International Law' (2009) 4 Climate and Carbon Law Review 459

Craig, RK, 'Marine Biodiversity, Climate Change, and Governance of the Oceans' (2012) 4 Diversity 224

Galland, G, E Harrould-Kolieb and D Herr, 'The Ocean and Climate Change Policy' (2012) 12 Climate Policy 764

Gattuso, J-P and L Hanson (eds), *Ocean Acidification* (OUP, 2011)

Goodwin, EJ, *International Environmental Law and the Conservation of Coral Reefs* (Routledge, 2011)

IGBP, IOC, SCOR, *Ocean Acidification Summary for Policymakers – Third Symposium on the Ocean in a High-CO_2 World* (International Geosphere-Biosphere Programme, 2013)

IPCC, 'Summary for Policymakers' in *Climate Change 2013: The Physical Science Basis. Contribution of Working Group I to the Fifth Assessment Report of the Intergovernmental Panel on Climate Change* (CUP, 2013)

IPCC, 'Ocean Systems', in *Climate Change 2014: Impacts, Adaptation, and Vulnerability. Part A: Global and Sectoral Aspects. Contribution of Working Group II to the Fifth Assessment Report of the Intergovernmental Panel on Climate Change*, chapter 6

IPCC, 'The Ocean', in *Climate Change 2014: Impacts, Adaptation, and Vulnerability. Part A: Global and Sectoral Aspects. Contribution of Working Group II to the Fifth Assessment Report of the Intergovernmental Panel on Climate Change*, chapter 30

IPCC, 'Summary for Policymakers' in *Climate Change 2014: Impacts, Adaptation, and Vulnerability. Part A: Global and Sectoral Aspects. Contribution of Working Group II to the Fifth Assessment Report of the Intergovernmental Panel on Climate Change* (CUP, 2014)

IPCC, 'Coastal Systems and Low-Lying Areas' in *Climate Change 2014: Impacts, Adaptation, and Vulnerability. Part A: Global and Sectoral Aspects. Contribution of Working Group II to the Fifth Assessment Report of the Intergovernmental Panel on Climate Change* (CUP, 2014) chapter 5

Lockwood, M et al., 'Marine Biodiversity Conservation Governance and Management: Regime Requirements for Global Environmental Change' (2012) 69 Ocean and Coastal Management 160

Rayfuse, R, 'Climate Change and the Law of the Sea' in R Rayfuse and S Scott (eds), *International Law in the Era of Climate Change* (Edward Elgar Publishing, 2012) 147

Scientific Committee on Ocean Research and International Oceanographic Committee, 'The Ocean in a High CO_2 World' (2004) 17 Oceanography 72

Turley, C et al., 'Future Biological Impacts of Ocean Acidification and their Socioeconomic-policy Implications' (2012) 4 Current Opinion in Environmental Sustainability 278

6. Broad-spectrum efforts to enhance the conservation of vulnerable marine ecosystems
Edward J. Goodwin

1. INTRODUCTION

This chapter seeks to provide an overarching insight into the initiatives being pursued by international law for advancing the conservation of marine ecosystems that are perceived to be in a vulnerable position. As will be described, humankind's mark upon marine ecosystems is so extensive that arguably there is no part of the ocean that is not vulnerable in some way and to some degree, in the short or long term. This fact demands wide-reaching broad-spectrum action to address the underlying drivers. Nevertheless, the author has observed elsewhere[1] that a necessary complement to this is that scientists and governments need to have systems in place to identify instances of marine ecosystems that are suffering particularly badly, so that targeted treatment can be deployed. Such focused responses are complementary and significant. For instance, international support in response to such events might provide increased capacity – in terms of knowledge, expertise or financing. Furthermore, international engagement in developments might raise the priority given to the conservation of a site in the face of competing claims upon limited national resources.

Whilst the author therefore regards legal initiatives as falling along a continuum, from macro to micro level action, this chapter takes advantage of the opportunity presented to focus upon, and look in more detail at, the broad-spectrum responses of international law. That said, mention will still be made of the systems available to catch emergency cases. As to the macro responses, conservation practices designed to secure ecosystem functioning will be considered – most notably marine protected areas. In addition, the principal drivers of vulnerability will be explored, such as land-based sources of pollution and fishing. Climate change and ocean acidification will only be mentioned briefly, given that more on this is provided in the chapter by Professor Rayfuse. The chapter will also highlight the principal future challenges in this field, as well as highlighting possible subsequent research directions for legal academics.

2. VULNERABLE MARINE ECOSYSTEMS

Although contrasting in their 'depth of field', the unifying theme of macro and micro measures remains a desire to identify vulnerability and to secure the continued

---

[1] EJ Goodwin, 'Threatened Species and Vulnerable Marine Ecosystems' in DR Rothwell and others (eds), *The Oxford Handbook of the Law of the Sea* (OUP, 2015), 799.

provision of marine ecosystem services and values. Therefore, before embarking upon this study, it is important to be more precise in the framing of the subject matter. This will be done by clarifying certain key terms and concepts, whilst also describing the principal drivers of vulnerability.

2.1 Marine Ecosystems

The qualifier 'marine' might be expected to limit consideration to areas of seawater. The problem with this is that to adopt such a course would exclude brackish water systems, such as salt-marshes or mangroves. As a result, this chapter intends to frame its inquiry using Park and Allaby's definitions, which means that reference to marine ecosystems is to those systems operating by the interaction of a community of living organisms in conjunction with non-living components, one of which must be saltwater.[2]

Early life on earth blossomed in such marine ecosystems, and that life in turn drove the development of conditions capable of supporting ever more diverse and complex life forms, both under water and on dry land.[3] Today, marine ecosystems continue to abound with life[4] although some are more diverse than others. For example, coastal ecosystems display the greatest biological diversity in tropical waters, such as those off the coast of the Philippines.[5] Further, shallow warm water coral reefs are known to account for 100,000 species, including 4,000–5,000 species of fish (40 per cent of the known marine fish species) even though such reefs only amount to 0.2 per cent of the ocean.[6]

Healthy, bio-diverse marine ecosystems are best placed to provide maximum long-term benefits for humankind. These beneficial ecosystem services[7] were classified into four categories under the Millennium Ecosystem Assessment.[8] The first of these services was provisioning, which most obviously relates to providing food, but also preservatives such as salt, energy and pharmaceuticals. Indeed, global marine capture

---

[2] This definition combines those for 'ecosystem', 'marine ecosystem' and 'saltwater' in C Park and M Allaby, *A Dictionary of Environment and Conservation* (2nd edn., OUP, 2013), 135, 258 and 378.

[3] *See* C Roberts, *Ocean of Life* (Penguin, 2013), chapter 1.

[4] In 2010 it was estimated that there were 250,000 confirmed species of marine life, with expectations being that there were at least a further 750,000 unconfirmed species; JH Ausubel and others (eds), *First Census of Marine Life 2010: Highlights of a Decade of Discovery* (Census of Marine Life, 2010), 11.

[5] *Ibid*, 15. *See also* T Austin and others, *The Exploitation of Coral Reefs* (Field Studies Council, 1996), 3.

[6] WWF, *The Value of Our Oceans: The Economic Benefits of Marine Biodiversity and Healthy Ecosystems* (WWF Germany, 2008), 15.

[7] Ecosystem services are those conditions and processes that sustain and fulfil human life; G Daily, 'Introduction: What are Ecosystem Services?' in G Daily (ed), *Nature's Services: Societal Dependence on Natural Ecosystems* (Island Press, 1997), 3.

[8] Millennium Ecosystem Assessment, *Ecosystems and Human Wellbeing: A Framework for Assessment* (Island Press, 2003), 5.

fisheries of 79.9 million tonnes were recorded by the Food and Agriculture Organization (FAO) in 2012.[9] In economic terms, such servicing can be significant. In the United Kingdom in 2004, 654,000 tonnes of fish were landed, worth US$1,027 million at first point of sale.[10]

The Millennium Ecosystem Assessment also recognised that there are cultural services provided by ecosystems. Indeed, the variety of environments within which people have settled has driven diverse human cultures and ways of life.[11] Cultural services also extend to education, ecotourism and recreation.[12] Whilst it is difficult to directly attribute a value to these, some choices as to where to spend leisure time can be clearly linked to marine ecosystems. For example, US$625 million was spent in order to dive on Caribbean coral reefs in 2000, whilst whale watching in California was estimated as being worth US$25 million per year in 2008.[13]

Ecosystems provide a third benefit to humankind in that they are significant in regulating many natural processes that support life. Marine ecosystems are particularly influential when it comes to maintaining the earth's climate, water quality (where mangroves help to filter sediments from freshwater outflow into the ocean) and coastal zone protection from storm damage. Indeed, mangroves and coral reefs can absorb 90 per cent of wave energy,[14] and it has been estimated that in the Caribbean the coastal defence services provided by such reefs are worth between US$775 million to US$2.2 billion per year.[15]

The final form of ecosystem service recognised by the Millennium Ecosystem Assessment is that of support. This relates to the way in which an ecosystem is a necessary contributor to (rather than a regulator in) the functioning of another ecosystem through, for example, producing oxygen as a product of photosynthesis, or nutrient recycling.[16]

Placing an accurate economic value on these four marine ecosystem services is difficult, and, for some, objectionable in principle.[17] Nevertheless much effort continues to be expended in producing figures.[18] One significant study published by Robert Costanza and others in 1997 suggested that marine systems account for 63 per cent of the entire service value to humankind provided by all ecosystems.[19] Destroying or diminishing this service is something we simply cannot afford.

---

[9] FAO, *The State of World Fisheries and Aquaculture 2014* (FAO, 2014), 5 available at <http://www.fao.org/3/a-i3720e.pdf>.
[10] WWF (n 6), 14.
[11] Millennium Ecosystem Assessment (n 8), 58.
[12] *Ibid.*, 58–59.
[13] WWF (n 6), 6.
[14] UNEP/WCMC, *In the Front Line: Shoreline Protection and Other Ecosystem Services From Mangroves and Coral Reefs* (UNEP/WCMC, 2006), 5.
[15] WWF (n 6), 25.
[16] Millennium Ecosystem Assessment (n 8), 59–60.
[17] *See* A Gillespie, *International Environmental Law, Policy and Ethics* (2nd edn., OUP, 2014).
[18] For example, 'The Economics of Ecosystem Benefits' initiative available at <http://www.teebweb.org>.
[19] R Costanza and others, 'The Value of the World's Ecosystem Services and Natural Capital' (1997) 387 Nature 253, 259.

2.2 Vulnerability in Ecosystems

Turning to the concept of vulnerability, a suitable starting point is that ecological theory suggests that ecosystems can have alternative stable states.[20] Thus a coral reef can change from one stable state to another if a perturbation of sufficient enormity and type occurs.[21] This event then pushes that ecosystem over the tipping point between the basins within which these stable states exist.[22] Whilst ecosystems might therefore seem perpetual, the concern surrounding such transitions is that the new stable state is impoverished in terms of the quality of the four ecosystem services it delivers, and the richness of biological diversity it supports.

Vulnerability might therefore be regarded as a propensity to shift states when perturbed, in contrast to resilience that results in sliding back to the stable point within the basin in which it originally existed. Given the predominant uncertainty over the nature, quality and value of that alternate state, the precautionary principle would suggest that such transitions ought to be avoided, especially where the perturbation is anthropocentric in origin.

This conception of vulnerability is reflected in some multilateral regimes for the oceans, albeit that it has been predominantly employed in maritime discourse on deep-sea fisheries in areas beyond national jurisdiction ('ABNJ'). Thus the FAO defines vulnerability in this context as:

> ... related to the likelihood that a population, community, or habitat will experience substantial alteration from short-term or chronic disturbance, and the likelihood that it would recover and in what time frame. These are, in turn, related to the characteristics of the ecosystems themselves, especially biological and structural aspects. [Vulnerable Marine Ecosystem] features may be physically or functionally fragile ...[23]

An important component of vulnerability is, therefore, the likelihood that an ecosystem will be forced to shift from one stable state to another. Likelihood, and therefore vulnerability, is consequently a product of ecosystem fragility, the force and/or number of perturbations and the capacity for the ecosystem to arrest its transition to an alternate state because of its resilience, capacity to recover, or humankind's ability to intervene and aid recovery.[24] Thus vulnerability will vary as a product of ecosystem features and factors external to the functioning of that ecosystem. Rather self-evidently, and by way of illustration, deep-water coral reefs might be fragile and slow to regrow if damaged, but if perturbations are either weak or unlikely to occur in a given location, then those ecosystems are not vulnerable.

---

[20] JA Estes and others, 'Trophic Downgrading of Planet Earth' (2011) 333 Science 301, 301.
[21] RC Lewontin, 'The Meaning of Stability' (1969) 22 Brookhaven Symposia in Biology 13, 15.
[22] Estes (n 20).
[23] FAO, *International Guidelines for the Management of Deep-Sea Fisheries in the High Seas* (FAO, 2009), 4.
[24] *Ibid.*, 4.

A further important observation is that there is no reason why such vulnerability should only be encountered (and perhaps only considered relevant) in ABNJ. This jurisdictional division is the product of State consensus rather than one dictated by nature. Furthermore, many coastal ecosystems are known to be vulnerable to alteration when subjected to many of the well-known marine habitat threats described below. This chapter is therefore premised upon the belief that vulnerable marine ecosystems (VME) should be expected to be found, and therefore framed as a concern requiring policy development, in multiple maritime zones. Indeed the UN General Assembly (UNGA) recently refused to limit itself to conceiving of VMEs as an issue restricted to the high seas.[25]

2.3 Vulnerabilities

Marine ecosystems are vulnerable to five well-known perturbations – dubbed the five horsemen of the apocalyptic loss of biodiversity by Callum Roberts.[26] These are: (1) overkill by hunting or fishing; (2) habitat loss; (3) pollution; (4) climate change; and (5) invasive 'alien' species.[27] Acting alone, or more often side by side, these threats have led to, amongst other things, just close to 30 per cent of fish stocks being over-exploited (a proportion that has been increasing since the mid-1970s).[28] Further, because of climate change and ocean acidification, 70 per cent of coral reefs are expected to collapse,[29] even before factoring in other threats to these bio-diverse habitats.[30] Finally, by 2003, between 35 per cent and 86 per cent of mangroves had been lost, principally through clearance for fish farms, urbanisation and coastal landfill.[31]

Two factors make mobilising effort to address these threats particularly difficult. First, much of this goes on out of sight underwater, so gaining public and, consequently, political support for action is a problem. In 1998, 80–90 per cent of coral reefs died in parts of the Indian Ocean due to coral bleaching,[32] yet only marine scientists really raised the alarm.[33] Second, later generations forget the abundance and diversity enjoyed by their predecessors and can, therefore, only reflect upon loss relative to their own experiences. These shifting baselines mean that conservation aims

---

[25] UNGA Res 68/70 (2013), Preamble, [153], [217].
[26] Roberts (n 3), 177.
[27] *Ibid.*
[28] FAO (n 9), 11.
[29] K Frieler and others, 'Limiting Global Warming to 2°C Is Unlikely to Save Most Coral Reefs' (2013) 3 Nature Climate Change 165, 165.
[30] *See further* Secretariat of the Convention on Biological Diversity, *Global Biodiversity Outlook 4* (2014), 78–81 available at <http://www.cbd.int/gbo4/>.
[31] NC Duke and others, 'A World Without Mangroves?' (2007) 317 Science 41.
[32] C Wilkinson, 'Executive Summary' in C Wilkinson (ed), *Status of Coral Reefs of the World: 2000* (Australian Institute of Marine Science Cape Ferguson, 2000), 11.
[33] Roberts (n 3), 2.

and expectations operate at levels that previous generations would find unacceptable, and even if achieved leave marine ecosystems in a relatively impoverished state.[34]

2.4 A Role for Public International Law

The above observations and clarifications on definitions lead to some queries about the appropriate role for international law. International law has traditionally been anticipated to respond where there are high levels of interdependence, trans-boundary impacts from independent State decision-making, uncertainty, and the possibility of mutual gains from interstate cooperation.[35] Such forces certainly underpin a number of treaties mentioned herein and concluded for marine environment purposes. Nevertheless, it is also recognised that ecosystems are confronted by grave threats that endanger the survival of present and future generations.[36] The protection of such fundamental aspects of life on earth is recognised as a 'common concern of humankind'.[37]

As Duncan French explores in detail in his chapter in this handbook, common concern of humankind justifies international laws being adopted for threats that are not trans-boundary, and for ecosystems that are located entirely within the sovereign territory of a State. This is important for VME's such as mangroves, sea grass beds and warm water coral reefs, which are predominantly located within the territorial waters of coastal States. Further, in substance the principle implies a sophisticated form of cooperation: a sense of standing,[38] whereby 'the international community can hold them accountable for compliance with their obligations through institutions such as the Conference of the Parties'.[39] In return, given such ecosystems are enjoyed by all humankind, other States are obliged to assist in those conservation efforts.[40]

We can, therefore, expect international law to have grounds for action to protect all VMEs. However, it must also be recognised that the body of law that contributes to the

---

[34] Published research on historic abundance levels is a small step towards addressing this phenomenon; see for example, C Roberts, *The Unnatural History of the Sea* (Gaia, 2007).

[35] RS Dimitrov and others, 'International Non-regimes: A Research Agenda' (2007) 9 International Studies Review 230, 235; *See also* RO Keohane, *After Hegemony: Cooperation and Discord in the World Political Economy* (Princeton University Press, 1984), 79; OR Young, *International Cooperation: Building Regimes for Natural Resources and the Environment* (Cornell University Press, 1989), 198–202.

[36] *The Implications of the 'Common Concern of Mankind' Concept on Global Environmental Issues (Note of the Executive Director to UNEP, Dr. Mostafa K Tolba, to the Group of Legal Experts Meeting Malta December 13–15, 1990)* [1]–[2] available at <http://www.juridicas.unam.mx/publica/librev/rev/iidh/cont/13/doc/doc27.pdf>.

[37] A Kiss, 'The Common Concern of Mankind' (1997) 27(4) Environmental Policy and Law 244. Although the principle underpins earlier treaties, express recognition was first included in the Convention on Biological Diversity (n 41). The preamble to that treaty affirms 'that the conservation of biological diversity is a common concern of humankind'.

[38] A Boyle, 'The Rio Convention on Biological Diversity' in MJ Bowman and C Redgwell (eds), *International Law and the Conservation of Biological Diversity* (Kluwer, 1996), 40.

[39] P Birnie and others, *International Law and the Environment* (3rd edn., OUP, 2009), 131–132.

[40] As Duncan French says in his chapter in this handbook, '[the] raison d'être of common concern is thus the collective responsibility to act'.

current condition of the marine environment is often motivated by multiple concerns. For example, the 1992 Convention on Biological Diversity (CBD)[41] addresses multiple aspects of biodiversity including access to biotechnology and the equitable sharing of the benefits that arise from the utilisation of genetic resources.[42] The 1972 Convention Concerning the Protection of the World Cultural and Natural Heritage[43] (WHC) was introduced to coordinate international capacity building for the long-term protection of properties dotted around the world that were of such wonder and significance that they were to be preserved for humankind as a whole.[44] In the case of natural sites, this significance could be because they contained 'superlative natural phenomena or areas of exceptional natural beauty and aesthetic importance'.[45] Similarly, the goal of maintaining sites of significance, rather than exclusively because of their parlous condition, can also be discerned under the 1971 Convention on Wetlands of International Importance (Ramsar Convention).[46]

This does not rule out the fact that many treaties, like the CBD, are intended as a response to the numerous human threats to the functioning of ecosystems. It simply means that, given the variety of other objectives pursued by the same MEAs, this chapter must seek to distinguish and extract the rules that tackle the principal drivers that render an ecosystem vulnerable.

3. MULTILATERAL MEASURES

In this section, the general, broad-spectrum responses of international law will be considered. These measures seek to minimise the threats posed by overkill, habitat loss, pollution, climate change and invasive 'alien' species.

3.1 Modern Maritime Zones

The legal reconditioning of the oceans beginning in the mid-twentieth century and culminating in provisions of the 1982 United Nations Law of the Sea Convention (LOSC),[47] can be seen as one of the foundational measures taken towards enhancing the conservation of VMEs. The historical background to this is well known, with the original position being encapsulated in the seventeenth-century writings of Hugo Grotius. Grotius asserted that the sea was *res communis*; in other words incapable of

---

[41] (1992) 31 ILM 818.
[42] CBD, Article 1.
[43] (1972) 11 ILM 1358.
[44] WHC, Preamble. *See also* F Francioni, 'Preamble' in F Francioni and F Lenzerini (eds), *The 1972 World Heritage Convention: A Commentary* (OUP, 2008), 11.
[45] UNESCO, *Operational Guidelines for the Implementation of the World Heritage Convention* (July 2013) (hereafter 'WHC Guidelines'), [77(vii)] available at <http://whc.unesco.org/archive/opguide13-en.pdf>.
[46] (1971) 996 UNTS 245. Ramsar is the Iranian town where the convention text was adopted.
[47] (1982) 21 ILM 1245.

private ownership but open to all.[48] In contrast, he said natural resources (such as fish) were *res nullius*, which meant capable of private ownership once captured.[49] Given the wide acceptance of this position, beyond the maritime areas regarded as sovereign territory of the coastal State (the territorial sea), for much of the twentieth century the model of rights over marine resources was proprietorial.[50] States' jurisdiction only extended to vessels flying their flag, or over appropriated resources. The ensuing conditions made environmental management difficult, and produced conditions ripe for a 'tragedy of the commons'.[51]

Based upon a belief that allocating exclusive jurisdiction to one State would be more effective for managing natural marine resources, the preferred response was for a transition towards, principally, coastal State stewardship of natural resources.[52] However, as explained below, the form of resource allocation ultimately adopted no longer fits neatly into the language of *res nullius* or *res communis*.

The LOSC employs differential settings concerning the jurisdiction and rights of coastal, land-locked and flag States within seven maritime zones, these being internal waters, an expanded territorial sea, contiguous zone, exclusive economic zone (EEZ), continental shelf, the area and the high sea. For example, 40 per cent of the ocean fell within the EEZ. The freedom for any nation to fish in these, the most productive, parts of the sea[53] was thereby replaced, so that almost 90 per cent of commercial fisheries fell within single State control.[54] Access to these resources is no longer open to all, but instead to a group of States selected, and regulated, by the coastal State, with property allocated on an appropriation basis thereafter. However the natural resources were not fully appropriated to the coastal State, which instead had exclusive competence to manage these fisheries, almost as an agent of the international community, subject to (albeit generous) internationally agreed rules and policies.[55] The latter, arguably, therefore has a closer affinity to the common concern of humankind than any of the traditional freedoms associated with the sea.

The handling of high seas natural resources has also changed. The proprietorial principles of *res nullius* have been restrained. The LOSC's designation of resources found in the area[56] as the common heritage of humankind, and the appointment of the International Seabed Authority as manager of these resources, ensured that a principle

---

[48] H Grotius, *De Jure Praedae Commentarius* (GL Williams trans) (OUP, 1950), 231.
[49] *Ibid.*, 232.
[50] AV Lowe, 'Reflections on the Waters: Changing Conceptions of Property Rights in the Law of the Sea' (1986) 1(1) International Journal of Estuarine and Coastal Law 1, 4.
[51] *See* G Hardin, 'The Tragedy of the Commons' (1968) 162(3859) Science 1243.
[52] Lowe (n 50), 9.
[53] The EEZ is an area that a State may elect to claim and which is (like the territorial sea) measured in a seaward direction from its maritime baseline for up to 200 nautical miles; LOSC Article 57. Given the greater range, it is an area beyond and adjacent to the territorial sea; LOSC Article 55.
[54] E Hay, 'The Fisheries Provisions of the LOS Convention' in E Hay (ed), *Developments in International Fisheries Law* (Brill 1999), 27.
[55] Lowe (n 50), 9.
[56] The area means 'the seabed and ocean floor and subsoil thereof, beyond the limits of national jurisdiction'; LOSC Article 1(1).

154 *Research handbook on biodiversity and law*

of non-appropriation prevailed.[57] Further, the adoption of subsequent agreements concerning highly migratory species and straddling fish stocks, alongside provisions and other agreements for named high seas species, also displaced the 'free-for-all' for many marine stocks.[58] Conversely, the conditions for appropriating the remaining high seas species are far less stringent. Nevertheless, the LOSC remains imbued with the language of cooperation and shared responsibility,[59] which again suggests an element of collective common concern for high seas biodiversity, and a transition away from a pure sense of *res nullius*.[60] Nevertheless, with only rudimentary provisions remaining for these high seas resources and in practice poor implementation and enforcement by States, a *de facto* open access regime continues to operate in these waters.[61] Thus open access competition has been concentrated into a smaller area, which has increased the risks to VMEs and threatened species[62] and of a continuing tragedy of the commons. High seas ecosystems therefore remain highly vulnerable to the open competition to appropriate marine resources in these waters.

3.2 Marine Protected Areas and the Reconditioned Maritime Zones

Despite the progressive reallocation of jurisdiction within the seven previously mentioned maritime zones under the LOSC, doubts about competences can be raised, most pertinently those surrounding marine protected areas (MPAs).[63] MPAs are widely regarded as pivotal for the conservation of VMEs,[64] yet it is important to establish whether the LOSC allocated appropriate jurisdiction to deploy them.

3.2.1 Internal waters

In all but the high seas, there appears to be coastal State jurisdiction to introduce MPAs, although this conclusion can only be reached after careful consideration. The internal waters (those areas lying on the landward side of a coastal State's maritime baseline) are treated as akin to that State's land territory. Here, full sovereign powers apply,[65] leading to the freedom to deal with these waters as a State chooses. It is important to note that these coastal State powers are rarely tempered by the rights of others. Unless permitted by prior agreement (usually bilateral), as a general rule vessels from other States may not enter internal waters, nor demand access to a coastal State's

---

[57] *See* further D French's chapter in this handbook.
[58] For coverage, *see* DR Rothwell and T Stephens, *The International Law of the Sea* (Hart Publishing, 2010), 303–319.
[59] *See* LOSC, Article 87(2), Part VII, Section 2.
[60] *See further* Rothwell and Stephens (n 58), chapter 19.
[61] R Barnes, 'The Convention on the Law of the Sea: An Effective Framework for Domestic Fisheries Conservation?' in D Freestone, R Barnes and D Ong (eds), *The Law of the Sea: Progress and Prospects* (OUP, 2006), 240.
[62] *Ibid.*, 241.
[63] *See further* the 1946 International Convention for the Regulation of Whaling's authority to set quotas for whales within EEZs; A Gillespie, *Whaling Diplomacy: Defining Issues in International Environmental Law* (Edward Elgar Publishing, 2005), 289–294.
[64] Roberts (n 34), 377.
[65] LOSC, Article 2(1).

ports. Indeed such vessels may only enter a port when the vessel is in distress and there is a danger to human life in accordance with customary law.[66]

Thus the legal condition of the internal waters of a coastal State imposes no real practical or legal constraints upon the authority and ability to designate and manage an MPA. This is significant for certain vulnerable ecosystems, such as mangroves, and coastal fringing reefs. The same, however, is not the case as a coastal State looks to deploy MPAs on the seaward side of its baseline.

3.2.2 The territorial sea

The territorial sea[67] allocates sovereignty to the coastal State over the sea, airspace, seabed and subsoil.[68] This authority is only subject to obligations imposed upon the State under international law (treaty or customary), and also the freedom of other States to engage in innocent passage through such waters. Passage is restricted to navigation through the territorial sea in order to traverse it, or to travel to and from the internal waters, and it must be continuous and expeditious.[69] Such passage will not be innocent if, amongst other things, it includes fishing, research and surveying activities, as well as wilful or serious pollution, and unloading people against coastal State regulation.[70] Indeed, the LOSC limits innocent passage to activities having a direct bearing upon passage as such.[71]

The act of designating an MPA is, therefore, arguably permitted pursuant to the coastal State's sovereign powers over the territorial sea, whilst in practice the activities likely to be targeted during management, and therefore regulated, do not form a fundamental part of innocent passage i.e. fishing, prospecting, tourism. What cannot, apparently, be unilaterally managed is navigation as passage simpliciter. Nevertheless, the LOSC describes the laws and regulations that coastal States may still enact with respect to the innocent passage of vessels. In particular, such laws and regulations may relate to the conservation of the living resources of the sea, prevention of the infringement of the fisheries laws and regulations, and the preservation of the marine environment of the coastal State.[72] Fabio Spadi suggests that these provisions are enough to entitle the coastal State to exclude or limit navigation in particular areas of the territorial sea.[73]

A further basis for effectively excluding vessels may exist under LOSC Article 22, which allows coastal States to require vessels to engage in innocent passage through

---

[66] *See* for example, *Kate A. Hoff Case (the Rebecca)* (1929) IV RIAA 444, 447. One further limited exception is provided in LOSC Article 8(2), according to which vessels may pass through internal waters where a baseline drawn to reflect a coastline characterised by heavy indentation or fringing islands encloses waters that were not previously regarded as internal.

[67] A zone extending 12 nautical miles in a seaward direction from the coastal State's baseline.

[68] LOSC, Article 2.

[69] LOSC, Article 18.

[70] LOSC, Article 19(2).

[71] LOSC, Article 19(2)(l).

[72] LOSC, Article 21(1).

[73] F Spadi, 'Navigation in Marine Protected Areas: National and International Law' (2000) 31 Ocean Development and International Law 285, 289.

defined sea lanes in order to ensure safety of navigation. In designating these sea lanes, the coastal State is not entitled to ultimately hamper the innocent passage of vessels[74] and must take account of the factors listed in Article 22(3), e.g. any recommendations with a bearing upon sea lanes from the International Maritime Organization (IMO), and the density of traffic which will use the route.[75] The coastal State may not, therefore, designate sea lanes so as to exclude passage throughout the breadth of the territorial sea, and strictly speaking the aim of deploying sea lanes should be to protect vessels from danger caused by VMEs, rather than vice versa. Nevertheless, IMO has increasingly adopted routing measures beyond the territorial waters for environmental reasons and this could encourage coastal States to adopt a similar approach in the territorial sea, where the rights of third-party States are comparatively weaker.[76]

3.2.3 The EEZ

The EEZ differs significantly from the internal and territorial waters in that here the coastal State does not enjoy territorial sovereignty. Instead, it exercises sovereignty and jurisdictional competence over particular issues or activities. Thus the LOSC in Article 56 initially provides that the coastal State exercises full sovereign rights over:

> exploring and exploiting, conserving and managing the natural resources, whether living or non-living, of the waters superjacent to the seabed and of the seabed and its subsoil, and with regard to other activities for the economic exploitation and exploration of the zone, such as the production of energy from the water, currents and winds.[77]

This seems to cover a number of activities beyond the obvious exercise of sovereign rights over the fisheries of the EEZ. For example, tourist activities focused upon interest in marine ecosystems, such as coral reefs, represent a form of exploitation of both the living and non-living natural resources. The LOSC goes on to allocate jurisdiction to the coastal State within the EEZ over particular issues which are dealt with under other parts of the convention. Specifically and under Article 56(1)(b), the coastal State exercises jurisdiction over scientific research, as well as over the protection and preservation of the marine environment. Taken together the LOSC appears to afford the coastal State full sovereign rights over a number of activities that MPAs might seek to manage or control. But can a coastal State deploy MPAs as part of managing these activities?

The LOSC only specifically refers to the creation of MPAs in the EEZ in Article 211(6). This article allows for the creation of MPAs with the approval of IMO in special circumstances where general rules and standards for reducing, controlling and preventing pollution from vessels are deemed inadequate for a given area of the EEZ. This provision is, of course, restricted to the problem of vessel-source pollution. As noted earlier, MPAs are intended to manage a far more diverse range of activities and to meet other aims beyond simply pollution control, prevention and reduction. The question therefore becomes whether or not the LOSC gives coastal States the authority

---

[74] LOSC, Article 24.
[75] LOSC, Article 22(3).
[76] Spadi (n 73), 290.
[77] LOSC, Article 56(1)(a).

to designate MPAs in order to control resource exploitation, tourist activity and scientific research or to generically protect VMEs.

Article 56 (referred to above) does not seek to restrain coastal States by prohibiting particular management techniques for controlling the activities described, and seems to implicitly approve of enclave approaches in Article 62, which requires all other States to respect a coastal State's laws and regulations which regulate specific areas in which fishing is permitted or prohibited.[78] Further, and more generally, Article 56(1)(b)(iii) provides that jurisdiction over the protection and preservation of the environment lies with the coastal State. The State thus has the authority and responsibility to enact and enforce legislation for protecting and preserving the environment of the EEZ. By the same token, jurisdiction over scientific research within the EEZ has been allocated to the coastal State. It therefore seems to the author that the LOSC gives the coastal State jurisdiction over environmental protection and preservation as well as scientific research, in such a manner that the coastal State could choose to enact legislation which creates MPAs in the EEZ.

There is, however, one major restraint upon this authority enshrined in Article 58, which is that its exercise must not affect the freedoms of other States to engage in navigation (a notably wider activity than innocent passage), over-flight and the laying of submarine cables. This does not outlaw the establishment of MPAs, but may have serious repercussions for their management. As Tullio Scovazzi says: 'It could thus be asked whether there is any use in establishing a protective regime for an area where a particularly fragile marine ecosystem is located, if foreign super tankers or ships carrying hazardous wastes are expected to move around the area.'[79]

It remains generally accepted that coastal States may not unilaterally prohibit navigation within particular areas of the EEZ. This is generally reflected in State practice.[80] Nevertheless, the freedom of navigation may be restricted, though only with the consent of IMO. This is based on two arguments, the first of which has already been noted. Coastal States have the power to establish MPAs in the EEZ pursuant to Article 211(6), i.e. where IMO agrees that the preconditions relating to vessel-source pollution prevention, control and reduction have been met. The State may then adopt rules and regulations for the area, including those regulating navigational practices.

---

[78] LOSC, Article 62(4)(c).

[79] T Scovazzi, 'Marine Specially Protected Areas Under International Law' in T Scovazzi (ed.), *Marine Specially Protected Areas* (Kluwer, 1999), 17, 18.

[80] For a snapshot of national legislation reflecting the attitudes of States towards legislating in a manner which restricts or prevents navigation, *see* Spadi (n 73), 286–289. *See also contra* such an assumed and practised restraint upon coastal State authority, Scovazzi who argues that the supremacy of freedom of the seas as formulated in the seventeenth century must now be balanced with interests that have a collective character since they belong to the international community. The protection of the environment and sustainable development are such interests. He therefore concludes that there can be no predetermined solution to the conflict between the interests of States exercising their freedoms, and a coastal State's wish to create and manage an MPA. Factors such as the fragility of the ecosystem in question as well as the practical disruption that would be caused to navigation, he suggests, would lie at the heart of determining the appropriate balance between these competing interests; *ibid.*, 18–20.

The second mechanism involves having an area recognised by IMO as a Particularly Sensitive Sea Area (PSSA), as has already happened around the Florida Keys, Great Barrier Reef and the Galapagos.[81] IMO may declare PSSAs if a site meets criteria, including containing habitat that is fragile and susceptible to degradation.[82] An additional requirement is that the values of an area must be vulnerable to damage from international shipping activities.[83]

It is possible that designation as a PSSA alone might be valuable,[84] for example in heightening awareness of vulnerable ecosystems whilst navigating in a particular area of the sea. However, associated protective measures (APMs) are key to delivering a wider range of management measures in order to reduce the risk to an area from international shipping. The difficulty is that such APMs must be approved separately from the recognition of an area as a PSSA, albeit often in parallel as part of the same application. The latter poses some obstacles to seeking PSSA status. Foremost is that there must be a legal basis for IMO's adoption of these measures, which means the APM:

(a) is available under an existing IMO instrument (e.g. routing measures under safety at sea regulations);
(b) could become available if IMO were to amend or adopt an instrument (in such instance the APM only applies once IMO makes such amendments or adopts such an instrument); or
(c) could be adopted by a coastal State in its territorial sea or under LOSC Article 211(6).[85]

Legal basis (c) is by far the most wide ranging. Whilst, admittedly, Article 211(6) is limited to preventing pollution[86] from vessels in the EEZ, the reference to the territorial sea in (c) offers more scope. It has already been noted that Article 21(1) enables the coastal State to subject navigation in such territorial waters to regulation in order to, inter alia, preserve the environment. As Markus Kachel therefore notes, the PSSA may draw upon such authority as a legal basis, irrespective of the maritime zones it actually covers. Thus a basis for regulation that would be available within the territorial sea can be used to control navigation within a section of a PSSA that falls within the EEZ.[87]

---

[81] For details of these and other PSSAs, *see* <http://pssa.imo.org/#/globe>.
[82] The conditions for establishing a PSSA fall into three categories: (i) ecological, (ii) social, cultural and economic, and (iii) scientific and educational criteria: IMO Res 982 (24) (2006), [4].
[83] *See further* MJ Kachel, *Particularly Sensitive Sea Areas* (Springer, 2008), chapter 7, and T Henriksen, 'Conservation of Marine Biodiversity and the International Maritime Organization' in C Voigt (ed), *Rule of Law for Nature* (CUP, 2013), 331.
[84] Kachel, *ibid.*, 179–181.
[85] IMO Res 982 (24) (2006), [7.5.2.3].
[86] It is, of course, possible (as explained at the start of Section 3.4 below) to give pollution a rather wider interpretation than many academics or stakeholders realise, going beyond common perceptions of, for example, oil spill incidents or dumping, to include bottom trawling, anchor damage from ships, blast fishing techniques and the introduction of sound energy.
[87] Kachel (n 83), 187–188.

Despite the fact that this unlocks the potential to control navigation through the EEZ with the assistance of IMO, making designation of MPAs possible legally and practicably, Tore Henriksen has revealed resistance to compulsory routing measures in decisions concerning APMs attached to PSSAs, the preference being for recommendatory APMs.[88] This clearly has the potential to undermine the effectiveness of management initiatives.

3.2.4 The high seas

Establishing MPAs in the high seas is difficult and an important emerging challenge for the law of the sea. Here MPAs currently need to be pursued in multilateral settings. Such multilateral efforts fall into two groupings. First, there are functional groupings. Thus fisheries management falls under the responsibility of FAO and regional fisheries management organisations (RFMOs), whilst seabed mineral exploitation is part of the purview of the International Seabed Authority. Finally, international shipping is regulated by IMO.

Second, and beyond functional groupings, efforts have been channelled through the various regional seas initiatives. However, gaps and problems persist. The regional seas groupings do not cover all regions and rarely areas of the high seas.[89] Even where they do have jurisdiction over the high seas, initiatives to declare MPAs may be circumscribed by State consensus to exclude fishing and maritime transport regulation, presumably in deference to functional division of competences assigned to other bodies.[90] Therefore under current legal conditions, establishing enclaves in ABNJ requires complex coordination between regional and functional bodies, and even then leaves certain gaps in terms of coverage.[91]

Significantly, in 2013 UNGA resolved to begin work on formulating a proposal on the scope, parameters and feasibility of an international instrument on the conservation and sustainable use of marine biodiversity in ABNJ.[92] A series of meetings organised by the designated Ad Hoc Open-ended Informal Working Group in April and June 2014 discussed a variety of issues under this mandate but they agreed that one issue needing to be addressed in negotiations for an agreement included area-based management, including MPAs. Already identified as a potential stumbling block is determining who may declare MPAs in the high seas.[93] With resistance to a new institutional body being

---

[88] Henriksen (n 83), 338–342. For further discussion on controlling navigation in the territorial sea and the EEZ, see J Roberts, 'Protecting Sensitive Marine Environments: The Role and Application of Ships' Routing Measures' (2005) 20(1) International Journal of Marine and Coastal Law 135.

[89] KM Gjerde and A Rulska-Domino, 'Marine Protected Areas Beyond National Jurisdiction: Some Practical Perspectives for Moving Ahead' (2012) 27 International Journal of Marine and Coastal Law 351, 356–358.

[90] 1992 Convention for the Protection of the Marine Environment of the North-East Atlantic (1993) 32 ILM 1068 (OSPAR), Annex 5, Article 4.

[91] *Ibid.*; KN Scott, 'Conservation on the High Seas: Developing the Concept of High Seas Marine Protected Areas' (2012) 27 International Journal of Marine and Coastal Law 849.

[92] UNGA Res 68/70 (2014), Preamble, [197]–[201].

[93] UNGA A/69/177, *Letter dated 25 July 2014 from the Co-Chairs of the Ad Hoc Open-ended Informal Working Group to the President of the General Assembly*, [64].

160   *Research handbook on biodiversity and law*

evident from State representatives participating in these early talks, it was recognised that coastal States did not have authority to declare enclaves, which left flag States with the exclusive jurisdiction to declare such MPAs.[94] If an institutional body were to be created for this purpose, then MPAs designated by this body would only bind those States party to any future agreement.[95] The working group ultimately came to the conclusion that an implementing agreement under LOSC should be negotiated and proposed a timetable for UNGA to adopt; the negotiation phase being projected to begin after 2017.[96]

Additionally, the CBD has initiated a programme for 'Ecologically or Biologically Significant Sea Areas' in ABNJ aimed at guiding the international community in identifying such sites, thus facilitating designation of MPAs.[97] One criterion for selecting suitable sites for protection is the presence of VMEs.[98] At the most recent CBD COP, the results of regional workshops to explore methodologies, and ultimately identify such ecologically and biologically significant marine areas, were recorded.[99] This list of sites and methodology is being made available and recommended for use by the aforementioned Ad Hoc Open-ended Working Group; the CBD's initiative being viewed as complementary to these significant developments in UNGA.[100] Therefore both initiatives might ultimately make enclaves in the high seas more practicable, although this is certainly an area of the law undergoing significant change requiring monitoring and further research.

3.3 Unsustainable Fisheries

Roberts reports that overfishing is

> one of the biggest soluble environmental problems in the world. We know what must be done and it takes no more than a sentence to say it. We have to fish less, waste less, use less destructive methods to catch what we take, and provide safe havens ... So why hasn't it happened?[101]

In large measure these solutions have not happened due to political failures, for example where scientific advice is ignored when setting catch limits in order to appease

---

[94] *Ibid.*
[95] *Ibid.*
[96] *Recommendations of the Ad Hoc Open-ended Informal Working Group to study issues relating to the conservation and sustainable use of marine biological diversity beyond areas of national jurisdiction to the sixty-ninth session of the General Assembly*, 23 January 2015 available at <http://www.un.org/Depts/los/biodiversityworkinggroup/biodiversityworkinggroup.htm>.
[97] CBD, Decision VIII/24.
[98] CBD, Decision IX/20.
[99] CBD, Decision XII/22.
[100] *Ibid.*, [9].
[101] Roberts (n 3), 278.

industry and meet political goals.[102] Nevertheless, criticisms can also be levelled at international law.

It has already been noted that the law of the sea during the twentieth century made important changes to the condition of the oceans, whether by way of ever more sophisticated zoning of the sea or evolution from seventeenth-century concepts concerning ownership of marine resources. One of the most significant moves was to allocate to coastal States authority and control over fisheries management through the enlargement of the territorial sea or formal recognition of the EEZ and archipelagic waters. This eliminated multi-State conflict and complexity by giving one State the decision-making authority and competence over all marine resources and habitats. Whilst this has curtailed multi-State freedom of fishing and therefore, supposedly, helped to avert the environmental problems associated with the 'tragedy of the commons', the extension of jurisdiction has so far failed to prevent over-exploitation of marine resources. This has been known to cause disturbances of magnitudes that can lead to ecosystem instability,[103] as well as direct damage to marine habitats. Some of the factors behind this are explored below.

3.3.1 Archipelagic and coastal State authority

Within the internal waters, territorial sea and any archipelagic waters, the coastal State exercises territorial sovereignty. With the exception of navigational rights and the traditional fishing rights of States neighbouring archipelagic nations,[104] this means that coastal and archipelagic States enjoy full authority to enact and enforce their own fisheries laws and regulations. Access by foreign enterprises requires authorisation.

There are few constraints upon the coastal and archipelagic States' fisheries policies in these zones. In contrast to the EEZ, there is no obligation under the LOSC demanding that the State conserve or optimally utilise the resources in this area.[105] It is true that the obligation under Article 192 to 'protect and preserve the marine environment' would apply in this context,[106] but the generality of this provision does little to guide and contain conduct without further elaboration. As a result, the only other international legal constraints placed upon these States are those that flow from voluntary ratification or accession to environmental treaties, such as the CBD, although these provide few specifics as to what sort of catch levels and gear should be deployed.

Moving away from the territorial sea and archipelagic waters, it has already been observed that, with the recognition of the EEZ, the coastal State was allocated sovereignty over the exploitation, exploration, conservation and management of marine living resources.[107] This left the coastal State to set allowable catch levels within the

---

[102] DG Webster, 'The Irony and the Exclusivity of Atlantic Bluefin Tuna Management' (2011) 35 Marine Policy 249; Roberts (n 3), 51.
[103] *See*, for example, the experience of Jamaica in JW Nybakken and MD Bertness, *Marine Biology: An Ecological Approach* (6th edn., Benjamin Cummings, 2005), 453.
[104] LOSC, Article 51(1).
[105] Hay (n 54), 20; C de Klemm, 'Fisheries Conservation and Management and the Conservation of Marine Biological Diversity' in E Hay (ed), *Developments in International Fisheries Law* (Kluwer, 1999) 423, 433.
[106] SB Kaye, *International Fisheries Management* (Kluwer, 2001), 90–91.
[107] LOSC, Article 56.

162 *Research handbook on biodiversity and law*

zone,[108] subject to partial constraint. More specifically they must, inter alia, take proper conservation and management measures to ensure the resources are not over-exploited and populations of harvested species are maintained or returned to a level that supports maximum sustainable yield (MSY).[109]

In addition, the coastal State is supposed to give access to fishers from other countries if their own needs do not exhaust the annual catch limit.[110] However, this need not create undue demands or undermine any environmental initiatives voluntarily pursued by a coastal State. As Burke has observed, 'allowable catch does not mean ... a quantity of fish determined by objective, scientific criteria, leading to the maximum harvest of fish'.[111] Article 61(3) allows catch levels to reflect economic and environmental factors, such that it has been observed that the coastal State has so much discretion that practically any level of catch can be legitimately set.[112] And of course, setting limits according to broader environmental policies is supported by the sovereign right of States in the EEZ to implement such policies in this zone.[113]

Nevertheless, there is little international legal control of fisheries policy unless the coastal State undertakes additional duties under other MEAs. Additional treaties and soft law guidelines have followed on from the LOSC provisions on fishery regulation but reservations exist surrounding their provisions, as explained below.

3.3.2 TACs

After this realignment of jurisdiction over natural marine resources, it has already been mentioned that under the LOSC there was also a setting of objectives for managing these resources in the EEZ and high seas. Thus in the EEZ and high seas, States (coastal or flag) were to formulate conservation plans or TACs (Total Allowable Catch) for their vessels that would maintain or return stocks to levels that could support MSY.[114] This reflected persistence with a well-established concept even though in 1982 MSY was already subject to criticism.[115] Since then humankind's understanding of the marine environment has developed further. Ecosystem thinking has come to define approaches to biodiversity conservation.[116] Focusing on just the recruitment and mortality rate of a stock of one species is now considered erroneous as it misses the inter-connected nature of life in the oceans. It ignores the effect on predators, species caught as by-catch and the knock-on effect for ecosystem functioning and human

---

[108] LOSC, Article 61(1).
[109] LOSC, Articles 61(2) and 61(3).
[110] LOSC, Article 62.
[111] WT Burke, *The New International Law of Fisheries: UNCLOS 1982 and Beyond* (OUP, 1994), 44.
[112] RR Churchill and AV Lowe, *The Law of the Sea* (3rd edn., Manchester University Press, 1999), 289.
[113] *See* Section 3.2.3.
[114] LOSC, Articles 61(3), 119(1)(a). For a definition of MSY *see* P Birnie, A Boyle and C Redgwell, *International Law and the Environment* (3rd edn., OUP, 2009), 590–591.
[115] *Ibid.*
[116] CBD, Decision VII/11.

communities.[117] Moving fisheries regulation away from the narrow focus of MSY has been a challenge.

It is not readily apparent whether the drafting of the LOSC allows for the introduction of ecosystem management of marine fisheries. As has already been mentioned, Article 61(3) qualifies MSY calculations in the EEZ by the inclusion of additional variables, these being 'environmental and economic factors, including economic needs of fishing communities and the special requirements of developing States'. Further, under Article 61(4) the effects on associated or dependent species must be considered. Similar provisions can be found covering high seas stocks.[118] Barnes doubts that this is sufficient to demand ecosystem-sensitive catch levels, however, since the unconstrained qualifications might justify higher TACS than MSY, and are in any event only to be taken into account, suggesting their subordination to use.[119] Ultimately the convention remains species led, rather than conditioned by the viability of ecosystems.[120]

Following the adoption of the CBD in 1992, the ecosystem approach has gained greater prominence. Subsequent fisheries agreements and soft law guidelines reflect this. The 1995 Fish Stocks Agreement (FSA)[121] replicates the LOSC measures but also injects a precautionary approach into the management of (predominantly) high seas fish stocks and highlights the role of excess capacity in overfishing.[122] Ecosystem management is also woven into the management of fisheries through Article 5, which provides that States 'should adopt, where necessary, conservation and management measures for species belonging to the same ecosystem', although the qualification 'where necessary' seems to subordinate the viability of ecosystems once again.[123]

The non-binding FAO *Code of Conduct for Responsible Fisheries* (1995)[124] (FAO Code of Conduct) also demands attention since it seeks to offer guidance applicable to domestic fisheries, as well as those operating in the EEZ and high seas. The code sets as the goal of management maintaining or restoring stocks to MSY as qualified by environmental and economic factors, whilst enumerating these factors so as to include: avoiding excess fishing capacity; that economics serve to promote responsible fishing rather than undermine it; duly protecting fishers, habitats and ecosystems as well as endangered species; and minimising by-catch and impacts on other species.[125]

The above suggests that global multilateral fisheries regulation is showing increasing awareness of more holistic approaches to TACs, but the obvious problem with the

---

[117] R Hilborn, *Overfishing: What Everyone Needs to Know* (OUP, 2012), 117.
[118] LOSC, Article 119.
[119] Barnes (n 61), 243–244.
[120] *Ibid.*
[121] 1995 Agreement for the Implementation of the Provisions of the United Nations Convention on the Law of the Sea of 10 December 1982 Relating to the Conservation and Management of Straddling Fish Stocks and Highly Migratory Fish Stocks (1995) 34 ILM 1542.
[122] Barnes (n 61), 247.
[123] *Ibid.*, 248–249; there is some debate as to whether these terms implement the LOSC or amend it, with a bearing upon those States that have not become a party to the FSA.
[124] FAO (Fisheries and Aquaculture Department), *Code of Conduct for Responsible Fisheries* (1995), available at <http://www.fao.org/docrep/005/v9878e/v9878e00.htm>.
[125] *Ibid.*, Articles 7.2.1, 7.2.2.

current position is that gaps continue to exist in coverage. For instance, non-migratory marine resources in the high seas receive no additional legal protection beyond the LOSC. Incorporating ecosystem management is thus largely dependent upon either political will or regional multilateral agreements.

With respect to the latter, regional fisheries management organisations (RFMOs) both pre- and post-date the FSA and FAO Code of Conduct. Further, they either look to manage multiple fish stocks (straddling, migratory and/or confined to the high seas) within a geographical zone, or focus upon a single species. Problematically, MSY persists, particularly in early arrangements, and an effort has been made to initiate substantive reviews to bring these agreements in line with best practice.[126] More recent agreements attempt realignment, with varying success. The 2012 Convention on the Conservation and Management of High Seas Fisheries Resources in the North Pacific Ocean[127] calls for individual or collective actions to ensure stocks are capable of producing MSY, albeit stating that account needs to be taken of dependent species.[128] The same convention sets as its objective the long-term conservation and sustainable use of fisheries whilst protecting the marine ecosystems in its area, and certainly contains more provisions drawing attention to adopting an ecosystem approach and impacts upon VMEs of fishing methods.[129] Nevertheless, it leaves the tension and priority between MSY and the ecosystem approach unresolved.

Notable older and recent exceptions to MSY targets can be found. The 1980 Convention on the Conservation of Antarctic Marine Living Resources pursues rational use rather than MSY and an ecosystem approach to conservation.[130] The 2009 Convention on the Conservation and Management of High Seas Fishery Resources in the South Pacific Ocean[131] sets itself the objective of the long-term conservation and sustainable use of fisheries but specifically through the application of a precautionary and ecosystem approach.[132] MSY is notably absent from the text, with, instead, fishing being undertaken in a manner commensurate with the sustainable use of resources.[133] As if to emphasise the point: 'An ecosystem approach shall be applied widely to the conservation and management of fishery resources through an integrated approach under which decisions in relation to the management of fishery resources are considered in the context of the functioning of the wider marine ecosystems in which

---

[126] *See*, for example, the account of NAFO in MJ Bowman, PGG Davies and C Redgwell, *Lyster's International Wildlife Law* (2nd edn., CUP, 2010) ('*Lyster*'), 135–139.

[127] For the adopted text *see* the President of the USA's transmission of the text to the Senate on 22 April 2013, available at <http://docs.lib.noaa.gov/noaa_documents/NOAA_related_docs/Fish_Wildlife/Convention_conservation_management_high_seas_fisheries_td113_2.pdf>.

[128] *Ibid.*, Article 3(b).

[129] *Ibid.*, Articles 2, 3(c) and 3(e).

[130] Convention on the Conservation of Antarctic Marine Living Resources (1980) 19 ILM 837, Article II.

[131] Text currently available via United Nations Treaty Series Collection Online <https://treaties.un.org/doc/Publication/UNTS/No%20Volume/50553/Part/I-50553-0800000280363a44.pdf>.

[132] *Ibid.*, Article 2.

[133] *Ibid.*, Article 3(1)(a)(ii).

they occur …'.[134] Recent developments therefore suggest some hope for appropriate TAC setting, but regional practice remains inconsistent.

3.3.3 Methods

Bottom fishing methods, i.e. those where fishing gear is likely to have contact with the seabed as part of normal operation (e.g. trawling), are well-known threats to VMEs. The LOSC's general obligations to protect and preserve the marine environment,[135] and to take measures to protect and preserve rare or fragile ecosystems,[136] when coupled to the above provisions regarding the precautionary and ecosystem approaches, form a rudimentary framework for the control of fishing methods.

Beyond this the FAO Code of Conduct calls for fishing to be managed so that ecosystems are conserved, and that by-catch of fish and non-fish species are minimised.[137] The Code goes on to suggest this might include phasing out gear that is inconsistent with responsible fishing, gear restrictions or closed areas.[138] Whilst the code has enjoyed positive reviews about its influence there is little to suggest that this has extended beyond policy development tackling dynamite fishing and poisoning, so as to include trawling.[139]

Efforts have been made within UNGA to ban the destructive practice of bottom trawling, particularly in the high seas where seamounts and deep-water coral ecosystems catalyse life. However, one such initiative was blocked by a handful of fishing States in 2006.[140] A resolution adopted in its place reminds States to protect VMEs, including seamounts, hydro-thermal vents and cold-water corals, from destructive fishing methods, and calls upon RFMOs to work on identifying VMEs, assess the impacts of bottom fishing upon them, and to close areas of VMEs until suitable management can be introduced.[141] To its credit, this resolution does, therefore, attempt to introduce a greater element of precaution into bottom trawling activities.

There is evidence that the above-mentioned resolution's approach is, at least, being incorporated. For example, the Commission to be established under the North Pacific Ocean agreement mentioned previously is required to:

> adopt conservation and management measures to prevent significant adverse impacts on vulnerable marine ecosystems in the Convention Area, including but not limited to:

---

[134] *Ibid.*, Article 3(2)(b).
[135] LOSC, Article 192.
[136] *Ibid.*, Article 194.
[137] FAO (n 124), [7.2.2].
[138] *Ibid.*, [7.6.4], [7.6.9].
[139] FAO, Fisheries and Aquaculture Department, *Circular No. 1038* (2009), 22.
[140] BBC News, 'Ban on Brutal Fishing Blocked' *BBC Online* (London, 24 November 2006), available at <http://news.bbc.co.uk/1/hi/sci/tech/6181396.stm>. With debate being held in closed sessions, it is difficult to uncover the positions taken in opposition to a well-known cause of habitat loss. Ultimately, the requirement for consensus was enough for a tiny number of States to prevent the interim ban's adoption. It is thought those States who were opposed were Russia, Iceland, South Korea, China and Japan; 'Blame Iceland' *Washington Post* (Washington, 3 December 2006), available at <http://www.washingtonpost.com/wp-dyn/content/article/2006/12/02/AR2006120200937.html>.
[141] UNGA Res 61/105 (2007), [80], [83]; *see also* UNGA Res 64/72 (2010), [112]–[130].

166 *Research handbook on biodiversity and law*

(i) measures for conducting and reviewing impact assessments to determine if fishing activities would produce such impacts on such ecosystems in a given area;
(ii) measures to address unexpected encounters with vulnerable marine ecosystems in the course of normal bottom fishing activities; and
(iii) as appropriate, measures that specify locations in which fishing activities shall not occur.[142]

Clearly in the absence of a simple ban on trawling, identifying VMEs at risk becomes the priority. As noted in the above provision, scientific methods need to be developed which are operable by fishing boats. One central approach developed by RFMOs has been closed areas and to introduce move-on rules. For example, the North Atlantic Fishery Organisation closed the Newfoundland, New England and Corner Seamounts in 2006, along with a number of coral and sponge protection zones in 2007.[143] Further, a number of RFMOs require fishing boats to stop trawling and move on two nautical miles upon catching 60 kg of corals or 800 kg of sponges.[144] But problems persist. Move-on rules, it has been argued, are set too high, and do not reveal soft-bodied species, which are pulverised and extruded in nets before reaching the surface.[145] Furthermore, the precautionary approach is often preferable and this demands that impact assessments be conducted before fully opening an area to fishing.[146]

3.4 Pollution

Responding to concerns about pollution of the sea, the LOSC begins by defining marine pollution as 'the introduction by man, directly or indirectly, of substances or energy into the marine environment, including estuaries, which results or is likely to result in such deleterious effects …'.[147] The principal types of pollution[148] and those most often implicated in ecosystem disturbance are: sewage, industrial waste and agricultural run-off (referred to as a group as land-based sources of pollution (LBSP)); alien invasive species; and carbon dioxide, which is still predominantly of land-based origin, although shipping remains a contributor.[149] These will be considered in subsequent sections. Before doing so, there is merit in pausing to note the potential width of the LOSC definition beyond the obvious examples of oil spill incidents or dumping. David Ong suggests that the kinetic energy released if a vessel strikes a

---

[142] (n 127), Article 7(1)(e).
[143] NAFO FC/DOC.06/5 (September 2006); FC/DOC.07/18 (September 2007).
[144] PJ Auster and others, 'Definition and Detection of Vulnerable Marine Ecosystems on the High Seas: Problems with the "Move-On" Rule' (2011) 68(2) ICES Journal of Marine Science 254, 258.
[145] *Ibid.*
[146] *Ibid.*, 261.
[147] LOSC, Article 1(4).
[148] Group of Experts on the Scientific Aspects of Marine Environmental Protection, *The State of the Marine Environment: UNEP Regional Seas Reports and Studies No. 115* (UNEP, 1990), [363].
[149] In 2007 shipping contributed 3.3 per cent of global CO_2 emissions; *see* IMO, Ø Buhaug et al, *Second IMO GHG Study* (2009), 6 available at <http://www.imo.org/blast/blastDataHelper.asp?data_id=26046&filename=4-7.pdf>.

marine habitat arguably leads to such incidents satisfying this definition of pollution, which simply requires the introduction by man of energy into the environment with deleterious effects.[150] If this is so, then a wider range of activities which cause damage to VMEs would meet the definition. Interestingly, bottom trawling could then be seen as a type of collision between a man-made structure deliberately introduced into the marine realm and deep-sea reefs. To the author's knowledge, though, this is not an argument that has been pursued.

3.4.1 LBSP

LBSP create significant problems for coastal habitat. For example, shallow coral reef ecosystems exist in a state of careful balance between coral species and marine algae. Corals can only maintain a competitive edge over algae when nutrient levels are low.[151] As a result, the introduction of more nutrients by humankind's action tips the balance in favour of fleshy algae. Under such favourable conditions, algae will overgrow and kill the coral, as well as repelling the dispersal and establishment of coral larvae in new areas.[152]

LOSC Article 194(1) states contracting parties 'shall take, individually or jointly ... all measures ... that are necessary to prevent, reduce and control pollution of the marine environment from any source'. Article 207 then obliges States to 'adopt laws and regulations to prevent, reduce and control pollution of the marine environment from land-based sources'. This commitment has been pursued primarily through regional initiatives. However, and reflecting the fact that LBSP remained a problem 10 years after the adoption of the LOSC, Agenda 21 called for the United Nations Environment Programme (UNEP) to convene an intergovernmental meeting on LBSP.[153] The resulting global conference adopted the 'Global Programme of Action for the Protection of the Marine Environment from Land-Based Activities' (GPA).[154] The programme is international in scope, but legally non-binding.

Under the GPA, States are encouraged to draw up national action programmes (NAPs) that integrate land-use, river basin and coastal management.[155] A number of recommended actions are set out in the GPA for producing these programmes, including assessing and identifying areas of concern such as vulnerable ecosystems.[156] The GPA proceeds to highlight types of LBSP, and sets out ways to prevent harm, drawing attention to many VMEs, such as mangroves.[157]

---

[150] D Ong, 'The Relationship Between Environmental Damage and Pollution: Marine Oil Pollution Laws in Malaysia and Singapore' in MJ Bowman and A Boyle (eds), *Environmental Damage in International and Comparative Law* (OUP, 2002), 194.
[151] CRC Sheppard and others, *The Biology of Coral Reefs* (OUP, 2009), 224.
[152] MD Spalding and others, *World Atlas of Coral Reefs* (University of California, 2001), 57.
[153] UNGA, *Report of the United Nations Conference on Environment and Development*, UNCED A/CONF.151/26 (1992), [17.25], [17.26].
[154] UNEP, Global Programme of Action for the Protection of the Marine Environment from Land-Based Activities, UNEP(OCA)/LBA/IG.2/7 (1995).
[155] *Ibid.*, [19].
[156] *Ibid.*, [21(e)(i)].
[157] *Ibid.*, [97(b)(iii)], [133], [149], [152].

168 *Research handbook on biodiversity and law*

Since its adoption in 1995, three intergovernmental meetings have been convened to assess the programme's progress and to adopt work plans. These meetings have uncovered mixed progress. They have highlighted a lack of political will on the part of some States, a steady increase in NAP adoption, and most recently resistance to calls to set targets for the reduction of nutrient concentrations, which would have helped with monitoring implementation.[158] Given States' evident discomfort with mechanisms that could engender accountability, their preference lies in looser global partnerships for improving water quality where responsibility and success can be shared.[159]

Ultimately there is little evidence to suggest that the GPA is stemming the negative impacts of LBSP. Sewage is a growing problem linked to population rises but sanitation facilities remain prohibitively expensive.[160] Given that it is a non-binding arrangement, the unavoidable reality is that the GPA is dependent upon political will and this has been more evident regionally, rather than at the global level across which the programme looks to operate.

Under Article 197 of the LOSC:

> States shall cooperate on a global basis and, as appropriate, on a regional basis, directly or through competent international organizations, in formulating and elaborating rules, standards and recommended practices and procedures consistent with this Convention, for the protection of the marine environment, taking into account characteristic regional features.

Given that 10 regional agreements had been concluded by 1982, this provision can be seen as the codification of an existing practice.[161] It supports externally concluded MEAs between States, enabling the creation of more focused and detailed obligations to supplement those portions of the LOSC that were only adopted as a framework.

A handful of regional seas groupings have progressed beyond memoranda of cooperation (which set out the broad principles for cooperation and action) to the point where an agreement has been concluded which is dedicated to LBSP.[162] Of particular note is the protocol on LBSP concluded for the Caribbean region.[163] Like the majority of regional agreements on this threat, the Caribbean protocol begins with a general obligation for parties to take appropriate measures to prevent, reduce and control

---

[158] *See* International Institute for Sustainable Development, 'Summary of the Third Intergovernmental Review Meeting on the Implementation of the Global Programme of Action for the Protection of the Marine Environment From Land-Based Activities: 25–26 January 2012' (2012) 25(77) Earth Negotiations Bulletin 1–8.

[159] *Ibid.*, 6–7.

[160] DL VanderZwaag and A Powers, 'The Protection of the Marine Environment from Land-Based Pollution and Activities' (2008) 23 International Journal of Maritime and Coastal Law 423, 440.

[161] CO Okidi, 'Protection of Marine Environment through Regional Arrangements' in AHA Soons (ed), *Implementation of the Law of the Sea Convention Through International Institutions: Proceedings of the 23rd Annual Conference of the Law of the Sea Institute* (Law of the Sea Institute, 1990), 474.

[162] *See*, for example, OSPAR, Annex I.

[163] 1999 Protocol concerning Pollution from Land-Based Sources and Activities to the Convention for the Protection and Development of the Marine Environment of the Wider Caribbean Region.

pollution from LBSP.[164] There are, however, a number of significant points of divergence from regular practice. The Caribbean agreement envisages that annexes will be developed that contain effluent and emissions limits and/or management practices (together with timetables for achieving these) for a list of particular LBSP.[165] These sources are split according to priority of concern,[166] and in a reversal of priorities compared to some regions, domestic sewage and agricultural non-point source pollution are specified as a main concern.[167]

Worth highlighting for their innovative approach are the detailed provisions for tackling domestic wastewater, including sewage.[168] Coastal waters likely to be affected by domestic wastewater are divided into two classes. Class I waters contain ecosystems vulnerable to domestic wastewater. This includes coral reefs and areas located within MPAs implemented under the region's enclave protocol.[169] Waters that are not so sensitive fall within Class II. For Class I waters the Annex commits parties to ensuring that domestic wastewater discharges do not exceed specific concentration levels, which are 80 per cent lower than those permitted for Class II waters.[170] These targets are to be achieved according to timescales that vary depending upon the size of the community using the wastewater service.[171] The larger the community, the longer the contracting party has for providing compatible wastewater systems.

The Wider Caribbean protocol on LBSP arguably reflects some of the best practice for protecting VMEs against LBSP and as such ought to be promoted around many regions with vulnerable coastal ecosystems.

3.4.2 Carbon dioxide and other greenhouse gases

The most recent reports of the Intergovernmental Panel on Climate Change record that global carbon dioxide (CO_2) emissions are 54 per cent higher than they were in 1990.[172] This reflects our reliance upon the burning of fossil fuels to power homes, industry and transportation, as well as the level of cement production, gas flaring and deforestation. The effect has been anthropogenic increases in the emission of CO_2. This causes two dangerous responses. First, large quantities of CO_2 influence the pH of seawater as it is absorbed. The Intergovernmental Panel on Climate Change (IPCC) estimates that almost 28 per cent of CO_2 emissions have been taken up by the oceans.[173] This absorption leads to the acidification of seawater with potentially

---

[164] *Ibid.*, Article III(1).
[165] *Ibid.*, Article IV(1).
[166] *Ibid.*, Annex I.
[167] *Ibid.*
[168] *Ibid.*, Annex III.
[169] *Ibid.*
[170] *Ibid.*
[171] *Ibid.*
[172] IPCC, *Climate Change 2013: The Physical Science Basis. Contribution of Working Group I to the Fifth Assessment Report of the Intergovernmental Panel on Climate Change* (2013), 12.
[173] *Ibid.*

catastrophic effects for calcifying organisms such as corals and shelled marine life.[174] Greater detail on the legal response is provided in Professor Rayfuse's chapter in this handbook.[175]

The second consequence is that it is 'extremely likely'[176] that the unprecedented levels of anthropogenic CO_2, in association with methane, nitrous oxide and a raft of halocarbons, have caused the current warming of the global climate.[177] As such, CO_2 is often referred to as one of the greenhouse gases (GHG), capable of having a radiative forcing effect.[178] For the oceans, this affects water temperatures, sea levels and the number (and intensity) of storms.[179]

As already mentioned, the legal regime in place for addressing climate change is discussed in greater depth elsewhere in this handbook, however some brief observations can be made here for the sake of completeness. The 1982 UN Framework Convention on Climate Change sets as the ultimate objective of the regime keeping GHGs in the atmosphere at a level that would 'prevent dangerous anthropogenic interference with the climate system'.[180] As formally confirmed at the 2010 Cancun plenary meeting of States, political opinion has coalesced around the view that limiting climate change to a rise of 2°C would be in keeping with the objective.[181] This has confirmed previous commitments under the Kyoto Protocol[182] as being entirely inadequate since the 5 per cent quantifiable emissions reduction target accepted by the group of States subject to mitigation obligations are drastically below what would likely achieve this upper limit.[183]

Acceptance of deeper cuts has stalled over concerns that the largest future emitters of GHGs cannot be subject to quantified mitigation targets without diverging from the framework convention's terms. The regime's future is therefore somewhat uncertain.

---

[174] *See* EPOCA, D Laffoley and JM Baxter (eds), *Ocean Acidification: The Facts* (2009), available at <http://www.epoca-project.eu/dmdocuments/OA.TF.English.pdf>.

[175] As a short observation, there is a gap in the law. Attempts to deal with ocean acidification under the climate change framework are second rate as it is calibrated and oriented towards the radiative (not acidifying) potential of gases, of which CO_2 is just one; R Baird, M Simons and T Stephens, 'Ocean Acidification: A Litmus Test for International Law' (2009) 4 Carbon and Climate Law Review 459.

[176] A term used by the IPCC to denote a 95–100 per cent likelihood; IPCC, *Climate Change 2014: Synthesis Report Summary for Policymakers* (2014), 2 (fn 1).

[177] *Ibid.*, 4.

[178] 'Radiative forcing is a measure of the influence a factor has in altering the balance of incoming and outgoing energy in the Earth-atmosphere system'; IPCC, PK Pachauri and A Reisinger (eds), *Climate Change 2007: Synthesis Report* (2007), available at <http://www.ipcc.ch/pdf/assessment-report/ar4/syr/ar4_syr.pdf>,36.

[179] IPCC (n 172), 3, 6 and 9.

[180] (1982) 31 ILM 851, Article 2.

[181] FCCC/CP/2010/7/Add.1, Decision 1/CP.16 (The Cancun Agreements: Outcome of the work of the Ad Hoc Working Group on Long-term Cooperative Action under the Convention), [4].

[182] 1997 Protocol to the Framework Convention on Climate Change (1998) 37 ILM 22.

[183] *See* M Meinshausen, 'What Does a 2°C Target Mean for Greenhouse Gas Concentrations?' in HJ Schellnhuber and others (eds), *Avoiding Dangerous Climate Change* (CUP, 2006), 265.

The Copenhagen Accords[184] (a non-binding arrangement whereby all States gave an indication of the mitigation cuts they would make), and negotiations under the Ad Hoc Working Group on the Durban Platform for Enhanced Action, have taken steps to address such obstacles to meaningful commitments, but until the key COP to be held in Paris in late 2015 concludes, it is difficult to predict what any future agreement on reducing GHG emissions will look like.

From the perspective of VMEs, and as the author has considered at length elsewhere,[185] shallow coral reef ecosystems are perhaps the most immediately vulnerable and threatened by climate change. Recommendations from marine biologists are that GHG concentrations of 450 ppm threaten widespread destruction of reefs with only remnants left of reduced diversity.[186] The call is therefore for stabilisation at 350 ppm, which is based upon historic concentrations known to have caused negative responses amongst corals.[187] The problem is that the IPCC is currently promulgating advice to the effect that maximising GHG concentrations of 450ppm by 2100 will likely limit temperature increase to a rise of 2°C (and this demands cuts of between 40 and 70 per cent of 2010 GHG emissions).[188] Without significant inflation of State ambition and commitments to mitigation, the regime looks likely to leave one VME in a devastated condition.

3.5 Alien Invasive Species[189]

The impact of alien invasive species[190] upon marine ecosystems can be illustrated with the case study of the Indo-Pacific Lionfish (*Pterois volitans*). This species is native to the Red Sea, western Pacific and Indian Ocean. However, it is believed that around a dozen specimens were introduced into the waters off the Florida coast (most likely because of the aquarium trade) in the early to mid-1990s.[191] Reproductive adults release 2 million eggs per year and have few predators in this region, and consequently their distribution has now spread from Rhode Island, USA southwards to the coast of

---

[184] Decision 2/CP.15, UN Doc FCCC/CP/2009/11/Add.1 (30 March 2010).
[185] See EJ Goodwin, *International Environmental Law and the Conservation of Coral Reefs* (Routledge, 2011), chapter 9.
[186] JEN Veron and others, 'The Coral Reef Crisis: The Critical Importance of <350ppm CO_2,' (2009) 58 Marine Pollution Bulletin 1428, 1430.
[187] *Ibid.*
[188] IPCC, *Climate Change 2014 Mitigation of Climate Change: Working Group III Contribution to the Fifth Assessment Report of the Intergovernmental Panel on Climate Change* (2013), 10.
[189] See further Peter Davies' chapter in this handbook.
[190] It is important to limit concerns to invasive alien species, since modern society relies upon many non-native species to provide food and drink. For example the Holstein Friesian Cow originates from the Netherlands and northern Germany and has since been introduced to dairy farms around the globe. Humankind has been able to control the range and impact of such species, so that they are not invasive and threatening to native ecosystems.
[191] D Wilson Freshwater and others, 'Mitochondrial Control Region Sequence Analyses Indicate Dispersal from the US East Coast as the Source of the Invasive Indo-Pacific Lionfish *Pterois volitans* in the Bahamas' (2009) 156 Marine Biology 1213, 1213–1214.

Colombia, with their centre of concentration being in the northern Caribbean.[192] Density in the latter area far exceeds those in native ranges; greater than 390 Lionfish per ha^{-1} in the Bahamas compared to about 80 per ha^{-1} in the Red Sea.[193] With experiments suggesting these alien species can reduce reef fish recruitment by 80 per cent, the knock-on effects for maintaining the fragile balance between competing corals and algae are feared to be disastrous.[194]

The vectors by which invasive species colonise the marine environment are varied, including international trade (as evidenced above) and, more problematically, due to the expulsion of ballast water from ships engaged in international shipping which require water for stabilisation (especially when un-laden). The international legal response has tended to focus upon these pathways, rather than a comprehensive treatment. That said, the CBD contains some provisions of relevance. It has already been highlighted that the convention aims to conserve biological diversity in situ, further to which Article 8(h) calls on States to 'prevent the introduction of, control or eradicate those alien species which threaten ecosystems, habitats or species'. State policy development has been encouraged in accordance with various guidelines, most recently on best practice covering trade in species as pets, aquarium and terrarium species, and as live bait and live food.[195]

The LOSC offers a rather uncertain approach to invasive species in terms of whether or not to characterise such introductions as marine pollution. It has already been observed that the convention's definition of pollution includes the introduction by humankind of substances into the marine environment, which results in, or is likely to result in, deleterious effects including harm to marine life.[196] It is not immediately obvious that a living creature falls into the category of 'substance'. However, one reading of Article 196(1) might confirm[197] that such introductions are a form of pollution:

> States shall take all measures necessary to prevent, reduce and control pollution of the marine environment resulting from the use of technologies under their jurisdiction or control, or the intentional or accidental introduction of species, alien or new, to a particular part of the marine environment, which may cause significant and harmful changes thereto.

This provision adds detail to the general obligations of parties to protect and preserve the environment,[198] to protect rare and fragile ecosystems,[199] and to prevent, reduce and

---

[192] WJ Sutherland and others, 'A Horizon Scan of Global Conservation Issues for 2010' (2010) 25(1) Trends in Ecology and Evolution 1, 4.
[193] Ibid.
[194] See H Hoag, 'Bounty Hunters' (2014) 513 Nature 294, 295.
[195] CBD, Decision XII/16. The latter requires a degree of coordination with the 1975 Convention on International Trade in Endangered Species (993 UNTS 243) to which end see Resolution Conf. 13.10 (Rev. CoP14) adopted under the 1975 agreement.
[196] LOSC, Article 1(4).
[197] The debate would turn upon whether a colon ought to be read into the drafting after 'from' or the first occurrence of 'control'.
[198] LOSC, Article 192.
[199] LOSC, Article 194(5).

control pollution.[200] Nevertheless, the article does not elaborate upon the necessary measures for achieving this.[201] Furthermore, Briony MacPhee observes that the reference to alien invasive species in a provision also covering technologies is not an obvious alliance; Article 211 concerning pollution from the operating of vessels being a more obvious location, which would have had the knock-on benefits associated with that provision's comprehensive guidelines for preventing pollution.[202]

Against this background, more specific rules have been sought focusing upon the vectors to colonisation used by these species. In the marine sphere this has principally been caused by the expulsion of ballast waters by international shipping, which has led, for example, to invasions of Zebra Mussels from the Black Sea into western and northern Europe, and ultimately into freshwater bodies of water in the Great Lakes region of the USA and Canada.[203] As a result, in 1992 Agenda 21 declared that States:

> ... acting individually, bilaterally, regionally or multilaterally and within the framework of IMO and other relevant international organizations ... should assess the need for additional measures to address degradation of the marine environment (a) from shipping by: ... (vi) Considering the adoption of appropriate rules on ballast water discharge to prevent the spread of non-indigenous organisms ...

Following on from their interim guidelines,[204] IMO met this challenge in 2004 with the adoption of the International Convention for the Control and Management of Ship's Ballast Water and Sediments (BWC).[205]

Space constraints prevent a detailed discussion of the convention, but a few general points ought to be made.[206] In broad terms the convention seeks to 'prevent, minimize and ultimately eliminate the transfer of Harmful Aquatic Organisms and Pathogens'. To achieve this objective the vast majority of vessels[207] are expected to conform to specified ballast water exchange standards and carry certification from the flag State of their conformance with these standards. Such certificates and logs of ballast water management activities must be made available for inspection by port State authorities.

The Annex to the BWC stipulates that ballast exchange should take place in waters no less than 50 nautical miles (and whenever possible 200 nautical miles) from shore and in waters of at least 200 metres in depth.[208] Thereafter the Annex presents two

---

[200] LOSC, Article 194.
[201] B MacPhee, 'Hitchhikers' Guide to the Ballast Water Management Convention: An Analysis of Legal Mechanisms to Address the Issue of Alien Invasive Species' (2007) 10 Journal of International Wildlife Law and Policy 29, 39.
[202] *Ibid.*
[203] For other examples visit <http://globallast.imo.org/poster4_english.pdf>.
[204] IMO MEPC Resolution 50(31).
[205] IMO Doc. BWM/CONF/.
[206] For greater detail, *see* M Tsimplis, 'Alien Species Stay Home: The International Convention for the Control and Management of Ship's Ballast Water and Sediments 2004' (2005) 19(4) International Journal of Marine and Coastal Law 411.
[207] Exceptions exist for, inter alia, vessels that do not employ ballast water systems, vessels navigating solely within a coastal State's jurisdiction (as long as it will not cause trans-boundary harm, and government non-commercial shipping (e.g. warships); *see* BWC, Article 3.
[208] BWC, Annex Regulation B-4.

174 *Research handbook on biodiversity and law*

standards regarding the processing of ballast water during management. The first standard is to be immediately applicable and calls for water to be pumped through the ballast tanks three times or alternatively that 95 per cent of the volume of water has been exchanged.[209] The intention behind this appears to be more concerned with flushing residues from the tank, rather than having a bearing upon the presence of alien species.[210] The second standard is to be phased in for ships constructed before 2009, with these vessels being subject to the second standard as from 2014. For those ships constructed after 2009, the second standard applies immediately. The second standard prohibits exceeding the stipulated concentrations of viable plankton per cubic metre of water, as well as given concentrations of microbes such as cholera and E-Coli.[211]

At the time of adoption there was uncertainty as to whether the specified standards could be met due to the absence of suitable technology to achieve them, as well as scientific uncertainty whether the concentrations would prevent colonisation of invasive species. The Annex therefore allows for these regulations to be adjusted, up or down, according to various findings and concerns. These include safety concerns for the ship and crew, environmental acceptability and cost-effectiveness.[212] The latter hints at one of the early criticisms of the adopted text, this being that with the presence of many exceptions to these regulations, there was undue concern not to inconvenience the shipping market and ship-owners, thereby securing more engagement but at the expense of environmental improvements.[213] As Michael Tsimplis observes:

> From the environmental point of view one can only express unhappiness with the way in which shipping and financial interests are given unambiguous priority to any environmental concerns … Thus although it is arguably correct that human life and safety cannot be compromised for the benefit of environmental concerns, it is not as clear that any delay or deviation of the ship for the purpose of improved environmental standards should have such priority.[214]

Despite this generosity to shipping and economic interests, a further problem for the convention is its failure to garner sufficient support in 10 years so as to enter into force. Article 18(1) provides that entry into force will take place 12 months after the date on which no fewer than 30 States, representing a combined total of at least 35 per cent of the gross tonnage of the world's merchant shipping, have ratified, accepted, approved or acceded to the convention. At the time of writing, major shipping nations such as Greece, China and the USA had not ratified the BWC; indeed 44 States are currently parties, representing 32.86 per cent of the world's gross tonnage.[215]

---

[209] BWC, Annex Regulation D-1.
[210] Tsimplis (n 206), 428.
[211] BWC, Annex Regulation D-2.
[212] BWC, Annex Regulation D-5.
[213] Tsimplis (n 206), 444.
[214] *Ibid.* This criticism stems from an exception to the 50/200 nautical miles from shore and 200 m depth of water regulation concerning the location for ballast exchange activity; *see* BWC, Annex Regulation B-4(3).
[215] IMO, *Status of multilateral Conventions and instruments in respect of which the International Maritime Organization or its Secretary-General performs depositary or other*

3.6 Habitat Protection and Capacity

Ecosystems are rarely being pushed to the limits of their stable state by a single threat. Instead, multiple, varied and cumulative threats – perhaps small incidents that chip away at components – can amount to a perturbation of sufficient force to push an ecosystem out of one stable state into another. This produces certain consequences, foremost of which is a move away from sectoral thinking towards holistic management and VME protection through national measures, given that this is the scale of most threatening incidents.[216] This calls for capacity building in terms of suitable policy development and implementation, appointment of trained personnel, and nurturing channels of communication between stakeholders and governments. In a marine context, this has been an objective of the CBD, the 1971 Ramsar Convention (Ramsar)[217] and the 1972 WHC.[218] The initiatives to this end under each are considered below.

3.6.1 CBD

The CBD's objectives are established in Article 1 as the conservation of biodiversity, the sustainable use of its components, and the fair and equitable sharing of the benefits arising from the use of genetic resources, which includes access to genetic resources and technology transfer. The loss of ecosystems and the services they provide falls within this ambit.[219] One of the principal ways in which the CBD seeks to achieve this objective is through the development and implementation of government policy covering habitats and ecosystems. It provides that States shall 'develop national strategies, plans or programmes for the conservation and sustainable use of biological diversity' (or adapt existing ones) so as to reflect the suggested measures agreed under the convention.[220] Such plans are referred to as National Biodiversity Strategies and Action Plans (NBSAP), and the idea is that they should not remain static documents, but be periodically updated as needs must. NBSAPs have been drawn up by 95 per cent of parties.[221] However, with new initiatives and targets being agreed in 2010, only 42 of

---

functions as at 12 February 2015, 500 available at <http://www.imo.org/About/Conventions/StatusOfConventions/Documents/Status%20-%202015.pdf>.

[216] Of course, it ought to be noted that, in contrast to national measures, the International Seabed Authority, which has jurisdiction over the area and exploitation of the mineral resources found therein, has developed a mining code for prospecting and extraction of these resources, aimed in part at protected VMEs in the area; *see* International Seabed Authority, 'Mining Code' (2013), available at <http://www.isa.org.jm/en/mcode>.

[217] 996 UNTS 245.

[218] (1972) 11 ILM 1358.

[219] CBD, Article 2 defines biological diversity as: 'the variability among living resources including, *inter alia*, terrestrial, marine and other aquatic ecosystems and the ecological complexes of which they are part; this includes diversity within species, between species and of ecosystems.'

[220] CBD, Article 6(a).

[221] CBD, CBD Secretariat, 'National Biodiversity Strategies and Action Plans' (2014), available at <https://www.cbd.int/nbsap/default.shtml>.

the 194 States Parties have updated or produced NBSAPs that are consistent with these new developments.[222]

Whilst the CBD reiterates the sovereign right of States to exploit their natural resources and formulate their own conservation policies,[223] the convention still seeks to steer the content of NBSAPs. This is achieved through treaty provisions, thematic programmes and adopting periodic strategic plans. As to the first, the treaty encourages States to provide research and training,[224] public education programmes,[225] and community involvement in conservation initiatives.[226] Of particular note is an emphasis on in-situ conservation, especially Article 8(a) under which contracting parties shall 'as far as possible and as appropriate' establish a system of protected areas.

Given the focus of this chapter upon VMEs, significant developments have happened under a dedicated marine and coastal biodiversity programme. Thus, at the request of the first COP, the CBD's scientific committee produced a recommendation for conserving marine and coastal habitats focused around five actions: implementing integrated coastal zone management, establishing and maintaining MPAs, managing living resources in a sustainable manner, ensuring that mariculture is conducted sustainably, and controlling or eradicating harmful alien species.[227] The second COP, in 1995, adopted the Jakarta Mandate on Marine and Coastal Biodiversity, in which the previous COP's recommendation was supported, subject to further development, and stated its belief that the recommendation was a solid basis for future action.[228] Such further development has continued, for example, and as previously mentioned, concerning the identification of 'Ecologically or Biologically Significant Sea Areas' in ABNJ. Other recommendations have focused upon coral reef bleaching, to which end parties have been encouraged to integrate resilience to bleaching principles into MPA management and design.[229] Over-fishing and water quality are specifically highlighted as factors in determining resilience.[230]

The above has highlighted the principal treaty provisions and recommendations that have sought to shape NBSAPs. In terms of developing strategic plans to the same end, a number of these have been adopted, and they have, in part, looked to set measurable goals that tie into the thematic programmes. In this way strategic plans go beyond mere encouragement of policy formulation towards measurable targets. The current plan for 2011–2020 contains what is known as the Aichi Biodiversity Targets.[231] It challenges States to, inter alia, halve habitat loss, manage fish stocks sustainably and using an

---

[222] *Ibid.*
[223] CBD, Article 3.
[224] CBD, Article 12.
[225] CBD, Article 13.
[226] CBD, Article 8(j).
[227] CBD, Subsidiary Body on Scientific, Technical and Technological Advice ('SBSTTA'), Recommendation I/8 (1995).
[228] CBD, Decision II/10.
[229] CBD, Decision VII/5, *Specific Work Plan on Coral Bleaching*.
[230] For the most recent general assessment of the work on biodiversity and climate change, see CBD/SBSTTA, *In-Depth Review of the Work on Biodiversity and Climate Change: Note by the Executive Secretary* (UNEP/CBD/SBSTTA/14/6).
[231] CBD, Decision X/2 (2010).

ecosystem approach, and establish MPAs for 10 per cent of coastal and marine areas.[232] This is to be achieved by 2020. Thus the work of the CBD demonstrates a sound understanding of the threats to VMEs. Whilst much is rightfully left to action at national level, this does bring the international community closer to being able to judge whether the appropriate steps are actually being taken by contracting parties.

3.6.2 Ramsar

The Ramsar Convention is primarily designed to encourage national wetland policies, practices and procedures so that these critical habitats receive heightened consideration in the welter of competing social and economic developments. Thus the Preamble to Ramsar recognises that wetlands are important regulators of water regimes, and habitats supporting characteristic flora and fauna. Therefore, the convention expresses its objective as being 'to stem the progressive encroachment on and loss of wetlands now and in the future'.[233] Ramsar's jurisdiction extends to wetlands within the territories of contracting parties. Wetlands are defined in Article 1(1) as: 'areas of marsh, fen, peatland or water, whether natural or artificial, permanent or temporary, with water that is static or flowing, fresh, brackish or salt, including areas of marine water the depth of which at low tide does not exceed six metres'. Wetlands are, therefore, geomorphological areas, involving various types of substrate occasionally covered or saturated with water.[234] In practice this still affects an extensive range of wetland habitats, and their associated ecosystems, including some shallow water marine and coastal wetlands such as coral reefs,[235] mangroves and sea grass beds.[236]

Significant to Ramsar is its employment of an inventory as part of its conservation strategy. Thus, wetlands falling within Article 1(1) may be inscribed on the Ramsar List of Wetlands of International Importance; a smaller sub-category of special wetlands to which additional obligations apply. Nominations are automatically accepted, making designation to the List a unilateral act by the contracting party.[237] Sites qualify for various reasons including, notably for VMEs, that they support 'vulnerable, endangered, or critically endangered species or threatened ecological communities'.[238]

States parties are then obliged to formulate and implement plans that promote the use of all wetlands wisely and to conserve listed wetlands.[239] Whilst these plans might apparently need to aim at different standards, use and interpretation under the regime

---

[232] *Ibid.*
[233] Ramsar, Preamble.
[234] GVT Matthews, *The Ramsar Convention on Wetlands: Its History and Development* (Ramsar Convention Bureau, 1993), 42–43.
[235] Although not obviously in relation to coral reefs; Goodwin (n 185), 147–154.
[236] Ramsar, Article 1(1), and Recommendation 4.7.
[237] *See,* in contrast, the next section on the WHC. That said, the decision to list has been reviewed by the COP in the past in relation to some sites in Pakistan. These were suspected as never having met the criteria and so, with the involvement of Pakistan, they have been removed or joined with other wetland areas that together do meet the requisite standard; *see* Lyster (n 126), 413–414.
[238] Strategic Framework and Guidelines for the Future Development of the List of Wetlands of International Importance of the Convention on Wetlands, Criterion 2.
[239] Ramsar, Article 3(1).

has assimilated the two, so that wise use and conservation both bear a close relationship to sustainable use, allowing humankind to extract benefit from wetlands provided their ecosystem services are conserved.[240] For example, wise use of wetlands is taken to mean 'the maintenance of their ecological character, achieved through the implementation of ecosystem approaches, within the context of sustainable development'.[241]

One of the perceived strengths of the regime is its elaboration of core commitments through the regular issuance and updating of guidelines and handbooks designed to help stakeholders implement the agreement's provisions.[242] Central to the idea that governments need to be engaged in sensitive policy implementation, the *Guidelines for the Implementation of the Wise Use Concept* observe it 'is desirable, in the long term, that all contracting parties should have comprehensive national wetland policies, formulated in whatever manner is appropriate to their national institutions'.[243]

In terms of steering appropriate policy, the treaty calls for enclaves to be used, research to be conducted and personnel to be trained.[244] Channels of communication are secured through the National Ramsar Committee initiative whereby governments are encouraged to establish dedicated bodies that can be the focus of national implementation and allow discussions between government departments, individuals and NGOs.[245] In a developed system, problems with local wetlands can then be addressed nationally. This leaves the plenary meetings under Ramsar to draw attention to threatened wetlands of global significance or as a communication channel for those operating in a State with undeveloped national mechanisms for dialogue.

The convention therefore promotes good practice for general conservation of VMEs, and has been successful in encouraging States to apply its terms to coastal habitats. Over 700 marine wetlands have been inscribed on the Ramsar List, of which almost 500 are for reasons that include the presence of vulnerable, endangered or critically endangered species or threatened ecological communities.[246] Obviously this is not always sufficient to reverse significant harm to some wetlands.[247] Wetlands loss persists usually due to the higher priority given to economic development. As a result the latest Strategic Plan[248] aims to produce a shift from sectoral, demand-driven approaches to an ecosystem-based approach to policy and decision-making, which secures the wise use of wetlands. This will require greater integration of wetland concerns within government planning and coordination with local stakeholders. The projected introduction of a high-level ministerial segment to future COPs ought to go some way towards achieving this goal.

---

[240] *Lyster* (n 126) 414–416. Although note, wise use of wetlands is qualified by 'as far as possible', unlike the obligation to conserve.
[241] Ramsar, Resolution IX.1, Annex A, [22].
[242] The numerous guides and 21 handbooks can be accessed via <http://www.ramsar.org/library>.
[243] As adopted under Ramsar, Recommendation 4.10 (Annex), 6, and supplemented by Resolution 5.6 (Introduction).
[244] Ramsar, Article 4.
[245] Ramsar, Resolution 5.7.
[246] For data visit <https://rsis.ramsar.org/ris-search/>.
[247] *See* reports on the Danube-Oder-Elbe Canal.
[248] Ramsar, Resolution X.1 adjusted by XI.3.

Before moving on, it is worth noting the micro measures offered by Ramsar, since it is one of the few instances where areas of marine habitat can receive such individualised attention. To this end, Ramsar obliges States to inform its bureau about any adverse ecological changes to listed wetlands.[249] A formal list of sites undergoing such change was introduced[250] – now known as the 'Montreux Record' – although States must consent to their wetlands being included. Of the 48 wetlands currently on the Montreux Record, 20 contain marine ecosystem elements.[251] Whilst inclusion in the Record might attract unwanted publicity for a party, this is counter-acted by prioritising allocation of Ramsar-controlled funding for remediation efforts. Furthermore, Ramsar Advisory Missions may be sent to the site in question, thus providing free assessment and advice for States with limited capacity.[252]

3.6.3 WHC

Although the WHC is often recognised as one of the principal environmental treaties for the conservation of habitat and species, its relevance to broad-spectrum measures to protect VMEs is negligible. This stems from the parameters of its jurisdiction. The WHC operates under a founding principle that there exist around the world sites of such importance that they need to be preserved for humankind as a whole.[253] The convention defines such natural heritage as:

- natural features consisting of physical and biological formations or groups of such formations, which are of outstanding universal value from the aesthetic or scientific point of view;
- geological and physiographical formations and precisely delineated areas which constitute the habitat of threatened species of animals and plants of outstanding universal value from the point of view of science or conservation;
- natural sites or precisely delineated natural areas of outstanding universal value from the point of view of science, conservation or natural beauty.[254]

Endowed States are subject to exacting conservation standards[255] designed to maintain and present the 'outstanding universal value' of these properties. Furthermore, the WHC maintains an inventory of properties. This is the well-known 'World Heritage List'. Admission of a site to the inventory depends upon a variety of criteria being met – including evidence of the presence of 'outstanding universal value' – the deliberate result being that the list contains only the most outstanding properties.[256]

---

[249] Ramsar, Article 3(2).
[250] Ramsar, Recommendation 4.8.
[251] <http://www.ramsar.org/cda/en/ramsar-documents-montreux/main/ramsar/1-31-118_4000_0__>.
[252] *Lyster* (n 126), 445–446.
[253] WHC, Preamble.
[254] WHC, Article 2.
[255] *See* EJ Goodwin, 'The World Heritage Convention, the Environment and Compliance' (2009) 20(2) Colorado Journal of International Environmental Law and Policy 157.
[256] UNESCO, *Operational Guidelines for the Implementation of the World Heritage Convention* (July 2013) ('WHC Guidelines'), [52].

180 *Research handbook on biodiversity and law*

The key concept of 'outstanding universal value' is generally defined in the Guidelines to the convention as: 'natural significance which is so exceptional as to transcend national boundaries and to be of common importance for present and future generations of all humanity. As such, the permanent protection of this heritage is of the highest importance to the international community as a whole.'[257] Furthermore, the Guidelines go on to state the criteria that the regime's governing body regards as indicating outstanding universal value. In relation to nature these are:

> ... (vii) to contain superlative natural phenomena or areas of exceptional natural beauty and aesthetic importance;
>
> (viii) to be outstanding examples representing major stages of earth's history, including the record of life, significant on-going geological processes in the development of landforms, or significant geomorphic or physiographic features;
>
> (ix) to be outstanding examples representing significant on-going ecological and biological processes in the evolution and development of terrestrial, fresh water, coastal and marine ecosystems and communities of plants and animals;
>
> (x) to contain the most important and significant natural habitats for in-situ conservation of biological diversity, including those containing threatened species of Outstanding Universal Value from the point of view of science or conservation.[258]

The cumulative effect of this stance is that VMEs will not necessarily find protection under the convention, since vulnerability of ecosystems (in contrast to vulnerability of species) is not a criterion of outstanding universal value. This differs from what was encountered in relation to Ramsar, where vulnerability of ecosystems was a recognised value signifying international importance.

Of course, threatened species may benefit from the strong conservation provisions pursuant to criterion (x) above, and these species may coincidentally inhabit VMEs. Thirty-three sites containing habitat of threatened marine species have thus been inscribed on the World Heritage List. The threatened species benefiting include coral and sea grass, but this does not necessarily mean that these ecosystems are particularly vulnerable.

Where the WHC therefore really has a role to play is in terms of micro measures – responding to significant perturbations at individual marine sites. First, the WHC operates a reactive monitoring system to review worsening conditions of heritage sites.[259] The system depends upon receipt of information to trigger an investigation, and whilst this ought to come from the endowed State, non-State actors may make submissions.[260] Time is allocated at committee meetings to review such reports and any proposals from missions sent to sites in response.

Furthermore, the WHC maintains a second list of world heritage properties, these being those regarded as in danger.[261] The dangers faced by properties may be either

---

[257] *Ibid.*, [49].
[258] *Ibid.*, [77].
[259] WHC Guidelines, Part IV.A.
[260] *Ibid.*, [174].
[261] WHC, Article 11(4); WHC Guidelines, Part IV.B.

'ascertained', that is, 'specific and proven imminent danger', or 'potential', meaning there are 'major threats which could have deleterious effects on its inherent characteristics'.[262] Further, the danger must be one that can be corrected by human action.[263] Inclusion of a property on this 'danger list' is a formal recognition of a state of affairs that calls for safeguarding measures, and is also a way to secure resources since sites in danger enjoy a degree of priority when it comes to allocating funds under the WHC.[264]

This mechanism offers a micro-level response to the growing vulnerability to an area of outstanding universal value, and four properties have been inscribed on the danger list because of threats to their marine ecosystem components. These are the Belize Barrier Reef Reserve, East Rennell, the Everglades National Park, and the Galapagos Islands. All but the last remain on the list. To illustrate the potential of the system, unregulated tourist development had been a factor in the 2007 danger listing of the Galapagos, along with poor controls over human immigration and introduction of species to a site of high endemism.[265] By 2010 monitoring missions continued to feel that Ecuador was falling short in terms of: implementing proposals to limit points of entry and improve biosecurity measures; empowering the National Park Service; and formulating a tourism plan.[266] Nevertheless the site was removed on the basis of existing progress.[267] Fortunately, progress has continued in the three years since, suggesting the danger listing has kept the Ecuadorian government focused upon the recommended measures to reduce risk to the island's vulnerable ecosystems.

The danger list is an important component in the WHC's set-up, which contributes to its ability to pull States towards action for protecting marine sites. However, unlike the potential scope of Ramsar's system, it only affects those few properties that are able to meet the criteria for listing as world heritage in the first place, and, as has been highlighted, the fact that an area is a VME is not a valid basis for such initial inclusion.

4. CONCLUSIONS AND FUTURE CHALLENGES

The international regulation of the threats to VMEs is both varied and extensive, going far beyond simply efforts to address vulnerable areas in the high seas. The techniques range from encouraging policy development, controlling fishing intensity, endorsing particular fisheries practices, enabling conservation techniques such as MPAs, to setting standards for the emissions of pollutants. The task for stakeholders, policy-makers and academics is to try to coordinate all of these strands, nationally and internationally. This demands due attention, but being aware of them and drawing many of them together in one place, as this chapter has attempted to do, is a significant step.

Looking at the various sectors, pressing future challenges abound. Controlling fishing has long been at the forefront of this. Greater deployment of effective protected

---

[262] WHC Guidelines [180].
[263] *Ibid.*, [181].
[264] *Ibid.*, [236].
[265] On the original listing, *see* Goodwin (n 255), 175–178.
[266] WHC-13/37.COM/7A.Add (2013), 35–40.
[267] WHC, Decision 37COM7A (2013), 15.

spaces (such as MPAs or PSSAs) is needed to reduce anthropogenic stress upon biodiversity. Also, more gear restrictions (particularly a ban on bottom trawling) need to be introduced and enforced. Further, the narrative of sustainable development (rather than short-term economic profit) needs to be given greater influence since marine waters are so significant for food supply. A failure to sustainably manage these resources (whether through wild caught or aquaculture) places additional burdens upon terrestrial production and the stability of multiple ecosystems. In relation to this, there are two significant areas for future monitoring and research that merit highlighting at this point. Aquaculture offers much potential, but current practices are having such a negative effect on VMEs such as mangroves that it is clear that remodelling of this fishery to ensure optimum sustainability is needed. Second, the UNGA process surrounding an international instrument on the conservation and sustainable use of marine biodiversity in ABNJ could have a major impact upon the availability of conservation techniques, fishing methods and TACs. However, the multiplicity of interests involved in the process already suggests these negotiations will be difficult.

Pollution remains a considerable challenge to VMEs, not least as it undermines conservation measures such as MPAs. The challenges here are multiple and include gaining sufficient support for agreements so that they enter into force and raising political ambition and commitment to reducing such pollution in the face of competing economic demands. Furthermore, one challenge lies in integrating marine environmental concerns across multiple government agencies. Some sources of pollution fall within traditional law of the sea competences, such as the release of alien invasive species through ballast water. However, others fall outside of marine managers' and ministries' exclusive remits, such as LBSP and CO_2 emissions. Integration of marine concerns across government departments and planning therefore remains a basic tenet of conservation. However, ambitious reductions in GHG emissions, and (somehow) CO_2 in particular, are perhaps the keystones to securing marine biodiversity. All of this requires political will but ultimately this currently seems to be in short supply.

SELECT BIBLIOGRAPHY

Auster, PJ and others, 'Definition and Detection of Vulnerable Marine Ecosystems on the High Seas: Problems with the "Move-On" Rule' (2011) 68(2) ICES Journal of Marine Science 254
Baird, R, M Simons and T Stephens, 'Ocean Acidification: A Litmus Test for International Law' (2009) 4 Carbon and Climate Law Review 459
Barnes, R, 'The Convention on the Law of the Sea: An Effective Framework for Domestic Fisheries Conservation?' in D Freestone, R Barnes and D Ong (eds), *The Law of the Sea: Progress and Prospects* (OUP, 2006), 240
Bowman, MJ, PGG Davies and C Redgwell, *Lyster's International Wildlife Law* (2nd edn., CUP, 2010), chapter 5
Gjerde, KM and A Rulska-Domino, 'Marine Protected Areas Beyond National Jurisdiction: Some Practical Perspectives for Moving Ahead' (2012) 27 International Journal of Marine and Coastal Law 351
Goodwin, EJ, *International Environmental Law and the Conservation of Coral Reefs*, (Routledge, 2011)
Henriksen, T, 'Conservation of Marine Biodiversity and the International Maritime Organisation' in C Voigt (ed), *Rule of Law for Nature* (CUP, 2013), 331
Hilborn, R, *Overfishing: What Everyone Needs to Know* (OUP, 2012)
Kachel, MJ, *Particularly Sensitive Sea Areas* (Springer, 2008)
Roberts, C, *Ocean of Life* (Penguin, 2013)

Roberts, J, 'Protecting Sensitive Marine Environments: The Role and Application of Ships' Routing Measures' (2005) 20(1) International Journal of Marine and Coastal Law 135

Scott, KN, 'Conservation on the High Seas: Developing the Concept of High Seas Marine Protected Areas' (2012) 27 International Journal of Marine and Coastal Law 849

Scovazzi, T, 'Marine Specially Protected Areas Under International Law' in T Scovazzi (ed.), *Marine Specially Protected Areas* (Kluwer, 1999), 17

Spadi, F, 'Navigation in Marine Protected Areas: National and International Law' (2000) 31 Ocean Development and International Law 285

Tsimplis, M, 'Alien Species Stay Home: The International Convention for the Control and Management of Ship's Ballast Water and Sediments 2004' (2005) 19(4) International Journal of Marine and Coastal Law 411

VanderZwaag, DL and A Powers, 'The Protection of the Marine Environment From Land-Based Pollution and Activities' (2008) 23 International Journal of Maritime and Coastal Law 423

7. Alien invasive species: is the EU's strategy fit for purpose?

Peter Davies

1. INTRODUCTION

The spread of alien species throughout the world is a phenomenon which gathered momentum in colonial times as more trade routes opened up, and has proliferated in today's world of integrated economies and markets.[1] International trade has particularly facilitated the movement of alien species as has our ability to travel far more easily than ever before.[2] Having arrived in its new environment, a given alien species might lack the capacity to adapt and survive. Others may continue to exist with minimum impact on the ecosystem in question, and may indeed bring benefits to a given ecosystem.[3] However, some alien species become 'invasive' in that their introduction or spread presents a serious threat to the native wildlife and biodiversity generally. In Europe alone it is estimated that 10 per cent of the 12,000 alien species are invasive,[4] and it is believed that there has been an increase of 76 per cent in the number of invasive alien species in this region since the 1970s.[5] The cost of invasions by invasive alien species (IAS) in Europe has been estimated to be at least €12 million each year, whilst the total cost to date in Australia, Brazil, India, South Africa, the United Kingdom and the USA is thought to be in the region of US$300 billion.[6]

The Parties to the 1992 Convention on Biological Diversity (CBD)[7] have defined 'alien species' as 'a species, subspecies or lower taxon, introduced outside its natural past or present distribution; includes any part, gametes, seeds, eggs, or propagules of

---

[1] C Shine, N Williams and L Gundling, *A Guide to Designing Legal and Institutional Frameworks on Alien Invasive Species* (IUCN, 2000) 4.

[2] *Ibid.*

[3] Pearce takes the view that certain alien species should be regarded as 'ecological saviours' due to their conservation value; F Pearce, 'No Trespassers?' *The Independent* (London, 14 April 2015) 31. In an interesting shift in perspective from the traditional approach to alien species, the same author has stipulated that he has 'become convinced that alien species are part of the solution to nature's current crisis, rather than part of the problem': F Pearce, 'Loving the Alien' (2015) 33(7) Wildlife 71, 72.

[4] Regulation (EU) 1143/2014 of the European Parliament and of the Council on the prevention and management of the introduction and spread of invasive alien species, OJ 2014 317/35, recital 1.

[5] European Commission, *Invasive Alien Species: a European Response* (European Commission, 2014), 7.

[6] European Environment Agency (EEA), *The Impacts of Invasive Alien Species in Europe* (EEA, 2012), 7.

[7] (1992) 31 ILM 818.

such species that might survive and subsequently reproduce'.[8] The word 'introduction' in this context is defined as 'the movement by human agency, indirect or direct, of an alien species outside of its natural range (past or present). This movement can be either within a country or between countries or areas beyond national jurisdiction.'[9] An organism which is therefore transported due to human action into an area beyond the species' natural distribution is defined as 'alien' in nature. But what of an *invasive* alien species? This is defined under the CBD regime as 'an alien species whose introduction and/or spread threaten biological diversity'.[10] A key distinction must therefore be made between an 'alien species' and an 'alien *invasive* species'. It is only the latter that threatens biodiversity and requires human intervention to halt its introduction and spread.[11] After habitat destruction and fragmentation, the spread of IAS is believed to be the most pressing cause of biodiversity loss worldwide.[12]

This chapter first seeks to address the means by which IAS have spread globally before discussing the negative impact of such invasions. An introduction to the existing international and regional legal response to IAS will then be given before discussion turns to the recently adopted EU Regulation on the Prevention and Management of the Introduction and Spread of Invasive Alien Species (the 'Regulation').[13] The latter is the foundation of the EU's new strategy on IAS and entered into force on 1 January 2015. The chapter's key focus is to provide an analysis of the adequacy of the Regulation by primarily addressing the extent to which it operates in line with key international guidance provided by the CBD.

2. THE SPREAD OF INVASIVE ALIEN SPECIES

Alien species can be introduced by humans either intentionally or unintentionally.

2.1 Intentional Introduction

The numerous reasons as to why humans have seen fit to introduce alien species include the following: to develop commercial forestry; to promote commercial or sport fishing; to facilitate the pet trade; to develop fur farming; to provide a biological control in agriculture; for ornamental reasons; and with a view to breeding specimens

---

[8] CBD COP 6, Decision VI/23 Alien Species that Threaten Ecosystems, Habitats or Species, Annex, footnote 57. Defining exactly what is meant by 'IAS' has at times proved controversial. For reasons of space this chapter will not address such controversies. However, *see* S Riley, 'A Weed by any Other Name: Would the Rose Smell as Sweet if it were a Threat to Biodiversity?' 22 (2009–2010) Georgetown International Environmental Law Review 157.

[9] CBD COP 6, Decision VI/23 Alien Species that Threaten Ecosystems, Habitats or Species, Annex, footnote 57.

[10] *Ibid.*

[11] *See* letter by M Lambertini and others, 'Invasives: A Major Conservation Threat' 333 (6041) Science (2011) 404–405.

[12] *See*, for example, Bern Convention on the Conservation of European Wildlife and Natural Habitats, *European Strategy on Invasive Alien Species* (Council of Europe, 2003), 'Introduction'.

[13] Regulation (EU) 1143/2014 (n 4).

in captivity for scientific or commercial reasons.[14] Indeed, many such introductions have proved advantageous to society as a whole. However, some of these alien introductions have proved far from positive in that the species in question has become invasive. The cane toad provides a notorious example of an alien species intentionally introduced for pest control, but which has impacted beyond the target species to become a major predator of a range of other native species.[15] Naturally occurring in Central America and the tropical regions of South America, cane toads (*rhinella marina*) were introduced into North Queensland (Australia) from Hawaii in the mid-1930s as a biological control to reduce the damage to sugar cane occasioned by scarab beetles.[16] This large and resilient toad has now spread rapidly throughout north-eastern Australia. Shine has noted that they 'eat a wide variety of prey, have greater fecundity than native anurans, and develop rapidly in tropical regions'[17] and, as such, 'colonizing cane toads attain very high densities'.[18]

Poisonous throughout the entirety of their life cycle, the introduction of the cane toad has certainly had an undesirable effect on Australian biodiversity. For example, cane toads have had a significantly negative impact on certain populations of native species including the bluetongue lizard, the northern death adder, the northern quoll, and some freshwater crocodile populations.[19] It is also believed that a number of bird species in Kakadu Park might be at risk from their presence.[20] Competition between this IAS and native frogs is thought additionally to have had a negative impact on the latter.[21] The Australian government has recognised the ongoing threat posed by the cane toad to Australian wildlife and noted that 'the range of cane toads has expanded through Australia's northern landscape and they are now moving westward at an estimated 40 to 60 km per year'.[22]

2.2 Unintentional Introduction

Increased trade and tourism has facilitated the unintentional transportation of alien species across the world. Alien species can, for example, attach to the hulls of ships and be transported vast distances before being introduced into a new environment.[23] Border and quarantine requirements have often been needed to ensure that those alien species hidden within consignments of traded products are detected in time.[24] Moreover, it is estimated that around 650 million tourists cross international boundaries each

---

[14] Shine, Williams and Gundling (n 1) 5–6.
[15] *Ibid.*, 5.
[16] See <http://australianmuseum.net.au/Cane-Toad>.
[17] R Shine, 'The Ecological Impact of Invasive Cane Toads in Australia' 2010 85(3) The Quarterly Review of Biology 253, 254.
[18] *Ibid.*
[19] *Ibid.*, 258–260 and 263.
[20] *Ibid.*, 269.
[21] *Ibid.*, 261.
[22] See <http://www.environment.gov.au/biodiversity/invasive-species/publications/factsheet-cane-toad-bufo-marinus>.
[23] Shine, Williams and Gundling (n 1), 7.
[24] *Ibid.*

year and in doing so can inadvertently introduce infectious agents that are harmful to humans or to agricultural production.[25]

New infrastructure projects can also facilitate and reinforce the introduction of alien species. The Suez Canal, for example, was opened for traffic in 1869 and has since allowed the migration of more than 300 tropical species to the Mediterranean 'causing major changes to composition and structure of native flora and flora'.[26] The inadvertent inclusion of alien species in vessels' ballast water can additionally lead to the spread of alien marine organisms. Ballast water is commonly utilised to stabilise ships and is eventually discharged, often into a new environment. The Global Ballast Water Management Programme (established by the Global Environment Facility (GEF), United Nations Development Programme (UNDP) and the International Maritime Organization (IMO)) has noted that

> the potential for species transfer is compounded by the fact that almost all marine species have planktonic stages in their life-cycle, which may be small enough to pass through a ship's ballast water intake ports and pumps. This means that species with adult stages that are large or attached to the seabed, may still be transported in ballast water.[27]

Numerous species of plankton dinoflagellates are believed to have been transported large distances by ballast water.[28] Such algae can be absorbed by shellfish, such as oysters and scallops, and then release toxins. The subsequent consumption of contaminated shellfish has led to paralysis and even mortality.[29] It is interesting to note that records of such poisonings had until the 1970s been very largely limited to the consumption of shellfish from European, North American and Japanese waters.[30] By 1990 the geographical pattern of shellfish poisoning had spread not only further afield in the Northern Hemisphere (to include India, Thailand and the Philippines), but also throughout the Southern Hemisphere (such as Australia, South Africa, and New Zealand).[31] It is thought that the transportation of dinoflagellates within ballast water

---

[25] JA McNeely and others (eds), *A Global Strategy on Invasive Alien Species* (IUCN, 2001), 9 and 12.

[26] Shine, Williams and Gundling (n 1), 7.

[27] Global Ballast Water Management Programme (GBWMP), *Stopping the Ballast Water Stowaways!* (GBWMP, 2001), 2. Note also the 2004 International Convention for the Control and Management of Ships' Ballast Water and Sediments (available at http://www.imo.org/ – not in force). On the latter see KN Scott 'Defending the World Below the Brine: Managing Alien Invasive Species under the 2004 Ballast Water Convention – a New Zealand Perspective' (2008) 14(4) Journal of International Maritime Law 307, and J Firestone and JJ Corbett, 'Coastal and Port Environments: International Legal and Policy Responses to Reduce Ballast Water Introductions of Potentially Invasive Species' (2005) 36(3) Ocean Development and International Law 291.

[28] GBWMP (n 27), 3.

[29] *Ibid.*

[30] G Hallegraeff, 'Transport of Toxic Dinoflagellates via Ships' Ballast Water: Bioeconomic Risk Assessment and Efficacy of Possible Ballast Water Management Strategies' (1998) 168 Marine Ecology Progress Series 297.

[31] *Ibid.*

may well have been the reason for the increased geographical spread of such incidents.[32]

3. THE IMPACT OF INVASIVE ALIEN SPECIES

The spread of IAS can negatively impact upon biodiversity, ecosystem services, human health and economic activities.[33] Some examples will serve to provide informative illustrations.

3.1 Biodiversity

Adverse impact occasioned by IAS can take the form of competition to native species or the predation of such species, the transmission of disease or harm to native organisms, and the hybridisation of IAS and native species.[34] Of the 395 species critically in danger of extinction in Europe in 2011, 110 were at risk due to the impact of IAS.[35] The American mink (*neovison vison*) provides an example of an IAS which out-competes a native species, the threatened European mink.[36] Originally introduced into Europe by fur farmers, the American mink has also proved to be a major predator of European water voles, and a number of birds that nest on the ground such as the common tern and the black-headed gull.[37]

A further example of an IAS threatening biodiversity is afforded by the Chytrid fungus (*batrachochytrium dendobatidis*), which has transmitted a lethal disease known as chytridiomycosis to approximately 500 amphibians in around 40 countries.[38] The disease reduces the ability of amphibians to respire through their skins resulting in heart attack, and also contributes to the thickening of their skins, which can lead to suffocation. A final illustration is provided by the American Ruddy duck (*oxyura jamaicensis*), which was brought by wildfowl collectors to the UK in the 1940s. Either released from captivity or having otherwise escaped, the Ruddy duck has threatened the existence of the closely related White-headed duck (*oxyura leucocephala*) through hybridisation in the UK and continental Europe.[39]

3.2 Ecosystem Services

The contribution of ecosystems to human interests can be significantly affected by the introduction and spread of IAS. They can, for example, modify the quality of soil or prove to be a hindrance to crop pollination. The Spanish slug (*arion vulgaris*), which finds food not only in vegetable gardens. but also in agricultural fields growing maize,

---

[32] *Ibid.*
[33] European Environment Agency (n 6), 11.
[34] *Ibid.*, 10.
[35] *Ibid.*, 8.
[36] *Ibid.*, 19.
[37] *Ibid.*
[38] *Ibid.*, 35.
[39] *Ibid.*, 40.

rape and sunflowers,[40] is a good example. It is believed that these slugs were unintentionally introduced into much of Europe by contaminated soil and also as stowaways in gardening equipment.[41] Each slug can produce 400 eggs and the species is thought to have been responsible for a 50 per cent reduction in yield in Norwegian strawberry fields.[42] The pontic rhododendron (*rhododendron ponticum*) also provides an illustration of the negative impact of a species on ecosystem services in that it shades out plants that grow beneath it, and its leaves host poisonous chemicals which reduce the ability of vegetation around the plant to survive.[43] Introduced in the UK around 250 years ago as an ornamental plant, it also acts as a host for the Sudden Oak Death disease, thereby impacting on woodland ecosystems as well as timber production.[44]

3.3 Human Health

Invasive alien species can also have a negative effect on human health. For example, the common ragweed (*ambrosia artemisiifolia*), which is native to North America, has spread throughout Europe in the last 25 years, having been introduced as a contaminant in agricultural products and construction materials.[45] Its pollen is known to induce hay fever and rhinoconjunctivitis, and the plant itself contains oils that act as a skin irritant.[46] Additionally, the giant hogweed (*heracleum mantegazzianum*) negatively impacts not only on native plants by reducing the amount of light they receive, but also on human health by causing a burning effect on contact with skin potentially causing serious skin lesions.[47]

3.4 Economic Activities

The coypu (*myocastor coypus*) was introduced to various parts of the world from South America during the course of the last century, and has subsequently escaped from fur farms. The rodent is known for its burrowing activities, which have had a major impact on man-made infrastructures such as dykes, levees and riverbanks.[48] In the period 1995–2000 the rodent damaged riverbanks in Italy at a cost of more than EUR 10 million even though an eradication programme had removed more than 200,000 from the environment.[49] Moreover, rose-ringed parakeets (*psittacula krameri*), which have often escaped from aviaries, are known to impact on agricultural activities in India and Pakistan by eating cereals, pulses and oil seeds. They have also damaged plantations in

---

[40] *Ibid.*, 58.
[41] *Ibid.*, 60.
[42] *Ibid.*, 58.
[43] *Ibid.*, 55.
[44] *Ibid.*, 56.
[45] *Ibid.*, 80.
[46] *Ibid.*
[47] *Ibid.*, 82.
[48] *Ibid.*, 84.
[49] *Ibid.*

Australia by stripping bark, thereby causing the death of affected trees.[50] A final example is provided by the Japanese Knotweed (*fallopia japonica*), which can damage buildings and negatively impact on biodiversity. Found to be present across 10 acres of the London Olympic Games site, it was finally eradicated but only at a cost of £70 million.[51]

4. THE DEVELOPING INTERNATIONAL AND REGIONAL LEGAL RESPONSE

A large number of international treaties now address the spread of IAS,[52] as have numerous international guidelines and codes of conduct.[53] Particularly significant

---

[50] *Ibid.*, 100.

[51] H Wallop, 'Japanese Knotweed: How Do We Tackle This Scourge?' *The Telegraph* (London, 1 April 2014) available at <http://www.telegraph.co.uk/finance/property/10737159/Japanese-knotweed-How-do-we-tackle-this-scourge.html>.

[52] For a brief introduction to some of these regimes *see* <www.cbd.int/invasive/done.shtml>. *See* also Shine, Williams and Gundling (n 1), Appendix I. Examples of international regimes which address IAS include the following: the 1994 Agreement on the Application of Sanitary and Phytosanitary Measures (available at https://www.wto.org/english/docs_e/legal_e/15sps_01_e.htm), which applies to 'additives, contaminants, toxins and disease-carrying organisms in food, beverages and feedstuffs' many of which are IAS; the 1979 Bonn Convention on Migratory Species (1980) 19 ILM 15, Article III(4)c of which notes that Parties that are range states of Appendix I listed migratory species 'shall endeavour … to prevent, reduce or control factors that are endangering or are likely to endanger the species, including strictly controlling the introduction of, or controlling or eliminating, already introduced exotic species'; the 2004 International Convention for the Control and Management of Ships' Ballast Water and Sediments (available at http://www.imo.org/ – not in force) under which Parties 'undertake to give full and complete effect to the provisions of this Convention and the Annex thereto in order to prevent, minimize and ultimately eliminate the transfer of Harmful Aquatic Organisms and Pathogens through the control and management of ships' Ballast Water and Sediments' (Article 2(1)); the 1982 UN Convention on the Law of the Sea (1982) 21 ILM 1261, which places Parties under an obligation to 'take all measures necessary to prevent, reduce and control pollution of the marine environment resulting from … the intentional or accidental introduction of species, alien or new, to a particular part of the marine environment, which may cause significant and harmful changes thereto' (Article 196(1)); the 1971 Convention on Wetlands of Strategic Importance (1972) 11 ILM 963 under which Parties have adopted a 2009–15 Strategic Plan which encourages Parties to 'develop a national inventory of invasive alien species that currently and/or potentially impact the ecological character of wetlands, especially Ramsar sites … [and to] develop guidance and promote procedures and actions to prevent, control or eradicate such species in wetland systems' (Strategy 1.9); the 2000 Cartagena Protocol on Biosafety (2000) 39 ILM 1027, which stipulates that the 'development, handling, transport, use, transfer and release of any living modified organisms are undertaken in a manner that prevents or reduces the risks to biological diversity, taking into account risks to humans' (Article 2); and the 1997 Convention on the Law of the Non-navigational uses of International Watercourses (1997) 36 ILM 719, which notes that States 'shall take all measures necessary to prevent the introduction of species, alien or new, into an international watercourse which may have effects detrimental to the ecosystem of the watercourse resulting in significant harm to other watercourse states' (Article 22).

[53] *See*, for example, the 1995 FAO Code of Conduct for Responsible Fisheries, which notes that 'States should, in order to minimize risks of disease transfer and other adverse effects on

bearing in mind its global importance is the CBD, Article 8(h) of which places an obligation on Parties to 'prevent the introduction of, control or eradicate those alien species which threaten ecosystems, habitats and species'. The EU itself and all its 28 Member States are party to the CBD and are therefore legally bound by this provision.[54] Providing 'a comprehensive, global approach to the protection of Earth's biodiversity previously lacking in international law',[55] the CBD regime regards IAS as a cross-cutting issue relevant to the entirety of the CBD's work. The tenth meeting of the CBD's Conference of the Parties in 2010 adopted an updated Strategic Plan for Biodiversity for the period 2011–2020. The plan includes the 'Aichi Biodiversity Targets', which incorporates a goal that 'by 2020, invasive alien species and pathways are identified and prioritized, priority species are controlled or eradicated, and measures are in place to manage pathways to prevent their introduction and establishment'.[56] But how might this ambitious objective be achieved? Key internationally respected 'soft law' guidance has been provided under the auspices of the CBD in the form of 15 'Guiding Principles for the Prevention, Introduction and Mitigation of Impacts of Alien Species that Threaten Ecosystems, Habitats or Species' (hereinafter 'CBD Guiding Principles'). The latter were adopted at the sixth Conference of the Parties in 2002,[57] and are designed to provide a structure within which governments and organisations can develop effective strategies to reduce the spread and impact of IAS. The 15 non-binding principles are therefore essential goals to be achieved and have set an important benchmark against which relevant strategies can be assessed.

Action within regional legal regimes has also been endorsed and promoted.[58] Activity under the 1979 Convention on the Conservation of European Wildlife and Natural Habitats (the 'Bern Convention') has arguably proved to be the most significant in this regard. Article 11(2)b of the Bern Convention obliges Parties to 'strictly control the introduction of non-native species' and the EU and its Member States are bound by this provision.[59] An IAS Experts' Group has been established by the Parties to the treaty and first met in 1993. This group played an essential part in the preparation of a

---

wild and cultured stocks, encourage adoption of appropriate practices in the genetic improvement of broodstocks, the introduction of non-native species, and in the production, sale and transport of eggs, larvae or fry, broodstock or other live materials' (para. 9.3.3). Also see the 2005 International Council for the Exploration of the Seas' Code of Practice on the Introductions and Transfers of Marine Organisms, available at <www.ices.dk/publications/Documents/Miscellaneous%20pubs/ICES%20Code%20of%20Practice.pdf>.

[54] In relation to EU membership *see* Council Decision 93/626/EEC concerning the conclusion of the Convention on Biological Diversity, OJ 1993 L309/1.

[55] MJ Bowman, PGG Davies and C Redgwell, *Lyster's International Wildlife Law* (2nd edn., CUP, 2010), 594.

[56] CBD Decision X2, Aichi target 9.

[57] *See* CBD Decision VI/23, Annex.

[58] *See*, for example, the 1995 Protocol Concerning Specially Protected Areas and Biological Diversity in the Mediterranean (Barcelona Convention); Article 6(d) notes that Parties 'shall take the protection measures required, in particular: the regulation of the introduction of any species not indigenous to the specially protected area in question'.

[59] In relation to EU membership *see* Council Decision 82/72/EEC concerning the conclusion of the Convention on the Conservation of European Wildlife and Natural Habitats, OJ 1982 L38/1.

key IAS-related recommendation which was adopted in 2003 by the Standing Committee; the 'European Strategy on Invasive Alien Species' is designed to foster greater cooperation in this area as well as the adoption of effective national policies and legislation.[60] This strategy was, of course, introduced one year after the adoption of the CBD's Guiding Principles and sought to facilitate the latter's implementation.

The Bern Convention's IAS Expert Group has also engaged in other IAS-related activity, including analysis of national measures taken by Parties, the drafting of codes of conduct on IAS as well as guidelines on eradication, and also the identification of those sensitive habitats (including islands) particularly susceptible to invasion. Moreover, the Bern Convention's Standing Committee has deliberated over the spread of a variety of IAS within its 'case file' monitoring system and has subsequently adopted specific recommendations relating to the protection of particular species,[61] as well as those relating to the eradication or control of IAS.[62] Standing Committee recommendations have additionally endorsed a range of relevant technical codes of conduct,[63] and numerous IAS-related reports have been produced.[64]

5. THE REGULATION

At the EU level, whilst the control of some IAS had been addressed or was capable of being addressed under existing binding measures,[65] a comprehensive system

---

[60] Recommendation No. 99 (2003) on the European Strategy on Invasive Alien Species.

[61] *See*, for example, Recommendation No. 78 (1999) on the conservation of the Red squirrel in Italy; Recommendation No. 18 (1989) on the protection of indigenous crayfish in Europe; Recommendation No. 61 (1997) on the conservation of the White-headed Duck; Recommendation No. 114 (2005) on the control of the Grey squirrel and other alien squirrels in Europe.

[62] *See*, for example, Recommendation No. 124 on progress in the eradication of the Ruddy Duck. Other Standing Committee recommendations related to the eradication or control of IAS include Recommendation No. 126 on the eradication on some invasive alien plant species and Recommendation No. 91 (2002) on Invasive Alien Species that threaten biological diversity in Islands and geographically and evolutionary isolated ecosystems.

[63] *See*, for example, Recommendation No. 154 (2011) on the European Code of Conduct on Pets and Invasive Alien Species, Recommendation No. 160 (2012) on the European Code of Conduct for Botanical Gardens on Invasive Alien Species, and Recommendation No. 161 (2012) on the European Code of Conduct for Zoological Gardens and Aquaria on Invasive Alien Species.

[64] *See*, for example, C de Klemm, 'Introduction of Non-native Organisms into the Natural Environment' Nature and Environment series no. 73 (1996).

[65] For example, under Article 22(b) of the Habitats Directive (Directive 92/43/EEC, OJ 1992 L206/7) Member States shall 'ensure that the deliberate introduction into the wild of any species which is not native to their territory is regulated so as not to prejudice natural habitats within their natural range or the wild native fauna and flora and, if they consider it necessary, prohibit such introduction'. Moreover, Article 11 of the Birds Directive (Directive 2009/147/EC, OJ 2009 L20/7) stipulates that 'Member States shall see that any introduction of species of bird which do not occur naturally in the wild state in the European territory of the Member States does not prejudice the local flora and fauna'. Some IAS, such as the American Ruddy duck and the painted turtle, are regulated by the CITES Regulation (Regulation 338/97, OJ 1997 L61/1). Note also Council Directive 2000/29 on protective measures against the introduction into the

specifically designed to regulate IAS was lacking. Moreover, national responses by EU Member States to the spread of IAS had been criticised; a 2006 review of frameworks in 27 countries (now all EU Member States) came to the conclusion that 'although most Member States have some regulations in place relating to IAS … the fragmented measures in place are unlikely to make a substantial contribution to lowering the risks posed by IAS to European ecosystems'.[66] A 2011 study further concluded that 'some Member States are more advanced than others in their initiatives to tackle IAS, and approaches to the IAS issue differ'.[67] There was additionally evidence that action taken by Member States was often simply reactive to IAS detection rather than preventive.[68] A more comprehensive and co-ordinated cross-border response was therefore required and the EU's 2020 Biodiversity Strategy, adopted in May 2011, noted that the European Commission ('Commission') would 'fill policy gaps in combating IAS by developing a dedicated legislative instrument by 2012'.[69] In fact the Commission's proposal saw the light of day in September 2013,[70] and the Regulation was finally adopted on 22 October 2014.

The EU Regulation's definitions of 'alien species', 'introduction', and 'invasive alien species' are broadly in line with those used by the CBD. However, particular types of IAS are in fact excluded from the remit of the measure due to the fact that other EU measures have already introduced applicable control regimes.[71] Moreover, the Regulation makes an exception – similar to that now endorsed under the Bern Convention – in that it excludes species moving outside their natural range in response to 'changing ecological conditions and climate change'.[72] This particular exception seeks to clarify a

---

Community of organisms harmful to plants or plant products and against their spread within the Community (OJ 2000 L169/1), those measures designed to prevent the introduction of animal diseases (such as Council Directive 97/78/EC laying down the principles governing the organisation of veterinary checks on products entering the Community from third countries (OJ 1998 L24/9)), and Regulation 708/2007 on the use of alien and locally absent species in aquaculture (OJ 2007 L168/1).

[66] C Miller, M Kettunen and C Shine, *Some Options for EU Action on Invasive Alien Species: Final Report for the European Commission* (Institute for European Environmental Policy, 2006), 50.

[67] European Commission, *A Comparative Assessment of Existing Policies on Invasive Species in the EU Member States and in selected OECD countries* (European Commission, 2011), 21.

[68] European Commission, 'Proposal for a Regulation on the prevention and management of the introduction and spread of invasive alien species', COM(2013) 620 final, 2.

[69] European Commission, 'Our Life Insurance, Our Natural Capital: an EU Biodiversity Strategy to 2020' COM (2011) 244, Action 16.

[70] European Commission (n 68).

[71] Article 2. IAS excluded from the Regulation include genetically modified organisms regulated under Directive 2001/18/EC, those pests harmful to plants regulated by Directive 2000/29/EC, species used in aquaculture and controlled by Regulation 708/2007, and microorganisms used in plant protection products and biocidal products duly authorised or being assessed under Regulation 1107/2009 and Regulation 528/2012 respectively.

[72] Article 2(a) Recommendation No. 142 (2009) of the Bern Convention's Standing Committee interpreting the CBD's definition of IAS had taken into account climate change in noting concern that 'native species moving to neighbouring areas may be considered alien due to the fact that climate change is the result of human action and that such species may be

concern that such species might otherwise be regarded as invasive in nature having indeed moved as a consequence of human action into a new habitat outside their natural range – the climate change phenomenon is, of course, in part the consequence of human activity. This exclusion of such species from regulation can be justified by the fact that such species have no alternative but to move from their original habitats due to climatic changes. In effect, the species in question may have nowhere else to go, whilst, by contrast, the IAS subject to the Regulation have no such difficulties surviving within their own original natural range.

The Regulation envisages the establishment of a list ('the List') which will note those IAS for which concerted action at the EU level is required. The List is to be updated over time by either adding to it or removing certain species that no longer fulfil the Regulation's criteria for listing. Species will be listed, and hence become 'IAS of Union Concern', if they inter alia are likely to have significant adverse impact on biodiversity or the related ecosystem services, and may also have an adverse impact on human health or the economy. When adopting the List or adding to it, the Commission will apply the applicable criteria for listing 'with due consideration to implementation costs for Member States, the cost of inaction, the cost-effectiveness and the socio-economic aspects'.[73] A scientific forum will provide the Commission with advice on scientific questions, particularly those relating to the listing process and the application of any emergency measures.[74]

An IAS is to be listed only following a risk assessment, which will provide an analysis of the risks involved with the introduction of the species in question. This assessment is to be carried out in accordance with common criteria, thereby ensuring a uniformity of approach which was generally lacking across the Member States prior to the Regulation. In line with the principle of subsidiarity, the risk assessment must inter alia demonstrate that 'concerted action at Union level is required to prevent their introduction, establishment or spread'.[75] A committee of experts will assist the Commission by evaluating the risk assessments.[76]

If listed, the IAS in question becomes subject to an import restriction in that the species cannot be intentionally brought into the EU.[77] Other restrictions include the fact that the listed species cannot be intentionally kept, bred or transported to, from or within the EU.[78] Neither can the listed species be placed on the market, allowed to reproduce, grow or be cultivated or released into the environment.[79] Additionally,

---

unnecessarily controlled' and recommended that the term 'alien species' be interpreted 'as not including native species naturally extending their range in response to climate change'.

[73] Article 4(6). The List will be adopted by means of an implementing act. The draft List is to be examined by a committee composed of representatives of Member States; see Article 27. The List will be adopted only if the committee delivers a positive opinion on the draft List in accordance with the examination procedure noted in Article 5 of Regulation 182/2011 (OJ 2011 L55/13).

[74] Article 28.
[75] Article 4(3)d.
[76] Article 27.
[77] Article 7(1)a.
[78] Article 7(1)b–d.
[79] Article 7(1)e–h.

Member States must take steps to prevent the unintentional introduction of listed species.[80] The Regulation also inter alia establishes a surveillance system to detect IAS,[81] allows emergency action to be taken in specific circumstances, and places obligations on Member States to introduce appropriate preventive, eradication and management measures.[82]

6. ANALYSING THE REGULATION IN THE LIGHT OF CBD GUIDING PRINCIPLES

6.1 Precautionary Approach – Guiding Principle 1

The first CBD Guiding Principle notes that:

> [g]iven the unpredictability of the pathways and impacts on biological diversity of invasive alien species, efforts to identify and prevent unintentional introductions as well as decisions concerning intentional introductions should be based on the precautionary approach, in particular with reference to risk analysis.

Furthermore, the precautionary approach is defined as 'that set forth in principle 15 of the 1992 Rio Declaration[83] ... and in the preamble to the [CBD]'.[84]

Additionally, a precautionary approach should be applied in giving consideration to 'eradication, containment and control measures in relation to alien species that have become established. Lack of scientific certainty about the various implications of an invasion should not be used as a reason for postponing or failing to take appropriate eradication, containment and control measures.'[85] In similar vein, Parties to the Bern Convention have acknowledged that the

> potential impact of a new alien species can only be predicted with a high degree of uncertainty. Unknown variables include the likelihood that an organism will survive transport to, establish and spread in a given location and the possible time lag before an introduced species shows invasive characteristics. For these reasons, precaution is particularly relevant to alien species issues. Precaution with regard to IAS has been described as 'guilty until proven innocent'.[86]

---

[80] Article 7(2).
[81] Article 14.
[82] Articles 17 and 19.
[83] Principle 15 of the Rio Declaration notes that '[w]here there are threats of serious or irreversible damage, lack of full scientific evidence shall not be used as a reason for postponing cost-effective measures to prevent environmental degradation'.
[84] The preamble to the CBD notes 'that where there is a threat of significant reduction or loss of biological diversity, lack of full scientific certainty should not be used as a reason for postponing measures to avoid or minimize such a threat'.
[85] The CBD Guiding Principle 12 further notes that 'mitigation measures should take place in the earliest possible stage of invasion, on the basis of the precautionary approach'.
[86] Bern Convention, Contribution to the 6th meeting of the Subsidiary Body on Scientific, Technical and Technological Advice of the CBD (Montreal, 12–14 March 2001), 8 [document T-PVS (2001) 12 revised].

196 *Research handbook on biodiversity and law*

In effect, the CBD's first Guiding Principle indicates that the identification and prevention of unintentional introductions, decisions concerning intentional introductions and consideration of mitigation measures should be based on the precautionary approach, an approach also expressly endorsed by the Bern Convention's 'European Strategy on Invasive Alien Species'.[87]

Express reference to the need for such a precautionary approach regarding IAS in the Regulation is, however, limited to one instance in its operative provisions and another in the measure's preambular recitals. In relation to the former, Article 8(1) allows some restrictions placed on intentionally introduced IAS of Union Concern to be lifted to carry out research or *ex situ* conservation pursuant to a permit issued by a Member State. Moreover, where the use of products obtained from IAS of Union Concern 'is unavoidable to advance human health' a permit issued by a Member State can also allow for the scientific production and subsequent medicinal use of such products. In relation to the possible withdrawal of such a permit, Article 8(5) then proceeds to note that Member States

> shall empower their relevant competent authority to withdraw the permit at any point in time, temporarily or permanently, if unforeseen events with an adverse impact on biodiversity or related ecosystem services occur. Any withdrawal of a permit shall be justified on scientific grounds and, *where scientific information is as yet insufficient, on the grounds of the precautionary principle* and having due regard to national administrative rules [emphasis added].

This, therefore, is a very limited circumstance in which the text of the Regulation expressly provides for a precautionary approach to be taken and would in itself fall a long way short of complying with the CBD guidance.

However, the Regulation's recitals (rather than text) do make reference to the precautionary principle in relation to the need for emergency measures in certain circumstances noting that 'Union level emergency measures would equip the Union with a mechanism to act swiftly in case of presence or imminent danger of a new invasive species in accordance with the precautionary principle'.[88] As such, under Article 10(4) of the Regulation the Commission enjoys the ability to adopt emergency measures for the Union as a whole when it receives notification from a Member State or has other evidence of the presence or imminent introduction of an IAS not as yet included on the List of IAS of Union Concern, but which is deemed likely by the Member State concerned and the Commission to meet the criteria for listing '*on the basis of preliminary scientific evidence*'. The notifying Member State can also take emergency measures in such circumstances which must be repealed or amended if the Commission adopts EU-wide emergency measures.[89] Conclusive scientific evidence is not therefore required to introduce emergency measures and the latter could as a result potentially be introduced in accordance with a precautionary approach in these limited circumstances relating to an IAS not as yet on the List. However, emergency measures cannot last indefinitely. The provisions of Article 10(3) oblige the Member State in

---

[87] Bern Convention (n 12), 3.4.
[88] Recital 20.
[89] Article 10(1) and (5).

question to carry out a risk assessment within two years of the introduction of its emergency measures, and then place the onus on the Commission to determine whether or not the species is to be formally listed taking into account the results of the assessment. If it is not subsequently included in the List, emergency measures would be withdrawn.

The decision to list or not to list a given species is the critical element of the regime established by the Regulation – only if the species is listed will the full general restrictions and control measures established by the Regulation apply. The crucial question must therefore be asked as to whether the criteria to be applied in the listing process are in line with the precautionary approach advocated by CBD guidance. The initial draft List of IAS of Union Concern is to be drawn up by the Commission within a year of the Regulation's entry into force. The draft list must therefore be compiled by the end of 2015. However, at no stage in the process established by the Regulation is a precautionary approach expressly endorsed in relation to the drafting and adoption of that List. This is surprising given the importance of the precautionary approach in CBD guidance in relation to the making of decisions on IAS. In the absence of express provision for a precautionary approach, the following three arguments might, however, be made that such an approach *could* nonetheless be applied by the Commission. By contrast, a fourth and final line of reasoning contends that it will be very unlikely such an approach will in practice be utilised in the listing process.

6.1.1 Interpretation of 'available scientific evidence'

The criteria for listing stipulated in Article 4(3)[90] inter alia note the need for a finding that the IAS in question is alien to the Union, capable of establishing a viable population and spreading in one biogeographical region shared by more than two Member States or one marine subregion. In addition, the species would be deemed 'likely to have significant adverse effects on biodiversity or the related ecosystem services' as well as a possible adverse impact on human health or the economy. All such findings must be 'based on *available* scientific evidence' (emphasis added). The required risk assessment must also be carried out with regard inter alia to the need for

---

[90] Article 4(3):

Invasive alien species shall only be included on the Union list if they meet all of the following criteria:

(a) They are found, based on available scientific evidence, to be alien to the territory of the Union excluding the outermost regions;
(b) They are found, based on available scientific evidence, to be capable of establishing a viable population and spreading in the environment under current and in foreseeable climate conditions in one biogeographical region shared by more than two Member States of one marine subregion excluding their outermost regions;
(c) They are, based on available scientific evidence, likely to have significant adverse impacts on biodiversity or the related ecosystem services, and may also have an adverse impact on human health or the economy;
(d) It is demonstrated by a risk assessment performed pursuant to Article 5(1) that action at Union level is required to prevent their introduction, establishment and spread;
(e) It is likely that the inclusion in the list will effectively prevent, mitigate or mitigate their adverse impact.

'an assessment of the potential future impacts having regard to *available* scientific evidence' (emphasis added).[91] There is therefore no reference in these conditions to the need for, say, *robust or conclusive or comprehensive* scientific evidence, but consistently only to *available* scientific evidence. It might therefore be argued, albeit very tentatively, that the adoption of a precautionary approach in the listing of IAS would implicitly be appropriate and in line with Principle 15 Rio Declaration, which notes that '[w]here there are threats of serious or irreversible damage, lack of full scientific evidence shall not be used as a reason for postponing cost-effective measures to prevent environmental degradation'. However, this argument would be much stronger if the text had also explicitly referred to the fact that findings should be based on the available scientific evidence even where that falls short of being conclusive. Alternatively, the condition might have been expressed in terms of requirement for 'preliminary scientific evidence' as is required for the introduction of emergency measures. No such references are in fact present in the Regulation with regard to the listing process.

6.1.2 Provisions of the Treaty on the Functioning of the European Union (TFEU)

Article 191(1) TFEU notes that Union policy on the environment shall contribute to the pursuit of environmental protection and preservation, as well as to the protection of human health. Furthermore, Article 191(2) TFEU indicates that the Union's environmental policy is 'based on the precautionary principle'. Bearing these two treaty provisions in mind in the specific context of protecting human health, the European Court of Justice (ECJ) in *Gowan Comercio Internacional e Servicos* indicated that '[i]t follows from the precautionary principle that, where there is uncertainty as to the existence or extent of risks to the health of consumers, *the institutions may take protective measures without having to wait until the reality and the seriousness of those risks become fully apparent*' [emphasis added].[92]

By analogy, it is submitted that the Commission would also be in a position to add an IAS to the List where there is uncertainty as to the existence of or extent of risk to either the environment or human health posed by the species in question. If one accepts this argument, a precautionary approach *could* therefore be utilised by the Commission in making any decisions on the listing of IAS. However, the Commission would not be obliged to adopt such an approach.

6.1.3 Risk assessment methodology

It has already been noted that Article 4(3) stipulates the need for a risk assessment in the listing process and that this risk assessment must demonstrate that 'concerted action at Union level is required to prevent their introduction, establishment or spread'. Article 5(1) of the Regulation notes that a risk assessment is to have regard to certain elements

---

[91] Article 5(1)f (emphasis added).
[92] Case C-77/09 *Gowan Comercio Internacional e Servicos v Ministero della Salute* European Court Reports (2010) para. 73. Furthermore '[w]here it proves to be impossible to determine with certainty the existence or extent of the alleged risk because of the insufficiency, inconclusiveness or imprecision of the results of studies conducted, but the likelihood of real harm to public health persists should the risk materialise, the precautionary principle justifies the adoption of restrictive measures, provided they are non-discriminatory and objective': para. 76.

which inter alia include the range of the species, its reproduction and spread patterns, any potential pathways of introduction, the risks of introduction, a projection as to likely future distribution, and the potential costs of damage. Furthermore, the European Commission is empowered to adopt by means of delegated legislation a detailed description of the application of these factors.[93] This description is to include the methodology to be applied in risk assessments. In the absence of an express endorsement of the precautionary approach in the listing criteria set out in the Regulation, the methodology endorsed by the Commission in this respect will be of much importance as it could potentially endorse the taking of a precautionary approach. However, this eventuality is improbable when one bears in mind that, although the precautionary principle is specifically mentioned in the particular situations already highlighted, such an approach is not expressly endorsed by the Regulation itself in relation to the listing process.

6.1.4 Application of World Trade Organization (WTO) rules

It is submitted that it is unlikely that the Commission would opt to pursue any of the three above-mentioned avenues which would arguably have allowed it to apply the precautionary approach to the listing process. The key reason for this submission is that the EU and all its Member States are members of the WTO and therefore bound by its trade rules. Those rules as they relate to the protection of human, animal or plant life or health from particular identified risks (such as pests and diseases) are to be found in the 1995 WTO Agreement on Sanitary and Phytosanitary Measures ('SPS Agreement'). Trade is, of course, a key pathway for the introduction into the EU of IAS. It will be recalled that once listed under the Regulation, border and transportation controls must inter alia be applied to ensure that the IAS of Union Concern are not intentionally or unintentionally brought into the EU. The SPS Agreement potentially applies to such regulatory measures and aims to ensure that their application is not an arbitrary or unjustified restriction on trade. The relevant articles of the SPS Agreement in this context are Articles 2.2, 5.1 and 5.7. Article 2.2 provides that:

> Members shall ensure that any sanitary or phytosanitary measure is applied only to the extent necessary to protect human, animal or plant life or health, is based on scientific principles *and is not maintained without sufficient scientific evidence, except as provided for in paragraph 7 of Article 5* (emphasis added).

By virtue of Article 5.1 any sanitary or phytosanitary measure must also be based on a risk assessment of the threat to human, animal or plant life or health, as is indeed similarly required under the Regulation.[94] As a general rule, therefore, any measure introduced for sanitary or phytosanitary reasons cannot be maintained under WTO rules unless there is 'sufficient scientific evidence' to support it. This begs the question, what exactly amounts to 'sufficient scientific evidence'? In *US – Poultry (China)* the WTO's Appellate Body noted that for a measure to be maintained with 'sufficient' scientific

---

[93] Article 5(3).
[94] Recital 11 of the Regulation notes that the criteria for listing 'should include a risk assessment pursuant to the applicable provisions under the relevant Agreements of the World Trade Organization on placing trade restrictions on species'.

evidence 'the scientific evidence must ... be sufficient to demonstrate the existence of the risk which the measure is supposed to address'.[95] The need for such evidence would seemingly raise a significant question mark over whether a precautionary approach could be applied in relation to the introduction of an import restriction and still be in line with the provisions of Article 2.2.

However, could a precautionary approach be adopted in these circumstances under another provision of the SPS Agreement? Article 5.7 encouragingly notes that:

> In cases *where relevant scientific evidence is insufficient, a Member may provisionally adopt sanitary or phytosanitary measures on the basis of available pertinent information*, including that from the relevant international organizations as well as from sanitary or phytosanitary measures applied by other Members. In such circumstances, Members shall seek to obtain the additional information necessary for a more objective assessment of risk and review the sanitary or phytosanitary measure accordingly within a reasonable period of time (emphasis added).

The WTO's Appellate Body in *US/Canada – Continued Suspension* referred to the application of Article 5.7 as a 'temporary "safety valve" in situations where some evidence of risks exists but not enough to complete a full risk assessment, thus making it impossible to meet the rigorous standards set by Articles 2.2 and 5.1'.[96] In *Japan – Agricultural Products II* it was further made clear that the adoption and maintenance of any such provisional measure under Article 5.7 must comply with the following requirements:

> [be] (1) imposed in respect of a situation where 'relevant scientific information is insufficient'; and (2) adopted 'on the basis of available pertinent information'. Pursuant to the second sentence of Article 5.7, such a provisional measure may not be maintained unless the Member which adopted the measure: (1) 'seek[s] to obtain the additional information necessary for a more objective assessment of risk'; and (2) 'review[s] the ... measure accordingly within a reasonable period of time'.[97]

All four of these requirements must be met.[98] Moreover, the insufficiency of the scientific evidence must be such that an adequate risk assessment (as required by Article 5.1 of the SPS Agreement) is not possible.[99] Bearing this Article 5.7 jurisprudence in mind, the Regulation's provisions as they relate to emergency measures potentially appear to be in line with Article 5.7 SPS Agreement. The WTO's Appellate Body has indeed expressly acknowledged in *EC – Hormones* that '[t]he precautionary principle ... finds reflection in Article 5.7 of the SPS Agreement'.[100] However, it should be noted that any emergency measure adopted under the Regulation would of

---

[95] WTO Appellate Body Report, *US – Poultry (China)* (2010), para. 7.200.
[96] WTO Appellate Body Report, *US/Canada – Continued Suspension* (2008), para. 678.
[97] WTO Appellate Body Report, *Japan – Agricultural Products II* (1999), para. 89. What amounts to a 'reasonable period of time' in this context was discussed in the same case; *see* para. 93.
[98] *Ibid.*
[99] WTO Appellate Body Report, *Japan – Apples* (2003), para. 184.
[100] WTO Appellate Body Report, *EC – Hormones* (1998), para. 124.

course need to be reviewed later and that, once emergency measures have been applied, an active obligation is placed on the Commission and the Member State concerned to seek to obtain a sufficiency of information to enable a full risk assessment to be carried out.

The Appellate Body in *EC – Hormones* did nonetheless note that:

> [a] Panel charged with determining, for instance, whether 'sufficient scientific evidence' exists to warrant the maintenance by a Member of a particular SPS measure may, of course, and should, bear in mind that responsible, representative governments commonly act from perspectives of prudence and precaution where risks of irreversible, e.g. life-terminating, damage to human health are concerned.[101]

However and importantly, the Appellate Body additionally stated that the precautionary principle 'has not been written into the SPS Agreement as a ground for justifying SPS measures that are otherwise inconsistent with the obligations of members set out in particular provisions of that Agreement'.[102] The Appellate Body's deliberations in this case have prompted two highly distinguished WTO legal experts to conclude that '[t]he practical effect ... is to limit the relevance of the precautionary principle under the SPS Agreement to the situation covered by Article 5.7. The precautionary principle can thus not be relied upon to add flexibility to the scientific disciplines in Articles 2.2 and 5.1 of the SPS Agreement.'[103] In sum, the precautionary principle can legitimately apply under WTO rules to the introduction of *emergency* measures, but it is very doubtful indeed that it could otherwise be applied to justify a more permanent measure as the latter would very probably fail the test for 'sufficient' scientific evidence under the provisions of Article 2.2. To conclude, the Regulation is very arguably in line with both the CBD guidance and the SPS Agreement in relation to the applicability of a precautionary approach to the introduction of emergency measures. However, no express mention is made as to the application of such an approach to the listing of IAS of Union Concern. It is likely that a precautionary approach in the listing process will not take place as it is very arguably not allowed under the SPS Agreement by which the EU and its Member States are legally bound. It is submitted that the EU in adopting the approach to the listing of IAS of Union Concern in the Regulation has chosen to side with the general line endorsed by the 'hard law' SPS Agreement rather than the 'soft law' CBD guidance.

6.2 Three-stage Hierarchical Approach – Guiding Principle 2

The CBD guidelines advocate that priority should be given to preventing the introduction of IAS. Prevention is cost-effective and regarded as more 'environmentally desirable than measures taken following introduction and establishment of an invasive

---

[101] *Ibid.*
[102] *Ibid.*
[103] P Van den Bossche and W Zdouc, *The Law and Policy of the World Trade Organization* (3rd edn., CUP, 2013), 932. Peter van Den Bossche is a current member of the WTO's Appellate Body, while Werner Zdouc is the current Director of the Appellate Body Secretariat.

alien species'.[104] It is, of course, true that any action designed to prevent introduction will negate the need for the significant human and economic resources otherwise required to eradicate IAS once they have been released in to the environment. Prevention is therefore the first approach to be applied where possible. However, bearing in mind that such an approach will not always be sufficient, there must also be a system of 'early and rapid action ... to prevent [the] establishment' of IAS that have already been introduced.[105] In this respect, the preference would be to eradicate the species as soon as possible.[106] It is, for example, believed that had the Zebra mussel (*dreissena polymorpha*) invasion of the Ebro Delta (Spain) been tackled earlier by means of eradication there would not now be a need for an ongoing effort costing more than 4 million Euro each year to limit the damage caused by the invasion and prevent the further spread of these alien mussels.[107] Early and rapid action to include detection and eradication is therefore the second approach to be applied in terms of hierarchy. Nonetheless, the CBD guidance also acknowledges that a programme of eradication is not always possible or that there may be a lack of resources to implement such an approach. As such, efforts should then be taken to contain the spread of the species,[108] and introduce long-term control measures.[109] Action to contain and control is therefore the third stage of the hierarchical approach endorsed by the Guiding Principles. To what extent does the Regulation endorse such a three-stage hierarchical approach?

6.2.1 Prevention – stage one

The preamble of the Regulation fully endorses the notion that priority should indeed be given to preventing the introduction of invasive alien species, thereby halting IAS establishment from the outset; this preventive approach is also recognised as being both cost-effective and desirable from an environmental point of view in line with CBD guidance.[110] The application of a preventive approach, wherever possible, is indeed also generally endorsed in the text of the Regulation. Those aforementioned restrictions, for example, which are placed on IAS of Union Concern, are in line with such a general approach.[111] Indeed, where derogations from such restrictions are allowed under Article

---

[104] Guiding Principle 2(1).
[105] Guiding Principle 2(2).
[106] *Ibid. See* further Principle 13.
[107] European Commission, *Invasive Alien Species: a European Response* (European Commission, 2014), 11.
[108] Guiding Principle 14.
[109] Guiding Principle 15.
[110] Preamble, paragraph 15.
[111] These restrictions are noted in full in Article 7:

1. Invasive alien species of Union concern shall not be intentionally:
   (a) brought into the territory of the Union, including transit under customs supervision;
   (b) kept, including in contained holding;
   (c) bred, including in contained holding;
   (d) transported to, from or within the Union, except for the transportation of species to facilities in the context of eradication;
   (e) placed on the market;
   (f) used or exchanged;

8 by way of permit to facilitate research, *ex situ* conservation, or the development of medicinal products, tight conditions are to be applied to ensure the proper containment and surveillance of the species concerned, thereby limiting the chances of escape or removal into the wider environment. Contingency measures – including eradication plans – in anticipation of possible escape are also required.[112] Two further provisions in the Regulation can also be said to be generally in line with the preventive approach: first, a Member State's ability to introduce its own emergency measures in relation to IAS not as yet on the List but with the potential to meet the criteria for such listing;[113] and, secondly, the obligation placed on Member States to perform an analysis of the pathways of unintentional introduction of IAS of Union Concern, which should include an identification of those pathways of priority concern due to their potential for introducing either large numbers of species or species particularly damaging to the environment or public health.[114]

It is also of interest to note that the CBD guidance notes the need to take action to prevent the spread of IAS not only between States, but also within States. In this regard, the Regulation gives Member States the opportunity to establish their own national list of IAS of concern ('national list of invasive alien species of Member State concern'), and apply national restrictions that are compatible with the TFEU.[115] The ability to draw up such a national list and apply national restrictions is an acknowledgement of the fact that not all invasive species will necessarily require control at the EU level, but might need to be regulated within a given Member State. The hedgehog (*erinaceus europaeus*) is an example of such a species; although native to western and some parts of northern Europe (including mainland Scotland), it is non-native and invasive on the isles of North Uist, South Uist and Benbecula, which form part of Scotland's Western Isles.[116] The ability to create national lists of IAS is in line with the CBD's preventive approach.

The Regulation can nevertheless be criticised for widening the ability to derogate too far. By virtue of Article 9 Member States may 'in exceptional circumstances' allow permits to be given to establishments to carry out activities other than those provided for in Article 8 for reasons of 'compelling public interest, including those of a *social or*

      (g) permitted to reproduce, grown or cultivated, including in contained holding; or
      (h) released into the environment.
2. Member States shall take all necessary steps to prevent the unintentional introduction or spread, including, where applicable, by gross negligence, of invasive alien species of Union concern.

[112] Article 8.
[113] Article 10.
[114] Article 13(1).
[115] Article 12(1). An 'Invasive alien species of Member State concern' is defined as 'an invasive alien species other than an invasive alien species of Union concern, for which a Member State considers on the basis of scientific evidence that the adverse impact of its release and spread, even where not fully ascertained, is of significance for its territory, or part of it, and requires action at the level of that Member State': Article 3(4).
[116] *See* 'Why hedgehogs are not welcome in the Hebrides'; <http://www.bbc.co.uk/nature/22093131>.

economic nature.[117] This significantly broadens the potential for derogations and was included as a consequence of negotiations between the European Parliament and the Council on the Commission's original proposal. Any such derogation would need to be vetted and approved by the Commission,[118] and, if approved, the latter can establish conditions to be included in any permit issued by a Member State.[119] These conditions may well include the need to contain the IAS in question in specified establishments to prevent escape or unlawful release.[120] Nonetheless, the provisions of Article 9 allow for the potential authorisation of certain commercial activities which would not otherwise be allowed.[121]

6.2.2 Early and rapid action: detection and eradication – stage 2

It will be recalled that the CBD Guiding Principles call for a system of 'early and rapid action … to prevent [the] establishment' of IAS that have already been introduced, and that a preference is expressed to eradicate at the earliest opportunity.[122] In relation to the need for early and rapid action the Regulation places a clear obligation on all Member States to establish a surveillance system within 18 months of the adoption of the List,[123] which is to be 'sufficiently dynamic to detect rapidly the appearance … of any invasive alien species of Union Concern, whose presence was previously unknown'.[124] Moreover, once an IAS of Union Concern has been detected the Member State concerned must inform the Commission 'without delay' as well as other Member States,[125] and within three months of such notification 'apply eradication measures'[126] utilising methods which 'are effective in achieving the complete and permanent removal of the population'.[127] The Regulation therefore broadly complies with the CBD's Guiding Principles in this regard. An early response is duly recognised as being the most cost-effective and efficient manner in which to tackle the spread of an IAS.

---

[117] Article 9(1).
[118] *Ibid.*
[119] Article 9(6).
[120] Article 9(1).
[121] Note should be made that the Regulation also includes certain transitional provisions for commercial stocks (Article 32). These inter alia include the ability of keepers of a commercial stock of specimens of IAS of Union Concern to sell such specimens for a period of one year after inclusion of the species in question on the List to non-commercial users. It is also of interest to note that, by virtue of Article 31, non-commercial owners of specimens of IAS on the List are allowed to keep them until they die if they are 'companion animals' as long as they are kept in contained holding and measures are put in place to make sure reproduction and escape are not possible.
[122] Principle 2(2).
[123] Article 14(1).
[124] Article 14(2).
[125] Article 16.
[126] Article 17(1).
[127] Article 17(2).

6.2.3 Containment and control: management of established IAS of Union Concern – stage 3

It has already been noted that the Guiding Principles acknowledge that a programme of eradication is not always possible or that there may be a lack of resources to implement such an approach. In these circumstances steps should be taken to contain the spread of the species to within a given area and to ensure long-term control measures are in place. Reflecting the essence of this third stage of the hierarchical approach the Regulation does indeed specify that Member States must have effective management measures in place so that the impact of widely spread IAS of Union Concern are minimised within 18 months of an IAS being placed on the List;[128] such national measures will include not only eradication programmes but also steps aimed at the population control or containment of a given population.[129] The need to have in place such management measures is to be welcomed. However, although not out of line with CBD guidance, the time limit of 18 months within which such measures are to be put into place is less than ideal in facilitating the effective control of the species in question.

6.3 Ecosystem Approach – Guiding Principle 3

This CBD Guiding Principle notes that '[m]easures to deal with invasive alien species should, as appropriate, be based on the ecosystem approach, as described in decision V/6 of the Conference of the Parties'. Decision V/6 describes such an approach as 'a strategy for the integrated management of land, water and living resources that promotes conservation and sustainable use in an equitable way'.[130] This ecosystem approach also focuses on the 'structure, processes, functions, and interactions among organisms and their environment',[131] and underlines the importance of the 'conservation of ecosystem structure and functioning, in order to maintain ecosystem services'.[132]

The Regulation's provisions relating to rapid eradication of IAS at an early stage of invasion could be said to be generally in line with this approach as methods of eradication should be used but 'with due regard to human health and the environment, especially non-targeted species and their habitats'.[133] Furthermore a Member State may in fact decide not to apply eradication measures at all if they 'have serious adverse impact on human health, the environment and other species'. Additionally, applicable management measures envisaged by the Regulation 'shall include actions applied to the

---

[128] Article 19(1).
[129] Article 19(2).
[130] CBD Decision V/6, para. 1.
[131] *Ibid.*, Principle 5.
[132] *Ibid.*, para. 1 and Principle 5.
[133] Article 17(2).

receiving ecosystem aimed at increasing its resilience to current and future invasions',[134] and are to be applied in Member States so that the impact of IAS on biodiversity and related ecosystem services are minimised.[135]

Another key principle of the ecosystem approach noted in CBD Decision V/6 is that 'ecosystem managers should consider the effects of their activities on adjacent and other ecosystems'. The impact of management measures by a given ecosystem manager aimed at the eradication, population control or containment of IAS can clearly have a negative effect on adjacent or other ecosystems. Such ecosystems may span territorial boundaries and it is therefore of interest to note that the Regulation seeks to foster consultation and cooperation between Member States by, for example, applying jointly agreed management measures where there is a risk that a given IAS is likely to spread to another Member State.[136]

What of the need to restore a given ecosystem adversely affected by the spread of an IAS? CBD Decision V/6 acknowledges the need to not only conserve but also, where appropriate, restore ecological interactions and processes to preserve ecosystem structure and functioning. This could be said to be in line with the obligation placed on Member States in the Regulation to take 'appropriate restoration measures to assist the recovery of an ecosystem which has been degraded, damaged, or destroyed' by IAS of Union Concern.[137] It is of interest to note, however, that the final text of the Regulation following European Parliament and Council discussion introduced a proviso that such restorative measures would not be required when a cost-benefit analysis indicated that 'the costs of those measures will be high and disproportionate to the benefits of restoration'.[138] This represents a significant watering down of the obligation to take restorative action as one would anticipate that restoration costs in many instances may well be substantial, or that a given Member State will at least claim they are prohibitive when justifying a decision not to take restorative action.

6.4 The Role of States – Guiding Principle 4

Guidance from the CBD underlines that States:

> should recognize the risk that activities within their jurisdiction or control may pose to other States as a potential source of invasive alien species, and should take appropriate individual and cooperative actions to minimize that risk, including the provision of any available information on invasive behaviour or invasive potential of a species.[139]

---

[134] Article 19(2).
[135] Article 19(1).
[136] Article 19(5).
[137] Article 20(1). Under the Regulation restoration measures will include as a minimum steps not only to 'increase the ability of an ecosystem exposed to disturbance caused by the presence of [IAS of Union Concern] to resist, absorb, accommodate to and recover from the effects of disturbance', but also measures which 'support the prevention of reinvasion following an eradication campaign': Article 20(2).
[138] Article 20(1).
[139] Principle 4(1).

Furthermore, an onus is placed on States to 'identify, as far as possible, species that could become invasive and make such information available to other States' with a view to minimising the spread and impact of IAS.[140]

A number of the Regulation's provisions could be said to be in line with this approach. For example, the fact that a given Member State is allowed to take emergency measures if it is aware of the imminent introduction in its territory of an IAS which is not yet included on the List, but which it finds is likely to meet the criteria for listing.[141] Such emergency measures may include a ban on the transit to and from other EU States. In that respect, the Member State is endeavouring to reduce the risk posed to a neighbouring Member State. Any emergency measure imposed must also be made known to all other Member States together with accompanying evidence as to the risk involved.[142] A further example is provided by the fact that any Member State can submit to the Commission a request that an IAS be added to the List.[143] In so doing, they must provide certain information relevant to the case, including a risk assessment and evidence that inter alia the IAS concerned is likely to have significant adverse impacts on biodiversity or related ecosystem services. As such they would indeed be identifying species which 'could become invasive' and would also ultimately be making information available to other States (via the Commission) in accordance with the CBD Guiding Principles.

6.5 Research and Monitoring – Guiding Principle 5

The Parties to the CBD have seen fit to endorse the need for research on and the monitoring of IAS 'to develop an adequate knowledge base to address the problem'. More particularly, research on an IAS should document '(a) the history and ecology of invasion (origin, pathways and time-period); (b) the biological characteristics of the invasive alien species; and (c) the associated impacts at the ecosystem, species and genetic level and also social and economic impacts, and how they change over time'. Apart from an obligation on Member States to carry out a 'comprehensive analysis of the pathways of unintentional introduction and spread of [IAS of Union Concern], at least, in their territory',[144] there is surprisingly no express obligation placed on Member States in the Regulation to carry out this type of general research. The Regulation is therefore deficient in this regard when compared with the CBD Guiding Principles. This is particularly disappointing as recognised IAS experts have recently concluded that 'managers lack appropriate risk assessment methods to prioritise invasion threats because few general models or "rules of thumb" exist on which to predict the occurrence and impacts of IAS',[145] and have indeed endorsed the need to 'target the

---

[140] Principle 4(3).
[141] Article 10(1).
[142] Article 10(2).
[143] Article 4(4).
[144] Article 13(1).
[145] JM McCaffrey and others, 'Tackling Invasive Alien Species in Europe: the Top 20 Issues' (2014) 5(1) Management of Biological Invasions 1, 10.

R&D needed to increase confidence levels in risk assessment methods'.[146] The Regulation could have been improved by specifically highlighting the need for general research in a number of areas such as the ecology and biology of IAS, the susceptibility of European ecosystems to invasions, and the wide-ranging impact of IAS for society.[147]

As far as monitoring is concerned, the measure is more comprehensive. Reflecting the CBD guidance, the Regulation stipulates that 'surveillance systems offer the most appropriate means of early detection of new invasive species',[148] and that such monitoring should include both targeted and general surveys.[149] More particularly, Member States are placed under an obligation to establish the surveillance system of IAS of Union Concern referred to earlier that records data on the occurrence of IAS (both new and already established species) within 18 months of the List being adopted.[150] This system is also to be utilised to monitor the effectiveness of any eradication, population control or containment programmes,[151] and to contribute to the early detection of IAS.[152] The obligation to monitor is encouraging, especially as the majority of Member States had not established a monitoring system specifically to identify the presence of IAS prior to adoption of the Regulation.[153] In implementing their monitoring obligations, it is submitted that Member States must seek to target the most potentially important points of entry (such as airports, train stations and ports), as well as other relevant locations (such as zoos and horticultural establishments). An effective surveillance system is made all the more important when one bears in mind that, once an IAS is within the EU's borders, there will be no border checks between Member States in line with the functioning of the internal market.

6.6 Education and Public Awareness – Guiding Principle 6

Stressing that 'raising the public's awareness of the invasive alien species is crucial to the successful management of invasive alien species', this Guiding Principle underlines the importance of promoting 'education and public awareness of the causes of invasion and the risks associated with the introduction of alien species'. Certainly a general lack of awareness of the problems caused by IAS creates a considerable impediment to addressing IAS spread. The Regulation acknowledges this and would seem to be in line with CBD guidance by, for example, placing Member States under an obligation to establish and implement action plans to address the key pathways of unintentional introduction of IAS of Union Concern into its territory; these plans will include measures designed to 'raise awareness'.[154] It is suggested that action to increase

---

[146] Ibid., 4.
[147] A number of these areas of potential research are highlighted in the Bern Convention's 'European Strategy on Invasive Alien Species' (n 12), 2.2 at Box 4.
[148] Recital 22.
[149] Ibid.
[150] Article 14(1) and (2).
[151] Article 19(4).
[152] Article 14(2)b.
[153] European Commission (n 67), 93.
[154] Article 13(4).

awareness of the causes and risks of IAS must not only seek to enlighten the general public, but go beyond the CBD guidance by also educating key decision-makers, NGOs and particular organisations within, for example, the tourist, fisheries, transport, horticultural and agricultural industries.[155] Member States would do well to take notice of some existing programmes which have sought to raise awareness within a local population or target group. These include the 'Weedbuster' programme in Australia, which encourages the public to take part in events in their local area to eradicate weeds (including invasive alien species),[156] and the AlterIAS ('ALTERnatives to Invasive Alien Species') project, designed to improve awareness of IAS within the horticultural industry.[157]

CBD guidance additionally notes that raising awareness is particularly important when mitigation measures are required. At this point 'education and public-awareness-oriented programmes should be set in motion so as to engage local communities and appropriate sector groups in support of such measures'. In this respect, the Regulation stipulates that the public must be given an opportunity to participate in the drawing up of relevant action plans and in the establishment of management measures to address IAS of Union Concern,[158] thereby 'contributing to public awareness of environmental issues and support for the decisions taken'.[159] It is clear that some plans to eradicate IAS have failed through lack of support. Public opposition to the removal of, for example, the grey squirrel (*sciurus carolinensis*) in Italy, the hedgehog from the Western Isles, and of the population of coypu (*myocastor coypus*) from a lake in Sicily have marked the end of planned eradications.[160]

6.7 Border Control and Quarantine Measures – Guiding Principle 7

This aspect of the CBD guidance stipulates that States should introduce 'border controls and quarantine measures' for IAS that would ensure that intentional introductions are subject to appropriate authorisation and also that unintentional introductions are minimised. More particularly in relation to intentional introductions, the Regulation specifies the need for structures to be put in place by 2 January 2016 to perform 'official controls' to prevent introduction.[161] Such controls comprise 'documentary, identity and, where necessary, physical checks'.[162] No further detail is provided as to the nature of the controls but presumably they could indeed include quarantine

---

[155] This is also acknowledged in the Bern Convention's 'European Strategy on Invasive Alien Species' (n 12), 1 at Box 2.
[156] <http://www.daf.qld.gov.au/plants/weeds-pest-animals-ants/weeds/weedbusters>.
[157] <http://www.alterias.be/en/about-us-the-alterias-project/project-description>.
[158] Article 26.
[159] Recital 29.
[160] *See* P Genovesi, 'Eradications of Invasive Alien Species in Europe: A Review' (2005) 7 Biological Invasions 127, 130–131.
[161] Article 15(1).
[162] Article 15(3).

measures which are, of course, a type of 'physical check'.[163] With regard to unintentional introduction, the Regulation obliges Member States to include in their action plans on IAS pathways measures that 'ensure appropriate checks at the Union borders'.[164] The checks in question are to be made when goods commonly linked to the introduction of IAS are brought into the Union.[165] Checks are not, however, to be made when goods already within the Union pass between Member States. This is compliant with the notion of the free movement of goods within the EU's internal market, but is not in line with the CBD guidance in that it fails to impose checks when goods pass through national borders between EU Member States.

6.8 Exchange of Information – Guiding Principle 8

The need to share information on IAS is underlined in the CBD guidance:

> States should assist in the development of an inventory and synthesis of relevant databases, including taxonomic and specimen databases, and the development of information systems and an interoperable distributed network of databases for compilation and dissemination of information on alien species for use in the context of any prevention, introduction, monitoring and mitigation activities.

In this context it is acknowledged that there are existing databases established which already provide invaluable information on Europe's IAS. Arguably the most significant is the European Alien Species Information Network (EASIN), which is a project set up by the European Commission's Joint Research Centre and provides free information on IAS (such as location, taxonomy and applicable pathways) from a variety of existing databases.[166] Several are national databases, resources which have been established by some – but not all – Member States. Prior to adoption of the Regulation, for example, no information could be found in relation to the establishment of joint information systems on IAS in nine Member States.[167]

The Regulation importantly stipulates that the Commission is to establish 'progressively' an information support system, which, within a year of the entry into force of the Regulation, will include 'a data support mechanism interconnecting existing data systems [on IAS] paying particular attention to information on the [IAS of Union Concern]'.[168] Once an IAS of Union Concern has been detected the Member State

---

[163] The quarantine of certain animals is undoubtedly allowed under certain EU measures: *see*, for example, Directive 91/496/ EEC laying down the principles governing the organisation of veterinary checks on animals entering the Community from third countries, OJ 1991 L268/56; Article 12(1)b.
[164] Article 13(4).
[165] Article 15(3).
[166] <http://easin.jrc.ec.europa.eu>. Initiatives which provide information to EASIN include the important NOBANIS database, which covers IAS in North and Central Europe, and the DAISIE database, which covers IAS in all Member States. At the international level, *see* the Global Invasive Species Database: <www.issg.org/database/welcome>.
[167] European Commission (n 67), 58.
[168] Article 25(1) and (2).

concerned must inform the Commission and other Member States,[169] and apply eradication measures.[170] The information support system to be established by the Commission will from its infancy assist the Commission and Member States in the handling of Member State notifications of this nature.[171] Furthermore, the data support mechanism will within four years of entry into force of the Regulation be a system that allows information to be exchanged, which 'may' include information on 'invasive alien species of Member State concern, pathways, risk assessment, management and eradication measures, when available'.[172] It is unfortunate that the provision in the Regulation only refers to the fact that such information *may* rather than *must* be included, but it is to be hoped that over a period of time an EU-wide system of information support will be established which will comply with CBD guidance.[173]

CBD guidance further stipulates that the dissemination of information gathered on IAS should be facilitated through the CBD's Clearing-House Mechanism (CHM). The latter seeks to provide easy access to knowledge and information relating to the aims of the CBD, including that relating to the control of IAS. Whilst the CHM is not specifically mentioned in the Regulation, Article 22(1) places Member States under an obligation to ensure close coordination with 'existing structures arising under … international agreements', and Article 25(2) stipulates that the information support mechanism established by the Regulation will connect with existing data systems on IAS. Although specific reference to the internationally important CHM in the Regulation might have been helpful,[174] it is fair to assume that such information will be made available to it bearing in mind the EU and its Member States already contribute to the CHM in relation to a variety of biodiversity-related issues.[175]

6.9 Cooperation, Including Capacity-building – Guiding Principle 9

The Regulation is of course itself an example of cooperation between States to combat the spread of IAS of Union Concern. As such, the measure is certainly 'a cooperative effort between two or more countries', which, according to the CBD guidance, may be required in particular circumstances. The latter also places an emphasis on the sharing of information on IAS 'with a particular emphasis on cooperation among neighbouring countries, between trading partners, and among countries with similar ecosystems and

---

[169] Article 16.
[170] Article 17(1).
[171] Article 25(2).
[172] Article 25(3).
[173] After all, much of this type of information is to be routinely gathered by Member States – for example, eradication measures applied at the national level are to be monitored by the Member State concerned and information as to their effectiveness given to the Commission and other Member States (Article 17(3) and (4)). The sharing of this type of information either within the information support system of otherwise would be invaluable as it would shine a light on which particular eradication approach has proved effective and may thereby offer lessons to other Member States contemplating an eradication programme.
[174] The more general wording could, however, be useful if the CBD mechanism were to change its name in the future.
[175] *See* <http://www.cbd.int/chm/network/>.

histories of invasion. Particular attention should be paid where trading partners have similar environments.'[176] The sharing of information of this type is likely to form part of the information support system already discussed which is to be established by the Commission. Additionally, Member States are obliged by the Regulation to 'make every effort' to coordinate their efforts with other Member States, particularly where the States in question share the same biogeographical and marine regions, borders and river basins.[177] Cooperation of this nature is essential as a given habitat or ecosystem may well straddle national boundaries, thereby requiring effective coordination of preventive and/or mitigation activity between those Member States concerned. A good example of existing cross-border cooperation to control an IAS is the Bern Convention's Action Plan for eradication of the Ruddy Duck in the Western Palaearctic region.[178] Feral Ruddy Ducks initially spread from the UK to neighbouring countries (France, the Netherlands and Belgium), but went on to migrate to over 20 countries in the Western Palaearctic area. The UK population of these birds is believed to have been reduced to approximately 40 as of April 2014, and the 2011 Bern Convention Action Plan aimed to eradicate all Ruddy Ducks in the Western Palaearctic by 2015.[179]

Furthermore, Member States under the Regulation will 'endeavour' to cooperate with third countries to fulfil the Regulation's aims,[180] and also there is an obligation that 'based on best practices' the Commission and the Member States will 'develop guidelines and training programmes to facilitate the identification and detection of [IAS of Union Concern] and the performance of efficient and effective controls'.[181] In this way there will therefore be an opportunity to learn from Member States' experiences as to which controls are the more successful in preventing the intentional introduction of IAS of Union Concern.

6.10 Intentional Introduction – Guiding Principle 10

The CBD Parties agreed that the intentional introduction of invasive or potentially invasive alien species should not take place without prior authorisation, and that an 'appropriate risk analysis, which may include an environmental impact assessment, should be carried out as part of the evaluation process' before deciding whether to authorise the introduction.[182] The Regulation certainly requires official controls to be in place to prevent the intentional introduction of IAS of Union Concern into the EU

---

[176] Principle 9(a).
[177] Article 22(1).
[178] The initial Action Plan was adopted in 1999. A subsequent plan was adopted in 2011; *see* Bern Convention, *Eradication of the Ruddy Duck in the Western Palaearctic: a Review of Progress and a Revised Action Plan, 2011–2015* (Council of Europe, 2010); document T-PVS/Inf (2010) 21 revised.
[179] Animal Health and Veterinary Laboratories Agency, *UK Ruddy Duck Eradication Programme Project Bulletin* (April 2014).
[180] Article 22(2).
[181] Article 15(8). This is also generally in line with Guiding Principle 9(c), which encourages capacity-building which may include 'the development of training programmes' where States lack the requisite expertise.
[182] Principle 10(1).

without prior authorisation.[183] The EU measure also requires that a risk assessment is carried out and the results taken into account in making the decision whether or not to list a particular alien species.[184] The risk assessment includes the need to assess the adverse impact on biodiversity and related ecosystems, as well as on human health, safety and the economy.

However, the CBD Guiding Principles further stipulate that decisions

> concerning intentional introductions should be based on the precautionary approach, including within a risk analysis framework ... . Where there is a threat of reduction or loss of biological diversity, lack of sufficient scientific certainty and knowledge regarding an alien species should not prevent a competent authority from taking a decision with regard to the intentional introduction of such alien species to prevent the spread and adverse impact of invasive alien species.

As previously noted, there is limited express reference to a need for such a precautionary approach in the Regulation. Certainly the Regulation fails to provide an express basis for the application of such an approach in carrying out risk assessments and, in this regard, the measure – as earlier indicated – would appear to be out of line with the CBD Guiding Principles.

6.11 Unintentional Introductions – Guiding Principle 11

The CBD Guiding Principles note that States 'should have in place provisions to address unintentional introductions' which could include statutory and regulatory provisions as well as the formation or strengthening of relevant bodies. The Regulation for its part indicates that Member States 'shall take *all necessary steps* to prevent the unintentional introduction or spread of invasive species of Union Concern',[185] which could be said to be broadly in line with such CBD guidance as the highlighted words would allow for the application of numerous measures, including those specifically identified in the CBD guidance. The Regulation's recitals more particularly note that measures could include both voluntary and mandatory measures.[186] The Regulation interestingly and positively also provides an animal welfare consideration not included in the CBD guidance in that where animals are to be eradicated they should be 'spared any avoidable pain, distress or suffering'.[187] In this respect, best practices should be taken into account such as the 'Guiding Principles on Animal Welfare' developed by the World Organisation for Animal Health.[188] Adherence to such welfare issues may indeed increase the public's acceptance of eradication plans.

---

[183] Article 15(2).
[184] Article 5(2).
[185] Article 7(2) (emphasis added).
[186] Recital 21, which further notes that measures should 'build on the experience gained ... in managing certain pathways, including measures established through the International Convention for the Control and Management of Ships Ballast Water and Sediments adopted in 2004. Accordingly, the Commission should take all appropriate steps to encourage Member States to ratify that Convention.'
[187] Article 17(2).
[188] Recital 25.

214 *Research handbook on biodiversity and law*

According to the CBD guidance, pathways of unintentional introductions need to be identified. The Regulation complies with this guidance as far as IAS of Union Concern are concerned in that it obliges Member States to undertake a thorough analysis of such pathways within 18 months of the adoption of the List and to identify those pathways to be regarded as priority ('priority pathways') due to the high volume of IAS introductions and the potential damage that is caused by such species.[189] Action plans are then to be drawn up at the national level in relation to the priority pathways describing the measures that will be adopted to address them.[190] These pathways will include measures that must be implemented, but can also include the adoption of codes of conduct or good practice.[191] The types of measures will, for example, include those seeking to raise awareness and others seeking to minimise the contamination of goods by IAS.[192] Recognition of the need to identify relevant pathways is an important step forward when one bears in mind that no information on any such identification process could be found in 11 Member States in 2011.[193]

Arguably, however, the Regulation has not gone far enough in relation to the release of untreated ballast waters bearing in mind that '[a]t international and intra-EU levels, releases of untreated ballast water and hull fouling are by far the most significant vectors of unintentional introductions of alien species'.[194] It will be recalled that action under the Regulation in relation to pathways can include both voluntary and mandatory measures and, in relation to the former, the recitals to the Regulation specifically make mention of the voluntary guidelines in the IMO's Guidelines for the Control and Management of Biofouling.[195] In relation to the latter, the recitals to the Regulation additionally refer to the 2004 International Convention for the Regulation of Ships' Ballast Water and Sediments, and urge the Commission to take steps to encourage Member States to ratify it.[196] Bearing in mind the importance of this pathway, the argument for the introduction of mandatory, rather than simply voluntary, measures is strong. At the time of writing, whilst the 2004 treaty will surely enter into force shortly,[197] only seven EU Member States have ratified.[198] The EU may therefore need to adopt its own mandatory measures at a later date if further Member States fail to

---

[189] Article 13(1).
[190] Article 13(2).
[191] *Ibid.*
[192] Article 13(4).
[193] European Commission (n 67), 55.
[194] Institute for European Environmental Policy, *Assessment to support continued development of the EU strategy to combat invasive alien species* (IEEP, 2010), 111.
[195] Recital 21.
[196] *Ibid.*
[197] Article 18(1) of the treaty notes that it will enter into force 'twelve months after the date on which not less than thirty States, the combined merchant fleets of which constitute not less than thirty-five percent of the gross tonnage of the world's merchant shipping' have ratified. At the time of writing (May 2015) forty-four states had ratified the treaty representing approximately 32.86% of the world's merchant fleet gross tonnage. For further detail on this agreement see the chapter by Edward Goodwin.
[198] As of May 2015 the EU ratifying states are Croatia, Denmark, France, Germany, the Netherlands, Spain and Sweden.

ratify the 2004 international treaty or there is evidence of a lack of adherence to the IMO's voluntary guidelines.[199]

6.12 Mitigation of Impacts – Guiding Principle 12

This Guiding Principle stipulates that, once the presence of an IAS has been established, States should take steps – such as eradication, containment and control – to mitigate adverse impact. The need to mitigate the impact of IAS by such means is also stipulated in the Regulation.[200] What of the costs of taking mitigation measures? The Guiding Principles note that those responsible for the introduction of the IAS in question 'should bear the costs of control measures and biological diversity restoration where it is established that they failed to comply with the national laws and regulations'.[201] 'Control measures' are not defined in the CBD's guidance but are referred to under a section relating to the 'mitigation of impacts'. It might therefore be argued that the Regulation in fact endorses the polluter-pays principle to a greater extent than the international guidance as the EU measure notes that Member States shall aim to recover not only mitigation and restoration costs, but also the costs of preventive measures.[202]

6.13 Eradication, Containment and Control – Guiding Principles 13, 14 and 15

Whilst priority should indeed be given to preventing the introduction of IAS, thereby halting their establishment from the outset, this is not always possible. As mentioned earlier, eradication is then often the most appropriate course of action, the CBD guidance noting that '[t]he best opportunity for eradicating invasive alien species is in the early stages of invasion, when populations are small and localized'. In this regard, Member States are obliged under the EU measure to notify the Commission of the early detection of an IAS of Union Concern and within three months of this notification to apply eradication measures.[203] This can therefore be said to be in line with the application of an eradication programme in the early stages of an IAS invasion as noted

---

[199] The European Parliament's Committee on the Environment, Public Health and Food Safety in fact advocated that the Commission should report on Member States' implementation of the IMO's Guidelines for the Control and Management of Biofouling within three years of the Regulation entering into force and should 'if appropriate, submit legislative proposal to incorporate such measures into Union law'; *Report on the Proposal for a Regulation on the Prevention and Management of the Introduction and Spread of Invasive Alien Species* (European Parliament, 2014) [document A7-0088/2014], 18. This proposed amendment to the Regulation was not incorporated in the final text.

[200] *See*, for example, Article 19(2).

[201] This is in line with another aspect of the ecosystem approach noted in CBD Decision V/6 (n 130). Principle 4 of the latter notes that there is 'usually a need to understand and manage the ecosystem in an economic context' and that an applicable ecosystem-management programme should '[i]nternalize costs and benefits in the given ecosystem to the extent feasible'. Such internalisation would include imposing environmental costs on those who cause them.

[202] Article 21. Member States will determine the penalties for infringement of the Regulation; such penalties will be 'effective, proportionate and dissuasive' (Article 30(1) and (2)).

[203] Article 17(1).

in the CBD Guiding Principles. CBD guidance indicates that eradication should take place, but only 'where it is feasible'. This is also acknowledged in the Regulation in that Member States may decide that eradication is inappropriate when either it is technically infeasible, or when the costs will be exceptionally high.[204] Similarly, eradication may be deemed to be inappropriate when suitable methods of eradication are unavailable, or would have a particularly adverse impact on human health, the environment or other species.[205] When eradication is not appropriate, both the CBD guidance and the Regulation acknowledge that appropriate containment or population control measures should be introduced.[206]

7. SOME CONCLUSIONS

The regime established by the Regulation is certainly an important step forward in the establishment of a more comprehensive and co-ordinated approach to the problems posed by IAS in the EU. It is positive, for example, that the Regulation will introduce measures which will potentially regulate far more IAS than had previously been the case under EU law, and that it requires priority to be given to preventing the introduction of such species rather than merely allowing Member States to react to invasions once they have already taken place. The ecosystem approach is also generally endorsed and will inter alia require restorative action to be taken to assist the recovery of a damaged ecosystem, albeit when the costs of restorative action are not prohibitive. Additionally, the establishment of a surveillance system must be welcomed bearing in mind that most Member States did not have early-warning measures in place to detect and respond to invasions prior to the Regulation.[207]

Whilst the Regulations' provisions might have been more demanding with regard to the release of ballast water, important obligations nonetheless now apply to the intentional introduction of IAS and to priority pathways of unintentional introduction. It is also encouraging that the polluter-pays principle is to be applied, that efforts are expected from Member States to raise public awareness, and that a scientific forum will provide the Commission with advice in the listing process as well as in the application of emergency measures. Furthermore, the measure establishes the need for risk assessment utilising common criteria, which is important when one bears in mind that research had concluded that 'tools for assessing IAS risk are still relatively new and poorly developed' in EU Member States, and that there had previously been no common method of performing risk assessment.[208]

There is therefore much to be applauded in the measure's content and the Regulation clearly seeks to complement the work of international and regional treaty regimes by introducing an approach which will also benefit from the unique enforcement machinery of the EU. Taking the form of a 'Regulation', the measure is both binding in its

---

[204] Article 18(1)a and b.
[205] Article 18(1)c.
[206] Article 19(2).
[207] European Commission (n 67), 214.
[208] *Ibid.*

entirety and directly applicable in all Member States. Too often international legal regimes lack the ability to apply the appropriate level of political and legal pressure to bring about effective implementation and compliance. By contrast, should a Member State fail to live up to its obligations under the Regulation, the Commission can opt to take legal action before the ECJ to enforce its provisions under Article 258 TFEU. If the ECJ finds that there has indeed been a violation of the measure and then rules subsequently that the Member State in question has failed to comply with that initial judgment, the ECJ may also impose a lump sum or penalty payment upon the Member State in question.[209]

Whilst the Regulation must therefore be regarded as a significant move in the right direction, some of its provisions can rightly be criticised. In relation to the need for further research on IAS the measure should have placed far more of an onus on Member States to carry out such work and is not in line with CBD guidance in this regard. Additionally, it will, for example, be recalled that Member States may derogate from the measure's restrictions 'in exceptional circumstances', thereby allowing permits to be given to establishments to carry out activities for reasons of 'compelling public interest, including those of a *social or economic nature*' (emphasis added).[210] This appreciably widens the potential for derogations and in vetting such permits the Commission will undoubtedly play a key role in determining the extent to which derogations are allowed for economic and/or social reasons. Might those involved in the pet trade successfully apply for such a derogating permit in relation to a species on the List if it would otherwise mean the collapse of a lucrative economic market? Might an argument submitted by a company in the horticultural industry with regard to a number of IAS of Union Concern similarly be treated as amounting to 'exceptional circumstances' if it would otherwise lead to the demise of a key employer? It is submitted that Member States in granting such permits, and the Commission in the vetting of the same, must give particular and adequate consideration to the fact that any such derogation is indeed granted *only* in exceptional circumstances and also with appropriate regard being given to the susceptibility of all protected habitats and species to a possible unintentional release of the IAS in question from any commercial establishment.[211]

Whilst the provisions of the Regulation can be said to be generally in line with the CBD's Guiding Principles in many respects, there are of courses instances where the measure falls short in this regard. The most significant of these relates to the applicability of the precautionary principle. Whereas the Regulation can be said to endorse a precautionary approach in relation to the possible application of emergency measures, these measures cannot continue indefinitely. Bearing in mind the lack of express reference to the precautionary approach in the Regulation's provisions as they

---

[209] Article 260(2) TFEU.
[210] Article 9(1).
[211] Article 9(4)g notes that an application for an authorisation must include 'an assessment of the risk of escape ... accompanied by a description of the risk mitigation measures to be put in place'. Furthermore the preamble stipulates that 'particular attention should be paid to avoiding any adverse impacts on protected species and habitats, in accordance with relevant Union law' (para. 19).

relate to the listing process, as well as the applicable WTO rules under the SPS Agreement, it is submitted that it is very unlikely that a precautionary approach will be taken in the initial listing process, which must be completed by the end of 2015. If this submission is correct, IAS will therefore not be listed in a situation where there is insufficient scientific certainty as to the species' impact. This will also be the case in relation to decisions as to any future additions to the List regardless of whether or not the species in question was subject to prior emergency measures. The lack of a precautionary approach in this regard would be unsatisfactory, especially when one bears in mind that 'it is extremely difficult to predict accurately which introduced alien species will have benign effects and which may become invasive in a new habitat. Time factors make prediction even harder. While some alien species show their invasiveness quickly, others may have a long "lag" time.'[212] The lack of a precautionary approach in the listing process would not be in line with CBD guidance nor the Bern Convention's 'European Strategy on Invasive Alien Species', and would be far from a more satisfactory approach in which a non-native species is 'guilty until proven innocent'.

SELECT BIBLIOGRAPHY

Convention on Biological Diversity, 'Guiding Principles for the Prevention, Introduction and Mitigation of Impacts of Alien Species that Threaten Ecosystems, Habitats or Species' (2002; Decision VI/23)

European Commission, *LIFE and Invasive Alien Species* (2014) available at <http://ec.europa.eu/environment/life/publications/lifepublications/lifefocus/documents/life_ias.pdf>

C Miller, M Kettunen and C Shine, *Some Options for EU Action on Invasive Alien Species: Final Report for the European Commission* (2006)

A Monaco and P Genovesi, *European Guidelines on Protected Areas and IAS* (2013) [Bern Convention document T-PVS/Inf (2013) 22 prepared for the 33rd meeting of the Standing Committee].

S Riley, 'A Weed by any Other Name: Would the Rose Smell as Sweet if it were a Threat to Biodiversity?' (2009–2010) 22 Georgetown International Environmental Law Review 157

S Riley, 'Invasive Alien Species and Biodiversity' (part of the IUCN Academy of International Environmental Law's 'Essential Readings in International Environmental Law' available at <www.iucnael.org>)

C Shine, N Williams and L Gundling, *A Guide to Designing Legal and Institutional Frameworks on Alien Invasive Species* (IUCN, 2000)

---

[212] Shine, Williams and Gundling (n 1), 3.

8. Countering fragmentation of habitats under international wildlife regimes
Arie Trouwborst

1. INTRODUCTION

The focus of this chapter is on the threat of habitat fragmentation, and the corresponding challenge of connectivity conservation. In particular, the chapter aims to identify the varying extents to which international wildlife regimes are conducive to, or even require, the maintenance or achievement of an adequate degree of ecological connectivity.[1] The latter can be achieved, for instance, by maintaining or (re-)establishing corridors between protected areas, by equipping highways and other human infrastructure with wildlife overpasses, and other measures countering the fragmentation of habitats. Ensuring adequate connectivity was always an important element of wildlife conservation. Climate change is now adding to the challenge, as connectivity conservation is a central component of strategies to facilitate the adaptation of wild flora and fauna to climate change.

The structure of the chapter is as follows. The next section contains a concise introduction to fragmentation as a biodiversity conservation problem, and of the associated need for connectivity conservation (Section 2). The subsequent sections are then composed of analyses reviewing relevant international legal wildlife regimes from a connectivity conservation perspective. As it is not feasible to cover *all* relevant regimes, the chapter restricts itself to the following legal instruments:[2]

---

[1] This chapter builds on prior research by the author. *See*, in particular, B Lausche, D Farrier, JM Verschuuren, AGM La Viña, A Trouwborst, C-H Born and L Aug, *The Legal Aspects of Connectivity Conservation – A Concept Paper*, IUCN Environmental Policy and Law Series, Vol. 85-1 (IUCN, 2013); A Trouwborst, 'Conserving European Biodiversity in a Changing Climate: The Bern Convention, the European Union Birds and Habitats Directives and the Adaptation of Nature to Climate Change' (2011) 20 RECIEL 62; A Trouwborst, 'La Adaptación de la Flora y la Fauna al Cambio Climático en un Paisaje Fragmentado, y el Derecho Europeo sobre la Conservación de la Naturaleza' (2011) 2 Revista Catalana de Derecho Ambiental 1; A Trouwborst, 'Transboundary Wildlife Conservation in a Changing Climate: The Adaptation of the Bonn Convention on Migratory Species and its Daughter Instruments to Climate Change' (2012) 4 Diversity 258; A Trouwborst, 'Climate Adaptation and Biodiversity Law' in JM Verschuuren (ed.), *Research Handbook on Climate Adaptation Law* (Edward Elgar Publishing, 2013); and A Trouwborst, 'The Habitats Directive and Climate Change: Is the Law Climate Proof?' in C-H Born, A Cliquet, H Schoukens, D Misonne and G van Hoorick (eds.), *The Habitats Directive in its EU Environmental Law Context: European Nature's Best Hope?* (Routledge, 2014).

[2] A larger selection of instruments is discussed in Lausche and others, *ibid.*

- Ramsar Wetlands Convention[3] (Section 3);
- World Heritage Convention[4] (Section 4);
- Convention on Migratory Species (CMS)[5] and selected daughter instruments (Section 5);
- Convention on Biological Diversity (CBD)[6] (Section 6);
- African Convention(s) on the Conservation of Nature and Natural Resources[7] (Section 7);
- Bern Convention on European Wildlife and Natural Habitats[8] (Section 8);
- EU Habitats Directive[9] (Section 9).

This selection is deemed to represent a good mix between global and regional instruments and instruments of general and more restricted substantive scope. Although connectivity is also a relevant consideration in the planning of marine protected area networks, the above selection reflects the fact that habitat fragmentation is a much more significant problem on land. As regards the strong representation of European instruments, this is deemed appropriate with a view to the particularly daunting connectivity conservation challenges in Europe, with its heavily fragmented physical, ecological and political landscapes. The chapter ends with one or two concluding remarks (Section 10).

The chapter discusses pertinent (legally binding) treaty provisions that are of direct or indirect relevance, decisions by the parties (which are generally non-binding), and other guidance provided within the context of the instruments concerned. It should be borne in mind throughout the analyses below that treaty provisions are to be interpreted in light of the treaty's objectives and taking into account any 'subsequent agreements' or 'subsequent practice' by the parties regarding their interpretation and application, in conformity with Article 31 of the Vienna Convention on the Law of Treaties[10] and customary international law. As regards interpretation in light of the aims of wildlife conservation treaties, it is of significance that, both generally speaking and particularly as a consequence of climate change, effective conservation can hardly be achieved without maintenance or restoration of adequate connectivity (see Section 2 below). As the International Law Commission has clarified: 'When a treaty is open to two

---

[3] 1971 Convention on Wetlands of International Importance Especially as Waterfowl Habitat, 996 UNTS 245, in force 21 December 1975.

[4] 1972 UNESCO Convention Concerning the Protection of the World Cultural and Natural Heritage (1972) 11 ILM 1358, in force 17 December 1975.

[5] 1979 Convention on the Conservation of Migratory Species of Wild Animals (1980) 19 ILM 15, in force 1 November 1983.

[6] 1992 Convention on Biological Diversity, 1760 UNTS 79, in force 29 December 1993.

[7] 1968 African Convention on the Conservation of Nature and Natural Resources, 1001 UNTS 4, in force 16 June 1969; the revised version of this Convention, which was adopted in 2003, has yet to enter into force.

[8] 1979 Convention on the Conservation of European Wildlife and Natural Habitats, ETS No. 104, in force 1 June 1982.

[9] Council Directive 92/43/EC on the Conservation of Natural Habitats and of Wild Fauna and Flora [1992] OJ L206/7.

[10] 1969 Convention on the Law of Treaties, 1155 UNTS 331, in force 27 January 1980. On treaty interpretation generally, see R Gardiner, *Treaty Interpretation* (OUP, 2008).

interpretations one of which does and the other does not enable the treaty to have appropriate effects, good faith and the objects and purposes of the treaty demand that the former interpretation should be adopted.'[11] Interpretation with reference to 'subsequent agreements' or 'subsequent practice' means that decisions regarding habitat fragmentation and connectivity by Conferences of the Parties (COP) and similar treaty bodies, although themselves non-binding, may influence the interpretation of binding treaty obligations.[12]

2. FRAGMENTATION AND CONNECTIVITY

Habitat fragmentation is one of the major threats to biodiversity worldwide.[13] Fragmentation refers to the anthropogenic transformation of large areas of wildlife habitat to smaller, isolated patches. It is caused, for instance, by clear-cutting in forests, the conversion of natural habitats into agricultural landscapes or urban areas, and the construction of linear infrastructure such as roads, railways, canals and fences. The following metaphor by David Quammen eloquently grasps the impact fragmentation can have on wild fauna and flora:

> Let's start by imagining a fine Persian carpet and a hunting knife. The carpet is 12 feet by 18, say. That gives us 216 square feet of continuous woven material. We set about cutting the carpet into 36 equal pieces, each one a rectangle, two feet by three. ... When we're finished cutting, we measure the individual pieces, total them up – and find that, lo, there's still nearly 216 square feet of recognizably carpetlike stuff. But what does it amount to? Have we got 36 nice Persian throw rugs? No. All we're left with is three dozen ragged fragments, each one worthless and commencing to come apart. Now take the same logic outdoors and it begins to explain why the tiger, *Panthera tigris*, has disappeared from the island of Bali ... It suggests why the jaguar, the puma, and forty-five species of birds have been extirpated from a place called Barro Colorado Island – and why myriad other creatures are mysteriously absent from

---

[11] *Yearbook of the International Law Commission* (1966, Vol. II) 219. *See* also International Court of Justice, *Whaling in the Antarctic (Australia v Japan: New Zealand intervening)*, Judgment of 31 March 2014.

[12] On the potential of such decisions serving as 'subsequent agreement' or 'subsequent practice' in the context of treaty interpretation *see*, inter alia, RR Churchill and G Ulfstein, 'Autonomous Institutional Arrangements in Multilateral Environmental Agreements: A Little-Noticed Phenomenon in International Law' (2000) 94 AJIL 623, 641; JM Verschuuren, 'Ramsar Soft Law is Not Soft at All: Discussion of the 2007 Decision by the Netherlands Crown on the Lac Ramsar Site on the Island of Bonaire' (2008), available at <http://www.ramsar.org/pdf/wurc/wurc_verschuuren_bonaire.pdf>; A Wiersema, 'The New International Law-Makers? Conferences of the Parties to Multilateral Environmental Agreements' (2009) 31 Michigan Journal of International Law 231; MJ Bowman, PGG Davies and CJ Redgwell, *Lyster's International Wildlife Law* (2nd edn., CUP, 2010), 46; and Trouwborst, 'Conserving European Biodiversity in a Changing Climate' (n 1), 66–67.

[13] Perusal of the IUCN Red List at <http://www.iucnredlist.org> renders ample illustrations. The scientific literature documenting the effects of habitat fragmentation on populations, species and biodiversity at large is huge, and any attempt at comprehensive referencing would result in a footnote covering several pages.

myriad other sites. An ecosystem is a tapestry of species and relationships. Chop away a section, isolate that section, and there arises the problem of unraveling.[14]

In Europe these effects have been particularly severe, as confirmed in a 2011 study, which systematically measured and compared the rate of landscape fragmentation by transport infrastructure and urban areas in 28 European countries.[15] It reports that this fragmentation is having various adverse ecological impacts and is contributing appreciably to the decline and loss of populations of wild flora and fauna, and to the endangerment of species at a European scale.[16] The report highlights the imperative of addressing the problem through connectivity conservation and restoration measures: 'there is an urgent need for action'.[17]

There is ample scientific literature on the measures needed to counter and reduce habitat fragmentation.[18] A key term in this regard is 'connectivity', which refers to the 'ease with which organisms move between particular landscape elements'.[19] Connectivity can be enhanced through an array of tools, including corridors, stepping stones, highway wildlife crossings, and generally through biodiversity-friendly management of landscapes outside/between protected areas (including in designated buffer zones around protected areas), thus creating actual protected area *networks*.[20] One study describes ecological corridors as follows:

> Ecological corridors are defined functionally to indicate connectivity and as physical structures to indicate connectedness. They are functional connections enabling dispersal and migration of species that could be subject to local extinction and they are landscape structures (other than core areas) varying in size and shape from wide to narrow and from meandering to straight, which represent links that permeate the landscape and maintain natural connectivity.[21]

The conservation problems posed by the fragmentation of habitats are increasingly being exacerbated by climate change.[22] Plant and animal populations are responding

---

[14] D Quammen, *The Song of the Dodo* (Hutchinson, 1996), 9.
[15] European Environment Agency and Swiss Federal Office for the Environment, *Landscape Fragmentation in Europe* (Publications Office of the European Union, 2011).
[16] *Ibid.*, 7.
[17] *Ibid.*
[18] It suffices here to mention AF Bennet, *Linkages in the Landscape: The Role of Corridors and Connectivity in Wildlife Conservation* (2nd edn., IUCN, 2003); and GL Worboys, WL Francis and M Lockwood (eds), *Connectivity Conservation Management: A Global Guide* (Earthscan, 2010).
[19] Worboys and others, *ibid.*, xxxi. Whereas generally human infrastructure, e.g. roads, contributes to fragmentation, it should be noted that in certain landscapes (particularly in heavily modified ones) the carefully planned construction of linear infrastructure can actually promote connectivity conservation. For instance, a railway with 'nature-friendly' banks can provide a travel route for various creatures through otherwise impassable monoculture farmland.
[20] Generally, *see* Bennet (n 18) and Worboys and others (n 18).
[21] Worboys and others (n 18), xxxii (references omitted).
[22] A clear overview and discussion of impacts and adaptation measures is contained in A Campbell and others, *Review of the Literature on the Links between Biodiversity and Climate Change: Impacts, Adaptation and Mitigation* (Secretariat of the Convention on Biological

(or attempting to respond) to modifications in temperature, humidity and weather patterns, inter alia by shifting their distributions to higher latitudes and altitudes, and these transitions are expected to intensify. The crucial role of connectivity conservation in this respect is evident. For many species, adequate connectivity between populations is important in order to survive and recover from the adverse impacts of extreme weather events and other agents associated with climate change, such as storms, droughts, floods, temperature extremes, fires and disease. Fragmented populations have a significantly higher extinction risk in such situations than interconnected ones, and the latter recover much faster than the former. As a Dutch study explains:

> Extreme meteorological conditions will occur more frequently. Extreme weather like a cold and wet spring, or an extended hot period during summer, can lead to higher mortality or reproduction failure. This results in more pronounced highs and lows in population size, and increases the likelihood of extinction. The recovery of a population after a disturbance happens much quicker when the spatial coherence of the habitat is good. When the habitat is fragmented, recovery takes many times longer. An increased frequency of disturbances can then lead to the disappearance of species.[23]

In this light, and in view of the gradual latitudinal and altitudinal shifts of species distributions mentioned earlier, maintaining and/or restoring adequate connectivity is universally considered a crucial measure to facilitate the adaptation of biodiversity to climate change with minimal losses. The challenges to international wildlife instruments, and the need for international cooperation to address them, are obvious in this regard.[24]

Diversity, 2009). Attempts by the present author to summarise the scientific literature on the effects of climate change on biodiversity and on recommended adaptation measures have been made in A Trouwborst, 'International Nature Conservation Law and the Adaptation of Biodiversity to Climate Change: A Mismatch?' (2009) 21 JEL 419, 419–421 and 426–429.

[23] C Vos, M van der Veen and PFM Opdam, *Natuur en Klimaatverandering: Wat Kan het Natuurbeleid Doen?* (Alterra 2006), 7 (translation by the present author).

[24] This is reflected in the growing academic literature devoted to assessing the current capacity of international wildlife conservation regimes to facilitate the adaptation of species and ecosystems to climate change, and to examining ways of enhancing that capacity, which includes MJ Bowman, 'Global Warming and the International Legal Protection of Wildlife' in RR Churchill and DAC Freestone (eds), *International Law and Global Climate Change* (Graham & Trotman, 1991); GC Boere and D Taylor, 'Global and Regional Governmental Policy and Treaties as Tools Towards the Mitigation of the Effect of Climate Change on Waterbirds' (2004) 146 Ibis 111; K Wheeler, 'Bird Protection & Climate Changes: A Challenge for Natura 2000?' (2006) 13 Tilburg Foreign Law Review 283; D Hodas, 'Biodiversity and Climate Change Laws: A Failure to Communicate?' in MI Jeffery and others (eds), *Biodiversity, Conservation, Law and Livelihoods: Bridging the North-South Divide* (CUP, 2008); A Cliquet, C Backes, J Harris and P Howsam, 'Adaptation to Climate Change: Legal Challenges for Protected Areas' (2009) 5 Utrecht Law Review 158; Trouwborst (n 22); S Erens, JM Verschuuren and K Bastmeijer, 'Adaptation to Climate Change to Save Biodiversity: Lessons Learned from African and European Experiences' in BJ Richardson and others (eds), *Climate Law and Developing Countries: Legal and Policy Challenges for the World Economy* (Edward Elgar Publishing, 2009); D Schramm and A Fishman, 'Legal Frameworks for Adaptive Natural Resource Management in a Changing Climate' (2010) 22 Georgetown International Environmental Law Review 491; JM Verschuuren, 'Rethinking Restoration in the European Union's Birds and

3. RAMSAR WETLANDS CONVENTION

The Ramsar Convention was adopted in 1971 to 'stem the progressive encroachment on and loss of wetlands now and in the future'.[25] Contracting parties 'shall formulate and implement their planning so as to promote the conservation of the wetlands included in the List [of Wetlands of International Importance], and as far as possible the wise use of wetlands in their territory'.[26] The latter half of this obligation applies to all wetlands. The same is true of the duty to 'promote the conservation of wetlands and waterfowl by establishing nature reserves on wetlands, whether they are included in the List or not'.[27] Furthermore, parties are to consult with each other concerning the implementation of the Convention, especially with respect to transboundary wetlands.[28]

Rivers, streams and other wetlands (can) manifestly provide for connectivity. The obligations under the Ramsar Convention can and do therefore contribute to connectivity conservation. Arguably, not only conservation but also 'wise use' of wetlands – which is defined as 'the maintenance of their ecological character, achieved through the implementation of ecosystem approaches, within the context of sustainable development'[29] – ought to cater for sufficient degrees of conservation connectivity. This is well acknowledged by Convention parties. To illustrate, a resolution on wetlands and climate change adopted by the 10th COP in 2008 affirms that the 'conservation and wise use of wetlands enables organisms to adapt to climate change by providing connectivity, corridors and flyways along which they can move'.[30] As regards transboundary cooperation, one means of implementing the aforementioned duty of parties to consult with each other is the joint designation of wetlands extending across national

---

Habitats Directives' (2010) 28 Ecological Restoration 431; MJ Bowman, 'Conserving Biological Diversity in an Era of Climate Change: Local Implementation of International Wildlife Law Treaties' (2010) 53 German Yearbook of International Law 289; Trouwborst, 'Conserving European Biodiversity in a Changing Climate' (n 1); Trouwborst, 'La Adaptación de la Flora y la Fauna al Cambio Climático' (n 1); Trouwborst, 'Transboundary Wildlife Conservation in a Changing Climate' (n 1); and A Trouwborst, 'Bird Conservation and Climate Change in the Marine Arctic and Antarctic: Classic and Novel International Law Challenges Converging in the Polar Regions' (2013) 16 JIWLP 1.

[25] Preamble. On the Convention generally, see MJ Bowman, 'The Ramsar Convention on Wetlands: Has it Made a Difference?' (2002) 10 Yearbook on International Cooperation on Environment and Development 61; Bowman and others (n 12), 403–450; and CM Finlayson and others, 'The Ramsar Convention and Ecosystem-based Approaches to the Wise Use and Sustainable Development of Wetlands' (2011) 14 JIWLP 176.

[26] Article 3(1).
[27] Article 4(1).
[28] Article 5.
[29] Ramsar COP Resolution IX.1 on Additional Scientific and Technical Guidance for Implementing the Ramsar Wise Use Concept (adopted 15 November 2005).
[30] Ramsar COP Resolution X.24 on Climate Change and Wetlands (adopted 4 November 2008).

borders as Transboundary Ramsar Sites (TRS) – thus also reducing 'political fragmentation'.[31] Thirteen TRS have hitherto been established, one in Africa – the Niumi-Saloum Complex in Gambia and Senegal – and the remainder in Europe.[32] Also noteworthy is the development of so-called Ramsar Regional Initiatives, which may comprise international site networks.[33] One pertinent instance is the Partnership for the East Asian-Australasian Flyway. This informal, voluntary initiative aims inter alia for the development of an international Waterbird Site Network in the region in question.[34] It should, in any case, be noted that wetlands on the Ramsar List are probably most likely to constitute actual or candidate core areas of protected areas networks, and to a lesser extent connecting corridors.

4. WORLD HERITAGE CONVENTION

A substantial number of ecologically important sites around the globe qualify as 'natural heritage' according to the definition in Article 2 of the World Heritage Convention, and a number of these have been included in the World Heritage List authorised under the Convention.[35] Contracting parties are committed to doing everything within their power to ensure the 'identification, protection, conservation, presentation and transmission to future generations' of the natural heritage situated on their territories.[36] Moreover, to warrant that 'effective and active measures' are taken for the protection of the sites concerned, the Convention stipulates that each party 'shall endeavour, in so far as possible, and as appropriate for each country', to 'integrate the protection of that heritage into comprehensive planning programmes' and to 'take the appropriate legal, scientific, technical, administrative and financial measures necessary for the identification, protection, conservation, presentation and rehabilitation of this heritage'.[37]

As the case may be, it is possible to argue that this last obligation extends to connectivity measures. Indeed, the very occurrence on the World Heritage List of large wetlands like the Wadden Sea and mountain ranges such as the Canadian Rockies and the Volcanoes of Kamchatka is significant from a connectivity conservation perspective. It is noteworthy in the present context that the Convention's Operational Guidelines instruct parties to provide for an 'adequate buffer zone' wherever this is 'necessary for

---

[31] Article 5 and Ramsar COP Resolution VII.19 on International Cooperation (adopted 18 May 1999).

[32] *See* <http://www.ramsar.org>. All TRS involve two countries, except the 'Trilateral Ramsar Site Floodplains of the Morava-Dyje-Danube Confluence', which unites four pre-existing Ramsar Sites in three countries (two in Austria, one in the Czech Republic, and one in Slovakia).

[33] *See* inter alia Ramsar COP Resolution X.6 on Regional Initiatives (adopted 4 November 2008).

[34] *See* <https://www.eaaflyway.net>.

[35] On the Convention generally, *see* F Francioni (ed), *The 1972 World Heritage Convention: A Commentary* (OUP, 2008); and Bowman and others (n 12), 451–482.

[36] Article 4.

[37] Article 5.

the proper conservation' of the site involved.[38] Similar to the listed Ramsar sites, as mentioned above, most natural sites on the World Heritage List are likely to constitute actual or candidate core areas of protected areas networks, and to a lesser extent connecting corridors – unless they are covering large-scale biomes. Finally, in parallel to the TRS scheme under the Ramsar Convention, various transboundary sites have been designated under the World Heritage Convention. The World Heritage Committee's Operational Guidelines refer to these as 'transboundary properties'.[39] Examples of transboundary properties inscribed on the World Heritage List include the Wadden Sea (Germany-Netherlands), Waterton-Glacier (Canada-United States), and Mount Nimba Strict Nature Reserve (Côte d'Ivoire-Guinea).

5. CONVENTION ON MIGRATORY SPECIES AND DAUGHTER INSTRUMENTS

The Convention on Migratory Species (CMS or Bonn Convention) aims for a 'favourable conservation status' for migratory species.[40] With regard to the endangered migratory species listed in Appendix I, Article III(4) of the Convention stipulates that CMS parties 'shall endeavour':

(a) to conserve and, where feasible and appropriate, restore those habitats of the species which are of importance in removing the species from danger of extinction;
(b) to prevent, remove, compensate for or minimize, as appropriate, the adverse effects of activities or obstacles that seriously impede or prevent the migration of the species; and
(c) to the extent feasible and appropriate, to prevent, reduce or control factors that are endangering or are likely to further endanger the species … .

It is not difficult to imagine circumstances where connectivity conservation action by states parties would be essential in order to meet the requirements under (a), (b) and/or (c). To illustrate, one may consider the implications of Article III(4)(b) for roads, fences, wind farms, power lines and other infrastructure, which can impair connectivity in respect of migratory wildlife.

Comparable considerations apply to the species listed in Appendix II, which are migratory species with an unfavourable conservation status and other species which would significantly benefit from specific cooperation. Appendix II species are to be the subject of focused ancillary or 'daughter' instruments, which may be 'AGREEMENTS'

---

[38] Intergovernmental Committee for the Protection of the World Cultural and Natural Heritage, *Operational Guidelines for the Implementation of the World Heritage Convention*, update July 2012, WHC.08/01 (UNESCO World Heritage Centre, 2012), para. 103.

[39] Ibid., paras 134–136.

[40] On the CMS regime generally, *see* S Lyster, 'The Convention on the Conservation of Migratory Species of Wild Animals (The "Bonn Convention")' (1989) 29 Natural Resources Journal 979; R Caddell, 'International Law and the Protection of Migratory Wildlife: An Appraisal of Twenty-Five Years of the Bonn Convention' (2005) 16 Colorado Journal of International Environmental Law and Policy 113; and Bowman and others (n 12), 535–583.

under Article IV(3) or less formal 'agreements' under Article IV(4). Regarding AGREEMENTS, the Convention states that these should, 'where appropriate and feasible,' provide for:

(e) conservation and, where required and feasible, restoration of the habitats of importance in maintaining a favourable conservation status, and protection of such habitats from disturbances ... ;
(f) maintenance of a network of suitable habitats appropriately disposed in relation to the migration routes;
(g) where it appears desirable, the provision of new habitats favourable to the migratory species ... ;
(h) elimination of, to the maximum extent possible, or compensation for activities and obstacles which hinder or impede migration.[41]

Various CMS daughter instruments are discussed below.[42]

That the objectives of the Convention cannot be achieved without ensuring adequate conservation connectivity has been acknowledged by the CMS COP on several occasions. Various resolutions adopted by the COP in 2011 may be viewed to illustrate this. The first of these, Resolution 10.3, is specifically devoted to critical sites and ecological networks in the context of the Bonn Convention. Its Preamble recognises that 'habitat destruction and fragmentation are among the primary threats to migratory species, and that the identification and conservation of habitats, in particular the critical sites and connecting corridors, are thus of paramount importance for the conservation of these species'.[43] This point is amplified in several respects. It will be recalled that 'ecological connectivity can have multiple advantages, such as maintenance of viable populations and migration pathways, reduced risk of a population becoming extinct and higher resilience to climate change'. It is furthermore observed that 'ecological networks usually include core areas and corridors, and sometimes also restoration areas and buffer zones', and that 'networks of critical sites are needed in order to achieve connectivity and to protect migratory species along their entire migration route, and that corridors can occur in any habitat and should meet the requirements of the targeted species'.[44] Finally, it is emphasised that 'the practical approach to the identification, designation, protection and management of critical sites will vary from one taxonomic group to another or even from species to species, and that the flyway approach provides a useful framework to address habitat conservation and species protection for migratory birds along migration routes'.[45] The Preamble is explicit that such flyways qualify as 'a specific type of migration corridor'.[46] A number of the operative paragraphs of

---

[41] Article V.
[42] The texts of all such instruments may be located via the 'CMS Instruments' link on the Bonn Convention website at <http://www.cms.int/>.
[43] CMS COP Resolution 10.3 on the Role of Ecological Networks in the Conservation of Migratory Species (adopted 25 November 2011), Preamble.
[44] Ibid.
[45] Ibid.
[46] Ibid.

Resolution 10.3 are reproduced here in full because of their considerable significance for present purposes:

(1) *Requests* Parties to promote the identification of the most relevant sites and corridors for migratory species, with an emphasis on those that are transboundary and would benefit from international cooperation;
(2) *Invites* Parties to enhance the coverage, quality and connectivity of protected areas as a contribution to the development of representative systems of protected areas and coherent ecological networks that include all taxonomic groups of migratory species;
(3) *Urges* Parties to undertake habitat restoration and management at protected areas and critical sites in order to ensure habitat availability during the different stages of the life cycle of migratory species;
(4) *Urges* Parties to explore actively the potentially suitable areas for cooperation over transboundary protected areas, ensuring that barriers to migration are to the greatest possible extent eliminated or mitigated and that migratory species are managed under commonly agreed criteria;
(6) *Invites* Parties to undertake concerted efforts to integrate protected areas into wider landscapes and sectors, including through the use of connectivity measures such as the development of biological corridors, where appropriate, and the restoration of degraded habitats and landscapes in order to address the impacts of and increase resilience to climate change;
(9) *Encourages* Parties to explore the applicability of ecological networks and corridors to marine migratory species that are under pressure from human activities such as oil and gas exploration, overexploitation, fishing and coastal development.

A closely related resolution on flyway conservation, Resolution 10.10, requests parties to 'ensure that migratory bird habitat requirements are integrated into land-use policies, including protected areas but also especially outside protected areas'.[47] Furthermore, it calls for a review of 'the coverage and protection status of current site networks' and for parties to 'consider the resilience of sites to climate change, taking account of the potential for shifts in the range of species due to climate change, as well as other factors'.[48] In addition, parties are requested to 'ensure that key migratory stop-over sites are identified to form part of coherent site networks for migratory species' and to promote the 'development of flyway-scale site networks, especially where they are least developed, to include the widest possible range of available habitat for migratory birds'.[49] The third COP10 decision singled out here is Resolution 10.19 on climate change and migratory species.[50] It stresses the importance of implementing Resolution 10.3 on ecological networks in the context of climate change, specifically urging parties to 'strengthen the physical and ecological connectivity between sites, permitting dispersal and colonization when species distributions shift'.[51] A final example from the

---

[47] CMS COP Resolution 10.10 on Guidance on Global Flyway Conservation and Options for Policy Arrangements (adopted 25 November 2011), para. 4.
[48] *Ibid.*, para. 6.
[49] *Ibid.*, para. 7.
[50] CMS COP Resolution 10.19 on Migratory Species Conservation in the Light of Climate Change (adopted 25 November 2011).
[51] *Ibid.*, para. 8(b).

same COP, the title of which is sufficient to illustrate its relevance for present purposes, is Resolution 10.11 on Power Lines and Migratory Birds.[52]

Several CMS daughter instruments are (potential) vehicles for coordinated connectivity conservation on a comparatively detailed level. A few selected examples are considered here. The 2007 Gorilla Agreement lays down the following general duty: 'Parties shall take co-ordinated measures to maintain gorillas in a favourable conservation status or to restore them to such a status.'[53] This is supplemented by an obligation to 'take measures to conserve all populations of gorilla'.[54] To this end, states parties shall 'identify sites and habitats for gorillas occurring within their territory and ensure the protection, management, rehabilitation and restoration of these sites' and 'coordinate their efforts to ensure that a *network of suitable habitats* is maintained or re-established throughout the entire range of all species and sub-species, in particular where habitats extend over the area of more than one Party to this Agreement'.[55]

To illustrate this in some detail, it is shown here how connectivity forms part of the specific actions laid down in the Action Plans that have been concluded under the Gorilla Agreement for each gorilla sub-species. The performance of these actions is mandatory for parties to the Agreement.[56] The Western Lowland Gorilla Action Plan refers to a project concerning the tri-national (Congo/Gabon/Cameroon) transborder protected area complex Dja-Odzala-Minkebe (TRIDOM), the objective of which is 'to maintain the functions *and ecological connectivity* in the TRIDOM and ensure long term conservation of its protected area system'.[57] The following is one of the national actions specified for the Central African Republic: 'A corridor connecting Mbaére – Bodingué and Dzanga – Ndoki must be negotiated with logging companies.'[58] The national actions for Angola include carrying out a '[c]ensus of the Mayombe Forest in Cabinda to identify viable populations of gorillas and connectivity'.[59] For the Democratic Republic of Congo, the plan calls for '[c]ommon planning and integrated management for the transboundary gorilla population between Dimonika, Conkouati and the reserves *and corridors* still to create on the Bas Fleuve'.[60] One of the measures included in the Eastern Lowland Gorilla Action Plan is the maintenance of an 'ecological corridor between lowland and montane populations' in the Kahuzi-Biega National Park.[61] Finally, the Cross River Gorilla Action Plan calls for surveys of poorly known areas, 'especially within potential corridors connecting population nuclei',[62] and

---

[52] CMS COP Resolution 10.11 on Power Lines and Migratory Birds (adopted 25 November 2011).
[53] 2007 Agreement on the Conservation of Gorillas and their Habitats (in force 1 June 2008), Article II(1).
[54] Article III(1).
[55] Article III(2)(b) and (c) (emphasis added).
[56] See Article II(2).
[57] Western Lowland Gorilla Action Plan, 9 (emphasis added).
[58] *Ibid.*, 19.
[59] *Ibid.*, 23.
[60] *Ibid.*, 25 (emphasis added).
[61] Eastern Lowland Gorilla Action Plan, 7.
[62] Cross River Gorilla Action Plan, 2.

emphasises the importance of finding ways 'to protect the corridors connecting the sub-populations'.[63]

The African-Eurasian Waterbirds Agreement (AEWA), another CMS ancillary treaty, covers the entire African-Eurasian flyway for migratory waterbirds.[64] The general duty of parties to warrant a favourable conservation status for the species involved[65] and the obligations regarding habitat protection[66] employ a language very similar to the obligations under the Gorilla Agreement cited above. These are complemented by more detailed provisions on the protection of important sites in the annexed, mandatory Action Plan.[67] These do not, however, expressly mention connectivity. Noteworthy, at any rate, is the prescription that parties 'shall, as far as possible, promote high environmental standards in the planning and construction of structures to minimize their impact on populations' and 'should consider steps to minimize the impact of structures already in existence where it becomes evident that they constitute a negative impact for the populations concerned'.[68]

To further the adaptation of waterbird populations to climate change, the 4th AEWA Meeting of the Parties (MOP) resolved to 'designate and establish comprehensive and coherent networks of adequately managed protected sites as well as other adequately managed sites, to accommodate range-shifts and facilitate waterbirds' dispersal'.[69] Besides, parties undertook to 'provide wider habitat protection for species with dispersed breeding ranges, migration routes or winter ranges where the site conservation approach would have little effect, especially under climate change conditions'.[70] Some of the technical guidance, which has been produced under AEWA auspices to aid parties in the performance of their treaty obligations, is also of significance in the context of connectivity conservation, in particular the guidelines on infrastructural developments[71] and those on climate adaptation.[72]

---

[63] *Ibid.*, 3.

[64] 1995 Agreement on the Conservation of African-Eurasian Migratory Waterbirds (1995) 6 YIEL 907, in force 1 November 1999. *See* B Lenten, 'A Flying Start for the Agreement on the Conservation of African-Eurasian Migratory Waterbirds (AEWA)' (2001) 4 JIWLP 159; R Adam, 'Waterbirds, the 2010 Biodiversity Target, and Beyond: AEWA's Contribution to Global Biodiversity Governance' (2008) 38 Environmental Law Review 87; M Lewis, 'AEWA at Twenty: An Appraisal of the African-Eurasian Waterbird Agreement and its Unique Place in International Environmental Law' (2016) 19 JIWLP 22.

[65] Article II(1).

[66] Article III(2)(c) and (d).

[67] AEWA Annex 3, para. 3.

[68] *Ibid.*, para. 4.3.5.

[69] AEWA MOP Resolution 4.14 on the Effects of Climate Change on Migratory Waterbirds (adopted 19 September 2008), para. 4.

[70] *Ibid.*, para. 7.

[71] G Tucker and J Treweek, *Guidelines on How to Avoid, Minimize or Mitigate Impact of Infrastructural Developments and Related Disturbance Affecting Waterbirds*, AEWA Conservation Guidelines No. 11 (AEWA 2008).

[72] I Maclean and M Rehfisch, *Guidelines on the Measures Needed to Help Waterbirds Adapt to Climate Change*, AEWA Conservation Guidelines No. 12 (AEWA, 2008).

The Memorandum of Understanding (MoU) on the Bukhara Deer is an example of a pertinent non-legally binding CMS instrument.[73] The Bukhara deer is an endangered red deer sub-species, which is threatened by a combination of habitat destruction and degradation and poaching. The MoU's objective is 'regaining a favourable conservation status of the populations of Bukhara Deer and their habitat'.[74] Its four signatories – Kazakhstan, Tajikistan, Turkmenistan and Uzbekistan – are committed to 'identify, conserve and, where feasible and appropriate, restore those habitats of the species that are of importance in removing the sub-species from danger of extinction'.[75] The accompanying Action Plan, among other things, promotes the creation of 'an interstate econet (system of protected areas) which could support self-sustaining population development' of Bukhara deer.[76]

With regard to CMS daughters generally, it is of interest to note that the COP resolution on ecological networks, which was discussed previously, urges states 'to consider the network approach in the implementation of existing CMS initiatives and instruments such as the Sahelo-Saharan Antelopes Action Plan, the Monk Seal MoU, the West African Elephant MoU, the Gorilla Agreement, the Saiga Antelope MoU, the Bukhara Deer MoU, South Andean Huemul MoU and – as is already the case – in the work on flyways'.[77]

6. BIODIVERSITY CONVENTION

The provisions of the CBD itself do not address connectivity conservation in so many words.[78] Several provisions are nevertheless of relevance to the topic, particularly the following paragraphs of Article 8 on *in-situ* conservation:

Each Contracting Party shall, as far as possible and as appropriate:

(a) Establish a system of protected areas or areas where special measures need to be taken to conserve biological diversity;
(b) Develop, where necessary, guidelines for the selection, establishment and management of protected areas or areas where special measures need to be taken to conserve biological diversity;
(c) Regulate or manage biological resources important for the conservation of biological diversity whether within or outside protected areas, with a view to ensuring their conservation and sustainable use;
(d) Promote the protection of ecosystems, natural habitats and the maintenance of viable populations of species in natural surroundings;

---

[73] 2002 Memorandum of Understanding concerning Conservation and Restoration of the Bukhara Deer (*Cervus elaphus bactrianus*), in force 1 August 2002.
[74] *Ibid.*, Preamble.
[75] *Ibid.*, para. 1.
[76] Bukhara Deer Action Plan, 4.
[77] Resolution 10.3 (n 43), para. 7.
[78] On the Biodiversity Convention generally, *see* D Bodansky, 'International Law and the Protection of Biological Diversity' (1995) 28 Vanderbilt Journal of Transnational Law 623; MJ Bowman and C Redgwell (eds), *International Law and the Conservation of Biological Diversity* (Kluwer Law International, 1996); and Bowman and others (n 12), 587–629.

(e) Promote environmentally sound and sustainable development in areas adjacent to protected areas with a view to furthering protection of these areas;
(f) Rehabilitate and restore degraded ecosystems and promote the recovery of threatened species, inter alia, through the development and implementation of plans or other management strategies;
(l) Where a significant adverse effect on biological diversity has been determined pursuant to Article 7, regulate or manage the relevant processes and categories of activities.

It is significant from the current perspective that Article 8 refers to a 'system' of protected areas and *other* 'areas where special measures need to be taken to conserve biological diversity'. Further germane provisions include the duties of states parties to develop national biodiversity strategies or plans and, as far as possible and as appropriate, to integrate biodiversity conservation into other 'relevant sectoral or cross-sectoral plans, programmes and policies'.[79] The latter obligation must be deemed to apply, for example, to infrastructural and agricultural policies, which evidently have far-reaching implications for connectivity conservation.

These provisions from the Convention have come to be accompanied and informed by a growing set of non-binding commitments and guidelines adopted by the CBD COP, including with respect to climate change adaptation and protected area networks. These attach considerable significance to connectivity in the implementation of Convention obligations. For instance, the eleventh of the so-called Aichi Targets, laid down in the Strategic Plan for Biodiversity 2011–2020, reads:

> By 2020, at least 17 per cent of terrestrial and inland water areas, and 10 per cent of coastal and marine areas, especially areas of particular importance for biodiversity and ecosystem services, are conserved through effectively and equitably managed, ecologically representative and *well connected* systems of protected areas and other effective area-based conservation measures, and integrated into the wider landscapes and seascapes.[80]

This target builds, among other things, on the CBD Programme of Work on Protected Areas, which calls for the establishment and management of 'ecological networks, ecological corridors and/or buffer zones, where appropriate, to maintain ecological processes and also taking into account the needs of migratory species'.[81] Furthermore, according to the fifth Aichi Target, by 2020 the rate of loss of natural habitats should be 'at least halved and where feasible brought close to zero, and degradation *and fragmentation* [should be] significantly reduced'.[82]

In a 2010 COP Decision on protected areas, CBD parties resolved to '[e]nhance the coverage and quality, representativeness and, if appropriate, connectivity of protected areas' as a contribution to the establishment of 'representative systems of protected areas and coherent ecological networks'.[83] In the context of climate change, the same

---

[79] Article 6.
[80] CBD COP Decision X/2 on the Strategic Plan for Biodiversity 2011–2020 and the Aichi Biodiversity Targets (adopted 29 October 2010), Annex (emphasis added).
[81] CBD COP Decision VII/28 on Protected Areas (adopted 20 February 2004), Annex, Goal 1.2, para. 1.2.3; see also paras. 1.2.1, 1.2.4 and 1.2.5.
[82] CBD COP Decision X/2 (n 80), Annex (emphasis added).
[83] CBD COP Decision X/31 on Protected Areas (adopted 29 October 2010), para. 1(a).

decision calls for 'concerted efforts to integrate protected areas into wider landscapes and seascapes and sectors, including through the use of connectivity measures such as the development of ecological networks and ecological corridors, and the restoration of degraded habitats and landscapes in order to address climate-change impacts and increase resilience to climate change'.[84] Also in connection with ecosystem restoration, the Decision urges parties to employ, as appropriate, 'connectivity tools such as ecological corridors and/or conservation measures in and between protected areas and adjacent landscapes and seascapes'.[85]

Another 2010 COP Decision, addressing the adaptation of species and ecosystems to climate change, summons CBD parties to strengthen protected areas' networks 'including through the use of connectivity measures such as the development of ecological networks and ecological corridors and the restoration of degraded habitats and landscapes'[86] and to integrate biodiversity 'into wider seascape and landscape management'.[87] Earlier COP decisions on this topic had already called on parties to 'take measures to manage ecosystems so as to maintain their resilience to extreme climate events and to help mitigate and adapt to climate change';[88] to 'integrate climate change adaptation measures in protected area planning, management strategies, and in the design of protected area systems';[89] and to 'cooperate regionally in activities aimed at enhancing habitat connectivity across ecological gradients, with the aim of enhancing ecosystem resilience and to facilitate the migration and dispersal of species with limited tolerance to altered climatic conditions'.[90] Finally, detailed guidance for parties regarding conservation connectivity, whether or not in the context of climate change, has been provided in a large number of volumes published over the years in the CBD Technical Series.[91]

---

[84] *Ibid.*, para. 14(a).
[85] *Ibid.*, para. 26(a).
[86] CBD COP Decision X/33 on Biodiversity and Climate Change (adopted 29 October 2010), para. 8(d)(iii).
[87] *Ibid.*, para. 8(d)(iv).
[88] CBD COP Decision VII/15 on Biodiversity and Climate Change (adopted 20 February 2004), para. 12.
[89] Decision VII/28, above note 80, Annex, Goal 1.4, para. 1.4.5.
[90] CBD COP Decision VIII/30 on Biodiversity and Climate Change (adopted 31 March 2006), para. 4.
[91] This concerns in particular the following volumes (available at <https://www.cbd.int/ts>): 10: *Interlinkages between Biological Diversity and Climate Change* (2003); 15: *Biodiversity Issues for Consideration in the Planning, Establishment and Management of Protected Area Sites and Networks* (2004); 18: *Towards Effective Protected Area Systems* (2005); 23: *Review of Experience with Ecological Networks, Corridors and Buffer Zones* (2006); 24: *Closing the Gap: Creating Ecologically Representative Protected Area Systems* (2006); 29: *Emerging Issues for Biodiversity Conservation in a Changing Climate* (2007); 35: *Implementation of the CBD Programme of Work on Protected Areas: Progress and Perspectives* (2008); 41: *Biodiversity and Climate Change Mitigation and Adaptation: Report of the Second Ad Hoc Technical Expert Group on Biodiversity and Climate Change* (2009); 42: *Review of the Literature on the Links between Biodiversity and Climate Change – Impacts, Adaptation and Mitigation* (2009); 44: *Making Protected Areas Relevant: A Guide to Integrating Protected Areas into Wider Landscapes* (2010); 51: *Biodiversity and Climate Change: Achieving the 2020 Targets* (2010).

7. AFRICAN NATURE CONSERVATION CONVENTION(S)

The African Union's major nature conservation treaty, the African Convention on the Conservation of Nature and Natural Resources, was first adopted in 1968 (Algiers Convention). It was modernised in 2003, but the revised Convention (Maputo Convention) has yet to enter into force.[92] Both versions of the African Convention are relevant to the topic of connectivity conservation.

No express reference to connectivity occurs in the Algiers Convention. Various provisions are nonetheless of implicit relevance. According to the overarching obligation laid down in Article II, parties 'shall undertake to adopt the measures necessary to ensure conservation, utilization and development of soil, water, flora and faunal resources in accordance with scientific principles and with due regard to the best interests of the people'. Clearly, the 'measures necessary' will in many cases (have to) include connectivity conservation measures. Similar considerations apply to a number of other obligations. A provision concerning wild flora requires states parties to 'adopt scientifically-based conservation, utilization and management plans of forests and rangeland, taking into account [inter alia] the habitat requirements of the fauna'.[93] With a view to those habitat requirements, the plans referred to will, depending on the circumstances, need to make provision for corridors connecting animal populations. The need for connectivity conservation may likewise be read between the lines of the provision on protected species in which contracting parties 'recognize that it is important and urgent to accord a special protection to those animal and plant species that are threatened with extinction, or which may become so, and to *the habitat necessary to their survival*'.[94] A final example is the Algiers Convention's provision concerning 'conservation areas', which term comprises several different protected area types. It contains a duty to

> maintain and extend where appropriate ... the Conservation areas existing at the time of entry into force of the present convention and, preferably within the framework of land use planning programmes, assess the necessity of establishing *additional conservation areas* in order to [inter alia] *ensure conservation of all species* and more particularly of those listed or may be listed in the annex to this convention.[95]

Again, this duty can evidently be used to support the implementation of conservation connectivity initiatives.

Among the many significant changes and additions incorporated in the 2003 Maputo Convention, the more comprehensive provision on 'conservation areas' stands out:

---

[92] On the African Conventions generally, *see* IUCN, *An Introduction to the African Convention on the Conservation of Nature and Natural Resources* (2nd edn., IUCN, 2006), which incorporates the text of the 2003 Maputo revision; and Bowman and others (n 12), 262–296.
[93] Algiers Convention, Article VI.
[94] Article VIII (emphasis added).
[95] Article X (emphasis added).

(1) The Parties shall establish, maintain and extend, as appropriate, conservation areas. They shall, preferably within the framework of environmental and natural resources policies, legislation and programmes, also assess the potential impacts and necessity of establishing additional conservation areas and wherever possible designate such areas, in order to ensure the long term conservation of biological diversity, in particular to:
  (a) conserve those ecosystems which are most representative of and peculiar to areas under their jurisdiction, or are characterized by a high degree of biological diversity;
  (b) ensure the conservation of all species and … of the habitats that are critical for the survival of such species.
(2) The Parties shall seek to identify areas critically important to the goals referred to in sub paragraph 1(a) and 1(b) above which are not yet included in conservation areas, taking into consideration the work of competent international organisations in this field.
…
(4) The Parties shall, where necessary and if possible, control activities outside conservation areas which are detrimental to the achievement of the purpose for which the conservation areas were created, and establish for that purpose buffer zones around their borders.[96]

The consistent implementation of this provision would clearly be conducive to connectivity conservation in the region. Another provision of significance in the current context states that parties 'shall ensure that … in the formulation of all development plans, full consideration is given to ecological … factors'.[97] 'To this end, the Parties shall', inter alia, 'to the maximum extent possible, take all necessary measures to ensure that development activities and projects' – for instance the construction of roads or other infrastructure – 'are based on sound environmental policies and do not have adverse effects on natural resources and the environment in general'.[98]

8. BERN CONVENTION ON EUROPEAN WILDLIFE AND NATURAL HABITATS

The Bern Convention's aims are 'to conserve wild fauna and flora and their natural habitats, especially those species and habitats whose conservation requires the cooperation of several States, and to promote such cooperation' with a particular emphasis on endangered and vulnerable species, including migratory ones.[99] Article 2 states a general but unconditionally phrased obligation to take 'requisite measures' to 'maintain the population of wild flora and fauna at, or adapt it to, a level which corresponds in particular to ecological [and other] requirements'. Article 3 contains a more qualified duty to 'take steps to promote national policies for the conservation of

---

[96] Maputo Convention, Article XII.
[97] Article XIII(1).
[98] Article XIII(2)(a).
[99] Bern Convention, Article 1. On the Convention generally, *see* C Lasén Díaz, 'The Bern Convention: 30 Years of Nature Conservation in Europe' (2010) 19 RECIEL 185; Bowman and others (n 12), 297–345; and F Fleurke and A Trouwborst, 'European Regional Approaches to the Transboundary Conservation of Biodiversity: The Bern Convention and the EU Birds and Habitats Directives' in L Kotze and T Marauhn (eds.), *Transboundary Governance of Biodiversity* (Brill Nijhoff, 2014).

wild flora, wild fauna and natural habitats, with particular attention to endangered and vulnerable species, especially endemic ones, and endangered habitats'.[100] It also stipulates that each party 'undertakes, in its planning and development policies ... , to have regard to the conservation of wild flora and fauna'.[101] Article 4 on the protection of habitats is of particular importance from a connectivity conservation point of view:

(1) Each Contracting Party shall take appropriate and necessary legislative and administrative measures to ensure the conservation of the habitats of the wild flora and fauna species, especially those specified in Appendices I and II, and the conservation of endangered natural habitats.
(2) The Contracting Parties in their planning and development policies shall have regard to the conservation requirements of the areas protected under the preceding paragraph, so as to avoid or minimise as far as possible any deterioration of such areas.
(3) The Contracting Parties undertake to give special attention to the protection of areas that are of importance for the migratory species specified in Appendices II and III and which are appropriately situated in relation to migration routes, as wintering, staging, feeding, breeding or moulting areas.
(4) The Contracting Parties undertake to co-ordinate as appropriate their efforts for the protection of the natural habitats referred to in this article when these are situated in frontier areas.

The strong, result-oriented obligations in Article 4(1) and Article 2, read in light of the Convention's objectives, appear to entail a duty for the Bern Convention's parties to ensure the maintenance and/or creation of adequate conservation connectivity for different species groups.[102]

Recommendations adopted by the Convention's Standing Committee (the principal treaty body, in which all parties are represented), inter alia in the context of the 'Emerald Network' of Areas of Special Conservation Interest set up under the Convention,[103] reinforce this conclusion. The importance of connectivity was recognised by the Standing Committee early on, in a 1991 Recommendation on the conservation of natural areas outside protected areas proper.[104] The Recommendation invites parties to 'encourage the conservation and, where necessary, the restoration of ecological corridors in particular by taking the following measures':

1. *Rights of way of roads, railways and high-voltage lines*

    – Authorising agreements between nature conservation authorities and government or public bodies owning or responsible for such areas with a view to maintaining natural plant cover and preserving the sites of rare or endangered plant species, prohibiting or limiting the use of phytosanitary products and of fire in those areas, as well as restricting the use of machinery to the strict minimum necessary for safety reasons.

---

[100] Article 3(1).
[101] Article 3(2).
[102] See also Trouwborst, 'Conserving European Biodiversity in a Changing Climate' (n 1).
[103] As to which, see <http://www.coe.int/t/dg4/cultureheritage/nature/EcoNetworks/portal_en.asp>.
[104] Standing Committee Recommendation No. 25 (1991) on the Conservation of Natural Areas Outside Protected Areas Proper (adopted 6 December 1991).

- Taking measures to restore or to compensate for the loss of ecological corridors caused by the building of new roads and other constructions that prevent animals from migrating or interchanging. In these cases, the responsible authority has to safeguard such crossing routes, for example, by building special tunnels for otters and badgers, by building so-called cerviducts for deer, by closing roads during the spring migrational period for amphibians, or by any other appropriate means.

2. *Watercourses*

- Maintaining certain watercourses or parts thereof in their natural state, and where necessary restoring them, by prohibiting the building of dams, any straightening or canalisation work and the extraction of materials from their beds, and by maintaining or restoring vegetation along their banks. Ensuring that dredging operations, when they prove essential, do not harm the integrity of the aquatic ecosystem or of the banks.
- On other watercourses, limiting canalisation and straightening work to whatever is absolutely essential, providing fish passes across dams, maintaining a minimum flow in low-water periods as far as possible, limiting extraction of materials from the bed and maintaining vegetation along the banks.[105]

A central role for connectivity conservation is also reserved in a series of detailed Recommendations issued by the Standing Committee regarding the adaptation of flora and fauna to the effects of climate change.[106] A 2008 Recommendation, for instance, calls on parties to establish 'networks of interconnected protected areas (terrestrial, freshwater and marine) and intervening habitat mosaics to increase permeability and aid gene flow'.[107] Similarly, in a specific section on amphibians and reptiles, parties are called upon to '[f]acilitate in-situ adaptation and natural range shifts by redoubling efforts to maintain or restore large intact habitats and large-scale connectivity'.[108] Detailed and comprehensive climate adaptation guidance adopted in 2009 offers another good example where it proposes the following action on protected areas and connectivity generally:

(4) Ensure the development of a sufficiently representative and connected network of protected areas so as to allow for species dispersal and settlement in new suitable sites as a consequence of climate change. In a context of great uncertainty, such a network would constitute an insurance policy to provide protection for most endangered species and habitats. ...
(5) Connect protected areas into functional ecological networks to allow the movement of species between them. Techniques include, as appropriate, buffer zones, stepping stones, corridors, and measures to reduce habitat fragmentation.
(6) Carry out integrated management of the wider countryside to alleviate the overall pressure on biodiversity and facilitate movement of species between conservation areas,

---

[105] *Ibid.*, Appendix, Part III.
[106] For more detail, *see* Trouwborst, 'Conserving European Biodiversity in a Changing Climate' (n 1).
[107] Standing Committee Recommendation No. 135 (2008) on Addressing the Impacts of Climate Change on Biodiversity (adopted 27 November 2008), Appendix, para. II(3)(c).
[108] *Ibid.*, para. I(13).

as species dispersal is likely to be the most important mechanism of species adaptation to climate change.[109]

9. EU HABITATS DIRECTIVE

The Habitats Directive is the principal legally binding nature conservation instrument of the EU.[110] Together with the Wild Birds Directive,[111] the Habitats Directive constitutes the principal means for the implementation of the Bern Convention within the Union (the EU itself and all Member States are Bern Convention parties). It has often been asserted that the Habitats Directive is frail when it comes to countering habitat fragmentation.[112] To assess these assertions, an overview is provided below of Directive provisions that are of relevance to the issue, followed by an analysis of these provisions in light of the need for connectivity conservation.

Article 2 proclaims in general terms that all measures taken by Member States pursuant to the Directive 'shall be designed to maintain or restore, at favourable conservation status, natural habitats and species of wild fauna and flora of Community interest'. Such a status is to be aimed for at least at the national level, and perhaps even also at the level of individual protected areas.[113] According to the Directive, the status of a habitat qualifies as 'favourable' when, among other things, its range is 'stable or increasing' and the 'structure and functions which are necessary for its *long-term* maintenance exist and are likely to continue to exist for the *foreseeable future*'.[114] The conservation status of a species is deemed favourable when, inter alia, the species 'is maintaining itself on a *long-term* basis as a viable component of its natural habitats'

---

[109] Standing Committee Recommendation No. 143 (2009) on Further Guidance for Parties on Biodiversity and Climate Change (adopted 26 November 2009), para. 3(4)–(6).

[110] On the Directive generally, see JM Verschuuren, 'Implementation of the Convention on Biodiversity in Europe: 10 Years of Experience with the Habitats Directive' (2002) 5 JIWLP 251; JM Verschuuren, 'Effectiveness of Nature Protection Legislation in the EU and the US: The Birds and Habitats Directives and the Endangered Species Act' (2003) 3 Yearbook of European Environmental Law 305; Fleurke and Trouwborst (n 99).

[111] Directive 2009/147/EC of the European Parliament and of the Council on the Conservation of Wild Birds [2010] OJ L20/7; this is the codified version of Council Directive 79/409/EEC as subsequently modified.

[112] See, inter alia, Cliquet and others (n 24), 171; Erens and others (n 24), 217–218; and Verschuuren, 'Rethinking Restoration in the European Union's Birds and Habitats Directives' (n 24), 436.

[113] For a discussion of the level(s) at which a favourable conservation status ought to be achieved, see A Trouwborst, 'Living with Success – and with Wolves: Addressing the Legal Issues Raised by the Unexpected Homecoming of a Controversial Carnivore' (2014) 23 EEELR 89; Y Epstein and others, 'A Legal-Ecological Understanding of Favorable Conservation Status for Species in Europe' (2015) Conservation Letters (published online 28 September 2015).

[114] Article 1(e) (emphasis added).

and 'there is, *and will probably continue to be*, a sufficiently large habitat to maintain its populations on a *long-term* basis'.[115]

An important role in achieving these objectives is reserved for the site protection obligations of Member States in respect of species listed in Annex II and habitat types listed in Annex I of the Directive.[116] Following a multiple-step procedure, sites important to these species and habitats are to be designated as Special Areas of Conservation (SACs). The SACs, combined with the Special Protection Areas (SPAs) designated under the Wild Birds Directive, are to constitute the 'coherent European ecological network' of protected areas called Natura 2000.[117] With regard to SACs, Article 6(1) of the Habitats Directive requires Member States to take 'the *necessary* conservation measures' which 'correspond to the ecological requirements' of the habitats and species involved.[118] Article 6(2) states that Member States 'shall take appropriate steps to avoid, in the SACs, the deterioration of natural habitats'. Concrete projects and plans which are potentially harmful to the protected nature within SACs and SPAs[119] are subject to a restrictive authorization scheme, laid down in Article 6(3)–(4) of the Habitats Directive:

(3) Any plan or project not directly connected with or necessary to the management of the site but likely to have a significant effect thereon, either individually or in combination with other plans or projects, shall be subject to appropriate assessment of its implications for the site in view of the site's conservation objectives. In light of the conclusions of the assessment of the implications for the site and subject to the provisions of paragraph 4, the competent authorities shall agree to the plan or project only after having ascertained that it will not adversely affect the integrity of the site concerned and, if appropriate, after having obtained the opinion of the general public.

(4) If, in spite of a negative assessment of the implications for the site and in the absence of alternative solutions, a plan or project must nevertheless be carried out for imperative reasons of overriding public interest, including those of a social or economic nature, the Member State shall take all compensatory measures necessary to ensure that the overall coherence of Natura 2000 is protected. It shall inform the Commission of the compensatory measures adopted. Where the site hosts a priority habitat type and/or a priority species,[120] the only considerations which may be raised are those relating to human health or public safety, to beneficial consequences of primary importance for the environment or, further to an opinion from the Commission, to other imperative reasons of overriding public interest.

---

[115] Article 1(i) (emphasis added).
[116] Article 4.
[117] Habitats Directive, Article 3.
[118] Emphasis added.
[119] Article 7 of the Habitats Directive extends the reach of Article 6(3)–(4) to Birds Directive SPAs.
[120] Incidentally, it appears that the added layer of protection intended for priority habitats and species (which are marked in the Directive's annexes with an asterisk) does not always materialise in practice. In particular, the various opinions that have been issued by the European Commission under Article 6(4) seem to have been overly permissive, and have involved the approval of harmful linear infrastructure for economic reasons. For a critical discussion, *see* L Krämer, 'The European Commission's Opinions under Article 6(4) of the Habitats Directive' (2009) 21 JEL 59.

The ECJ has developed an extensive jurisprudence regarding the above rules on the designation and protection of sites under the Birds and Habitats Directives. Throughout this case law, the Court has tended to interpret the rules involved in such a way as to maximise their effectiveness in light of the Directives' objectives. For instance, this case law has made it abundantly clear that considerations of an economic nature, or concerning expected future management difficulties, are to play no part in the site designation process.[121] Another important example is the stance of the Court regarding the assessment and authorisation of plans and projects. The Court determined that under Article 6(3) of the Habitats Directive, plans or projects may in principle be authorised only 'where no reasonable scientific doubt remains as to the absence' of harmful impacts.[122] 'A less stringent authorisation criterion', the Court argued, 'could not as effectively ensure the fulfilment of the objective of site protection.'[123] Regarding Article 6(2), the prescription in the latter to take the 'appropriate steps to avoid ... the deterioration' of habitats in SACs or SPAs has repeatedly been interpreted by the Court as an obligation to 'do what it takes'. What the 'appropriate steps' are will depend on the problem at hand, but what ultimately counts is the result.[124] Clearly, effective measures are to be taken before adverse effects occur.[125] Moreover, to meet the requirements of Article 6(2), damage which already *has* occurred must be undone. For instance, a 2002 judgment in a case involving harm through overgrazing by sheep in an Irish SAC confirmed in this regard that 'it is necessary for the Irish authorities not only to take measures to stabilise the problem of overgrazing, but also to ensure that damaged habitats are allowed to recover'.[126] Similarly, the Court has affirmed that 'the protection of SPAs is not to be limited to measures intended to avoid external anthropogenic impairment and disturbance but must also, according to the situation that presents itself, include positive measures to preserve or improve the state of the site'.[127]

As concerns the generic protection of species – both within and outside Natura 2000 sites – Articles 12 and 13 of the Habitats Directive commit Member States to 'take the requisite measures to establish a system of strict protection' for the animal and plant species listed in Appendix IV of the Directive.[128] Article 12 requires the establishment of prohibitions on, inter alia, the killing, capturing and disturbing of individual animals belonging to species from Annex IV, and the 'deterioration or

---

[121] For example, Case C-355/90 *Commission v Spain* [1993] ECR I-4221, paras. 26–27; Case C-44/95 *Regina v Secretary of State for the Environment* [1996] ECR I-3805, para. 26; Case C-3/96 *Commission v Netherlands* [1998] ECR I-3031.

[122] Case C-127/02 *Waddenvereniging* [2004] ECR I-7405, para. 61.

[123] *Ibid.*, para. 58.

[124] For a particularly clear example, *see* Case C-117/00 *Commission v Ireland* [2002] ECR I-5335, paras. 26–33.

[125] This is apparent if only from the use of the term 'avoid' in Article 6(2). *See* also European Commission, *Managing Natura 2000 Sites: The Provisions of Article 6 of the 'Habitats' Directive 92/43/EEC* (European Commission, 2000), 24.

[126] Case C-117/00, para. 31.

[127] Case C-535/07 *Commission v Austria* [2006] ECR I-2755, para. 59; *see* also Case C-418/04 *Commission v Ireland* [2007] ECR I-10947, para. 154.

[128] Articles 12(1) and 13(1).

destruction of breeding sites or resting places',[129] exceptions to which may only be allowed under strict conditions.[130] The jurisprudence of the EU Court makes clear that Member States must not only prohibit the acts in question, but must also take all measures necessary to ensure that the prohibitions in question are not violated in practice.[131] According to the EU Court, Article 12(1) 'requires the Member States not only to adopt a comprehensive legislative framework but also to implement concrete and specific protection measures'.[132] Besides, the prescribed 'system of strict protection' of Annex IV species presupposes the 'adoption of coherent and coordinated measures of a preventive nature'.[133] A specific duty to monitor 'incidental capture and killing' – e.g., mortality through road traffic or bycatch in fishing gear – of Annex IV animals is laid down in Article 12(4). Member States are to take the conservation measures necessary to ensure that such killing does not have a 'significant negative impact' on the species involved.[134]

Last but not least, connectivity is addressed specifically in Articles 3(3) and 10 of the Habitats Directive:

> Where they consider it necessary, Member States shall endeavour to improve the ecological coherence of Natura 2000 by maintaining, and where appropriate developing, features of the landscape which are of major importance for wild fauna and flora.[135]
>
> Such features are those which, by virtue of their linear and continuous structure (such as rivers with their banks or the traditional systems for marking field boundaries) or their function as stepping stones (such as ponds or small woods), are essential for the migration, dispersal and genetic exchange of wild species.[136]

The permissively worded Articles 3(3) and 10 cited above seem to lack 'legal teeth'[137] and to leave the issue of connectivity conservation largely to the discretion of individual Member States. Indeed, in reality, the greater part of Natura 2000 appears to be 'not a network but a collection of isolated sites'.[138] Besides, the proposition in the European Commission's 2009 Climate Adaptation White Paper that '[i]n *future* it may be necessary to consider establishing a permeable landscape in order to enhance the interconnectivity of natural areas'[139] could be taken as an acknowledgement that the

---

[129] Article 12(1).
[130] Article 16.
[131] See Case C-103/00 *Commission v Greece* [2002] ECR I-1147; Case C-518/04 *Commission v Greece* [2006] ECR I-42; and Case C-221/04 *Commission v Spain* [2006] ECR I-4515.
[132] Case C-183/05 *Commission v Ireland* [2007] ECR I-137, para. 29.
[133] *Ibid.*, para. 30.
[134] Article 12(4).
[135] Article 3(3); *see* also Article 10(1).
[136] Article 10(2).
[137] Verschuuren, 'Rethinking Restoration in the European Union's Birds and Habitats Directives' (n 24), 436.
[138] *Ibid.*
[139] European Commission, *Adapting to Climate Change: Towards a European Framework for Action*, Communication COM(2009) 147 (1 April 2009), para. 3.2.3 (emphasis added).

current Natura 2000 regime fails to provide for adequate connectivity.[140] There is every reason to believe, nevertheless, that Articles 3 and 10 do not exhaust the legal relevance of the Directive in respect of connectivity.

Significantly, the fact that an adequate degree of connectivity *between* populations located in Natura 2000 sites appears necessary to safeguard a favourable conservation status for the species in question entails the applicability of Directive provisions aimed at the conservation and/or restoration of habitats and species *within* Natura 2000 sites, in particular Article 6. For example, in view of the aim of securing a favourable conservation status as defined in Article 1, establishing adequate connectivity *between* sites must in many cases be deemed obligatory as a result of the duty in Article 6(1) to take 'the necessary conservation measures involving, if need be, ... appropriate statutory, administrative or contractual measures which correspond to the ecological requirements of the natural habitat types in Annex I and the species in Annex II present on the sites'.[141] Similar considerations apply to Article 6(2), cited above. After all, although the scope of the 'appropriate steps' envisaged in this provision is limited to the species and habitats 'in the special areas of conservation', it is common ground that they 'may need to be implemented *outside* the SAC' as Article 6(2) 'does not specify that measures have to be *taken* in the SAC' but instead that measures are to avoid impacts in the areas in question.[142] Likewise, Article 6(3) may stand in the way of, for example, a highway planned to run between two separately located Natura 2000 sites, thus threatening to impair the connectivity between the respective populations within those sites. In respect of Annex IV species, Articles 12 and 13 on generic species protection – applying inside and outside of Natura 2000 sites – appear to prescribe connectivity measures as well:

> In a world where most annex IV species find their habitat fragmented, long-term persistence of populations cannot be ensured in local habitat sites, but requires that these sites can interact in a habitat network. Therefore, an effective implementation of the EU-Habitats Directive requires a landscape level approach: habitat networks.[143]

Finally, in cases where significant numbers of Annex IV animals end up as roadkill, Article 12(4) may require the construction of fauna crossings over highways and other human infrastructure, thus promoting connectivity. On a general note, it should be borne in mind that the obligations just reviewed are to be interpreted in conformity with the Bern Convention, as discussed above.

The above interpretation of Habitats Directive provisions seems to square with the observation in a connectivity guidance document composed for the Commission in 2007 that, in principle, connectivity measures 'should be implemented whenever they are necessary to maintain or restore FCS [favourable conservation status] of habitats or

---

[140] Verschuuren, 'Rethinking Restoration in the European Union's Birds and Habitats Directives' (n 24), 436.

[141] The nature of Article 6(1) as an obligation of result interlinked with Article 2 is underlined in the European Commission's 2000 publication (n 125), 17.

[142] European Commission, *ibid.*, 24 (emphasis as in original).

[143] P Opdam and others, *Effective Protection of the Annex IV Species of the EU-Habitats Directive: The Landscape Approach* (Alterra, 2002), 23.

species of Community interest'.[144] It is convenient to note in this context that the mere fact that Articles 3(3) and 10 of the Habitats Directive contain the most specific language on connectivity does not entail that they possess a monopoly on the issue, and that by consequence their ostensibly voluntary nature should overrule the mandatory requirements just distilled from a combination of Articles 1, 2, 6, 12 and 13 of the Directive.[145]

In two judgments of 2010 and 2011, both pronounced in non-compliance procedures involving Spain, the EU Court highlighted the importance of avoiding habitat fragmentation and ensuring adequate connectivity. In the first of these, concerning the Iberian lynx (*Lynx pardinus*), the Court affirmed that 'linear transport infrastructures may constitute a real barrier for certain species referred to in the Habitats Directive and, by thus fragmenting their natural range, promote endogamy and genetic drift within those species'.[146] The second, and for present purposes more consequential, case centres on brown bears (*Ursus arctos*) and capercaillies (*Tetrao urogallus*) threatened by open-cast coal mining activities in the Spanish Natura 2000 site 'Alto Sil'.[147] Specifically, these activities affected an ecological corridor connecting two sub-populations of bears. As the judgment notes, this 'corridor, with a width of 10 kilometres, is a transit route of great importance for the western population of the said species, allowing in particular communication between two very important pockets of reproduction'.[148] A report is cited establishing that 'the bears move 3.5 to 5 kilometres from the areas of impact of the noise and vibrations caused by mining operations' and that these operations 'will prevent access for the brown bear to that corridor, or make it much more difficult'.[149] The Court qualifies these impacts as 'disturbances' of the Natura 2000 site, 'which are significant having regard to the conservation of the brown bear', amounting to a violation of Article 6(2) of the Habitats Directive.[150] Comparable conclusions are reached in respect of the capercaillie, as the mining operations are deemed 'capable of producing a barrier effect likely to contribute to the fragmentation of the habitat of the capercaillie and the isolation of certain sub-populations of that species'.[151] It should be noted that the mines in question are located within the 'Alto Sil' site, and that the affected ecological corridors are also situated within that site and in adjacent Natura

---

[144] See M Kettunen and others, *Guidance on the Maintenance of Landscape Connectivity Features of Major Importance for Wild Flora and Fauna: Guidance on the Implementation of Article 3 of the Birds Directive (79/409/EEC) and Article 10 of the Habitats Directive (92/43/EEC)* (Institute for European Environmental Policy 2007), 7.

[145] To illustrate, a parallel can be drawn with Article 12(1) and 12(4) of the Directive. The fact that Article 12(4) appears to have been drafted specifically with, inter alia, incidental mortality of marine turtles or cetaceans in fishing gear in mind, does not cancel the applicability to such bycatch of the generic prohibition to kill Annex IV species from Article 12(1). See, e.g., the answer of EU Commissioner Dimas to parliamentary question E-5890/2007 on harbour porpoise (*Phocoena phocoena*) bycatch in recreational gillnet fisheries (11 February 2008), see [2008] OJ C191/150.

[146] Case C-308/08, *Commission v Spain* (20 May 2010), [2010] ECR I-4281, para. 25.
[147] Case C-404/09, *Commission v Spain* (24 November 2011).
[148] Ibid., para. 189.
[149] Ibid., para. 188.
[150] Ibid., para. 191.
[151] Ibid., para. 148.

2000 sites. Even so, it is likely that the Court would have arrived at a similar decision in an imaginary case with bear and capercaillie populations in two *non*-adjacent Natura 2000 sites connected through a corridor without that legal status. In such a case, Article 6(2) would require the adequate safeguarding of the corridor *outside* the Natura 2000 sites in order to safeguard the bear and capercaillie populations *within* those sites. The 'Alto Sil' judgment thus clearly bolsters the conclusions drawn above regarding the scope of Article 6(2) in the context of connectivity conservation.[152]

In sum, the Habitats Directive imposes legal obligations for EU Member States to counter habitat fragmentation and ensure adequate degrees of connectivity.

10. CONCLUSION

The problem of habitat fragmentation and the corresponding need for connectivity conservation are not addressed in so many words in the provisions of most of the legal instruments reviewed above. In many cases, however, the interpretation of relevant provisions in light of (i) the instruments' objectives, (ii) current scientific knowledge on fragmentation and connectivity, including regarding the needs of species in order to cope with climate change, and (iii) relevant decisions by the parties, renders the conclusion that measures to counter fragmentation and ensure connectivity are mandatory.

Whereas the present chapter paints a broad picture of the way habitat fragmentation and connectivity conservation are currently addressed in a number of legal instruments, there is ample scope for further research into the topic. Future research might usefully target instruments not dealt with here; go into more depth concerning the instruments that have been dealt with; and, in particular, focus on ways to improve the contribution of the various instruments to resolving the problem.

SELECT BIBLIOGRAPHY

B Lausche, D Farrier, JM Verschuuren, AGM La Viña, A Trouwborst, C-H Born and L Aug, *The Legal Aspects of Connectivity Conservation – A Concept Paper*, IUCN Environmental Policy and Law Series, Vol. 85-1 (IUCN, 2013)

GL Worboys, WL Francis and M Lockwood (eds), *Connectivity Conservation Management: A Global Guide* (Earthscan, 2010)

---

[152] *See* also Trouwborst, 'Conserving European Biodiversity in a Changing Climate' (n 1).

9. Armed conflict and biodiversity
Karen Hulme

1. INTRODUCTION

While many industrial and developmental activities, such as minerals extraction, logging and the damming of rivers, have devastating impacts on biodiversity, many states are also plagued by the environmentally destructive phenomenon of warfare. Often what sets wartime environmental destruction apart is its perceived needlessness, such as the Iraqi oil-well fires during the 1991 conflict, as well as the intentionality of unleashing weapons and tactics with devastating inter-generational effects. And as scientific understanding increases, and the benefits and value of healthy natural habitats, including biodiversity, become clearer, it is natural to seek to minimise all causes of harm, including that caused by warfare. The current chapter, therefore, hopes that in more precisely pinpointing how wartime harm to biodiversity occurs, lawmakers might be better equipped to prevent it in the future.

Specific to the situation of armed conflict is the legal regime of international humanitarian law (IHL). Of current interest, however, are recent legal developments in the work of the International Law Commission[1] (ILC) and international judicial decisions,[2] among others, which point to greater emerging importance of environmental treaty obligations during armed conflict. With the ILC's new work stream on the *Protection of the Environment in Relation to Armed Conflict*[3] it seems an opportune time to examine threats to biodiversity as a consequence of armed conflict. Adopting the ILC's temporal approach[4] of analysing legal obligations applicable (a) before conflict, (b) during conflict and (c) post conflict, also offers new insights into how existing obligations might be used to better effect, and where the gaps lie for new legal provision.

According to the 1992 Convention on Biological Diversity (CBD),[5] 'biodiversity' includes the 'variability among living organisms from all sources ... and the ecological

---

[1] 2011 Draft Articles on the Effects of Armed Conflicts on Treaties, with Commentaries (2011) II:2 Yearbook of the International Law Commission.

[2] *Legality of the Threat or Use of Nuclear Weapons*, Advisory Opinion, ICJ Reports (1996) 226, [25]; *Legal Consequences of the Construction of a Wall in the Occupied Palestinian Territory*, ICJ Reports (2004) 136, [106]; *Armed Activities on the Territory of the Congo (DRC v Uganda)*, ICJ Reports (2005) 168, [216].

[3] 2011 Recommendation of the Working-Group on the Long-term Programme of Work, A/66/10, Annex E.

[4] International Law Commission, Report on the work of its sixty-fifth session (6 May to 7 June and 8 July to 9 August 2013) General Assembly, Official Records, Sixty-eighth Session, Supplement No. 10 (A/68/10), Chapter IX.

[5] (1992) 31 ILM 822.

complexes of which they are part'[6] including, fundamentally, diversity at all three levels of 'within species, between species and of ecosystems'.[7] The culmination of decades of work recognising that the richness of species 'creates the capacity for resilience'[8] in the earth's ecosystems, and, hence, promotes not just ecosystem health but also 'ecosystem stability',[9] the CBD now enjoys near universal acceptance with 196 states parties.[10] Of course, the CBD does not work alone and forms part of a complex and comprehensive web of protections and conservation management obligations for species and habitats. Indeed, all environmental protection rules will benefit biodiversity simply by reducing emissions of pollutants, reducing the exploitation of certain species or conserving protected areas and species in situ. While IHL may indirectly be able to reduce certain pollutants, the principal focus of the chapter will be an analysis of the wartime context through the final approach listed, namely, of protecting species *in situ*, an obligation which forms the cornerstone of biodiversity protection in the CBD[11] and is implicit in all species and habitats treaties.

The chapter, therefore, will first explore more broadly the biodiversity–conflict relationship, before focussing specifically on the biodiversity protections offered by IHL. Later sections seek to explore new ways of approaching the 'protected areas' conservation obligations found in many multilateral environmental agreements (MEAs) and the possible need for new IHL provisions, as potential new tools in the wartime protection of the environment.

2. THE BIODIVERSITY/CONFLICT NEXUS

The ILC's temporal approach to analysing the law maps onto the key phases in which conflict occurs and helps to highlight a number of factors that influence the conflict–biodiversity relationship.

The first factor is the geographical location of the conflict, particularly the environmental resources present, their pre-existing vulnerability and their capacity for resilience. In recent decades armed conflict has plagued many biodiversity-rich and unique ecosystems, such as the forests of the Democratic Republic of the Congo (DRC) and Rwanda, the important European biodiversity corridor of the River Danube in the conflict in Serbia (home to some 43.3 per cent of all existing species in Europe[12]), and the 'megadiverse' environment of Colombia, which ranks first in the world for bird

---

[6] CBD, Article 2.
[7] *Ibid.*
[8] T O'Riordan and S Stoll (eds), *Biodiversity, Sustainability and Human Communities: Protecting Beyond the Protected* (CUP, 2002), 5; D Tilman, 'Causes, Consequences and Ethics of Biodiversity' (2000) 403 Nature 208, 208–209.
[9] KS McCann, 'The Diversity-stability Debate' (2000) 405 Nature 228.
[10] As at 30 December 2015. The most notable non-party is the US, which has also been engaged in major armed conflicts over the past decade.
[11] CBD, Article 8.
[12] Republic of Serbia, First National Report of the Republic of Serbia to the United Nations Convention on Biological Diversity, July 2010, 8 available at <http://www.cbd.int/countries/?country=rs>.

species and hosts almost 10 per cent of global biodiversity.[13] In its *Desk Study on the Environment in Iraq*,[14] the United Nations Environment Programme (UNEP) suggests the possible extinction of otter, bat and endemic barbel fish species in the Iraqi Mesopotamian Marshes caused, in part, by armed conflict.[15] Clearly, different environments will contain different degrees of diversity. For example, tropical forests, wetlands and marine ecosystems will generally contain richer biodiversity than deserts or grasslands and, thus, the bomb damage impact on biodiversity will differ from place to place.

More specific, conflict-related factors include the scale, intensity and length of warfare, the types of equipment, weapons and tactics used, and the terrain of conflict (e.g. urban, jungle, desert). Obvious causes of harm are weapons and the environmental impacts of the targeting of military and industrial facilities, such as the thousands of oiled birds which perished in the oil-polluted shores of the Persian Gulf in the 1991 Gulf Conflict.[16] Weapons that utilise or release a harmful toxic component present a more obvious, inter-generational biological or ecosystem-level threat through impacts at the genetic level or in the food web.[17] Risks to species survival are also caused by inadvertent side-effects of weapons use, such as the fragmentation of forest habitat, the replacement of diversity-rich flora with diversity-poor grasses,[18] and the destruction of biodiversity corridors by large-scale bombing.[19] Often it is simply the proliferation of, and ease of access to, guns among armed groups and local communities that lead to greater threats to species, such as the killing of endangered hippos, elephants, buffalo and mountain gorillas in the Virunga forests in the DRC,[20] and the Asian elephant and Siamese crocodile of Cambodia.[21]

When moving large numbers of soldiers into or across an environmental space, as with the manoeuvre of battleships, tanks and aircraft around a battle zone, there is an inevitable environmental footprint in terms of fuel pollutants, ground disturbance and waste generation. War's broader biodiversity footprint emanates simply from the consequences of civil disorder itself, such as resource plunder and the breakdown in conservation management, and the population's desperation to meet their survival needs. For example, an estimated 300 km² of forest were damaged by the day-to-day

---

[13] Available at <http://www.cbd.int/countries/?country=co>.

[14] (UNEP 2003) available at <http://www.unep.org/pdf/iraq_ds.pdf>.

[15] *Ibid.*, 44.

[16] J Brauer, *War and Nature: The Environmental Consequences of War in a Globalized World* (Altamira Press, 2011), 26.

[17] D Vidosavljević and others, 'Soil Contamination as a Possible Long-Term Consequence of War in Croatia' (2013) 63(4) Acta Agriculturae Scandinavica, Section B – Soil & Plant Science 322.

[18] AH Westing and EW Pfeiffer, 'The Cratering of Indochina' (1972) 226(5) Scientific American 59.

[19] L Gibson and others, 'Near-Complete Extinction of Native Small Mammal Fauna 25 Years After Forest Fragmentation' (2013) 341 Science 1508.

[20] A Plumptre, 'Lessons Learned From On-the-Ground Conservation in Rwanda and the Democratic Republic of the Congo' in SV Price (ed), *War and Tropical Forests: Conservation in Areas of Armed Conflict* (Food Products Press, 2003), 77–82.

[21] C Loucks and others, 'Wildlife Decline in Cambodia, 1953–2005: Exploring the Legacy of Armed Conflict' (2009) 2 Conservation Letters 82.

survival needs of the 850,000 refugees fleeing the 1994 genocide and civil war in Rwanda.[22] And war-torn Afghanistan, a state listed towards the bottom of the World Development Index for over 30 years,[23] admitted that in its search for food its population had 'no option but to exploit biodiversity unsustainably'.[24] Long gone are the days of species-rich buffer zones (refugiums)[25] between disputed territories. Instead the all too brief respite the environment enjoys from the suspension of large-scale economic development activities, such as mineral extraction activities and dam-building projects, is shattered by the long-term contamination by landmines and other toxic or explosive remnants of war. An oddity then is the biodiversity rejuvenation witnessed in the heavily mined, 60-year demilitarised zone separating North and South Korea.[26]

Finally, existing levels of environmental damage and, thus, pre-existing vulnerabilities will contribute to the depth or severity of biological impacts sustained during conflict and, hence, the duration of the necessary recovery period – which will also be influenced by the restoration measures implemented, specifically the speed of implementation and the ability to pay. For example, although an oil-rich country, it still reportedly took Kuwait seven years to restore the Kuwaiti National Park to the rich ecosystem that it had been before the damage caused in the 1991 Gulf Conflict.[27] Fortunately for the DRC, inclusion on the 'in Danger' list of the 1972 Convention for the Protection of the World Cultural and Natural Heritage[28] (World Heritage Convention or WHC) aided the cash-strapped state to secure vital funds to assess and restore its war-torn biodiversity across its five forests.[29] For remediation obligations and damage assessments,[30] the ILC's temporal approach again offers a new analytical approach, necessitating analysis beyond IHL.

---

[22] JA McNeely, 'Conserving Forest Biodiversity in Times of Violent Conflict' (2003) 37(2) Oryx 142, 146; J Kalpers, 'Volcanoes Under Siege: Impact of a Decade of Armed Conflict in the Virungas' (Biodiversity Support Program, 2001) copy on file with author, 14–17; A Lanjouw, 'Building Partnerships in the Face of Political and Armed Conflict' in SV Price (ed), *War and Tropical Forests: Conservation in Areas of Armed Conflict* (Food Products Press, 2003), 97.

[23] United Nations Development Programme, available at <http://hdr.undp.org/en/countries>.

[24] Islamic Republic of Afghanistan, 'Afghanistan's Fourth National Report' for the CBD, March 2009, 3, available at <http://www.cbd.int/doc/world/af/af-nr-04-en.pdf>.

[25] PS Martin and CR Szuter, 'War Zones and Game Sinks in Lewis and Clark's West' (1999) 13 Conservation Biology 36.

[26] Home to increasing populations of red-crowned cranes, golden eagles, and lynx, among others, see SD Lanier-Graham, *The Ecology of War: Environmental Impacts of Weaponry and Warfare* (Walker Publishing Company, Inc., 1993), 73; JA McNeely, 'War and Biodiversity: An Assessment of Impacts' in JE Austin and CE Bruch (eds), *The Environmental Consequences of War* (CUP, 2000), 365. McNeely also points out (at 368) the increase in North Atlantic fish stocks while fishermen were hampered in their task by the Second World War.

[27] SAS Omar and others, 'The Gulf War Impact on the Terrestrial Environment of Kuwait: An Overview' in JE Austin and CE Bruch (eds), *The Environmental Consequences of War* (CUP, 2000), 329.

[28] (1972) 11 ILM 1358.

[29] UNESCO, *Resource Manual: Managing Natural World Heritage* (UNESCO, 2010), 26, 45–46, available at <http://whc.unesco.org/en/managing-natural-world-heritage/>.

[30] Note the environmental claims to the F-4 Panel of the UN Compensation Commission set up by SCR 687 (1991) available at <http://www.uncc.ch/>; UNEP's Post-Crisis Environmental Assessment Group: UNEP, *The Kosovo Conflict: Consequences For the Environment and*

The biodiversity/conflict nexus, therefore, involves a complex interaction between a range of factors from the temporal perspectives of (1) the pre-conflict environmental management and disarmament obligations, (2) the environmental impacts permissible in conflict, and (3) any post-conflict legal restoration obligations or damage assessments. Consequently, this novel method of approaching the issues by the ILC Special Rapporteur may help to identify new insights into this area of law. First, however, it is imperative to focus on IHL in order to examine its protective value for biodiversity, and to pinpoint any inadequacies.

3. INTERNATIONAL HUMANITARIAN LAW

While the concept of 'biodiversity' is absent from IHL, this finding is unsurprising since the main legal provisions date to the aftermath of the Vietnam War with the adoption of two additional protocols; specifically the 1977 Protocol (I) Additional to the Geneva Conventions of 1949 and Relating to the Protection of Victims of International Armed Conflicts[31] (Additional Protocol I or API), and Protocol II for non-international armed conflicts.[32] Broader notions of 'environmental' protection are, however, not completely absent and this section will therefore analyse the main dimensions of warfare dangers to biodiversity through the conduct of hostilities, namely, (1) environment-specific protections, (2) the lawful targeting of military objectives and (3) weapons use.

3.1 Environment-Specific Protections

It was in 1977 in the aftermath of the Vietnam War that states negotiated API and specifically included protection for the 'natural environment' in two provisions, namely Articles 35(3) and 55.[33] Article 35(3) prohibits the use of tactics and weapons 'which are intended, or may be expected, to cause widespread, long-term and severe damage to the natural environment'. Practically redundant today in direct terms, it is the extremely high, cumulative threshold which has, in practice, eliminated almost all of the value of the provisions. Environmental damage resulting from oil-polluted coastlines, desert

---

Human Settlements (UNEP and United Nations Centre for Human Settlements (Habitats), 1999), 63–68; UNEP, *The Democratic Republic of the Congo Post-Conflict Environmental Assessment: Synthesis for Policy Makers* (UNEP, 2011), 40–43; UNEP, *Afghanistan's Environment 2008* (UNEP and National Environmental Protection Agency of the Islamic Republic of Afghanistan, 2008), 15–17, available at <http://postconflict.unep.ch/publications/afg_soe_E.pdf>.

[31] (1977) 16 ILM 1391.

[32] Protocol (II) Additional to the Geneva Conventions of 12 August 1949, and Relating to the Protection of Victims of Non-International Armed Conflicts (1977) 16 ILM 1442.

[33] Protocol II's draft article protecting the environment was among those deleted to reduce the content of the Protocol to attract more ratifications. In the comprehensive study of the customary laws of armed conflict it was 'arguable' that Protocol I's environmental provisions had become customary law also in non-international armed conflict; J-M Henckaerts and L Doswald-Beck, *Customary Humanitarian International Law, Volume I: Rules* (CUP, 2005), Rules 44 and 45.

soils and the atmosphere in the Persian Gulf was not recognised as fulfilling the '20 or 30 years'[34] definition of 'long-term', since initial fears of species' decline and climatic cooling were over-estimated.[35] Similarly, too, the toxic contamination of the Danube, which serves as a vital biodiversity corridor in Europe, by NATO's attack on the petrochemical and oil facilities at Pancevo, Serbia in the 1999 Kosovo Conflict, and attacks on fragile ecosystems in protected areas would also fail the provision's long-term test.[36] Yet, it is certainly not clear how the interpretation of long-term harm of 20 to 30 years is being applied in practice, and certainly not as regards the notion of biodiversity. Clearly, many weapons components will persist in the environment for decades, such as the dioxin contaminants in Agent Orange,[37] many are chemically toxic, such as the Uranium (U238) in depleted uranium ammunition, some will bioaccumulate, such as lead, mercury and other heavy metals often used in weaponry, and many chemicals will lead to cancers and species mutations. Species extinction through ecosystem changes or reduction in populations are, similarly, long lasting, as is, surely, damage that leads to ecosystem changes, which result in a lower-diversity ecosystem emerging.[38]

The other two terms were not as precisely defined. For 'severity' the German delegation referred to the need for 'a major interference with human life or natural resources',[39] and subsequent state practice appears to show acceptance of an area of 'several hundred square kilometres'[40] as denoting 'widespread' harm. Key is the notion that such harm should have been at least foreseeable by the state, yet complicating the damage assessment further is the fact that impacts in the environment are not uniform or linear. Thus, impacts on one species could mean that others thrive and new species dominate, or else the loss of key species may cause ecosystem collapse.[41] Ecosystem resilience and toleration of ecosystems for particular contaminants will also, therefore,

---

[34] Official Records of the Diplomatic Conference on the Reaffirmation and Development of International Humanitarian Law Applicable in Armed Conflicts, Geneva (1974–1977) CDDH/215/Rev.1, [27].

[35] Iraq was also not a party to API. A Roberts, 'Failures in Protecting the Environment in the 1990–91 Gulf War' in Peter Rowe (ed), *The Gulf War in English and International Law* (Routledge, 1993), 111; MN Schmitt, 'Green War: An Assessment of the Environmental Law of International Armed Conflict' (1997) 22 Yale Journal of International Law 1, 15–20.

[36] International Criminal Tribunal for the Former Yugoslavia (ICTY): Final Report to the Prosecutor by the Committee Established to Review the NATO Bombing Campaign Against the Federal Republic of Yugoslavia, 8 June 2000 (2000) 39 ILM 1257, [17]; UNEP Kosovo (n 30), 67.

[37] F Ramade, *Ecotoxicology* (John Wiley & Sons, 1987), 31.

[38] Brauer (n 16), 22–26.

[39] Manual of the German Armed Forces, cited in W Heintschel von Heinegg and M Donner, 'New Developments in the Protection of the Natural Environment in Naval Armed Conflicts' (1994) 37 German Yearbook of International Law 281, 286.

[40] The US is not bound by API but its operational practice refers to this definition, *2010 Operational Law Handbook*, International and Operational Law Department, Judge Advocate General's Legal Center and School, United States, 352, available at <http://www.loc.gov/rr/frd/Military_Law/pdf/operational-law-handbook_2010.pdf>.

[41] Brauer (n 16), 27–29.

influence the resulting level of harm.[42] Also a factor will be historical pollutant contributions. This was a key determinant in reducing potential compensation levels awarded by the UN Compensation Commission (UNCC), which had been tasked with assessing environmental damage contingent on Iraq's unlawful invasion of Kuwait in 1990.[43] Consequently, it is probably not that surprising that no post-Vietnam weapons use or tactic has been found to violate the threshold. And, despite more than 25 years of more informed, more scientifically advanced environmental thinking, this scale has not budged.

Article 55 of API requires that the parties take 'care' to protect the environment against widespread, long-term and severe damage. The care obligation is a broader, due diligence-based obligation, but in containing the same high, three-fold threshold of harm it too has little practical use.[44] An alternative formulation of the care obligation has been drawn from state practice by the esteemed authors of the 2005 Customary International Humanitarian Law Study.[45] Here it is suggested that a customary rule has emerged requiring 'due regard' for the environment.[46] Its main bonus is that it does not contain a specific threshold of harm, but its customary status is disputed.[47] Finally, worthy of mention are the complementary prohibitions against attacks upon 'objects indispensable to the survival of the civilian population'[48] (such as agricultural areas, crops, livestock, drinking water installations and supplies), and on such extremely hazardous polluting facilities as nuclear power plants.[49] Of course, all such protections are subject to abuse.

In summary, the environmental provisions have done little in practice to limit wartime damage, and, stymied by a seemingly unmoveable threshold, their role has instead been in galvanising support for improving environmental protection in wartime more broadly.

3.2 Targeting

The notion of targeting generally refers to the lawful attack of enemy military objectives, such as military persons, vehicles, buildings and other objects, which may

---

[42] *Ibid.*, 81–110.

[43] F-4 Panel claims (n 30); MT Huguenin and others, 'Assessment and Valuation of Damage to the Environment' in CR Payne and PH Sand (eds), *Gulf War Reparations and the UN Compensation Commission: Environmental Liability* (OUP, 2011), chapter 3. Note the UNCC basis was the breach of Article 2(4) United Nations Charter, not the API threshold since API was not applicable to the conflict.

[44] K Hulme, 'Taking Care to Protect the Environment against Damage: A Meaningless Obligation?' (2010) 92 (879) International Review of the Red Cross 675; M Bothe and others, 'International Law Protecting the Environment During Armed Conflict: Gaps and Opportunities' (2010) 92(879) International Review of the Red Cross 569, 575.

[45] Henckaerts and Doswald-Beck (n 33), Rule 44.

[46] *Ibid.*

[47] GH Aldrich, 'Customary International Humanitarian Law – An Interpretation on Behalf of the International Committee of the Red Cross' (2005) 76 British Yearbook of International Law 503, 515.

[48] API, Article 54 and APII, Article 14.

[49] *Ibid.*

include environmentally polluting facilities such as industrial complexes and oil refineries[50] and even the environment itself as 'a specific area of land',[51] such as the direct chemical attacks on mangroves and forests during the Vietnam War.[52] Clearly, any limitations on the attack of the environment itself, and of such polluting facilities, will reduce the level of potential harm to habitats, species and species diversity.

Since the environmental provisions in API have been found lacking, the legal protection for biodiversity is more fundamentally to be derived from customary law obligations requiring that the parties distinguish in their targeting between military objectives and civilian persons and objects, the latter of which includes the environment. The rule, thus, provides that 'Civilian objects shall not be the object of attack ... Civilian objects are all objects which are not military objectives ...'[53] Hence, as a prima facie civilian object the environment cannot be subject to attack.

Civilian status is not an absolute, however; it can be lost, for example, if the environment, such as a forest, is used by the military or armed groups as a base or as cover or concealment, for example by the Viet Cong in the Vietnam War and by FARC in Colombia, and the use of transboundary forests as escape routes in Nepal.[54] The definition of 'military objectives' ensures that only those facilities are attacked which, if destroyed or captured, would make an 'effective' contribution to military action (due to their location or use for example), and, dependent upon the ambient circumstances, would offer a 'definite' military advantage.[55] Thus, even where the environment is used by a military force, its attack is not a foregone conclusion and the civil wars in Colombia, the DRC and Sri Lanka do evidence this point.

While forests, wetlands and nature reserves could be subject to direct targeting via their use as cover, concealment or headquartering of military forces, the more likely scenario is harm caused to the environment as a by-product of the targeting of military forces, materiel or communications installations located within.[56] Here, the incidental or 'collateral' damage caused to the environment in the attack of the military objective is not ignored, but forms part of the rule of proportionality. Accordingly, Article 51(5)(b) of API stipulates that an attack is prohibited where it is 'expected to cause incidental loss of civilian life, injury to civilians, damage to civilian objects, or a combination thereof, which would be excessive in relation to the concrete and direct military advantage anticipated'. Unfortunately, the environmental credentials of the

---

[50] K Hulme, *War-Torn Environment: Interpreting the Legal Threshold* (Martinus Nijhoff, 2004).
[51] Statements of UK, Canada, Germany, and Italy upon ratification of Protocol I, available at <http://www.icrc.org/applic/ihl/ihl.nsf/States.xsp?xp_viewStates=XPages_NORMStatesParties&xp_treatySelected=470>.
[52] AH Westing, *Ecological Consequences of the Second Indochina War* (Almqvist and Wiskell International, 1976).
[53] API, Article 52(1).
[54] Bothe and others (n 44), 576.
[55] API, Article 52(2).
[56] Note also the prohibition on scorched earth policies, API, Article 54(5).

proportionality rule are generally rather poor.[57] This criticism is largely due to the general imbalance in the rule in favour of the military advantage, as well as the difficulty in assessing the potential threat to the environment, and consequently biodiversity, particularly in terms of its projected severity and timeframe of harm.

In applying the rule of proportionality, therefore, one must assess the indirect damage caused, for example, by chemically toxic air pollution or oiled seas caused in an attack on an oil refinery.[58] It also includes damage caused to protected areas when the targeted military installations are located nearby or inside reserve boundaries, such as attacks undertaken by NATO, during the 1999 Kosovo Conflict, upon the telecom tower in Lovchen protected area of Serbia, which was hit by cluster bombs,[59] the Iriški Venac telecom tower in the Frusùka Gora National Park which caused 64 craters damaging the rare, red-listed orchid species over an area of approximately 30 hectares of land,[60] and a bridge in Skadar National Park (a listed Ramsar[61] wetland area) targeted by a cruise missile which missed and landed outside the protected area.[62] Notably, there is no explicit prohibition on siting such military installations inside nature reserves or other protected areas, although it is certainly implied in Article 58 of Additional Protocol I, which requires states to take 'necessary precautions' to protect such civilian objects under their control 'against the dangers resulting from military operations'. The problem is that these installations are often built in peacetime without thought as to their potential future military use, and upon the outbreak of hostilities they are not dismantled, which would be a legitimate precautionary measure.

In targeting military objectives, biodiversity and 'biodiversity hotspots'[63] have no specific or enhanced protection above any other civilian object, and, in fact, the very 'location' or 'use' of environmental areas, such as forests and plant cover, often qualifies the environment as a direct target. When it is not a direct target, similar calculation and compliance difficulties are raised by the rule of proportionality as are raised by Article 35(3) of API, namely a general lack of awareness for environmental pathways of harm.

---

[57] APV Rogers, *Law on the Battlefield* (Manchester University Press, 1996), 17; C Droege and M-L Tougas, 'The Protection of the Natural Environment in Armed Conflict – Existing Rules and Need for Further Legal Protection' (2013) 82 Nordic Journal of International Law 21, 29–33.

[58] UNEP, Kosovo (n 30) 12–21; P Elmer-Dewitt, 'A Man-Made Hell on Earth' *Time Magazine* (New York, 18 March 1992), 32.

[59] UNEP Kosovo (n 30), 66.

[60] *Ibid.*, 64.

[61] 1971 Convention on Wetlands of International Importance Especially as Waterfowl Habitat (1971) 996 UNTS 245 (amended in 1982 and 1987; amended text available at <http://www.ramsar.org/cda/en/ramsar-documents-texts-convention-on/main/ramsar/1-31-38%5E20671_4000_0__>).

[62] UNEP Kosovo (n 30), 66.

[63] O'Riordan and Stoll (n 8), 10.

3.3 Weapons

Weapons clearly have a multitude of environmentally damaging effects, such as the heat and blast effects caused on impact, the fragmentation effect of bombs causing the scattering of metal shrapnel over wide areas, and the chemical and toxic pollutants of used and decaying weapons seeping into soils and water sources. Surprisingly, most scientific assessments of weapons effects, however, appear to be based on laboratory experiments, human epidemiological studies and ecological theory. Few studies, including the UNEP conflict assessment studies, go beyond contaminant detection to analyse the levels of actual harm caused at the species or biodiversity level.[64] Consequently, we are generally forced to discuss only the *expected* persistence, toxicity and radioactivity of weapons and their mutagenic or teratogenic effects. Depleted uranium (DU) ammunition provides a good example of this apparent lack of field data,[65] despite containing all the scientific hallmarks of causing a biodiversity-level impact[66] due to its chemically and radioactively toxic character and laboratory results indicating mutagenic effects in rats.[67] While there is no formal acceptance of a 'precautionary approach' in IHL or weapons use, many states have, nevertheless, renounced DU and others have quietly removed it from their arsenal,[68] thus possibly indicating that public perceptions of such dangers might, in time, be as valuable as a legal prohibition. A similar comment could be made on the use of phosphorus wedges, an incendiary device used as an obscurant by Israeli forces in the 2008 conflict in Gaza, which causes horrific burns on contact with flesh.[69]

First and foremost, states are obliged to undertake assessments of the environmental impacts of newly acquired weapons under Article 36 of API, but the robustness of domestic processes is variable.[70] Secondly, IHL limits the 'means and methods' (weapons and tactics) available to parties in a number of ways. From the perspective of threats to biodiversity, the most obvious weapons threats would be from nuclear, chemical and biological weapons, which have the capacity for genetic-level impacts,[71] and which are generally prohibited by the principal rule of IHL requiring distinction in

---

[64] Brauer (n 16), 62; UNEP post-conflict assessments (n 30).
[65] UNEP, *Depleted Uranium in Bosnia and Herzegovina: Post-Conflict Environmental Assessment* (UNEP, 2003).
[66] A McDonald, JL Kleffner and B Toebes (eds), *Depleted Uranium Weapons and International Law: A Precautionary Approach* (TMC Asser Press, 2008).
[67] JL Domingo, 'Reproductive and Developmental Toxicity of Natural and Depleted Uranium: A Review' (2001) 15 Reproductive Toxicology 603, 603.
[68] Germany, Italy, Portugal, Pakistan, and France demanded a moratorium on the use of depleted uranium ammunition, and Australia stopped stockpiling it.
[69] Human Rights in Palestine and Other Occupied Arab Territories: Report of the United Nations Fact-Finding Mission on the Gaza Conflict, Human Rights Council, A/HRC.12/48, 25 September 2009, 194–196; IJ MacLeod and APV Rogers, 'The Use of White Phosphorus and the Law of War' (2007) 10 Yearbook of International Humanitarian Law 75.
[70] WH Parks, 'Conventional Weapons and Weapons Reviews' (2006) 9 Yearbook of International Humanitarian Law 55, 76.
[71] JP Robinson and J Goldblat, 'Chemical Warfare in the Iran-Iraq War', May 1984, SIPRI Fact Sheet, 6, available at <http://www.iranchamber.com/history/articles/chemical_warfare_iran_iraq_war.php>.

targeting, specifically that 'Attacks shall be limited strictly to military objectives'.[72] Primarily, this rule (often referred to as the prohibition on indiscriminate attack)[73] prohibits weapons that are not sufficiently targetable (spatially and temporally) to a military objective, such as the indiscriminate nature of the gas or toxin released in chemical and biological weapons. Both chemical and biological weapons are subject to absolute treaty bans, even on their ownership,[74] which are also undoubtedly customary norms.[75] Nuclear weapons, on the other hand, are subject only to a ban on their testing,[76] as well as their acquisition by, and transfer to, non-possessor states and non-state actors.[77]

The law does little to regulate the more routine blast and heat damage from smaller conventional weapons, which can also cause harm to biodiversity and ecosystems through more localised damage to habitats. Such was the case in Vietnam, where the biodiversity-rich forests and mangroves were replaced with low-diversity grasslands and mudflats following US use of bulldozers and area bombing, and the cratering effect that resulted.[78] More common in very low-tech civil wars is the use of hand guns, land mines, improvised explosive devices (IEDs) and, sometimes, machetes. The common problem associated with such low-tech weapons is the side-effect of their proliferation among armed bands and local populations, leading to greater exploitation of wild species, such as occurred in Rwanda and the DRC.[79] While machetes and guns are not prohibited, cluster munitions, land mines and certain IEDs are subject to a whole host of regulations from those requiring precision targeting[80] and self-deactivation reliability

---

[72] API, Article 52(2).

[73] API, Article 51(4).

[74] This includes a prohibition on use even against the military, since poisons were banned as a treacherous and needless weapon, and note the rule prohibiting weapons which cause 'unnecessary suffering' at API, Article 35(2).

[75] 1993 Convention on the Prohibition of the Development, Production, Stockpiling and Use of Chemical Weapons and on Their Destruction (1993) 32 ILM 800; 1972 Convention on the Prohibition and Development, Production and Stockpiling of Bacteriological (Biological) and Toxin Weapons and Their Destruction (1972) 11 ILM 309. Note also Rule 73 in Henckaerts and Doswald-Beck (n 33) and for the law relating to herbicides, note Rule 76 also therein.

[76] 1963 Treaty Banning Nuclear Weapons Tests in the Atmosphere, in Outer Space and Under Water (1963) 480 UNTS 43. Note the 1996 Comprehensive Test Ban Treaty (1996) 35 ILM 1443 also prohibits testing underground but is not yet in force. Note the Nuclear Weapons Advisory Opinion (n 2). Note also the nuclear weapon-free zones, such as 1996 African Nuclear Weapon-Free-Zone Treaty (1996) 35 ILM 698.

[77] 1968 Treaty on the Non-Proliferation of Nuclear Weapons (1968) 729 UNTS 161; and Security Council Resolution 1540 (2004) S/RES/1540, *Non-Proliferation of Weapons of Mass Destruction*.

[78] Westing and Pfeiffer (n 18), 59; Westing (n 52).

[79] Kalpers (n 22), 17–18; Plumptre (n 20), 79–82.

[80] Articles 3(3), 4, 5 and 6 of Protocol II on Prohibitions or Restrictions on the Use of Mines, Booby-Traps and Other Devices (Mines Protocol II) to the 1980 United Nations Convention on Prohibitions or Restrictions on the Use of Certain Conventional Weapons Which May be Deemed to be Excessively Injurious or to Have Indiscriminate Effects (1980 UN CCW) (1980) 19 ILM 1523; Article 3, Amended Protocol II to the 1980 UN CCW (Amended Mines Protocol II) (1996) 35 ILM 1206.

standards,[81] to those imposing a comprehensive ban on use, transfer and stockpiling.[82] And by the inclusion of non-state actors (for example armed groups) in the ban, and prohibiting states parties from transferring such weapons to non-state actors,[83] the number of armed groups using mines has reduced vastly in the past 40 years.[84]

Weapons protections thus provide much needed implicit protection for the environment and, thus, biodiversity, especially in recent years with the increasing adoption of post-conflict removal obligations – which ties in with the ILC's temporal approach to the analysis of legal obligations. Both the 1997 Ottawa Convention on the Prohibition of the Use, Stockpiling, Production and Transfer of Anti-Personnel Mines and on Their Destruction (APM Treaty)[85] and the 2008 Convention on Cluster Munitions (CMC)[86] bind states parties to remove unexploded munitions (so-called 'explosive remnants of war' or ERW) within their own jurisdiction or control.[87] The CMC, however, goes one step further in that it also 'strongly encourage[s]' user states to provide assistance to facilitate the marking, clearance and destruction of cluster munition remnants even for historical uses – i.e. uses that pre-date the Convention.[88] From past conflicts in Vietnam, Laos, Cambodia,[89] Sri Lanka, Mozambique and Angola[90] it is clear that the lingering presence of cluster munition duds and landmines poses a long-term – and often a very widespread – threat to the environment, and hampers reconstruction and remediation efforts.[91] Indeed, in surveying for its post-conflict report in Kosovo, UNEP found that no assessment could be undertaken in Kosovo's protected areas due to the presence of cluster munition and landmine ERW.[92] Robust removal obligations that are designed to assist human victims and to return stability to the state, including in terms of attracting tourism revenue, will therefore also provide important tools to help in the recoverability of biodiversity and the broader environment.

---

[81] Mines Protocol II, Article 5(1)(b); Amended Mines Protocol II, Articles 5, 6 and 17; and Article 9 and the voluntary Technical Annex, Part 3, of Protocol V on Explosive Remnants of War (ERW) to the 1980 UN CCW, available at <http://www.icrc.org/applic/ihl/ihl.nsf/Treaty.xsp?documentId=22EFA0C23F4AAC69C1256E280052A81F&action=openDocument>.
[82] Article 1 of 1997 Ottawa Convention on the Prohibition of the Use, Stockpiling, Production and Transfer of Anti-Personnel Mines and on Their Destruction (APM Treaty) (1997) 36 ILM 1507; and Article 1 of 2008 Convention on Cluster Munitions (CMC) (2009) 48 ILM 357.
[83] CMC, Article 1(1)(b); APM Treaty, Article 1(1)(c).
[84] Some six groups according to the International Campaign to Ban Landmines, *Landmine Monitor 2012* (Mines Action Canada, 2012), 11.
[85] (n 80).
[86] *Ibid.*
[87] APM Treaty, Article 5; CMC, Article 4; ERW Protocol, Article 3.
[88] CMC, Article 4(4).
[89] Kingdom of Cambodia, 'National Biodiversity Strategy and Action Plan' for the CBD, April 2002, 57, available at <http://www.cbd.int/countries/?country=kh>. Note, as the report admits, while the state had been unable to create protected areas in unsecure zones since 1980, the sites could not be developed either.
[90] AP Volger, 'Landmine Clearance – Its Probable Effect on Angolan Biodiversity' (1999) 33(2) Oryx 88.
[91] Landmine Monitor (n 84).
[92] UNEP Kosovo (n 30), 64.

Weapons regulation tends to be reactionary, with prohibitions being imposed only after many years of use and only then with sufficient proof that the weapon breaches one of the rules, such as that prohibiting indiscriminate weapons.[93] Of course, there is often a heavy human and environmental toll before that assessment is finally made, and, while civil society will continue to monitor the effects of weapons and push for new limitations where needed, there may be other areas where such pressure would be valuable. The following section, therefore, will turn to analyse two approaches that could allow for greater and, possibly, more biodiversity-focussed protection; namely, (1) possible developments in 'place-based' protections under IHL, and (2) the continuation (or co-application) of MEAs, and the CBD obligations in particular, during armed conflict.

4. TOWARDS BETTER PROTECTION OF WAR-TORN BIODIVERSITY

4.1 Special Status for Environmental Protected Areas During Conflict

The rules on targeting and weaponry confer very general protections in and against attack, but could this be complemented by developing more specific protection? International humanitarian law already recognises two special regimes for 'place-based' protection, namely those governing 'cultural property' and 'demilitarised zones'. Could a similar, place-based protection be developed for other environmental protected areas, such as important wetlands, nature reserves and biodiversity hotspots? While this change would clearly not cover all of the planet's biodiversity, it would go a long way towards achieving that goal.

Taking cultural property first, in the detailed regime created by the 1954 Hague Convention for the Protection of Cultural Property in the Event of Armed Conflict[94] (1954 Hague Convention) and its two Protocols (1954[95] and 1999[96]), and the separate provision at Article 53 of API,[97] it is specifically prohibited both to use such property in support of the military effort and to commit acts of hostility against that property.[98] What is more, the regime provides for obligations of emergency planning in peacetime as preparation for such eventualities as armed conflict.[99] While the notion of cultural property largely centres on ancient and religious buildings, artefacts and art, the Convention definition refers principally to 'movable or immovable property of great

---

[93] API, Article 51(4).
[94] (1954) 249 UNTS 240.
[95] 1954 First Hague Protocol for the Protection of Cultural Property in the Event of Armed Conflict (1954) 249 UNTS 358.
[96] 1999 Second Protocol to the Hague Convention of 1954 for the Protection of Cultural Property in the Event of Armed Conflict (Second Hague Protocol) (1999) 38 ILM 769.
[97] Note also APII, Article 16.
[98] 1954 Hague Convention, Article 4(1).
[99] 1954 Hague Convention, Article 3; Second Hague Protocol, Article 5.

importance to the cultural heritage of every people' and offers 'archaeological sites'[100] as one example. Thus, so-called 'mixed' sites of both cultural and natural heritage, and those listed as 'cultural landscapes',[101] could be included provided they meet the 1954 Hague Convention definition.[102] It is clearly incumbent on states parties to categorise such sites within their territory that fit the definition and to ensure the relevant markings are displayed during conflict to warn the enemy of their protected status.[103] At present, however, there is no direct link between the 1972 WHC and the 1954 Hague Convention (and Protocols). This situation appears to be changing, albeit very slightly, with the recognition of the need to explore synergies between the WHC's listing procedure and the procedure under the 1999 Second Hague Protocol.[104]

Similar, holistic protection is afforded to the second specific regime of 'demilitarised zones', designated by the parties to the conflict under Article 60 of API.[105] Once designated these areas are off limits to all military activities, although protection ceases upon military use. These areas often provide an invaluable civilian corridor or safe area, or allow for the demilitarisation of an area to establish peace talks.[106] To date, however, they have not been used as an environmental protection tool, and with the crux of the provision resting on an agreement between the parties, it might be striking the wrong chord. Notably, it is easy to see why newly created wartime demilitarised zones would ordinarily require the agreement of the parties to the conflict, since these zones would not have had any particular status in peacetime to make their presence and geographical limits otherwise identifiable. Yet, the same is not generally true of 'environmentally protected spaces',[107] the perimeters of which are usually very clearly mapped in

---

[100] 1954 Hague Convention, Article 1(a).

[101] Such sites find protection under the WHC, where cultural landscapes are defined as 'cultural properties' that represent the 'combined works of nature and of man', with three categorisations including 'parkland landscapes constructed for aesthetic reasons', as well as landscapes which have evolved in association with a traditional way of life or due to cultural associations, such as a grouping of 40 villages providing 'remarkable testimony to rural life in late Antiquity and during the Byzantine period' in Syria. *Operational Guidelines for the Implementation of the World Heritage Convention* (UNESCO World Heritage Centre 2008), [86] available at <http://whc.unesco.org/en/guidelines/> and with respect to Syria at <http://whc.unesco.org/en/culturallandscape/#2>.

[102] 1954 Hague Convention, Article 1.

[103] 1954 Hague Convention, Article 6.

[104] *Guidelines for the Implementation of the 1999 Second Protocol to the Hague Convention of 1954 for the Protection of Cultural Property in the Event of Armed Conflict*, CLT-09/CONF/219/3 REV. 4, (22 March 2012), [36], available at <http://unesdoc.unesco.org/images/0018/001867/186742e.pdf>.

[105] Henckaerts and Doswald-Beck (n 33) – in their Rule 36 the authors conclude that Article 60 is customary in international and non-international conflicts, the attack of which is a grave breach.

[106] J-M Henckaerts and L Doswald-Beck, *Customary Humanitarian International Law, Volume II: The Practice* (CUP, 2005) – the authors note the practice evidencing Rule 36, for example the practice of Colombia §89. Note also United Kingdom Ministry of Defence, *The Manual of the Law of Armed Conflict* (OUP, 2004), §5.39.2.

[107] M Bothe, 'War and Environment', in R Bernhardt (ed), *Encyclopaedia of Public International Law (Vol. 4)* (Elsevier, 2000), 1344.

peacetime. Indeed, as originally drafted, Article 60 was not designed to cover environmental spaces; that role fell to Draft Article 48*ter*,[108] which was not adopted.

The text of Draft Article 48*ter* was succinct, and built in both pivotal aspects of (a) the need for markings and (b) the loss of status if the area became militarised. However, it also eliminated the need for agreement of the parties to such designation, thus: 'Publicly recognized nature reserves with adequate markings and boundaries declared as such to the adversary shall be protected and respected except when such reserves are used specifically for military purposes.'[109] Key to its ultimate rejection appears to be the lack of agreement on the designation of such areas. Specifically, states were not clear on what qualifications were necessary for such 'public recognition',[110] and which body would perform this role.[111] Consequently, a major problem was the potential for abuse by states, either in falsely designating areas or in listing too many areas, in order to expressly impede the military operations of the enemy.[112]

The rejection of Draft Article 48*ter* from API was likely also due to a general lack of appreciation in the 1970s of the true value of protected areas. Today, we easily appreciate the pivotal contribution that the conservation of biodiversity and protected areas makes to human survival in terms of ecosystem resilience, and in more anthropocentric terms of providing safe drinking water, food sources, clean air, medicines, carbon storage and flood prevention.[113] Consequently, it is suggested that the principle behind Draft Article 48*ter* might find more acceptance today. In 1994, for example, just two years after the adoption of the CBD, states appear to have found the protected area concept less objectionable when it was included in a soft-law instrument covering marine protected areas and naval operations, namely Article 11 of the 1994 San Remo Manual on International Law Applicable to Armed Conflicts at Sea.[114] Article 11 encourages the parties to the conflict 'to agree that no hostile actions will be conducted in marine areas containing: (a) rare or fragile ecosystems; or (b) the habitat of depleted, threatened or endangered species or other forms of marine life'. Again, the crux of protection is still only an 'encouragement to make mutual agreements' to avoid

---

[108] Proposed Amendment, Official Records of the Diplomatic Conference on the Reaffirmation and Development of International Humanitarian Law Applicable in Armed Conflicts, Geneva [1974–1977] CDDH/III/276; Y Sandoz, C Swinarski, and B Zimmermann (eds), *International Committee of the Red Cross Commentary on the Additional Protocols of 8 June, 1977, to the Geneva Conventions of 12 August, 1949: In Collaboration with Jean Pictet* (Martinus Nijhoff, 1987), 664.

[109] Proposed Amendment, *ibid*.

[110] The original intention was apparently that such designation would only emanate from governmental action, Report to Committee III on the Work of the Working Group Submitted by the Rapporteur, 3 February–18 April 1975, Official Records (n 34), CDDH/III/275, Volume 15, 370.

[111] Summary Record of the Thirty-Eighth Meeting, 10 April 1975, Official Records (n 34), CDDH/III/SR.38, Volume 14, 407 [24].

[112] Report to Committee III on the Work of the Working Group Submitted by the Rapporteur, 3 February–18 April 1975, Official Records (n 34), CDDH/III/275, Volume 15, 370.

[113] PM Wood, *Biodiversity and Democracy: Rethinking Society and Nature* (UBC Press, 2000), 35.

[114] L Doswald-Beck (ed), *San Remo Manual on International Law Applicable to Armed Conflicts at Sea* (Grotius Publications CUP, 1995).

warfare in such areas. Further support for a protected area rule came in the aftermath of the 1990 Iraqi oil-well fires,[115] and then later in 2000 in the adoption of the Draft International Covenant on Environment and Development by the International Union for the Conservation of Nature and Natural Resources (IUCN).[116] Here the IUCN proposed Draft Article 32, stipulating that:

> All Parties involved in armed conflicts shall take the necessary measures to protect natural and cultural sites of special interest, in particular sites designated for protection under applicable national laws and international treaties, as well as potentially dangerous installations, from being subject to attack as a result of armed conflict, insurgency, terrorism, or sabotage. Military personnel shall be instructed as to the existence and location of such sites and installations.[117]

Neither draft provision is perfect. Indeed, substantively, the two provisions go little further than restating the rule of distinction. Neither elaboration specifically prohibits attacks on the environmental areas. The IUCN version suggests 'protection' only in a rather vague way, whereas Draft Article 48*ter* indicates a rather broad formulation of the loss of protection, notably, when the reserve is used specifically for military 'purposes'. Clearly, had the provision been taken further by the API drafting committee, its wording may have been tightened up somewhat in order to demand a more substantial military supporting role before the protection was lost. Furthermore, neither provision specifically recognises the defender's obligation to take precautions against the effects of attack by not placing military objectives in the protected area, although this obligation is, to an extent, implicit in the IUCN draft.

Section 4.3 will pick up this discussion with suggestions for future legal changes. First, however, Section 4.2 will address the same issue from the different angle of the continuity of peacetime MEAs, and in particular those protecting designated environmental areas.

4.2 Continuity and 'Co-Application' of Environmental Treaty Obligations

The current section builds on discourse spanning the last two decades, notably, first, the possible continued applicability of MEAs during armed conflict,[118] culminating, in part, in the ILC's 2011 Draft Articles on the Effects of Armed Conflicts on Treaties,[119] and, secondly, the notion of 'parallel application' or 'co-application' as regards how those obligations might 'interact' with IHL.[120]

Very few environmental law treaties contain express stipulation regarding their application during armed conflict. One such treaty is the 2003 African Convention on

---

[115] Bothe (n 107).
[116] Second Edition, available at <http://data.iucn.org/dbtw-wpd/edocs/EPLP-031-rev.pdf>.
[117] *Ibid*.
[118] WD Verwey, 'Protection of the Environment in Times of Armed Conflict: In Search of a New Legal Perspective' (1995) 8 Leiden Journal of International Law 7, 19–21.
[119] (n 1).
[120] F Hampson, 'Other Areas of Customary Law in Relation to the Study' in E Wilmshurst and S Breau (eds), *Perspectives on the ICRC Study on Customary International Humanitarian Law* (CUP, 2007).

the Conservation of Nature and Natural Resources.[121] Interestingly, this Convention lowers API's cumulative threshold requirement by requiring parties to refrain from the threat or use of means and methods of combat causing 'widespread, long-term *or* severe' harm to the environment.[122] Even more interesting is Article XV(1)(a), which requires parties in armed conflict to 'take *every practical measure* to protect the environment against harm'.[123] This time the provision simply omits any threshold of harm, going far beyond API's care obligation or the 2005 Customary International Humanitarian Law Study's 'due regard' formulation. Yet, being an African environmental law treaty, its applicability is limited to African states parties only, and it is difficult to get too excited bearing in mind the challenges often faced by African states in complying with treaty law obligations, and the fact that the Convention has yet to gain the 15 ratifications needed to enter into force.

Most MEAs do not contain an express stipulation and their provisions must, consequently, be construed for continuity. Possible treaties that appear to contain provision for continuity in armed conflict include the WHC and the 1982 UN Convention on the Law of the Sea.[124] Under the WHC, for example, Articles 6(3) and 11(4) can be read so as to suggest the treaty's continuity in armed conflict. Article 6(3) stipulates that parties may not 'deliberately' cause damage to the natural and cultural heritage of other states, whether directly or indirectly. With the specific contemplation in Article 11(4) of listing sites endangered by armed conflict, these two provisions taken together suggest some level of continuity of obligations. As regards the CBD, Article 22(1) stipulates that the CBD will affect 'any existing international agreement … where the exercise of those rights and obligations would cause a serious damage or threat to biological diversity'. Furthermore, the jurisdictional scope of the CBD is more far reaching than most MEAs, to include within its remit 'activities' carried out under a state's jurisdiction or control 'regardless of where their effects occur'.[125] Thus, while it is not clear whether any of these treaty phrasings were originally intended by the parties to express the treaties' wartime continuation, such an interpretation could arguably be made for both the WHC and CBD.[126]

With the ground-breaking recognition by the ILC that the very subject matter of MEAs involves an 'implication' of continuity during armed conflict,[127] it would appear that an even stronger recognition for continuity is finally starting to emerge.[128] Intending to 'dispel any assumption of discontinuity'[129] for non-IHL treaties, in reality

---

[121] Revised version available at <http://www.au.int/en/sites/default/files/AFRICAN_CONVENTION_CONSERVATION_NATURE_NATURAL_RESOURCES.pdf>.

[122] *Ibid.*, African Convention, Article XV(1)(b), emphasis added.

[123] Emphasis added.

[124] (1982) 21 ILM 1261; Article 236 limits warship immunity by the obligation on parties to 'act in a manner consistent, so far as is reasonable and practicable, with the Convention'.

[125] CBD, Article 4.

[126] ILC (n 1), Draft Articles 3 and 4.

[127] *Ibid.*, Commentary to Draft Article 1 [5], Draft Article 7, and Annex; Bothe and others (n 44) 581–591.

[128] S Vöneky, 'A New Shield for the Environment: Peacetime Treaties as Legal Restraints of Wartime Damage' (2000) 9(1) RECEIL 20.

[129] ILC (n 1), Commentary to Draft Article 3 [1].

the notion of the automatic termination or suspension of treaty obligations in wartime was probably, however, already waning.[130] While state practice on this point is not clear,[131] an evolving approach has referred to the, admittedly problematic, 'compatibility' or 'consistency' tests between a treaty's provisions and a situation of armed conflict.[132] And as Schmitt observes, the problem is in identifying obligations in MEAs that offer 'normative guidance of direct relevance to warfare',[133] since peacetime environmental obligations tend to apply a negligence/due diligence approach. However, under the ILC's probably equally problematic approach, in the absence of a continuity clause the starting point is Draft Article 6. This provision provides, in determining whether a treaty continues during conflict, that first one should consider a number of factors related to the nature of the treaty, including its object and purpose and content, in relation to the context of the territorial extent, scale and intensity of the particular conflict.[134] Consequently, as for the WHC and CBD, for example, the argument for continuity also includes the fundamental nature of their object and purpose, namely the protection of 'world heritage of mankind as a whole'[135] and the conservation of global biodiversity for its intrinsic value and as a common concern of humankind.[136] For most MEAs, which do not even appear to contemplate the situation of armed conflict, the process will therefore solely concern Draft Article 6's object and purpose/content approach.

It has to be said, however, that it is not entirely clear whether states will accept the continuity of MEAs during armed conflict, since what the ILC approach contemplates is the application of environmental law treaty obligations concurrently with IHL. Indeed, the little guidance offered by the ILC's Commentary to the Draft Articles suggests that in favouring the legal stability offered by the continuation of treaties, the preference is to mitigate the impacts of a suspended treaty on those states not party to the armed conflict,[137] as opposed to the realistic ability of the warring state(s) to fulfil any treaty obligations with each other and with non-warring states. Thus, it is Hampson's 'co-application' question[138] that will prove most problematic, because simply saying that an environmental treaty continues to apply is only the first step – the second is understanding how the two sets of rules might 'interact'. This is an issue that on first glance the ILC study appears to have completely avoided. However, on closer inspection, the Draft Article 6 analysis of the treaty's object and purpose and content mirrors the existing compatibility test and thus arguably can be seen as covering both issues of continuity of obligations and their extent. Therefore, understanding how specific environmental treaty obligations might apply alongside IHL protections will

---

[130] As regards bilateral treaties the rule can be found in the 1947 US Supreme Court case of *Clark v Allen* (1947) 331 US 503; DP O'Connell, *International Law* (2nd edn, Vol.1, Steven & Sons, 1970), 268–271.
[131] Note oral arguments in the Nuclear Weapons Advisory Opinion (n 2).
[132] Bothe and others (n 44), 587–588; Schmitt (n 35), 47–51.
[133] Schmitt, *ibid*, 50.
[134] ILC (n 1).
[135] WHC, Preamble.
[136] Vöneky (n 128), 27.
[137] ILC (n 1), Commentary to Draft Article 6 [3].
[138] Hampson (n 120).

not simply be a question of continuity by implication as suggested by Draft Article 7, but will require a rule-by-rule analysis, similar to the work being carried out in the human rights sphere.[139] This analysis will be undertaken in the following section.

4.3 Enhancing Biodiversity Protection from the Effects of Armed Conflict

Analysing the 'area protection' regime contained in the Biodiversity Convention, WHC, and other habitats/species protection treaties, there is now a strong legal basis upon which to suggest that (a) the continuation in conflict of a 'protected area' regime would be workable alongside IHL rules, and (b) the increasing influence of such designation suggests the possible acceptance of a new IHL provision on this issue. Regarding co-application, the first stage might be to suggest the continuation of the designation or recognition of the borders of such protected areas in conflict, with the second stage then relating more specifically to the 'blending' of conservation and management functions with IHL rules, which also brings in point (b) and the possibility of new IHL rules. Clearly, in zones located away from the main throes of battle, including in occupied territory, IHL would not appear to impose rules inconsistent with peacetime conservation management obligations. Indeed, in 2003 in occupied Iraq the question arose of whether the US and UK occupation force could improve environmental protections in the Arab marshes, where the environmental protection laws of the occupied state were lacking.[140] Consequently, in these areas, as far as is practically possible, and clearly bearing in mind resource constraints, there is no reason to suggest that the peacetime protected area regimes do not continue.

As regards the 'hot' battlezone, and the conduct of hostilities in particular, the question relates to the possible 'blending' of targeting and weapons laws related to the environment with such conservation management obligations. Here, IHL already provides for the prima facie designation of protected areas as civilian property, and these are, thus, not open to direct attack unless and until they become military objectives, and this protection includes contemplation of the environmental impact in the proportionality calculations. However, co-application of CBD Article 8 and WHC Article 6(3), for example, may also suggest (1) a clear, workable prohibition on the use of designated 'environmental protected areas' in support of the military effort – in keeping with enhanced peacetime obligations of a state's own *in situ* conservation, (2) an associated prohibition on committing general acts of hostility against such designated areas (akin to the 1954 Hague Convention, Article 4(1)), and (3) a clause mandating the need for pre-conflict designation, including setting the geographical boundaries, of such designated areas, and, in particular, linking the provision to the WHC listing of natural heritage sites – or to a specially established (sub)committee such as the 'Committee for the Protection of Cultural Property in the Event of Armed

---

[139] C Droege, 'Elective Affinities? Human Rights and Humanitarian Law' (2008) 90(871) International Review of the Red Cross 501. Human rights obligations explicitly continue in armed conflict; API, Article 72 and APII, Preamble.

[140] 1907 Regulations Respecting the Laws and Customs of War on Land, to the 1907 Convention (IV) Respecting the Laws and Customs of War on Land (1910) UKTS 9, Cd.5030, Article 43.

Conflict' established under Article 24 of the 1999 Second Hague Protocol. While designation by such a new WHC Committee would not completely rule out the possibility of abuse or misuse by states, it could serve as a neutral institution serving the interests of all parties – and, thus, go towards negating the main criticism of Draft Article 48*ter*. Plainly, however, the usual rules of targeting would not absolutely forbid attack of military objectives, for example a communications mast, located inside a protected area – and co-application of the CBD would be unlikely to change this situation, although the value of surrounding biodiversity should weigh heavily in the conduct of hostilities, namely, the attacking state's targeting decisions of distinction and proportionality. This aspect would, thus, suggest a need for an additional provision requiring states to use the best available biodiversity and environmental information when determining compliance with the rule of proportionality and the rule contained within Article 35(3) of API. Other IHL rules, such as the principle of causing the least feasible damage in choosing weaponry and the need for constant care in the conduct of military operations to spare civilians and civilian objects, should also be emphasised for specific application as regards the environment.[141]

Ultimately, it can be suggested that these obligations should be underpinned by a new legal provision specifically protecting designated environmental protected areas in armed conflict. While previous attempts to define such a provision have eluded states, the key elements might look something like: 'Designated environmental protected areas with clearly marked boundaries, displaying the internationally protected emblem, and declared as such to the adversary, shall be protected and respected. Specifically, it is prohibited to attack such designated areas except when such areas are used in direct support of military action.' Designation should, it is suggested, occur by states through a new WHC Sub-Committee suggested above.[142] Admittedly, the suggested wording would impose a key limit on the provision, as designation as a 'wetland' on the Wetlands Convention list or otherwise as a site of special scientific interest would not suffice. The reasoning for this suggested limitation is ultimately as a compromise to military necessity to achieve the extension of protection to natural heritage sites, since they are generally perceived to be the most important sites in terms of their irreplaceability as natural features of outstanding universal value.[143]

Such a new provision could afford invaluable protection to designated natural heritage sites against attack from the start of the conflict, and without the need for agreement between the parties for a particular site. If such protected spaces could be designated as prima facie off limits in war, then, consequently, there would be a clear presumption in favour of the continuation of their peacetime protected status, and the

---

[141] API, Article 57. A Loets, 'An Old Debate Revisited: Applicability of Environmental Treaties in Times of International Armed Conflict Pursuant to the International Law Commission's "Draft Articles on the Effects of Armed Conflict on Treaties"' (2012) 21(2) RECIEL 127, 129.

[142] As suggested by EJ Goodwin in correspondence in preparing this chapter, care might be needed to isolate the positions adopted by this sub-committee from the determinations of the full World Heritage Committee concerning admission to the World Heritage List (note WHC, Article 12), as well as the process of identifying sites meeting the WHC's definition of heritage as conducted by states as of right under WHC Article 3.

[143] WHC, Article 2.

continued recognition of the governing body's role, such as that of the World Heritage Committee, in monitoring and ensuring that protection.[144] For it is widely recognised that the collapse of park management and monitoring mechanisms during conflict is a major cause of species and habitat decline in nature reserves.[145] To combat the perennial problem of poaching, a new provision, similar to the 1954 Hague Convention regime,[146] could also recognise the continued arming of park rangers within reserve boundaries without such persons being deemed to be directly participating in hostilities.[147] Environmental management and risk assessment tools could also be provided by the conservation staff to feed into wider policies, such as the siting of displacement camps.[148]

A potential further IHL suggestion is as regards state obligations under Article 36 of API relating to weapons testing. It was noted that the impacts of a particular weapon or pollutant on the environment will not be uniform, with much resting on the features of the particular ecosystem in which it is used. Consequently, weapons assessments which only test for impacts in certain environments may be missing key information that may ultimately lead the state to breach its treaty obligations. Thus, both the environment and people would benefit from damage assessments that focus on the more important dimension of biodiversity and, by implication, ecosystem services and functioning.

Turning to an analysis of the value of peacetime regimes more broadly, as the ILC's recently adopted temporal approach[149] seeks to do, while the value of post-conflict obligations of the clearance of explosive remnants of war and environmental remediation have been apparent for some time,[150] it appears that little attention has been paid to peacetime, pre-conflict obligations. Specifically under the CBD, for example, states' biodiversity surveying, management planning and reporting requirements in peacetime could be used to provide key information to all states in wartime as to the location and characteristics of biodiversity-rich or sensitive ecological areas. Similarly, management planning is a requirement under the Wetlands Convention,[151] the WHC,[152] the African

---

[144] B Sjöstedt, 'The Role of Multilateral Environmental Agreements in Armed Conflict: "Green-keeping" in Virunga Park. Applying the UNESCO World Heritage Convention in the Armed Conflict of the Democratic Republic of the Congo' (2013) 82(1) Nordic Journal of International Law 129.

[145] Plumptre (n 20), 73.

[146] 1954 Hague Convention, Article 8(4).

[147] A more controversial suggestion is the employment of private security personnel (so-called Green Berets), LA Malone, 'Green Helmets: A Conceptual Framework for Security Council Authority in Environmental Emergencies' (1996) 17 Michigan Journal of International Law 515.

[148] United Nations High Commissioner for Refugees, *Framework for Assessing, Monitoring and Evaluating the Environment in Refugee-Related Operations*, 28 August 2009, available at <http://www.unhcr.org/4a97d1039.html>.

[149] (n 4).

[150] On the removal of ERW note Protocol V to the 1980 UN CCW (n 81) and D Jenson and S Lonergan, *Assessing and Restoring Natural Resources in Post-Conflict Peacebuilding* (Earthscan, 2012).

[151] Article 3.

[152] Articles 4 and 5.

Nature Conservation Convention,[153] and, in the EU, for example, under the Birds Directive[154] and the Habitats and Species Directive.[155] Thus, such management and conservation requirements, and the consequent studies and reports generated, will provide warring parties with full knowledge of the location of designated biodiversity hotspots or 'critical conservation priorities'.[156] Furthermore, if conservation management has been undertaken seriously then the siting of military objectives in protected areas would be minimised. Consequently, a key obligation for MEA treaty bodies is to ensure that states are, in fact, designating representative sites within their borders more rigorously and that these are more than mere 'paper' designations.

Similar to the host state's obligations under the 1954 Hague Convention regime,[157] part of the peacetime conservation management and planning obligations for biodiversity under the CBD and other MEAs should also include preparation for emergencies, including armed conflict. Part of that planning could include the creation of inventories of species, the possible relocation of species, and clarification of the role of personnel during conflict.[158] More specifically, planning might include the designation of more limited biodiversity corridors or species evacuation plans. The key, in this regard, appears to be to develop local, well-trained staff, who can foster a community commitment to conservation, and who are more likely to remain during times of civil strife and conflict.[159] Pre-conflict protection may also, therefore, take the form of environmental education requirements towards the local population. Interestingly, Rwandan rangers opined that the reduced level of violence to the endangered Bonobos gorillas in the Rwandan conflict, as opposed to the harm caused in the DRC conflict, was because of the value which the local population had come to see in the gorillas.[160] A pre-conflict priority in such biodiversity hotspots might then be for local education or the instilling of pride in biodiversity, but this must also be sensitive to local needs, since natural resources are clearly themselves often a source of dispute and conflict.

Biodiversity education for the military is necessary too, similar to the 1954 Hague Convention regime's obligation for military training in a 'spirit of respect for the culture and cultural property of all peoples'.[161] As regards a broader educational role, Kearns proposes the need for a new international organisation to develop greater awareness of wartime biodiversity issues and, particularly, to reduce the military

---

[153] Article XII.
[154] European Parliament and Council Directive 2009/147/EC of 30 November 2009 on the conservation of wild birds (2010) OJ L20/7.
[155] Council Directive 92/43/EEC of 21 May 1992 on the conservation of natural habitats and of wild fauna and flora (1992) OJ L206/7 (Habitats Directive).
[156] O'Riordan and Stoll (n 8), 36.
[157] 1954 Hague Convention, Article 3; Second Hague Protocol, Article 5.
[158] Michael Bothe, comments made at Expert Meeting, Geneva, July 2013.
[159] T Hart and J Hart, 'Conservation and Civil Strife: Two Perspectives from Central Africa' (1997) 11(2) Conservation Biology 308, 308–309.
[160] Plumptre (n 20), 89.
[161] 1954 Hague Convention, Article 7; Second Hague Protocol, Article 30.

ignorance that surrounds the many ways in which their actions could harm biodiversity.[162] While he suggests that the International Committee of the Red Cross (ICRC) might be in a good position to take up this mantle, a better fit, however, might be the IUCN. With their mapping work on Key Biodiversity Areas[163] and the development of the Red List of Ecosystems,[164] in addition to their well-established work on the Red List of Endangered Species[165] and Protected Areas, the IUCN is already well placed to undertake most of the educational and training aspects of Kearns' proposal.

Moving to the post-conflict dimension, Dudley and his co-authors suggest that the 'effects of post-war ecological and social disruption are at least as significant a threat to wildlife and ecosystems as war itself',[166] citing the increase in rates of deforestation in post-conflict Cambodia – largely as a result of illegal exploitation by military leaders and breakdown in the rule of law. Thus, the continuation in conflict of the peacetime protected area regime could help to fill this management vacuum, and aid in reducing such unregulated post-conflict exploitation. Similarly, with regularly updated auditing of its biodiversity and the conservation status of its protected areas, a state would also have a clear base-line for post-conflict assessments and ecological restoration – an obligation which is explicit in the EU's Natura 2000 system of protected sites[167] and, at least, implicit in the CBD and others. States and researchers would thus know, more specifically, which are the keystone species and sensitive environments to monitor, and, possibly, which species are to be reintroduced if their numbers have declined beyond the point at which the species is no longer viable.

Similar to the exposure reconstruction maps (or Geographic Information Systems) devised for Agent Orange use in Vietnam – produced so as to study the health impacts of exposure in US veterans[168] – post-conflict environmental assessments and detailed remediation planning[169] could indicate priority areas of investigation and restoration – much like the UNEP post-conflict assessments at present. Based on their work for the UNCC, Huguenin and his co-authors add that such assessments need to be undertaken immediately, or as soon as possible, so as not to lose perishable data, and that responsibility for collecting the data needs to be clearly assigned.[170] However, feeding

---

[162] B Kearns, 'When Bonobos Meet Guerrillas: Preserving Biodiversity on the Battlefield' (2013) 24 Georgetown International Environmental Law Review 123, 157–158.

[163] Available at <https://www.iucn.org/about/union/secretariat/offices/iucnmed/iucn_med_programme/species/key_biodiversity_areas/>.

[164] Available at <http://www.iucn.org/about/work/programmes/ecosystem_management/red_list_of_ecosystems/>.

[165] Available at <http://www.iucnredlist.org/>.

[166] JP Dudley and others, 'Effects of War and Civil Strife on Wildlife and Wildlife Habitats' (2002) 16(2) Conservation Biology 319, 326.

[167] Habitats Directive (n 155), Article 6.

[168] JM Stellman and others, 'A Geographic Information System for Characterizing Exposure to Agent Orange and Other Herbicides in Vietnam' (2003) 111(3) Environmental Health Perspectives 321.

[169] CR Payne, 'Oversight of Environmental Awards and Regional Environmental Cooperation' in CR Payne and PH Sand (eds), *Gulf War Reparations and the UN Compensation Commission: Environmental Liability* (OUP, 2011), 132.

[170] Huguenin and others (n 43), 92.

into such studies could also be a legal obligation requiring complete disclosure of the type, volume and geographical location of all weapons used in the conflict. At present, such obligations only exist for delayed-action weapons, such as landmines and cluster munitions, and so there is definite room for improvement here in post-conflict weapons information provision – indeed, it could be argued that there is an obligation resulting from the co-application of the CBD provisions in armed conflict that states must record full details of their weapons usage, at least, in the enemy's protected areas. Equally, some of the weapons and other war debris will be hazardous, thus there should be a priority obligation for the mapping and clearance of so-called toxic remnants of war[171] to ensure that post-conflict environmental restoration does not miss latent environmental dangers.[172] A longer-term solution to regional peace-building and stability has been the creation of successful transboundary 'peace parks'.[173] Peace parks are generally biodiversity-rich areas near to state borders or are transboundary park areas where states can engage peacefully in nature conservation and cooperation, and more could be learned from past experience of these, with a view to possibly including them in future peace treaties or Security Council ceasefire terms.

Overall, with new protections for certain protected areas, and with a clearer sense of the pre-conflict environmental base-line data, together with timely post-conflict weapons, location information and ecological impact assessments, the reconstruction efforts of states would undoubtedly be better planned.

5. CONCLUSIONS

Clearly, warfare can impact on biodiversity anywhere. Yet, where it impacts on areas of especially high biodiversity, or spaces that we have chosen to dedicate specifically to nature protection, such as nature reserves, wetlands and sites of natural heritage, it will cause much greater suffering, especially in terms of heritage and ecosystem services, than in other, less special locations.

As more research is undertaken under the auspices of the ILC, among others, it will become clear that much can be done in the pre-conflict stage by way of environmental management that can make ecosystems more robust and resilient to shocks, such as conflict, and much environmental information that can be used from that stage to inform the post-conflict rehabilitation of the environment. Yet, clearly, the greatest impact that such pre-conflict management of ecosystems can have is by its continuance in armed conflict. Thus, this chapter has proposed the adoption of a specific new provision on protected areas in armed conflict, specifically for designated natural heritage sites, in addition to greater analysis of just how the CBD, and other MEAs, might co-apply during conflict. With greater awareness today of the fragility and

---

[171] Statement of Costa Rica to the First Committee of the General Assembly, 29 October 2013, available at <http://www.un.org/disarmament/special/meetings/firstcommittee/68/pdfs/TD_29-Oct_CW_Costa-Rica.pdf>.
[172] Jenson and Lonergan (n 150).
[173] JA McNeely, 'Biodiversity, War, and Tropical Forests' in SV Price (ed), *War and Tropical Forests: Conservation in Areas of Armed Conflict* (Food Products Press, 2003), 1, 16–18.

uniqueness of many of the world's wetlands, forests and savannahs, the adoption of a provision in both international and non-international armed conflict that continues their protected status during armed conflict would go a long way to preserving the planet's rich biodiversity. It is a tragedy that many biodiversity-rich ecosystems are also the location of armed conflict, and, hence, some of the world's greatest ecological assets are constantly in danger of being lost to the pollutants and environmentally destructive tactics of warfare, the privateering motives of armed groups exploiting conflict resources, and the consequent survival needs of a war-torn population. Thus, a place-based protection for natural heritage sites would be a good starting point in providing a vital level of protection to assist in the continued management of such resources in wartime.

SELECT BIBLIOGRAPHY

Austin, JE and CE Bruch (eds), *The Environmental Consequences of War: Legal, Economic, and Scientific Perspectives* (CUP, 2000)
Brauer, J, *War and Nature: The Environmental Consequences of War in a Globalized World* (Altamira Press, 2011)
Harvey, L, 'Nature Needs Half: A Necessary and Hopeful New Agenda for Protected Areas' (2013) 19(2) Parks 13, 15
Jenson, D and S Lonergan, *Assessing and Restoring Natural Resources in Post-Conflict Peacebuilding* (Earthscan, 2012)
Kalpers, J, 'Volcanoes Under Siege: Impact of a Decade of Armed Conflict in the Virungas' (Biodiversity Support Program, 2001)
Kearns, B, 'When Bonobos Meet Guerrillas: Preserving Biodiversity on the Battlefield' (2013) 24 Georgetown International Environmental Law Review 123
Koppe, EV, 'The Principle of Ambiguity and the Prohibition against Excessive Collateral Damage to the Environment During Armed Conflict' (2013) 82(1) Nordic Journal of International Law 53
Lanier-Graham, SD, *The Ecology of War: Environmental Impacts of Weaponry and Warfare* (Walker Publishing Company, Inc., 1993)
Payne, CR and PH Sand (eds), *Gulf War Reparations and the UN Compensation Commission: Environmental Liability* (OUP, 2011)
Price, SV (ed), *War and Tropical Forests: Conservation in Areas of Armed Conflict* (Food Products Press, 2003)
Sjöstedt, B, 'The Role of Multilateral Environmental Agreements in Armed Conflict: "Green-keeping" in Virunga Park. Applying the UNESCO World Heritage Convention in the Armed Conflict of the Democratic Republic of the Congo' (2013) 82(1) Nordic Journal of International Law 129
Westing AH, *Ecological Consequences of the Second Indochina War* (Almqvist and Wiskell International, 1976)

PART III

GENERAL PRINCIPLES OF INTERNATIONAL ENVIRONMENTAL LAW

10. The Convention on Biological Diversity and the concept of sustainable development: the extent and manner of the Convention's application of components of the concept[1]

Veit Koester

1. INTRODUCTION

The Convention on Biological Diversity (CBD)[2] was adopted on 22 May 1992 and entered into force on 29 December 1993. According to Article 1 the objectives of the CBD are: (1) conservation of biological diversity; (2) sustainable use of biological resources; and (3) fair and equitable sharing of the benefits arising out of the utilisation of genetic resources by means of access to genetic resources, access to or transfer of relevant technologies, and funding. The CBD is based on, inter alia, two principles/rules: States exercise *sovereign rights* over their natural resources (Article 15(1) and Preamble, subpara, 4), but the conservation and sustainable use of biodiversity are a *common concern of humankind* (Preamble, subpara. 3).

The substantive scope of the Convention is biological diversity (biodiversity), encompassing all plants, animals and microbial species and ecosystems and processes of which they are part. More specifically, biodiversity is defined as the variability among living organisms from all sources including, inter alia, terrestrial, marine and other aquatic ecosystems and the ecological complexes of which they are part, including diversity within species, between species and of ecosystems (Article 2). According to a decision by the governing body of the CBD, the Conference of the Parties (COP), human genetic resources are, however, not included within the framework of the Convention.[3] On the other hand, it

---

[1] The author previously published the article 'The Nature of the Convention on Biological Diversity and its Application of Components of the Concept of Sustainable Development' in XVI Italian Yearbook of International Law (Martinus Nijhoff, 2007), 57–84. The present chapter is based on some of the ideas presented in the article. That said, the chapter takes the issues raised further, and takes into account relevant developments in related fields, whilst additionally concentrating on some key decisions made by the Conference of the Parties including at its eleventh meeting in 2012. Since the chapter was completed before the twelfth meeting of the Conference of the Parties in 2014 it does not refer to decisions made at that meeting.

[2] (1992) 31 ILM 818.

[3] Decision II/11, para. 2. All references to Decisions in these footnotes are, unless otherwise stated, references to decisions made by the COP to the CBD. The decisions are available on the website of the Convention (www.biodiv.org). Decisions are enumerated by means of a Roman numeral indicating the number of the meeting of the COP at which a decision was made, followed by an Arabic numeral indicating the number of the decision taken at that meeting. Hence, Decision II/11 means decision number 11 of the second meeting of the COP.

has been recognised[4] that humans, with their cultural diversity, are an integral component of many ecosystems.

It is obvious that the CBD, by virtue of its subject, its coverage and the almost universal range of Parties,[5] ought to be one of the most important environmental agreements relating to the concept of sustainable development. The question is, however, how the Convention reflects the principal components of sustainable development, and, especially, how the COP is in practice handling those and possibly other components of the concept. In other words, is the CBD an instrument which is in reality applying (important) components of sustainable development? The present contribution provides an attempt to respond to that question.

After introducing the modus operandi of the CBD, focusing on its 'two-track approach' (Section 2), and presenting some comments on the general nature of the concept of sustainable development (Section 3), the contribution examines the contents of the CBD as related to sustainable development (Section 4). Section 5, the core section of this chapter, provides some comments on COP decisions relating to substantive biodiversity issues, and presents a more detailed analysis of some key decisions of a cross-cutting nature with a view to identifying components of sustainable development that are addressed in normative terms and by what means. Section 6 contains some conclusions. It is underscored that a series of key principles related to sustainable development are applied by COP decisions, reinforcing that the CBD is an instrument to be included in the body of international law on this topic . Finally, Section 7 addresses a number of questions which remain unanswered and might deserve further discussion or research.

2. THE TWO-TRACK APPROACH OF THE CBD

The CBD may be perceived as having pursued by its modus operandi a kind of two-track approach. One track is focused on the development of supplementary legal instruments for specific problem areas, while the other is centred on processes designed to provide or refine guidelines concerning the conservation and sustainable use of biodiversity.

Since the entry into force of CBD, 12 ordinary meetings of the COP have been held and an extraordinary COP with a view to adopting the Cartagena Protocol on Biosafety (1999/2000). CBD Secretariat, *The Handbook of the Convention on Biological Diversity* (3rd edn., Montreal, 2005), which is also available on the website, includes various important texts, provides a reference guide to decisions adopted by the COP 1–7 as well as a guide to activities in relation to particular Articles. The Handbook also contains all decisions adopted by COP 1–COP 7 (however, with the exception of retired decisions or retired parts of decisions) in the format of both the full text and of references under individual provisions of the Convention.

[4] Decision VII/11, Annex I, A, para. 2.
[5] The CBD has 193 Contracting Parties (as of 1 June 2014), including the EU, meaning that the Convention has an almost global coverage.

To the first track belong the three protocols under the CBD. The first protocol to be adopted was the 2000 Cartagena Protocol on Biosafety[6] deriving from Article 19(3) of the CBD, which called explicitly for the consideration of such a protocol. The Nagoya-Kuala Lumpur Protocol on Liability and Redress to the Cartagena Protocol was adopted on 15 October 2010, but has not yet entered into force.[7] The third protocol is the Nagoya Protocol on Access to Genetic Resources and the Fair and Equitable Sharing of Benefits Arising From Their Utilization to the Convention on Biological Diversity, adopted on 29 October 2010, which is also as yet not in force.[8]

The programme of work on Article 8(j) and its related provisions[9] may also be included in the above approach. Central issues of this programme are *sui generis* systems for the protection of the knowledge and the innovations of indigenous and local communities, and mechanisms to ensure benefit-sharing and prior informed consent, linking the programme closely to the Nagoya Protocol.[10]

Also, the work undertaken by the CBD on Alien Species that Threaten Ecosystems, Habitats or Species may be mentioned in this connection.[11] The issue of alien species has the potential of being addressed by means of a legally binding instrument, since it is considered as being one of the most important drivers of biodiversity loss.[12] So far,

---

[6] Entry into force 11 September 2003. 167 Contracting Parties (as of 1 June 2014), among them the EU.

[7] Entry into force requires adherence by 40 of the Parties to the Protocol on Biosafety (Article 18(1)). So far (as of 1 November 2015) the Protocol has only been joined by 33 Parties. The Protocol is a response to Article 27 of the Biosafety Protocol on elaboration of international rules and procedures in the field of liability and redress for damage resulting from transboundary movements of living modified organisms.

[8] After the Protocol had been joined by 50 Parties to the CBD it entered into force on 12 October 2014.

[9] Article 8(j) deals with the respect, preservation and maintenance (according to the chapeau: as far as possible and as appropriate and to the provision itself: '[s]ubject to ... national legislation') of knowledge, innovations and practices of indigenous and local communities relevant for the conservation and sustainable use of biological diversity as well as promoting their wider application with the approval of the holders and equitable sharing of the benefits arising from their utilization'. Related articles are Article 10(c) on protection of customary use of biological resources, Article 17(2) on exchange of information on, inter alia, indigenous and traditional knowledge, and Article 18(4) on cooperation for the development and use of technologies, including indigenous and traditional technologies.

[10] The latest CBD decision relating to the work on Article 8(j) is Decision XI/14. Decision X/42 on the Tkarihwaié:Ri Code of Ethical Conduct to Ensure Respect for the Cultural and Intellectual Heritage of Indigenous and Local Communities, is closely connected with the work.

[11] According to Article 8(h) Parties shall (as far as possible and as appropriate) prevent the introduction of, control or eradicate those alien species which threaten ecosystems, habitats or species. COP VI adopted by Decision VI/23, para. 4, Guiding Principles for the Implementation of Article 8(h), annexed to the decision. The *Aichi Biodiversity Targets* (*see* Section 5 below) includes as Target 9 a specific target on invasive alien species, as one of the main drivers of biodiversity loss as referred to in Decision XI/28, para. 25.

[12] *See* Peter Davies' contribution to this handbook, as well as A Long, 'Developing Linkages to Preserve Biodiversity' in OK Fauchald, D Hunter and X Wang (eds), *2010 Yearbook of International Environmental Law* (OUP, 2013) 41, 46 and 53.

however, there are no clear signs that the issue is going to be addressed in that manner, by contrast to climate change, another very important threat to biodiversity.[13]

Finally, the issue of liability and redress for damage to biological diversity (except where such liability is a purely internal matter) should be mentioned as a quite distinct, although extremely difficult, issue. The COP is, according to Article 14(2) of the Convention, under an obligation to examine this issue. However, not much progress has been achieved, and the question is perhaps rather whether this issue is at the point of being more or less abandoned altogether.[14]

The second track deals with processes designed to develop norms such as 'guiding principles', 'guidelines', 'principles', 'programmes' (with goals and operational objectives) and 'criteria', directly linked to conservation of and sustainable use of biodiversity. Considering the various COP decisions in this field, it seems rather clear that most of the subjects dealt with are not suitable for 'translation' into precise legal obligations. In respect of the key objectives of the CBD – to conserve biodiversity and to use its components in a sustainable manner – the Convention serves more or less as a means of cooperation, coordination, exchange of information, sharing of experience and provision of guidance. The fact that this work is undertaken under the umbrella of a legally binding instrument underscores that cooperation is based on agreement and promotes international comity.[15]

There is, of course, no clear distinction between the above two approaches. Furthermore, many of the issues dealt with under the two approaches are interrelated.

3. THE CONCEPT OF SUSTAINABLE DEVELOPMENT

The concept of sustainable development was firmly established more than 20 years ago at the 1992 UN Conference on Environment and Development in Rio de Janeiro, although the origins may be traced to the 1970s. Sustainable development was the subject of the 2002 World Summit on Sustainable Development in Johannesburg and again in 2012 at the UN Conference 'Rio+20'. The social dimension was added to the two other dimensions of sustainable development (economic development and environmental protection) in 1997 at 'Rio+5'.[16]

There is, however, no commonly accepted definition of what 'sustainable development' means. Often, reference is made to the characterisation of the 1987 World Commission on Environment and Development Report ('Brundtland Report') of sustainable development as 'development that meets the needs of the present without compromising the ability of future generations to meet their own needs'. This

---

[13] *See* Decision XI/21 on Biodiversity and climate change: integrating biodiversity considerations into climate change related activities, and, generally, Long, *ibid.*, which, however, dating from before COP 11, does not refer to that decision.

[14] By Decision IX/23, para. 4, the COP decided to consider at COP 10 the need for future work in this area. However, by Decision X/9, para. b(vii), it was decided that the issue could only *be considered* by COP 12 in 2014 or early 2015.

[15] G Palmer, 'New Ways to Make International Law' (1992) 86 AJIL 269.

[16] GA Res. S-19/2, 28 June 1997. For example, all the textbooks on international environmental law referred to below address the UN Conferences in 1992 and 2002.

characterisation may be deepened in various ways, such as: 'economic and social development, which meets at least the basic needs of all the presently existing individuals and peoples of this world, whilst at the same time protecting the world's natural resources and the environment so that needs of future generations can be fulfilled as well'.[17]

In spite of the rather long history of the notion of sustainable development and its inclusion in a number of instruments, among those many treaties, it remains an elusive and controversial concept. Scholars still disagree widely about the legal nature of the concept, its precise content and possible implications. Some authors believe that the concept does not belong to law at all, because it is ranging below the threshold of normative quality.[18] Other scholars argue that sustainable development is a principle with normative content,[19] or that it fits within traditional categories of normativity.[20] The present chapter applies the term 'concept' rather than referring to the 'principle' of sustainable development. This terminology corresponds to the term 'concept of sustainable development' applied by the International Court of Justice in the *Gabčíkovo-Nagymaros Dam Case* (Hungary/Slovakia).[21] Although this judgment was pronounced almost two decades ago, nothing of a substantial nature in that regard has been added by later international jurisprudence.

An analysis of the concept of sustainable development and its legal implications exceeds the scope of the present chapter. However, it is noteworthy that the question concerning which elements or components (i.e. legal principles or other legal tools) to include in the concept vary, to some extent, as seen through the looking glass of various textbooks on international environmental law. Thus, an analysis of four authoritative textbooks, published in 2004, 2009, 2011 and 2012 respectively, provides the following results.

Kiss and Shelton's textbook of 2004[22] refers to seven principles: (1) sustainable use of natural resources; (2) equity and eradication of poverty; (3) common but differentiated responsibilities; (4) precautionary approach; (5) public participation and access to information and justice; (6) good governance; and (7) integration and interrelationship,

---

[17] O Spijkers, 'Global Values in the United Nations Charter' (2012) LIX Netherlands International Law Review 386, 388.

[18] U Beyerlin and T Marauhn, *International Environmental Law* (Hart Publishing, 2011), 81. *See* also, J Klabbers, *International Law* (CUP, 2013), 264, arguing that 'the term "sustainable development" remains as opaque as when it was first introduced'.

[19] Franz Xaver Perriz, 'Book Review of Christina Voigt, *Sustainable Development as a Principle of International Law: Resolving Conflicts Between Climate Measures and WTO Law*' (2009) 18 RECIEL 354, while Duncan French, 'Sustainable Development' in M Fitzmaurice and others (eds), *Research Handbook on International Environmental Law* (Edward Elgar Publishing, 2010), 56 observes that there is a need for caution in that respect.

[20] V Barral, 'Sustainable Development in International Law: Nature and Operation of a Legal Norm' (2012) 22 EJIL 377, 378. The article also includes a summary of the theory (378) and a critical bibliography, including a list of studies focusing on sustainable development as a new branch of international law (399).

[21] (1997) ICJ Reports 7, para. 140. On the judgment, *see*, for example, P Sands and J Peel, *Principles of International Environmental Law* (3rd edn., CUP, 2012) 313ff.

[22] A Kiss and D Shelton, *International Environmental Law* (3rd edn., Transnational Publishers, 2004), 218.

in particular in relation to human rights and social, economic and environmental objectives, while stating,[23] that the concept of sustainable development is deemed to include the attainment of economic, social and cultural rights.

Birnie, Boyle and Redgwell's work of 2009[24] includes the elements: (1) integration of environmental protection and economic development; (2) right to development; (3) conservation and sustainable use of natural resources; (4) inter-generational equity; (5) intra-generational equity; and (6) procedural elements (Environmental Impact Assessment, access to information, participation in decision-making).

Next, Beyerlin and Marauhn's textbook[25] argues that there is much in favour of deducing some self-contained norms 'from the principle of sustainable development', such as inter-generational equity, sustainable use, equitable use and intra-generational equity.

Finally, Sands and Peel[26] emphasise as elements: (1) inter-generational and (2) intra-generational equity; (3) sustainable use of natural resources; and (4) integration of environment and development, while noting[27] that international law of sustainable development, apart from environmental issues, includes: (5) the social and economic dimension of development; (6) the participatory role of major groups; and (7) financial and other means of implementation.

The variation in respect of elements to be included in the concept of sustainable development is probably due to the very broad scope of the three dimensions,[28] i.e. economic development, social development and environmental protection. Commonly included, however, are integration, inter- and intra-generational equity, sustainable use, and often precaution and some procedural elements referred to in Section 4 below.

The analysis below of the interrelationship between the CBD and sustainable development includes, in the format of a working method, a number of elements that may be encompassed by the concept of sustainable development. Some of the elements are undoubtedly included, while others probably are or may be included. However, the aim of the analysis is not to contribute to the discussion of the definition of the concept of sustainable development or of its status in international law, including whether it, legally speaking, is more than a concept. Neither is the purpose to evaluate the concept of sustainable development and its inter-linkages with international environmental law. The aim is rather to examine to what extent the CBD contains and applies principles relevant in the context of sustainable development, and which legal principles or tools in those respects are applied regularly by the CBD in order to further or attain its objectives.

---

[23] Ibid., 17.
[24] P Birnie, A Boyle and C Redgwell, *International Law and the Environment* (3rd edn., OUP, 2009), 116.
[25] Beyerlin and Marauhn (n 18), 82.
[26] Sands and Peel (n 21), 206ff.
[27] Ibid., 10.
[28] The result of the United Nations Conference on Sustainable Development in June 2012 (RIO+20) (the Declaration: 'Outcome of the Conference: The Future We Want') refers mostly to the three dimensions as being equal and that they should be integrated. *See* E Rehbinder, 'Contribution to the Development of Environmental Law' (2012) 42 Environmental Policy and Law 210.

4. THE CONTENTS OF THE CBD AS RELATED TO (COMPONENTS OF) SUSTAINABLE DEVELOPMENT

Conservation of biodiversity and *sustainable use of biodiversity* (i.e. of components of biological diversity) is a key principle which appears in numerous provisions. From a conceptual point of view, conservation and sustainable use of biodiversity is part of environmental protection, which is one of the three essential dimensions of sustainable development, along with economic and social development. As defined by the Convention, conservation and sustainable use of biodiversity includes the principle of *inter-generational equity*.[29] Rooted in and defined and operationalised by the CBD, the obligation to ensure conservation and sustainable use of biodiversity should be regarded as a legal principle or maybe even as a legal rule.

While the notion of *sustainable development* only appears once in the text of the CBD,[30] *economic and social development* is referred to both in the Preamble and in an operative provision, each of which stipulates that economic and social development (and the *eradication of poverty*)[31] are the first and overriding priorities of developing countries.[32] The provision relates this statement to '[t]he extent to which developing country Parties will effectively implement their commitments under [the] Convention'.[33] Although the statement is linked also to 'the effective implementation by developed Country Parties of their commitments under [the] Convention related to

---

[29] Defined by Article 2 as 'use in a way and at a rate that does not lead to the long-term decline of biological diversity, thereby maintaining its potential to meet the needs of present and future generations'. Present and future generations are also referred to in the Preamble, subpara. 23. A survey of the element 'future generations' in foundational documents of sustainable development is provided by H Ward, 'Beyond the Short Term: Legal and Institutional Space for Future Generations in Global Governance' in OK Fauchald, D Hunter and X Wang (eds), *2011 Yearbook of International Environmental Law* (OUP, 2013) 3, 7ff.

[30] Article 8(e) on the promotion of environmentally sound and sustainable development in areas adjacent to protected areas.

[31] Inter alia, eradication of poverty and managing the natural resource base for economic and social development are mentioned as overarching objectives of, and essential requirements for, sustainable development in para. 11 of the Johannesburg Declaration (*Johannesburg Declaration on Sustainable Development* (2002)), while eradication of extreme poverty and hunger is the first of the six *Millennium Development Goals*, UNEP/CBD/COP/7/20/ADD1, Annex, 30 November 2003.

[32] Preamble, subpara. 19 and Article 20(4): 'The extent to which developing country Parties will effectively implement their commitments under this Convention will depend on the effective implementation by developed country, by the following: Parties of their commitments under this Convention related to financial resources and transfer of technology and will take fully into account the fact that economic and social development and eradication of poverty are the first and overriding priorities of the developing country Parties.'

[33] *See*, E Morgera and E Tsioumani, 'Yesterday, Today, and Tomorrow: Looking Afresh at the Convention on Biological Diversity' in OK Fauchald, D Hunter and X Wang (eds), *2010 Yearbook of International Environmental Law* (OUP, 2013), 27 (fn 144), referring to one author having interpreted this provision as a statement of facts, while other authors have argued that the effect of the provision is to subject developing countries' policies to international biodiversity policies, to the extent that they receive funding under the Convention. On finance-related decisions at COP 11, *see* generally Morgera and Tsioumani (*ibid.*), 27ff.

financial resources and transfer of technology', the provision seems, at first sight, to subordinate conservation and sustainable use of biodiversity (i.e. environmental protection) to economic and social development.[34] There is, however, no doubt that the three dimensions of sustainable development are, generally speaking, considered as being equal.[35]

The principle of *intra-generational equity* is, to some extent, reflected by the relationship between the conservation obligations as related to developing countries and the obligations of developed countries to provide new and additional financial resources to the financial mechanism for the use of developing countries.[36] These obligations may also be regarded as an affirmation of the statement of the Preamble that conservation of biological diversity is a *common concern of humankind*,[37] which, again, in combination with the obligations to provide new and additional financial resources, might be summarised as the principle of *common but differentiated responsibilities*; a principle not included as such in the CBD. The principle of intra-generational equity, however, is expressed explicitly by the notion of fair and equitable sharing of benefits, which in operative provisions of the Convention is linked, in particular, to utilisation of genetic resources, or to the protection of knowledge, innovations and practices of indigenous and local communities.[38] All of the above elements are directly or indirectly included in the objectives of the CBD.[39]

The element/principle of *integration* – i.e integrating conservation and sustainable use into other plans, programmes, and policies – is contained in various provisions; notably in Article 6(b) concerning integration (as far as possible and as appropriate) into relevant sectorial or cross-sectorial plans, programmes and policies, and Article 10(a) on integration (as far as possible and as appropriate) into national decision-making.

While the *polluter-pays principle* is not reflected in the CBD[40] and the *precautionary approach* is included in the Preamble only,[41] some other elements, especially procedural elements, relevant in the context of sustainable development, are reflected in a provision requiring the introduction of appropriate *Environment Impact Assessment Procedures* (EIA Procedures) with regard to projects that are likely to have significant adverse effects on biological diversity. The same provision requires, where appropriate,

---

[34] This seems to contradict Preamble, subpara. 20: 'Aware that conservation and sustainable use of biological diversity is of critical importance for meeting the food, health and other needs of growing world population, for which purpose access to and sharing of both genetic resources and technologies are essential.'
[35] *See* n 28.
[36] Article 20(2) and Article 21.
[37] Preamble, subpara. 3.
[38] Article 8(j) and Section 2 above. Intra-generational equity may also be linked to the preambular statement in subpara. 19 on poverty eradication.
[39] Article 2 and Section 2 above.
[40] COP decisions only rarely refer to this principle. Examples are found in Decision VII/28, Annex, Section II, Goal 1.5, para. 1.5.2 (the polluter-pays principle to be incorporated in liability and redress measures in relation to damages to protected areas), and Decision X/32, para. 2(i) (encouraging, inter alia, Parties to apply the principle). *See* also Section 6.3 below.
[41] Preamble, subpara. 9.

allowance for *public participation*.[42] This is the only provision on public participation in decision-making,[43] and there are no provisions on access to information. Thus, Article 13 on public education and awareness is applying a top-down approach. Hence, the difference between the CBD and the 2000 Cartagena Protocol on Biosafety with regard to participatory elements is quite striking, since Article 23 of the Protocol, which in addition to a general obligation to consult the public in the decision-making process regarding living modified organisms, includes a provision on access to information on these organisms. Furthermore, there are no provisions on access to justice. So, in respect of procedural rights the Convention itself is extremely weak.

Cultural aspects are mentioned alongside a number of other aspects in the first recital of the Preamble, in which sense, however, the CBD does not differ from some other biodiversity-related Conventions, as can be seen, for example in recital three of both the 1971 Convention on Wetlands of International Importance especially as Waterfowl Habitat (Ramsar Convention) and the 1979 Convention on Migratory Species of Wild Animals. Cultural aspects are also reflected, albeit merely indirectly, in the provisions of the CBD on knowledge, innovation and practices of indigenous and local communities[44] embodying traditional lifestyles, and on the protection of customary use of biological resources in accordance with cultural practices.[45] This is to some extent in contrast with the Johannesburg Declaration, which refers to the vital role of *indigenous peoples* in sustainable development[46] while the CBD refers to *indigenous communities* in Article 8(j) and quite often to indigenous communities or indigenous groups in COP decisions. The dissimilarities between the terminology of indigenous-related notions or concepts in the CBD and the 2007 UN Declaration on the Rights of Indigenous Peoples[47] have already been raised[48] and are likely to be addressed at COP 12. A strong demand from indigenous peoples' representatives for the COP to adopt the terminology 'indigenous peoples' instead of 'indigenous communities' constitutes a basis for the request for the meeting of the COP Working Group on Article 8(j) and related provisions with the Secretariat to study the implications of such terminology by asking the opinion of the UN Office of Legal Affairs and to submit the study and the

---

[42] Article 14(1)(a) including the chapeau that Parties shall 'as far as possible and as appropriate'.

[43] Article 8(j), though, requires the promotion of wider application of indigenous knowledge, innovations and practices with 'the approval and involvement of the holders of such knowledge'.

[44] Protection of *indigenous people* and local communities is included in Rio Declaration, Principle 22 (other elements referred to above are also included in the declaration, e.g. public participation in Principle 10, precautionary approach in Principle 15, and EIA in Principle 17).

[45] Article 8(j), and Article 10(c). Annex I of the CBD contains an indicative list of components of biological diversity to be identified and refers, inter alia, to ecosystems, habitats and species of *cultural importance*. Preamble, subpara. 1, is conscious of, inter alia, the *cultural values* of biological diversity, and subpara. 12 recognises the close and *traditional dependence* on biological resources of many indigenous and local communities embodying traditional lifestyles.

[46] Johannesburg Declaration (n 31), para. 25.

[47] UN General Assembly Resolution 61/295, adopted on 13 September 2007, Annex: United Nations Declaration on the Rights of Indigenous Peoples.

[48] Morgera and Tsioumani (n 33), 21ff.

282 *Research handbook on biodiversity and law*

legal opinion to the forthcoming COP in October 2014.[49] Interestingly enough, no definition of or guidance related to the concept of *local communities* is available. Nor does the concept, in spite of its rather elusive nature, seem to have raised any queries as to its meaning.

In order to complete the above analysis of the CBD it might be relevant to add a few comments on the contents of the two protocols to the CBD as related to sustainable development.[50] The 2000 Cartagena Protocol on Biosafety is primarily an instrument on *conservation and sustainable use* of biodiversity because of its objective to contribute to ensuring an adequate level of protection, based on *the precautionary approach*, in respect of living modified organisms that may have adverse effect on biodiversity.[51] The Protocol includes a preambular statement, although of a rather limited scope, on *sustainable development*.[52] The Nagoya Protocol, on the other hand, focuses on the third objective of the CBD, *fair and equitable sharing of benefits*, which, however, is also considered as a means to contributing to the *conservation and sustainable use of biodiversity*.[53] The Protocol may also be perceived as an instrument to further *intra-generational equity* by way of its focus on benefit-sharing with indigenous and local communities,[54] and the reference to *poverty eradication* or poverty alleviation.[55]

5. COP DECISIONS ON SUBSTANTIVE BIODIVERSITY ISSUES

5.1 Introductory Remarks

The COP must keep under review the implementation of the Convention. For this purpose the COP must, inter alia, undertake any action that may be required for the achievement of the purposes of the Convention, and review scientific, technical and technological advice on biological diversity[56] provided by the Subsidiary Body on Scientific, Technical and Technological Advice (SBSTTA).[57] Based on these provisions, the COP has adopted numerous decisions[58] relating to substantive biodiversity issues,

---

[49] Doc. UNEP/CBD/COP/12/5.11 November 2013, Report of the Eighth Meeting of the Working Group. Recommendation 8/6.

[50] Leaving apart the Liability and Redress Protocol (*see* Section 2 above) to the Biosafety Protocol.

[51] Article 1 and Article 15 with Annex III.

[52] Preamble, subpara. 9.

[53] Articles 1 and 7.

[54] For example, Articles 6(2), 7 and 12.

[55] Preamble, subparas 7 and 16, including references, respectively, to the Millennium Development Goals (n 31) and sustainable development.

[56] Article 23(4)(i) and (b). The COP shall also consider national reports; Article 23(4)(a) and Article 26.

[57] Article 25.

[58] The total number of decisions exceeds 330. Many of them, however, deal with administrative, budgetary and financial matters. Some decisions or elements of decisions have

and covering both approaches identified above.[59] It should, however, be observed that COP decisions, like most other Multi-lateral Environmental Agreements, are considered for the most part to constitute soft law and not therefore legally binding on Parties.[60] Furthermore, they are usually phrased in a non-mandatory language. When operative provisions are addressing Parties, the language clearly indicates their recommendatory character, 'urging', 'encouraging' or 'inviting' Parties (to take this or that action). Issues are commonly dealt with by means of a decision with an annex containing a work programme and/or guidelines, which Parties are invited to implement or use. It should also be noted that all important decisions by the COP are made under the umbrella of 'decisions', thus not distinguishing between, for example, resolutions and recommendations. Accordingly, the normative nature of decisions has to be identified by means of the vocabulary used in the operative provisions addressing the Parties in their individual capacities.[61]

The present intention is to adopt a similar approach to previous papers produced by the author examining substantive COP decisions with a view to identifying common trends, if any, and in order to work out whether and to what extent components of sustainable development are applied by those decisions.[62] While that output considers individual COPs, this chapter takes a wider perspective and analyses some decisions which may be considered as key decisions in respect of sustainable development, because they are dealing with broad, overarching or cross-cutting issues. Accordingly, these decisions (i.e. the decisions on the Strategic Plan, the Ecosystem Approach, and on sustainable use) have left, and/or are going to leave, footprints in numerous other COP decisions, and in particular, decisions relating to the thematic work programme of the COP.[63] Such footprints may also be found in decisions by the governing bodies of

---

been retired, but according to a rough assessment around 150 decisions, still fully or partly in force, deal with matters relating to substantive CBD provisions.

[59] Section 2 above.

[60] Leaving aside the nature of decisions on Rules of Procedure, Financial Rules and similar decisions, as well as the question of internal binding effect, and without taking any position in respect of, for example, the position of A Wiersema, 'The New International Law-Makers? Conference of the Parties to Multilateral Environmental Agreements' (2009) 31 Michigan Journal of International Law 231, on the influence of COP decisions on hard law. However, it cannot be excluded that some decisions may constitute '[a] subsequent agreement between the parties regarding the interpretation of the treaty or the application of its provisions', which, according to Article 31(3) of the 1969 Vienna Convention on the Law of Treaties, shall be taken into account for the purpose of the interpretation of a treaty, especially since all substantive decision-making by the COP is by consensus.

[61] Wiersema, *ibid.*, 263.

[62] V Koester, 'The Nature of the Convention on Biological Diversity and its Application of Components of the Concept of Sustainable Development' in R Pavoni (ed), *The Italian Yearbook of International Law XVI*, (Martinus Nijhoff Publishers, 2007) 57.

[63] Includes, inter alia, agricultural biodiversity, dry and sub-humid lands, forest biodiversity, inland water ecosystems biodiversity, island biodiversity, marine and coastal biodiversity, and mountain biodiversity.

284 *Research handbook on biodiversity and law*

the other global biodiversity-related Conventions,[64] and, beyond those, in relevant regional treaties.[65]

5.2 The Strategic Plan

In 2010 COP 10 adopted the 'Strategic Plan for Biodiversity 2011–2020 and the Aichi Biodiversity Targets' with the subtitle 'Living in harmony with nature'.[66] This replaces the former Strategic Plan, adopted by COP 6 in 2002, the objective of which was not met,[67] namely to achieve by 2010 a significant reduction in the rate of biodiversity loss.[68]

The purpose of the latest Strategic Plan is, inter alia:

> … to promote effective implementation of the Convention through a strategic approach, comprising a shared vision, a mission, and strategic goals and targets ('the Aichi Biodiversity Targets') that will inspire broad-based action by all Parties and stakeholders … and for enhancing coherence in the implementation of the provisions of the Convention and the decisions of the Conference of the Parties including the programmes of work …[69]

The vision of the Strategic Plan is a world of living in harmony with nature by 2050, while the mission reflects the three objectives of the nature and, interestingly enough, also explicitly refers to decision-making based on sound science and the *precautionary approach*.[70] Apart from the vision and the mission, the Strategic Plan includes 20 targets for 2015 or 2020 (the 'Aichi Biodiversity Targets') organised under five strategic goals,[71] of which Goal A is of particular relevance in the present context.

Goal A is to '[a]ddress the underlying causes of biodiversity loss by mainstreaming biodiversity across government and society' emphasised, inter alia, by the second target, that by 2020, at the latest, 'biodiversity values have been integrated into national and local development and poverty reduction strategies and planning processes'. This mirrors to some extent the reference in the 'Rationale for the Plan' to biodiversity contributing to 'economic development and … [being] essential for the achievement of the Millennium Development Goals, including poverty eradication' and the 'national

---

[64] 1971 Convention on Wetlands of International Importance especially as Waterfowl Habitat (Ramsar Convention), 1972 Convention Concerning the Protection of the World Cultural and Natural Heritage (WHC), 1973 Convention on International Trade in Endangered Species of Wild Fauna and Flora (CITES), and 1979 Convention on the Conservation of Migratory Species of Wild Animals (CMS).

[65] The author's earlier paper (n 62) also included an explanation of the working methods of the COP. Since, however, a detailed analysis of the *modus operandi* of the COP was included in a recent article, reference herein will be made to that analysis, namely Morgera and Tsioumani (n 33), 6ff.

[66] Decision X/2.

[67] Recognised by Decision X/2, para. 5.

[68] Decision VI/26, paras 4 and 11.

[69] Decision XI. Annex, Section I.

[70] Decision X/2. Annex, Sections II and III. On the precautionary approach, *see*, also, Section I, para. 6, that 'scientific uncertainty should not be used as an excuse for inaction'.

[71] Decision X/2. Annex, Section IV.

planning process [beoming] more effective in mainstreaming biodiversity and in highlighting its relevance for social and economic agendas'.[72]

Mainstreaming biodiversity reflects the *principle of integration*, since it means integrating or including actions to conservation and sustainable use of biodiversity in strategies relating to production sectors, such as agriculture, fisheries, forestry, mining and tourism. All those sectors have been addressed by various COP decisions. Biodiversity considerations might also be included in poverty reduction plans as referred to in the above Goal A and sustainable development plans, hence equally constituting a way of mainstreaming biodiversity. According to Article 6(a) of the CBD the overall instrument for mainstreaming biodiversity on the national level is the National Biodiversity Strategy and/or National Action Plan (NBSAP),[73] which is addressed by Goal E mentioned below.

Under Goal B, to reduce the direct pressure on biodiversity and promote sustainable use, reference is made to, inter alia, the application of the Ecosystem Approach to achieving this goal by 2020 (Target 6). Goal D has a clear economic and social perspective with its focus on enhancing 'the benefits to all from biodiversity and ecosystem services'.[74] Its targets relating to ecosystem restoration and resilience were further elaborated by a COP decision in 2010 on ecosystem restoration.[75]

The fifth Goal (E) is to '[e]nhance implementation through participatory planning, knowledge management and capacity building', linking the Goal, inter alia, to the target (Target 17) that by 2015 each Party will have commenced implementing 'an effective, participatory and updated national biodiversity strategy and action plan'. Participation in this respect, however, seems, according to Target 18, only to include participation of indigenous and local communities.

NBSAPs, the development of which Parties are committed to,[76] are according to the Strategic Plan:

> ... key instruments for translating the Strategic Plan to national circumstances, including through national targets, and for integrating biodiversity across all sectors of government and society. The participation of all relevant stakeholders should be promoted and facilitated at all levels of implementation. Initiatives and activities of indigenous and local communities, contributing to the implementation of the Strategic Plan at the local level, should be supported and encouraged.[77]

At least one important element or principle related to sustainable development is emerging from the above provisions: the *principle of integration*. This requires integrating environmental protection into other plans, programmes and policies.

---

[72] Decision X/2. Annex, part II, paras 3 and 10(c) respectively.
[73] *See* C Prip and others, *Biodiversity Planning. An Assessment of National Biodiversity Strategy and Action Plans* (United Nations University of Advanced Studies, 2010).
[74] Morgera and Tsioumani (n 33), 12.
[75] Decision XI/16, para. 1.
[76] CBD Article 6(a).
[77] Decision X/2. Annex, Section V on implementation, monitoring, review and evaluation, para. 14.

286 *Research handbook on biodiversity and law*

It is, however, less clear whether another important principle, the principle of *public participation in decision-making*, belonging to the procedural elements of sustainable development, is embodied in the Strategic Plan itself. Not only the Strategic Plan, but also a number of other COP decisions regularly apply the notion of '(all) relevant stakeholders', often in various combinations with indigenous and local communities,[78] which makes it rather obscure to what extent civil society at large, including NGOs, is included. Thus, it is quite striking that while indigenous and local communities are referred to in the goals and targets of the Strategic Plan and in numerous other provisions of the plan, NGOs are only referred to twice.[79]

The Strategic Plan was supplemented by a decision on biodiversity for poverty eradication and development, adopted by COP 11 in 2012.[80] This may be considered as a kind of response to the outcome document of the UN Conference on Sustainable Development (Rio+20), 'The Future We Want'. The decision, which, inter alia, stresses the relevance of the Strategic Plan for the development of sustainable development goals in the context of the UN development agenda beyond 2015, aims at 'moving towards a road map for integrating biodiversity into the social and economic dimensions of sustainable development, taking into account the outcomes of the Rio+20 Conference, in the context of the Strategic Plan'.[81]

Issues relating to the links between biodiversity and human well-being, livelihoods, poverty eradication and sustainable development will be discussed at future COPs 'for the purpose of recommending specific actions to implement the Strategic Plan'.[82] The Executive Secretary is requested to ensure, inter alia, that 'the issue of biodiversity for poverty eradication and development is regarded as a cross-cutting theme in all relevant programmes of work under the Convention, [and] is integrated into [NBSAPs]'.[83] The last part of the task of the Executive Secretary, however, probably exceeds the ability and power of any executive secretary.

The COP 11 decision on biodiversity for poverty eradication is a rather curious instrument, because poverty reduction is already included as a rationale for the Strategic Plan and is included in its mission. Furthermore, this element arguably is an important aspect of economic and social development,[84] and, especially, Millennium Goal 1 on combating poverty and hunger is often referred to in COP decisions.[85] The

---

[78] For example, Decision VIII/28 on Impact assessment: Voluntary guidelines on biodiversity-inclusive impact assessment, Annex para. 5: '… The effective participation of relevant stakeholders, including indigenous and local communities, is a precondition for a successful EIA …'
[79] Decision X/2. Annex, paras 17 and 24.
[80] Decision XI/22.
[81] *Ibid.*, para. 4.
[82] *Ibid.*, para. 10.
[83] *Ibid.*, para. 11f.
[84] *See*, for example, Decision VII/12. Annex II, para. 2 according to which 'sustainable use is an effective tool to combat poverty, and, consequently to achieve sustainable development'.
[85] For example, Decision VII/32 on the programme of work of the Convention and the Millennium Goals. A note by the Executive Secretary of CBD, Doc UNEP/CBD/COP/7/20, Add 1, 30 November 2003, analysing the relationship between the Millennium Development Goals

explanation is probably that the Rio+20 outcome document clearly separates sustainable development and eradication of poverty.[86] Additionally, the Strategic Plan is, at least in the eyes of the Secretariat, 'the overarching framework on biodiversity, not only for the biodiversity-related conventions, but for *the entire United Nations systems*' (emphasis added).[87] Be that as it may, the Strategic Plan remains over and above an instrument for further conservation and sustainable use of biodiversity by application of the *principle of integration* of economic development, social development and environmental protection.

5.3 The Ecosystem Approach

The 'Ecosystem Approach' is a strategy 'for the integrated management of land, water and living resources that promotes conservation and sustainable use in an equitable way ... [which] will help to reach a balance of the three objectives of the Convention'. This definition was endorsed by COP 5 in 2000 by its first decision on the approach,[88] and still stands, although the decision has been supplemented by two later decisions on the approach.[89] The COP definition of the Ecosystem Approach might be compared with other definitions that, contrary to the COP decision, include or emphasise the importance of scientific considerations.[90] That said, the precautionary approach, according to the Ecosystem Approach of the CBD as referred to below, also constitutes a feature in that framework. The COP definition contributes to the interpretation of the CBD, because it follows from the definition that conservation of biodiversity and sustainable use of biodiversity are not perceived as alternatives, i.e. as biological diversity which is either protected or non-protected. Instead they should be 'seen in context and the full range of measures is applied in a continuum from strictly protected to human-made ecosystems'.[91]

The Ecosystem Approach has a rather close relationship with the Strategic Plan, because the basic elements of the Ecosystem Approach forms an essential part of the mission of the Strategic Plan,[92] which also includes some references to the concept.

---

and the CBD, concludes that none of the targets for the eight goals explicitly recognises the role of biodiversity or the CBD, and that only two of the indicators are related to biodiversity.

[86] Rehbinder (n 28) 210, 211. The reason for the separation of sustainable development and poverty eradication may be due to the fact that UNGA Res. 64/236 (2009) establishing 'Rio+20' decided that the Conference should focus on two themes, one of which was a green economy in the context of sustainable development and poverty eradication.

[87] *Strategic Plan for Biodiversity 2011–2020, including Aichi Biodiversity Targets* (<http://www.cbd.int/sp/>).

[88] Decision V/6. Annex, para. 1.

[89] Decision VII/11 and Decision IX/7 respectively. Decision VII/11 may be considered as the main decision, and most of the references below belong to that decision. Some, in the present context irrelevant, provisions of the decisions on the Ecosystem Approach were, in accordance with Decision IX/29, para. 14, retired by Decisions X/14 and XI/12.

[90] *See*, generally, A Trouwborst, 'The Precautionary Principle and the Ecosystem Approach in International Law: Differences, Similarities and Linkages' (2009) 18 RECIEL 26.

[91] Decision VII/11. Annex I, Table 1, Principle 10, Rationale.

[92] *See also* M Ambrus, 'Through the Looking Glass of Global Constitutionalism and Global Administrative Law. Different Stories About the Crisis in Global Water Governance?' (2013) 6

However, the concept as detailed in the relevant decisions clearly demonstrates its relevance in the context of sustainable development. The approach is operationalised by means of 12 principles that are explained by way of 'annotations' and are supplemented by implementation guidelines. Almost all elements or principles that are, or may be, perceived as being relevant components of sustainable development are included in one way or another in the decision in general, and in the annotations or implementation guidelines in particular.

The *principle of integration* is clearly reflected, although in various formulations. For example, '[t]hat management of living component is considered alongside economic and social considerations'; '[t]here is a need to integrate the ecosystem approach into agriculture, fisheries, and other production systems that have an effect on biodiversity'; '[i]ncorporate social and economic values of ecosystem goals and services into National Accounts, policy, planning, education, and resource management'; and '[a]ssess the extent to which ecosystem composition ... can ... contribute to the delivery of goods ... to meet the desired balance of conservation, social and economic outcomes'.[93] The COP decision, adopted in 2008, as an outcome of an in-depth review of the application of the Ecosystem Approach, characterises the approach as '[a] useful normative framework for bringing together social, economic, cultural and environmental values'.[94]

Inter-generational equity is mentioned in that '[i]t is also necessary to ensure that the needs of future generations and the natural world are adequately represented'; '[t]he inter-generational obligation to sustain the provision of ecosystem goods and services to future generations should clearly be stated'; and as a concept that 'need[s] to be applied to considerations of the temporal scale'.[95] Both inter-generational and to some extent *intra-generational equity* are included in an implementation guideline to the effect that: 'adaptive management processes should include the development of long-term visions, plans and goals that address inter-generational equity, while taking into account immediate and critical needs (e.g. hunger, poverty, shelter)'.[96]

The *precautionary approach* is referred to in connection with the principle of managing ecosystems within the limits of their functioning, because the current understanding is insufficient to allow these limits to be precisely defined.[97] Also, in the same vein, the *principle of impact assessment* (EIA) (and strategic environmental assessment, SEA) is included in the Ecosystem Approach to be carried out for 'developments that may have substantial environmental impacts'.[98]

The *principle of public participation in decision-making* takes a rather prominent position in the Ecosystem Approach, in particular, because it is, to some extent at least,

---

Erasmus Law Review 31, 36 on the close link between the ecosystem approach (as applied by some water Conventions) and sustainable development.
   [93] Decision VII/11. Annex I, paras 3(a) and 19, and Table 1, Principle 4.8 and Principle 5.3, respectively.
   [94] Decision IX/7. Annex I, para. 3(a).
   [95] Decision VII/11. Annex I, Table 1, Principle 1, Annotations; Annex II, para. 7; and Annex I, Table 1, Principle 7.6 respectively.
   [96] Decision VII/11. Annex I. Table 1, Principle 8.1.
   [97] Decision VII/11. Annex, Table 1, Principle 6.2.
   [98] Decision VII/11. Annex, Table 1, Principle 3.3.

the object of one of the 12 principles, i.e. 'Principle 12: The ecosystem approach should involve all relevant sectors of society and scientific disciplines.' The annotations and the implementation guidelines, however, include the rather obscure notion of 'all relevant stakeholders', which is also found in the Strategic Plan. On the other hand, the guidelines are also using language which to some extent belongs to 'true' public participation instruments. They state that: '[p]rocedures and mechanisms should be established to ensure effective participation of all relevant stakeholders and actors during the consultation process, decision-making on management goals and actions, and, where appropriate, in implementing the ecosystem approach'.[99]

Other provisions of the Ecosystem Approach equally reflect participatory aspects, however, without clarifying its precise content. Reference is made, for example, to 'the full and effective participation of indigenous and local communities and other stakeholders' or to the principles that '[a]ll interested parties (particularly including indigenous and local communities) should be involved in the process'. This is one of the characteristics of '[g]ood decision-making processes' or that '[m]anagement should involve all stakeholders'.[100]

An important element of the procedural component of sustainable development is *access to (environmental) information*, which, however, as observed above, is not addressed in any of the provisions of the CBD. While this element seems to be completely neglected in respect of the Strategic Plan, it is mentioned in connection with the Ecosystem Approach in noting that '[g]ood decisions depend on those involved having access to accurate and timely information and the capacity to apply this information' and 'enhancing the access of stakeholders to information because the more transparent the decision-making is, based on information at hand, the better ownership of the resultant decisions between partners, stakeholder and sponsors'.[101]

On the other hand, there seems to be no trace of the notion of *NGOs* in the decisions on the Ecosystem Approach. NGOs are mentioned a couple of times in the Strategic Plan (see Section 5.2 above), although not in connection with public participation in decision-making. In addition to the qualifying word 'stakeholder' and even more by the language of 'relevant stakeholders', this is why the provisions of both instruments are rather obscure as related to public participation.

The *principle of good governance* is highlighted as being essential for the successful application of the Ecosystem Approach.[102] Good governance includes, according to this provision, inter alia, 'sound environmental, resource and economic policies and administrative institutions that are responsible to the needs of the people' and that '[d]ecision-making should account for societal choices, be transparent and accountable

---

[99] Decision VII/11. Annex I, Principle 12 and guideline 12.3. The 1998 Aarhus Convention on Access to Information, Public Participation in Decision-making and Access to Justice in Environmental Matters includes a provision on 'effective public participation' (Article 6(4)).

[100] Decision VII/11, respectively para. 10; Annex I, Table 1, Principle 1 Annotations; and Table 1, Principle 2, Rationale. Decision IX/7, preamble para. (d) refers to the 'active participation of all relevant stakeholders'.

[101] Decision VII/11. Annex I, Table 1, Principle 1, annotations, and Annex I, para. 16, respectively.

[102] Decision VII/11. Annex I, para. 18. Also the principle of integration (para. 19) is included as an element of good governance.

290 *Research handbook on biodiversity and law*

and involve society'.[103] These features correspond more or less to the good governance principles singled out by the EU Commission as a basis for EU governance, namely openness, participation, accountability, effectiveness and coherence.[104] The principles reflected in the COP decision, however, concern good governance on the national or local level, which should not be confused with the issue of good governance on the international level, inter alia, in relationship with the global biodiversity-related Conventions which are referred to in Section 6 below.

The preconditions for good decision-making processes in relation to the application of the Ecosystem Approach are as follows: (a) all interested Parties (particularly including indigenous and local communities) should be involved in the process; (b) it needs to be clear how decisions are reached and who the decision maker(s) is (are); (c) the decision-makers should be accountable to the appropriate communities of interest; and (d) the criteria for decisions should be appropriate and transparent, and decisions should be based on, and contribute to, inter-sectorial communication and coordination.[105] This guidance is unusually clear in the sense that it does not contain any kind of reservation. Typically, guidance of a similar nature includes the reservation 'inter alia', 'including' or reservations to the same effect, in order to make clear that the guidance is not meant to be exhaustive.

Generally, the Ecosystem Approach may, as the Strategic Plan shows, be perceived as an instrument of integration, focusing, however, on integrating the three objectives of the Convention. By focusing, inter alia, on environmental considerations, sustainable use and reflecting both inter- and intra-generational equity, the principle of the Ecosystem Approach undoubtedly demonstrates essential components of sustainable development. The principle also contains elements of public participation in decision-making, but these elements may be seen to reflect a rather narrow approach in that respect.

5.4 Sustainable Use

The decision on Sustainable Use, referring to CBD Article 10 on sustainable use of components of biological diversity, was adopted by COP 7 in 2004. The decision includes an adoption of the Addis Ababa Principles and Guidelines for Sustainable Use as contained in Annex II to the decision. Annex II includes a kind of introduction, an explanation of the underlying causes for sustainable use (part A), and in part B the practical principles, 1–14, each of them supplemented by a 'rationale' and operational guidelines. The decision,[106] which still stands,[107] was supplemented by a decision,

---

[103] Corresponding more or less to the components of sustainable development included in the concept of sustainable development by D Magraw and LD Hawke, 'Sustainable Development' in D Bodansky, J Brunnée and E Hey (eds), *The Oxford Handbook of International Environmental Law* (OUP, 2008) 613, 633ff.

[104] C Harlow, 'Global Administrative Law: The Quest for Principles and Values' (2006) 17 EJIL 187.

[105] Decision VII/11. Annex 1, Table 1, Principle 1, Annotations.

[106] Decision VII/12.

[107] A couple of – in the present context irrelevant – provisions, were retired by Decision XI/12.

adopted by COP 10 in 2010, after a review of the implementation of the original decision.[108]

Sustainable use of biodiversity constitutes one of the three objectives of the CBD. Hence, the commitment to use biodiversity sustainably belongs to the core provisions of the CBD, and the decision, deepening the commitment, may be considered as one of the most important decisions on issues of a cross-cutting nature. It is notable that the decision was made by the same COP that adopted the Ecosystem Approach. Thus, the two decisions are generally interwoven,[109] and, by references to the Ecosystem Approach, the Sustainable Use Decision includes some of the elements identified in Section 5.3 above. It is, equally, incorporating some of these elements in its own language. Thus, the *principle of integration* is reflected by the statements that, while sustainability of use depends on biological parameters, it is recognised 'that social, cultural, political and economic factors are equally important'.[110] Another instance is the integration of 'the values of biodiversity and ecosystem services into national policies, plans, and strategies for relevant economic sectors'.[111]

The Sustainable Use Decision includes a rather clear reference to *public participation*, which is not found in the instruments analysed in Sections 5.2 and 5.3 above, i.e. the notion of 'full public participation'.[112] However, the decision also uses language in that respect which resembles the qualifying language of the two other instruments, e.g. 'all relevant stakeholders' or 'stakeholders including indigenous and local communities'.[113] In general, indigenous and local communities take a prominent position in the Sustainable Use Decision. In addition to a principle which is fully devoted to these communities,[114] they are in one way or another referred to in several other provisions, to some extent corresponding to what is a general feature also of the two instruments examined in, respectively, Sections 5.2 and 5.3 above.

The issue of *intra-generational equity* is, apart from in the context of indigenous and local communities, addressed only sporadically,[115] while *precaution* is referred to in rather obligatory language:

> To ameliorate any potential negative long-term effects of uses it is incumbent on all resource users, to apply precaution in their management decisions and to opt for sustainable use management strategies and policies that favor uses that provide increased sustainable benefits

---

[108] Decision X/32. The two decisions are referred to below as the Sustainable Use Decision.
[109] There are several cross-references in Decision VII/12 to the Ecosystem Approach. *See*, Annex II, para. 5 and the references to the approach in several footnotes to the Annex.
[110] Decision X/12. Annex II, Practical Principle 9, Rationale. However, to make sense some words like 'to achieving sustainable development' should be added.
[111] Decision X/32, para. 2(b).
[112] Decision VII/12. Annex II, Practical Principle 7, Operational guidelines.
[113] Decision VII/12. Annex II, Practical Principle 9, Operational guidelines and Practical Principle 12, Operational guidelines.
[114] Decision VII/12. Annex II, Practical Principle 12.
[115] For example, Decision VII/12. Annex II, Practical Principle 6, Operational guidelines.

while not affecting biodiversity. Governments should be certain that licensed or authorized sustainable uses of biological diversity are taking such precaution in their management.[116]

Unlike the two instruments referred to above, the Sustainable Use Decision includes a reference to the *polluter-pays principle*, which Parties are encouraged to apply together with effective market-based instruments that have the potential to support sustainable use of biodiversity.[117]

The Sustainable Use Decision including the Addis Ababa Principles is foremost an instrument on sustainable use of biodiversity, one of the three objectives of CBD, and deepens the core obligations of CBD Article 10 in that respect. On the other hand, sustainable use may be considered as one of the key components of the concept of sustainable development being highlighted in that respect by a number of scholars.[118] The Decision, however, also reflects a number of other elements relevant in the context of sustainable development.

6. SOME CONCLUSIONS

Despite the fact that the CBD does not define itself explicitly as an instrument based on the concept of sustainable development, it contains, in addition to the principle of conservation and sustainable use of natural resources, a number of elements which are, or may be, considered components of that concept (e.g. inter-generational equity and integration). The COP decisions examined in Section 5 above clearly perceive the CBD within the broader concept of sustainable development. COP decisions have also explored legal principles included in the CBD itself and embodied in the concept of sustainable development (e.g. public participation and the precautionary approach).

The decisions have also, in combination with other decisions, touched upon or added other elements that might be encompassed by the concept, notably protection of cultural diversity.[119] This can be seen, for example, in the Ecosystem Approach stipulating that humans, with their cultural diversity, are an integral component of many

---

[116] Decision VII/12. Annex II, para. 8(f). Annex II includes under Practical 5, Operational guidelines, one of the – generally speaking – rather rare references to the principles of the Rio Declaration, i.e. Principle 15 on the precautionary approach.
[117] Decision X/32, para. 2(i).
[118] *See* earlier discussion in Section 3.
[119] On cultural heritage, including intangible cultural heritage, as an element of sustainable development, *see* generally, J Blake, 'UNESCO/World Heritage Convention: Towards a More Integrated Approach' (2013) 43 Environmental Policy and Law 8ff. *See* also S Atapattu, 'Sustainable Development, Myth or Reality? A Survey of Sustainable Development Under International Law and Sri Lankan Law' (2001) XIV Georgetown International Environmental Law Review 272, note 20 '[t]he modern thinking is to include social factors (like human rights issues) and cultural factors in the development process. In other words, development must be environmentally, socially, and culturally sound and acceptable to the community or country in question'; and A Rosencrantz, 'The Origin and Emergence of International Environmental Norms' (2003) 26 Hastings International and Comparative Law Review 310, referring to the norm of 'cultural diversity and the right of indigenous people to retain a separate cultural identity' as one of the 20 prevailing or rising norms of global environmental law he is

ecosystems. The cultural dimension of the CBD is underscored, inter alia, by its close relationship with indigenous and local communities. The cultural element, however, is not being given a wider interpretation. Suffice to mention, for example, 'the Akwé: Kon Voluntary Guidelines for the Conduct of Cultural, Environmental and Social Impact Assessment Regarding Developments to Take Place, or Which are Likely to Impact on Sacred Sites, and Lands, and Waters Traditionally Occupied or Used by Indigenous People and Local Communities'[120] and 'The Tkarihwaié:ri Code of Ethical Conduct to Ensure Respect for the Cultural and Intellectual Heritage of Indigenous and Local Communities',[121] both adopted by the COP. Such instruments are obviously closely related to issues regarding human rights.

Do the CBD and its achievements clarify the concept of sustainable development, including its legal implications? According to the analysis a series of elements are from a legal point of view important within the framework of conservation and sustainable use of biodiversity. It might be argued that since these elements are important in that context, they are equally important with regard to sustainable development, since conservation and sustainable use of biodiversity are components of sustainable development. As referred to above,[122] some scholars include in the concept of sustainable development the attainment of cultural rights. These rights, in fact, play an important role with regard to conservation and the sustainable use of biodiversity, but whether they are included in the concept ultimately depends on the definition of sustainable development. If the concept of sustainable development is defined or understood in the same way as those authors, and if the CBD is perceived as an instrument pursuing sustainable development, then the findings of the analysis correspond by and large to the concept of sustainable development as defined by these authors.

There is no doubt that the CBD is an instrument to be included in the international law of sustainable development, since its objective is conservation and sustainable use of biodiversity. Which, then, are the legal principles commonly applied by the CBD in order to attain its objectives, irrespective of the substance matter being addressed? According to the findings, key principles being applied are: integration, intergenerational and intra-generational equity, and the precautionary approach. As far as intra-generational equity is concerned, however, COP decisions linked directly to benefit-sharing (and the 2010 Nagoya Protocol referred to in Section 4 above) and decisions on funding are probably more relevant than those analysed in Section 5 above. Thus, it is remarkable that the decision on the Strategic Plan (Section 5.2) was accompanied by a strategy for resource mobilisation in support of the achievement of the three objectives of the CBD.[123]

---

identifying. This was indeed confirmed by the 2007 UN Declaration on the Rights of Indigenous Peoples (n 47).
[120] Decision VII/7. Annex, Section F.
[121] Decision XI/42. Annex.
[122] Note 110 and discussion in Section 3.
[123] Decision X/3.

294   *Research handbook on biodiversity and law*

Also various procedural elements (for example public participation and EIA)[124] are applied. In fact, public participation plays a role in CBD decisions exceeding the scope of the rather limited and weak provisions of the CBD itself, but provisions of COP decisions in that respect are often ambiguous and may be considered as approaching this element too narrowly. It should, of course, be fully acknowledged that indigenous and local communities play an important role in the context of the subject matter of the CBD, and that the weight the COP has attached to them is nothing but laudable.[125] It is, however, difficult to escape from the thought that the very focus on indigenous and local communities[126] tends to make the COP overlook the fact that participation in decision-making also includes participation by civil society as such, including NGOs. After all, a number of Parties do not hold what may be characterised as 'indigenous and local communities' in the sense used by the CBD. This includes Parties that are not Parties to the 1998 Convention on Access to Information, Public Participation in Decision-Making and Access to Justice in Environmental Matters (Aarhus Convention). Probably, however, the deficiency of clearly indicating that public participation in decision-making should include the public at large is remedied to some extent 'on the ground'.[127]

The apparent unwillingness of the COP to address the issue of public participation in decision-making by clear, non-ambiguous and consistent normative language may be perceived as a reflection of the resistance of UN conferences on sustainable development to consider this issue in a more substantive manner.[128] Hence, no progress in a global perspective has been achieved since Principle 10 of the 1992 Rio-Declaration was adopted. The attitude of the COP vis-à-vis NGOs may be characterised as ambivalent. NGOs are not often referred to in the context of public participation in decision-making, but, on the other hand, the COP has facilitated their participation in meetings of the COP.[129]

The other dimensions of 'environmental rights,' i.e. access to environmental information and access to justice, are more or less ignored, at least in the decisions analysed in Section 5 above. Access to information is only referred to in a couple of their numerous provisions, and access to justice is not mentioned at all. This deficiency

---

[124] On EIA, *see* Decision VIII/28 on Impact assessment: Voluntary guidelines on biodiversity-inclusive impact assessment; and Decision XI/28 on Marine and coastal biodiversity: sustainable fisheries and addressing adverse impacts of human activities, voluntary guidelines for environmental assessment on marine spatial planning.

[125] According to Morgera and Tsioumani the CBD has emerged as the 'primary instrument and the preferred international forum for indigenous and local communities to express their interest for the protection of their traditional knowledge'; Morgera and Tsioumani (n 33) 21.

[126] On the 'status' of indigenous and local communities at COP negotiations, *see* P Kohler and others, 'Informing Policy. Science and Knowledge in Global Environmental Agreements' in PS Chasek and LM Wagner (eds), *The Road from Rio. Lessons Learned from Twenty Years of Multilateral Environmental Agreements* (RFF Press, 2013) 59, 75.

[127] *See*, for example, Prip and others (n 73), 34ff.

[128] The Outcome Document of Rio+20, 'The Future We Want', para. 99, only encourages action in that respect at the regional, national, subnational and local levels.

[129] Decision IX/29, para. 17 and Annex on steps for admitting qualified bodies and agencies whether governmental or non-governmental as observers to meetings.

remains a pity, irrespective of whether or not to include the two elements in a procedural dimension of the concept of sustainable development.

To the above enumeration of components of sustainable development applied by COP decisions may be added the principle of good governance, not only at a national, but also at an international level.[130] It is, in any event, quite striking that cooperation, coordination and promotion of synergies at the international level play a very important instrumental role for achieving the objectives of the CBD and have been addressed in numerous decisions of the COP. It remains, however, an open question to what extent the endeavours have been fruitful.[131] The same applies generally to information-sharing. The polluter-pays principle is only rarely referred to, but it is equally rare that this principle is mentioned by scholars in the context of sustainable development.[132]

It should be emphasised that, although the decisions examined may be characterised as key decisions, they represent only a tiny part of the total number of substantive decisions taken by the COP.[133] An analysis of all decisions may result in other conclusions. On the other hand, and in spite of the limited scope of the analysis, it has clearly demonstrated the truth of the criticism of the 'convoluted, repetitious, and disorderly drafting' of the guidelines and principles.[134] Virtually all provisions of the decisions examined dealing with the same substantive matter (for example, the principle of integration) are formulated in different ways, which may diminish their normative value. To some extent this may be due to the very nature of the COP, the substantive decision-making of which has to be based on consensus, notwithstanding, however, the advantages of this manner of decision-making.

7. FURTHER RESEARCH

The above conclusions and other findings of the present chapter may deserve further attention, discussion or research. Thus, it is obvious that the uncertainties or diverse views relating to the concept of sustainable development (Section 3 above) are pertinent. How are we to promote this concept as a key concept of international environmental law when in reality nobody knows exactly what it means? Faced with such uncertainties, it is difficult to agree with those – although probably rather few – academics claiming that the concept has gained status as a principle of customary international (environmental) law.

Another relevant question might be whether the conclusions of the analysis hold true in a broader perspective. Would an analysis of other COP decisions (Section 6 above) reflect a similar pattern?

---

[130] Good governance, however, is not a very precise concept, and it may be argued that it includes some of the elements which might be considered as components of sustainable development, for example public participation and access to information.

[131] For some rather critical reviews, see Long (n 12) 41, and R Caddell, 'The Integration of Multilateral Agreements: Lessons from the Biodiversity-Related Conventions' in Fauchald, Hunter and Wang (n 29), 37.

[132] See discussion in Section 3.

[133] See n 3.

[134] Morgera and Tsioumani (n 33), 8.

Further discussion or research would also be needed in order to identify the underlying causes of the formulation of the same notions or ideas in different ways, as outlined, for example, in Section 5.3 in respect of the principles of integration and of public participation, and to assess the consequences, if any. Are the causes related to the very nature of soft law decision-making processes, in the sense that developing soft law in the minds of governments does not require consistency because it is not hard law? And how would such findings correspond with the fact that negotiations on, for example, COP decisions are quite often lengthy and sometimes very difficult? Are soft law decision-making processes inspiring delegations to try to put their fingerprints on draft decisions rather than compelling them to rational thinking? It is inconceivable that the roots are to be found in the preparatory work of the Secretariat, since precedents are at the centre of the preoccupations of any Convention Secretariat. In respect of possible consequences of inconsistencies it might be relevant to examine to what extent the COP decisions analysed in the present contribution (and other decisions) would qualify as supplementary means of interpretation of provisions of the CBD in the sense of Article 31(3)(a) of the 1969 Vienna Convention on the Law of Treaties. Might one or the other decision be perceived as a subsequent agreement between the Parties regarding the interpretation of the CBD or the application of its provisions?

Another issue is why access to information is not prioritised (Section 5.2) and why access to justice seems to be almost a taboo (Section 6).

A number of questions may also be posed in relation to the notion of local and indigenous communities, such as have indigenous groupings in reality displaced the NGOs in general (Section 3)? And if so, is that merely due to the fact that indigenous communities are referred to explicitly in provisions of the CBD or are there other reasons as well? Why is it that the very vague notion 'local communities' seems never to have caused any problems (Section 5.1)? Has the notion been absorbed by the concept of indigenous communities or does it in practice mean more or less the same as indigenous communities? Who is in reality representing local communities at COPs?

Ultimately the CBD, and its COP decisions, remain a challenge. This challenge is, of course, foremost related to the substance matter of the Convention and to what extent it can or does contribute to the conservation and sustainable use of biodiversity. Closely linked to that, however, is the normative role of the COP, including its decision-making processes and their outcomes, which undoubtedly will continue to raise legal questions, as well as questions related to political science.

11. Whaling and inter- and intra-generational equity
Malgosia Fitzmaurice

1. PRELIMINARIES

1.1 Introduction

The focus of this chapter will be on issues relating specifically to questions of intra- and inter-generational equity and whaling. Therefore the 1946 International Convention on the Regulating of Whaling (ICRW),[1] and the practice of the International Whaling Commission (IWC) established thereunder, will be analysed only in so far as they are relevant to these questions. There is a host of other very important issues relating to whaling (such as the register of whaling vessels, and national control of citizens and vessels), but these have only a marginal or indirect impact on intra- and inter-generational equity and for that reason they will not be discussed in this chapter.

The question of intra- and inter-generational equity is very pertinent for the regulation of whaling. The Preamble to the ICRW states as follows: '[r]ecognizing the interest of the nations of the world in safeguarding for future generations the great natural resources represented by the whale stocks ...'. The Preamble thus clearly refers to future generations in relation to the preservation of whaling stocks. That said, the question of the relations between groups and people within one generation (intra-generational equity) and whaling was not explicitly mentioned in the ICRW but this does not mean that such an issue is irrelevant. It may be said that a reference to intra-generational equity can be presumed from the preambular paragraph '[r]ecognizing that it is in the common interest to achieve the optimum level of whale stocks as rapidly as possible without causing widespread economic and nutritional distress', and also Article V(2)(d) of the Whaling Convention, which states that while amending the Schedule the IWC will take into consideration 'the interests of the consumers of whale products and the whaling industry'. Certainly, the diverse (frequently conflicting) interests existing within each generation with regard to whaling (commercial, scientific, and in particular indigenous/subsistence whaling) do give rise to considerations of intra-generational equity.

The chapter will be structured as follows. Initially, a very short introduction to the legal structure regulating whaling will be given. Thereafter, a description of both concepts (intra- and inter-generational equity) will be presented and explained, followed by discussion as to the manner in which these principles are reflected in the issue of whaling.

---

[1] 161 UNTS 72. Originally, there were 15 signatory States to the Whaling Convention but participation is open to all States. There are currently 89 States Parties to the ICRW; see <http://iwc.int/members>.

1.2 The 1946 International Convention on the Regulation of Whaling: An Outline of Legal Issues

The legal controversies concerning whaling are not recent. Whaling has been a subject of discussion within international fora for many years. Indeed, the 1946 ICRW is not the first international instrument concerning whaling. It was preceded by two conventions: the 1931 Convention for Regulation of Whaling (the Geneva Convention)[2] and the 1937 International Agreement for the Regulation of Whaling.[3] These conventions were supposed to remedy the unsustainable hunting of whales, which was leading to the significant depletion of their stocks. This depletion was due to technological developments in the whaling industry, which had witnessed the introduction of steam engines, exploding harpoon guns, and the use of factory ships, all of which allowed the hunting of whales to become more effective and intensive.[4] However, these conventions were not successful, as many whaling nations, such as the USSR, Japan, Germany and Chile, were not parties to them.[5] Furthermore, the aftermath of the Second World War increased demand for edible fat.[6] Ever dwindling whale stocks led to the conclusion of the 1946 ICRW.

The main objectives of the Convention were expressed in the Preamble as follows:

---

[2] 1931 Convention for the Regulation of Whaling, 24 September, 155 LNTS 349.

[3] 1937 International Agreement for the Regulation of Whaling, 190 LNTS 79. The problems concerning whaling and the protection of whales have been the subject of many excellent studies, in particular the following publications, which take a comprehensive approach to these issues: A Gillespie, *Whaling Diplomacy: Defining the Issues in International Environmental Law* (Edward Elgar Publishing, 2005); MJ Bowman, '"Normalizing" the International Convention for the Regulation of Whaling' (2008) 29 Michigan Journal of International Law 293–499; GJ Nagtzaam, 'The International Whaling Commission and the Elusive Great White Whale of Preservationism' (2009) 33 William & Mary Environmental Law & Policy Review 375 and *The Making of International Environmental Treaties* (Edward Elgar Publishing, 2009); PGG Davies, 'Cetaceans' in MJ Bowman, PGG Davies and CJ Redgwell, *Lyster's International Wildlife Law* (2nd edn., CUP, 2010). See also MC Maffei, 'The International Convention for the Regulation of Whaling' (1997) 12 International Journal of Marine and Coastal Law 287; PW Birnie, 'Are Twentieth-Century Marine Conservation Conventions Adaptable to Twenty First Century Goals and Principles? Part II' (1997) 12 International Journal of Marine and Coastal Law 488, 499–505; S Freeland and J Drysdale, 'Co-Operation or Chaos? Article 65 of United Nations Convention on the Law of the Sea and the Future of the International Whaling Commission' (2005) 2 Macquarie Journal of International and Comparative Environmental Law 1; WC Burns, 'The International Whaling Commission and the Regulation of the Consumptive and Non-Consumptive Uses of Small Cetaceans: The Critical Agenda for the 1990s' (1994) 13 Wisconsin International Law Journal 105; AM Ruffle, 'Resurrecting the International Whaling Commission: Suggestions to Strengthen the Conservation Effort' (2002) 27 Brooklyn Journal of International Law 639; MJ Bowman, 'Transcending the Fisheries Paradigm: Towards a Rational Approach to Determining the Future of the International Whaling Commission' (2009) 7 New Zealand Yearbook of International Law 85.

[4] Ruffle (n 3), 645.

[5] Nagtzaam, *The Making of International Environmental Treaties* (n 3), 162.

[6] *Ibid.*, 165 and S Oberthür, 'The International Convention for the Regulation of Whaling: From Over-Exploitation to Total Prohibition' (1998/1999) Yearbook of International Cooperation on Environment and Development, 31.

[to] establish a system of international regulation for the whale fisheries to ensure proper and effective conservation and development of whale stocks ... and ... to provide for the proper conservation of whale stocks and thus make possible the orderly development of the whaling industry.

These two objectives (the conservation of whale stocks and the 'orderly' development of the whaling industry) have proved to be quite challenging to reconcile.[7] The view has also been expressed that the formulation of the objectives of the ICRW indicates that the primary purpose was the conservation of whale stocks and the development of the whaling industry was secondary. Thus, Bowman is of the view that the development of the industry as such did not constitute an essential part of the Convention's objectives, the main one being the preservation of stocks for future generations through the imposition of order upon the whaling industry.[8]

The ICRW established a main body to regulate whaling, the IWC, which consists of Commissioners who are the representatives of all States Parties to the Convention.[9] The IWC has a number of responsibilities, including: coordinating the compilation of catch reports and other statistical and biological records; encouraging, coordinating and funding whale research; publishing the results of scientific research; and promoting studies into related matters such as the humaneness of the killing operations.[10] Nevertheless, one of its most significant functions relates to keeping under review and revising, as necessary, the measures laid down in the Schedule that was appended to the ICRW and which governs the conduct of whaling throughout the world.[11] This Schedule, inter alia, sets limits on the numbers and size of whales that may be taken; prescribes open and closed seasons and areas for whaling; and approves the methods of taking whales.[12] The Schedule has been amended periodically. While amendment of the ICRW itself requires the agreement of all member States, amendments to the Schedule require a three-quarters majority.[13]

Certain normative *lacunae* in the Schedule have allowed States to abuse, or at least to misinterpret, the Convention. For example, although the Schedule banned the taking of gray, humpback and right whales in certain areas, it allowed aboriginal whaling of these species. Further, the territorial scope of the ICRW is contested, i.e., the question whether it applies also to exclusive economic zones and territorial seas of the States Parties or only to the high seas.[14] Finally, the Convention does not include a formal definition of what is a whale (the Schedule simply lists various particular species of whales); this amounts to one of the most crippling omissions in the Convention, and

---

[7] G Rose and G Paleokrassis, 'Compliance with International Environmental Obligations: A Case Study of the International Whaling Commission' in J Cameron, J Werksman and P Roderick (eds), *Improving Compliance with International Environmental Law* (Earthscan/ Routledge, 1996), 147, 147.
[8] Bowman (n 3), 293.
[9] ICRW, Article III.
[10] See <http://iwc.int/iwcmain>.
[11] ICRW, Article V.
[12] ICRW, Article V(1).
[13] ICRW, Article III(2).
[14] See Davies (n 3), 155–156.

has led some States Parties to claim that the Convention does not cover medium and small-size cetaceans, only the great ones. Therefore, they argue, the IWC has no jurisdiction over the first two of the above-mentioned categories, leaving hunting for them unregulated internationally.

All these unclear legal issues concerning the application of the Convention have a limiting effect on the powers, and therefore the effectiveness, of the IWC. It may also be mentioned that the system of 'opting out' provided for in Article 5(3) of the ICRW, which gives the right to all States Parties to the Convention to evade the effect of any decision of the IWC by notifying an objection, weakens its efficacy to a certain degree. However, such a decision-making system is very common in international bodies.

Increased whaling, generally changing attitudes towards the environment after the 1972 Stockholm Conference on the Human Environment, and the inclusion of many species of whales under the 1973 Convention on International Trade in Endangered Species of Fauna and Flora (CITES),[15] resulted in the imposition of the 1982 moratorium on whaling under the ICRW. This was a victory for preservationists. Thus, the amended Paragraph 10 of the Schedule read as follows:

> Notwithstanding the other provisions of paragraph 10, catch limits for the killing for commercial purposes of whales from all stocks for 1986 coastal and the 1985/86 pelagic seasons and thereafter will be zero. This provision will be kept under review, based upon the best scientific advice, and by 1990 at the latest the Commission will undertake a comprehensive assessment of the effects of these decisions on whale stocks and consider modification of this provision and the establishment of other catch limits (Paragraph 10 (e), IWC Schedule, February 1983).[16]

By restricting itself to whaling for commercial purposes the moratorium made an exception regarding aboriginal subsistence whaling; something that allowed limited taking of bowhead whales, for instance. Further, some States (Norway, Iceland) have by one means or another objected to the moratorium, and one of the main whaling nations, Canada, withdrew from the Convention altogether. Finally, the imposition of the moratorium on commercial whaling led to the States that still wanted to hunt for whales perceiving the IWC as an anti-whaling body. General dissatisfaction of these States and their supporters with the anti-whaling policy of the IWC led to the adoption of the St. Kitts and Nevis Resolution, by the narrowest of margins at the 2006 Annual Meeting.[17] The States which supported the Resolution were of the view that certain Commissioners were unreasonable in continuing to support the moratorium, which was meant as a temporary measure; its permanent existence was regarded as unsustainable

---

[15] 993 UNTS 243.

[16] Annual Report of the International Whaling Commission 33, at 40 (1983); Chairman's Report of the Thirty-Fifth Annual Meeting, Appendix 2.

[17] IWC Resolution 2006-1. Thirty-three commissioners voted in favour, 32 against and one abstained. The Resolution was passed with the support of the following States: St Kitts and Nevis, Antigua & Barbuda, Benin, Cambodia, Cameroon, Cote d'Ivoire, Dominica, Gabon, Gambia, Grenada, Republic of Guinea, Iceland, Japan, Kiribati, Mali, Republic of the Marshall Islands, Mauritania, Mongolia, Morocco, Nauru, Nicaragua, Norway, Republic of Palau, Russian Federation, St Lucia, St Vincent and the Grenadines, Solomon Islands, Suriname, Togo, Tuvalu.

and against scientific findings. Therefore these States called for a 'normalising' of the functions of the IWC.

However, it may be said that commercial whaling and the moratorium are not the only contentious issues concerning whaling. Two other types of whaling: scientific (provided for in Article VIII of the ICRW) and indigenous whaling (provided for in the Schedule of the Convention), are also causes of a conflict between States Parties. Indeed the former has been the subject of a recent judgment by the International Court of Justice (ICJ), and will be discussed later in this chapter. All of these contentious issues were raised at the 55th meeting of the IWC in 2003 in Berlin and resulted in the establishment of the Conservation Committee and the adoption of the so-called Berlin Initiative on Strengthening the Conservation Agenda of the International Whaling Commission.[18] Some views were expressed that the Berlin Initiative, which was a step further towards conservation of whales (from their consumptive utilisation), managed to further pull apart the whaling and non-whaling nations.[19]

2 INTRA- AND INTER-GENERATIONAL EQUITY

2.1 Introduction

In this part of this chapter the issues of intra- and inter-generational equity will be discussed in relation to whaling. The work of the IWC will be analysed to determine whether the regulatory techniques adopted by the Commission have an effect on, or relate to, either of these concepts, the discussion of which may overlap to a certain degree. A definition of intra-generational equity helps to begin this analysis.

2.2 Intra-generational Equity

Intra-generational equity is related to fairness among the present generation, and this primarily concerns the relationship between developed and developing States, i.e., the South/North controversy, which is focused on issues of distributive justice, post-colonial legacies and unequal relations of power between developed and developing States.[20] However, it may be also said that intra-generational equity in industrially developed countries is related to the 'ability of nations, or communities, to develop economically while preserving their environment'.[21] In developing countries intra-generational equity in relation to sustainability assumes a broader meaning, which is best understood at various levels of human activity. For example, at the local level intra-generational equity and sustainable development refers to the ability of a

---

[18] IWC Resolution 2003-1.
[19] WCG Burns, 'The Berlin Initiative on Strengthening the Conservation Agenda of the International Whaling Commission: Towards a new Era for Cetaceans?' (2004) 13 RECIEL 72.
[20] See generally D Shelton, 'Equity' in D Bodansky, J Brunnée and E Hey (eds), *The Oxford Handbook of International Environmental Law* (OUP, 2007) 32, 53.
[21] I Voinovic, 'Intergenerational and Intragenerational Equity Requirements for Sustainability' (1995) 22 Environmental Conservation 223, 225.

government and other agencies to alleviate urban and rural pressure points that are major problems in developing countries. This can be achieved through providing people with adequate shelter, food, drinking water, safe and effective sewage and waste disposal, effective transportation and food. However, it can be said that conservation is incompatible with absolute poverty. Therefore, it is an indispensable requirement that intra-generational equity has to be aimed at satisfying basic survival needs of people in the less-developed nations.[22] As to the use of scarce natural resources in relation to intra-generational equity, this will be based on either trade-offs between intra-generational requirements or the redistribution of resources among the current and future generations,[23] a rather complex issue. It is assumed that existing natural resource stocks are neither able to meet the needs of the current generation nor to meet the needs of future generation, so trade-offs will have to be made.[24]

In the context of whaling there is a tension between inter- and intra-generational equity: on the one hand the preservation of whaling stocks for future generations (which would be fostered by a continuing moratorium on commercial whaling); on the other, a possibility of providing food for persons in developing countries, which would support the case for the resumption of commercial whaling, thus fostering the principle of intra-generational equity. Additionally, the issue of whales' consumption of fish and other marine organisms could be regarded as a threat to food security. However, this needs to be balanced against the positive contribution of whales in the recycling of nutrients through such phenomena as 'whale falling'.[25]

2.2.1 Indigenous (subsistence) whaling[26]

The question of intra-generational equity is plainly relevant to the issue of indigenous whaling, as reflected in the relationship between indigenous peoples' subsistence

---

[22] *Ibid.*
[23] *Ibid.*
[24] *Ibid.*
[25] 'Whale falling' occurs when whales die and sink. The whale carcasses, or whale falls, provide a sudden, concentrated food source for organisms in the deep sea. Different stages in the decomposition of a whale carcass support a succession of marine biological communities. Scavengers consume the soft tissue in a matter of months. Organic fragments, or detritus, enrich the sediments nearby for over a year. The whale skeleton can support rich communities for years to decades, both as a hard substrate (or surface) for invertebrate colonisation and as a source of sulphides from the decay of organic compounds of whale bones. Microbes live off the energy released from these chemical reactions and form the basis of ecosystems for as long as the food source lasts. At deep sea levels this forms a new food web and provides energy to support single- and multi-cell organisms and sponges, thus adding to the ocean's food chain. *See* National Oceanic and Atmospheric Administration, 'What is a Whale Fall?' available at <http://oceanservice.noaa.gov/facts/whale-fall.html>.
[26] See generally M Fitzmaurice, 'Indigenous Whaling and Environmental Protection' (2012) 55 German Yearbook of International Law 419. In 2014 at its 65th meeting in Slovenia the IWC adopted Resolution 2014-1 to 'work to improve the process for ASW [Aboriginal Subsistence Whaling] in the future through a more consistent and long-term approach. This will include an expert workshop to assist the ASW Sub-Committee and the Commission with respect to improved procedures for considering catch limits, with a focus on consideration of need';

whaling and other States Parties to the ICRW which lack indigenous peoples. The ICRW does not itself include any special provision regulating aboriginal subsistence whaling; however, the Schedule to the Convention recognises its special position by excluding it from the definition of, and the provisions relating to, commercial whaling. In 1979, the IWC Anthropology Panel adopted an unofficial definition of 'subsistence whaling' as comprising:

(1) the personal consumption of whale products for food, fuel, shelter, clothing, tools, or transportation by participants in the whale harvest;
(2) the barter, trade, or sharing of whale products in their harvestable form with relatives of the participants in the harvest, with others in the local community or with persons in locations other than the local community with whom local residents share familial, social, cultural or economic ties. A generalised currency is involved in this barter and trade, but the predominant portion of the products from each whale is ordinarily directly consumed or utilised in their harvested form within the local community; and
(3) the making and selling of handicraft articles from whale products, when the whale is harvested for the purposes (1) and (2).[27]

'Aboriginal subsistence whaling' was again defined in 1981, this time by the IWC's Technical Committee Working Group on Development of Management Principles and Guidelines for Subsistence Catches of Whales by Indigenous (Aboriginal) Peoples, as comprising whaling conducted for 'purposes of local aboriginal consumption carried out by or on behalf of aboriginal, indigenous or native people who share strong community, familial, social and cultural ties related to a continuing traditional dependence on whaling and on the use of whales'.[28] These definitions are very problematic. For instance, the interchangeable use of the terms 'aboriginal', 'native' and 'indigenous' is in itself confusing, as in many indigenous communities these carry different meanings. For example, doubts have been raised as to whether whaling in Greenland can qualify as aboriginal.

The 1979 definition of 'subsistence use of whale products' coined by cultural anthropologists is more restrictive as to the area in which the distribution of whale products is permitted and does not recognise the distribution of whale products that involve cash, as in aboriginal subsistence whaling.[29] It may be added that the report proposing the definitions was not very consistent as it said that in some cases products are distributed to and used by communities away from the coastal areas where whaling is actually conducted, and in some areas the practice of trading to meet subsistence need has emerged. Further, the IWC ad hoc Working Group stated that it was arguable

---

Summary of Main Outcomes, Decisions and Required Actions from the 65th Annual Meeting of the IWC, available at <http://www.ifaw.org/sites/default/files/IWC_65-Main%20outcomes.pdf>.

[27] Report of the Panel Meeting of Experts on Aboriginal Subsistence Whaling, Special Issue 4 (IWC, Cambridge), 7.

[28] GP Donovan, 'The International Whaling Commission and Aboriginal Subsistence Whaling' April 1979 to July 1981; GP Donovan, *Aboriginal/Subsistence Whaling (with special reference to the Alaska and Greenland fisheries)*, Reports of the International Whaling Commission – Special Issue No. 4 (1982), 79–86.

[29] H Hamaguchi, 'Aboriginal Subsistence Whaling Revisited' (2013) 84 Senri Ethnological Studies 81, 86.

whether there is a difference in principle between the sale of whale products in order to buy essential goods and the direct exchange of whale products for such goods. According to Hamaguchi this is indicative of the fact that even the ad hoc Working Group's definition did not completely deny for all cases the extensive distribution of whale products or their distribution involving cash.[30]

At the outset, confusion has resulted from the lack of any conclusive definition of what constitutes 'commercial' whaling, which makes it difficult to differentiate between 'aboriginal' and 'commercial' whaling. The ad hoc Working Group attempted to distinguish between commercial and aboriginal whaling. These two forms were considered to be different in respect of two elements: management and catching. The main objective of the management of aboriginal subsistence whaling was to maintain individual stocks at the highest possible level, and the main purpose of aboriginal subsistence whaling was to fulfil nutritional and cultural needs. The main objective of commercial whaling on the other hand was to maximise yields from individual stocks, and the main purpose of catching whales commercially was to sell their products. Hamaguchi observes that 'these differences indicate that aboriginal subsistence whaling prioritizes quality (the cultural aspect) and commercial whaling prioritizes quantity (the economic aspect)'.[31]

The term 'indigenous' is one that grew in prominence, especially after the adoption of the 2007 United Nations Declaration on the Rights of Indigenous Peoples, as a means to describe aboriginal peoples in an international context. However, it may have contentious connotations since internationally, and in the United Nations context, it may define groups primarily in relation to their colonisers.[32]

According to the IWC, the objectives in regulating aboriginal subsistence whaling are as follows: (i) to ensure that the risk of extinction is not seriously increased (this being the objective with the highest priority); (ii) to enable harvests in perpetuity and appropriate to cultural and nutritional requirements; (iii) to maintain stocks at the highest net recruitment level (which is when the population is at its maximum sustainable yield level), and, if they fall below that, to ensure they move towards it.[33] In general the IWC has identified four specific whaling operations as qualifying for the status of aboriginal subsistence whaling, and therefore permitted: (i) minke and fin whales (formerly also humpback whales) in Greenland; (ii) humpback whales in the Lesser Antilles (specifically at the island of Bequia, St. Vincent and the Grenadines); (iii) bowhead whales (and formerly also gray whales) in the US (Alaska); and (iv) gray and bowhead whales in Russia (Chukotka).[34] It is the responsibility of national

---

[30] Ibid., 86.
[31] Ibid., 86.
[32] Strategic Alliance of Broadcasters for Aboriginal Reflection, *Key Terminology Guidebook for Reporting on Aboriginal Topics* available at <http://www.sabar.ca/wp-content/uploads/2012/06/SABAR-Glossary-English-Final.pdf>.
[33] See JG Cooke, 'A Review of Some Implications of Environmental Variability for the Management of Baleen Whale Populations, available at <http://www.researchgate.net/publication/255592686_The_influence_of_environmental_variability_on_baleen_whale_sustainable_yield_curves>.
[34] The IWC quotas for 2014–2018: agreed at IWC54 in Slovenia, and available at <http://uk.whales.org/issues/aboriginal-subsistence-whaling>:

governments to provide the Commission with evidence of the cultural and subsistence 'needs' of their aboriginal peoples, while the Scientific Committee provides scientific advice on safe catch limits for such stocks, as two necessary requirements for the allocation of quotas.[35]

Intra-generational equity is conceptualised in the principle of common but differentiated responsibilities, formulated in Principle 7 of the 1992 Rio Declaration on Environment and Development:

> States shall cooperate in a spirit of global partnership to conserve, protect and restore the health and integrity of the Earth's ecosystem. In view of the different contributions to global environmental degradation, States have common but differentiated responsibilities. The developed countries acknowledge the responsibility that they bear in the international pursuit of sustainable development in view of the pressures their societies place on the global environment and of the technologies and financial resources they command.

This principle may be interpreted in relation to indigenous whaling in the following way: States in whose territory indigenous peoples live should acknowledge their right to subsistence whaling. Indigenous peoples should, however, contribute to the protection and preservation of whale stocks, having regard to the principle of common but differentiated responsibilities. The principle of intra-generational equity (as understood in this way) is also an expression of solidarity in relation to environmental problems facing the international community as a whole, such as climate change and biodiversity decline.[36]

Whaling by Makah Indians is one of the most controversial cases of aboriginal whaling. It has been the subject of much discussion at the IWC and was also the subject of many legislative regulations and legal debates within the US. It still, however, remains controversial in a number of respects:

(i) it is an instance of a claim (there are others) to resume aboriginal whaling after a period during which, for different reasons, the practice had been abandoned; in this case, a claim by the Makah Indians to resume whaling after a 70 year hiatus;

The number of fin whales struck from the West Greenland stock in accordance shall not exceed 19 in each of the years 2015, 2016, 2017 and 2018.

The number of minke whales from the Central stock shall not exceed 12 in each of the years 2015, 2016, 2017 and 2018.

The number of minke whales struck from the West Greenland stock shall not exceed 164 in each of the years 2015, 2016, 2017 and 2018.

The number of bowhead whales struck from the West Greenland shall not exceed 2 in each of the years 2015, 2016, 2017 and 2018.

The number of Humpback whales struck off West Greenland shall not exceed 10 in each of the years 2015, 2016, 2017 and 2018.

[35] Cooke (n 33).
[36] P Cullet, 'Common but Differentiated Responsibilities' in M Fitzmaurice, D Ong, and P Merkouris (eds), *Research Handbook on International Environmental Law* (Edward Elgar Publishing, 2010), 169.

(ii) there were also doubts whether their whaling was purely aboriginal subsistence whaling or was also commercial;
(iii) the claim raised questions concerning the ethical nature of the resumption of aboriginal whaling;
(iv) it also raised the issue as to whether aboriginal whaling constitutes a cultural exemption.[37]

In general, the US first incorporated the IWC's regime into domestic law in the 1971 Pelly Amendment to the Fisherman's Protective Act of 1967. On the basis of this amendment, when the Secretary of Commerce determines that the nationals of a foreign country are diminishing the effectiveness of an international fishery conservation programme (such as the IWC's programme), the Secretary shall certify this fact to the President. The President then has the discretion to ban importation of fishing products from the offending country. For example, President Barack Obama informed the US Congress on 15 September 2011 of his decision not to apply trade measures to Iceland for commercial whaling in defiance of the IWC ban on such activity.[38] The US also governs aboriginal whaling by Alaska Natives via the Marine Mammal Protection Act (MMPA) and the Endangered Species Act (ESA).

Whaling by aboriginal peoples is allowed under US law to the extent that it is approved by the IWC.[39] The ICRW limits how many bowhead or gray whales aboriginal groups may harvest. However, no domestic law restricts harvest numbers on whales except specific regulations under the ESA or MMPA (provided the harvest is for non-wasteful subsistence use).[40] The Makah Indian tribe is the only indigenous group in the US with a treaty specifically reserving the right to hunt whales. According to the US Supreme Court ruling in *US v. Dion*, a treaty right cannot be extinguished by the absence of the exercise of this right.[41] For the Makah, the inclusion of this right in a treaty means that, even if no whales are taken, the right remains enforceable (providing that the treaty is still in force and the right has not been abrogated).

The Makah indigenous people live in the State of Washington and traditionally hunted for gray whales for centuries (commencing 1,500 years ago).[42] They ceded certain lands to the US on the basis of the 1855 Treaty of Neah Bay between the Makah and the US government. This treaty also guaranteed their rights to hunt for seals and whales and to fish in a reservation they were relocated to. Around the turn of the

---

[37] L Heinamaki, 'Protecting the Rights of Indigenous Peoples – Promoting the Sustainability of the Global Environment?' (2009) 11 International Community Law Review, 3, 46–52.

[38] Humane Society International, 'Iceland, Whaling and the Pelly Amendment', available at <http://www.hsi.org/issues/whaling/facts/iceland_whaling_and_pelly.html>.

[39] J Firestone and J Lilley, 'Aboriginal Subsistence Whaling and the Right to Practice and Revitalize Cultural Traditions and Customs' (2005) 8 Journal of International Wildlife Law and Policy 177, 197–200.

[40] Ibid.

[41] US Supreme Court, *US v. Dion*, 476 U.S. 734 (1986) (cited in J Sepez, 'Treaty Rights and the Right to Culture, Native American Subsistence Issues in US Law' (2002) 14(2) Cultural Dynamics 143,150).

[42] JJ Brown, 'It's in Our Treaty: the Right to Whale', available at <http://nativecases.evergreen.edu/docs/Makah_Case_Study_rev7_25_08.doc>.

20th century, the Makah people voluntarily ceased hunting for gray whales due to the depletion of their stocks, caused mainly by commercial hunting. The absence of hunting for whales had an adverse impact on the economy of the Makah people. When the eastern Pacific gray whale was removed from the endangered species list in 1994, the Makah tribe began preparations to resume hunting, as they alleged, for nutritional and cultural reasons in consultation with the National Marine Fisheries Service.[43] They also claimed that the resumption of whaling would enable them 'to instil in the tribe the values traditionally associated with whaling'.[44] However, the plan to resume this activity after a 70-year hiatus was met with fierce opposition in the IWC and in the US Congress; and also by some of the Makah indigenes themselves (acting as observers at the meeting of the IWC), who argued that the resumption of whaling had not been supported by all Makah people. Furthermore, due to the change in the IWC policy (following the bowhead crisis)[45] from one of a general exemption for aboriginal subsistence to one based on the requirement of 'needs', the Makah Indians' rights to aboriginal whaling were not automatically granted by the IWC[46] and the IWC initially decided that the Makah request did not fulfil the requisite conditions (i.e., subsistence needs and continuing traditional dependence) for aboriginal subsistence whaling.

In 1997, the US government brought the case before the IWC for a second time on behalf of the Makah. This time, however, the US submitted a joint request together with the Russian Federation (which was acting on behalf of the Chukotka people). It was in fact a trade-off between the US and the Russian Federation. The US argued the case for the resumption of Makah whaling on the basis of the rights granted by the 1855 Treaty of Neah Bay. The IWC agreed to this renewed request and set the limit of 620 catches for the period 1998 to 2002. Their resumption of hunting, however, was strongly objected to by environmental NGOs, which contended that, after the resumption of hunting for whales, the Makah people would engage in commercial

---

[43] Sepez (n 41), 149.
[44] Firestone and Lilley (n 39), 185.
[45] In the 1970s bowheads were considered an endangered species. At the height of the bowhead crisis, the IWC adopted Resolution 1979-4 on 'Bering Sea Bowhead Whales' (Report of the International Whaling Commission 30; 31st IWC Meeting 1979):

THE COMMISSION INTENDS that the needs of the aboriginals of the United States shall be determined by the Government of the United States of America. This need shall be documented annually to the Technical Committee, and shall be based upon the following factors:

1. importance of the bowhead in the traditional diet,
2. possible adverse effects of shifts to non-native foods,
3. availability and acceptability of other food sources,
4. historical take,
5. the integrative functions of the bowhead hunt in contemporary Eskimo society, and the risk to the community identity from an imposed restriction on native harvesting of the bowhead; and
6. to the extent possible, ecological considerations.

[46] Sepez (n 41), 149 *et seq.*

activities by selling meat to Japan (allegations which were strongly denied by the Makah people).[47]

The opponents of the resumption of whaling further argued that whaling was not necessary for subsistence of the Makah, noting that they had lived for a long period of time without whale meat. Again, the Makah opposed this contention, insisting that whale meat was a necessary part of their subsistence, despite the almost 100-year break.[48] There was also opposition to the argument that the resumption of whale hunting would contribute to the cultural revitalisation of the Makah. In this respect, the Makah argued that resumption of whaling would enable them 'to instil in the tribe the values traditionally associated with whaling'.[49] Opponents of the Makah's proposed hunt also felt that if the Makah claim were accepted, other claims would be made on the same basis. Additionally, some argued that such whaling would give a boost for Norwegian and Japanese claims for support of their own whaling traditions.

Opposition to the Makah Indians' resumption of whaling was also based on environmental grounds. For example, two NGOs[50] wrote to the US Department of Commerce (DOC) and to the National Oceanic and Atmospheric Administration (NOAA) stating that their departments had breached the National Environmental Policy Act (NEPA) because they authorised Makah whaling without first applying NEPA or making an Environment Impact Statement (EIS) and Environmental Assessment (EA). In response, a draft EA was issued and a new agreement between NOAA and the Makah was entered into; and the NOAA issued a final EA indicating a finding of 'No Significant Impact'.[51]

The coalition opposed to the resumption of whale hunting by Makah Indians is led by the Sea Shepherd Society; altogether, approximately 250 animal welfare organisations and 27 conservation organisations are involved. Some of these organisations filed a suit in the US courts. The decision of the Court stated that the EA was made too late in the decision-making process.[52] The NOAA had to abandon the agreement with the Makah in the light of the Court's decision. In 2001 a new draft EA was issued; and the same year the NOAA established the quota of five landings of gray whales for 2001 and 2002. This decision was again challenged by the anti-whaling lobby.[53] The dispute in essence was over the localised effect on the whole whale population in the area of the hunt. The Court required a full application of the EIS protocol in the light of the ambiguity and uncertainty in this respect, i.e., regarding the failure of the EIS to address fully the effect of the whaling permit on other Native Americans that might wish to hunt, and also of the effect on other IWC members.

---

[47] M Weinbaum, 'Makah Native Americans vs. Animal Rights Activists', available at <http://www.umich.edu/~snre492/Jones/makah.htm>.
[48] Firestone and Lilley (n 39), 186.
[49] *Ibid.*, 185.
[50] 'Australians for Animals' and 'BEACH Marine Protection'.
[51] Firestone and Lilley (n 39), 198.
[52] *Metcalf v. Daley* 214 F.3d 1135 (9th Cir. 2000). On this and other Makah-related cases (*Anderson v. Evans* 314 F.3d 1006 (9th Cir. 2002) and *Anderson v. Evans*, 371 F.3d 475 (9th Cir. 2004)), *see* generally Brown (n 42).
[53] *Anderson v. Evans*, 371 F.3d 475 (9th Cir. 2004).

According to other court decisions, the situation concerning the Makah whale hunting rights has still not yet been resolved.[54] The claim of the Makah tribe was supported by some scholars[55] who were of the view that Makah should not be deprived of their right to hunt whales. The latest decision of the US government is to scrap a seven-year-old draft environmental study on the impact of Makah tribal whaling and write a new impact statement in light of substantial new scientific information. The new information is that the gray whales which the tribe wants to hunt off the Washington coast may need to be managed separately from the overall gray whale population that migrates up and down the West Coast.[56]

Arguably, one of the causes of the unresolved situation concerning the Makah tribe's right to hunt gray whales is the concern over environmental issues; and, indeed, in this case environmental considerations (the preservation of gray whales stocks) clashed to some extent with the right to cultural diversity. It would, however, be both imprudent and simplistic to attempt to draw general conclusions based on this single case. According to the US courts, environmental obligations were not fully implemented, therefore it can be said that the right of the Makah people was overruled on the basis of a legal technicality (the failure to correctly apply EA), rather than on any informed discussion balancing the right to cultural diversity (whaling) and obligations stemming from the duty of environmental protection. The question thus arises, what would the outcome of the case have been if the application of the EA had been properly followed? It is possible that, if there had been a favourable result concerning application of the EA, Makah hunting for gray whale would have been allowed.

Whaling rights of aboriginal peoples are supported by Article 27 of the 1966 International Covenant on Civil and Political Rights (ICCPR).[57] This Article actually relates to minorities generally, but its application to indigenous peoples is of fundamental importance. In 1994, the Human Rights Committee (HRC) adopted General Comment No. 23, which referred explicitly to the applicability of this Article to indigenous peoples.[58] Indigenous rights to culture were also commented on by the Committee

---

[54] Firestone and Lilley (n 39), 201–207.
[55] Sepez (n 41), 143–159.
[56] This decision is based on a study by Canadian scientists (Tim Frazier and Jim Darling) who state that a separate, genetically distinct Pacific Coast Feeding Group of about 200 whales regularly feeds in areas that include waters between northern California and south-eastern Alaska during the summer and autumn. See P Gottlieb, 'Restart of whaling study disappoints Makah chairman, Peninsula News', available at <http://www.peninsuladailynews.com/article/20120525/news/305259989/restart-of-whaling-study-disappoints-makah-chairman>.
[57] 999 UNTS 171.
[58] 'With regard to exercise of the cultural rights protected under Article 27, the Committee observes that culture manifests itself in many forms, including a particular way of life associated with the use of land resources, especially in the case of indigenous peoples. That right may include such traditional activities as fishing and hunting and the right to live in reserves protected by law. The enjoyment of these rights may require positive legal measures of protection and measures to ensure the effective participation of members of minority communities in decisions which affect them'; HRC, General Comment No. 23 (50th Session), UN Doc. HR1/Gen/1/rev.3, para. 7, available at <http://www1.umn.edu/humanrts/gencomm/hrcom 23.htm>.

310 *Research handbook on biodiversity and law*

on the Elimination of Racial Discrimination (CERD),[59] when it stated that States must support the indigenous way of living.[60] Most importantly, the indigenous lifestyle is at the core of the 2007 UN Declaration on the Rights of Indigenous Peoples,[61] which states in its Preamble that 'respect for indigenous knowledge, culture and traditional practices contribute to sustainable and equitable development and proper management of the environment'.[62] Mention must also be made of the 1989 ILO Convention No. 169 concerning Indigenous and Tribal Peoples in Independent Countries,[63] which links the cultural identity of indigenous peoples with environmental protection, as shown in the Convention's Preamble, which refers to the 'distinctive contributions of indigenous and tribal peoples to cultural diversity and social and ecological harmony of humankind and to international co-operation and understanding'. Article 4 of that Convention also provides that 'special measures shall be adopted in appropriate form safeguarding the persons, institutions, property, labour, cultures and environment of the peoples concerned'. Finally, cultural diversity also finds support in several conventions and soft law instruments of the United Nations Educational, Scientific and Cultural Organization (UNESCO), such as the 2001 Universal Declaration on Cultural Diversity[64] and the 2005 Convention on the Protection and Promotion of the Diversity of Cultural Expressions.[65]

In conclusion, it can be said that there is support in various conventions, soft law instruments and practice of the IWC for indigenous subsistence whaling based on the right to cultural diversity. On the other hand, there is also a very persuasive and large opposition, led particularly by civil society, which finds requests by States to increase aboriginal whaling quotas by the IWC harmful for the preservation of whale stocks. During the 2012 meeting of the IWC, the Danish delegation walked out in protest against the refusal by the Commission to increase the quotas of humpback whales for Greenland's Inuit peoples and refused to accept any aboriginal quotas at all.[66] As noted above in the context of US litigation, the same conflicts can exist in national law.

Therefore it may be said that the principle of intra-generational equity is accepted both by the IWC and in the practice of States in relation to aboriginal whaling. However, its implementation is not without controversies. In the view of this author, aboriginal whaling will remain contentious due to the growing importance of environmental issues (especially protection of biodiversity) and the very strong position of civil

---

[59] The Committee was established under the 1966 International Convention on the Elimination of All Forms of Racial Discrimination, 660 UNTS 195.
[60] CERD, General Recommendation No. 23: Indigenous Peoples, UN Doc. A/52/18, Annex V (1997), para. 4. See also A Fodella, 'Indigenous Peoples, the Environment and International Jurisprudence' in N Boschiero and others (eds), *International Courts and the Development of International Law* (TMC Asser Press, 2013), 351–353.
[61] UNGA Res 61/295, Annex.
[62] Preamble, 11th recital. See further Articles 25 and 29 of the Declaration.
[63] (1989) 72 International Labour Office Official Bulletin 59.
[64] Available at <http://portal.unesco.org/en/ev.php-URL_ID=13179&URL_DO=DO_TOPIC&URL_SECTION=201.html>.
[65] 2440 UNTS 311.
[66] <http://www.tumblr.com/tagged/international%20whaling%20commission>. However, succeeded with the same request in 2014 meeting of the IWC, <http://www.bornfree.org.uk/campaigns/marine/marine-news/article/?no_cache=1&tx_ttnews[tt_news]=1670>.

society. In this context mention may also be made of the so-called small-type coastal whaling, propagated by Japan. Since the ban on commercial whaling Japan has continued to seek a whaling quota from the IWC to provide 'emergency relief' to four coastal towns that it argues are suffering financial hardship and cultural disintegration as a direct result of the ban (Abashiri, Ayukawa, Wada and Taiji).[67] Japan argues that the IWC's ban on minke whaling 'directly caused cultural disintegration and financial hardship in these towns; and that only the IWC could alleviate their problems, by allowing them to conduct a minke whaling operation in their coastal waters'.[68] The Whale and Dolphin Conservation Society has, however, argued that only Abashiri and Ayukawa have a sustained history of hunting minke whales in their coastal waters; from 1933 for Abashiri and from 1948 for Ayukawa. Further, there are no records of such operations by Taiji and Wada.[69] Moreover, it is claimed that, since the moratorium, each of the four towns has maintained hunts for short-finned pilot whales, Risso's dolphins and Baird's beaked whales in their coastal waters.[70] The Whale and Dolphin Conservation Society also argues that whaling is now of little importance, either culturally or financially, to the town of Abashiri, which does not even have its own active whaling vessel. It shares ownership of a vessel based in Ayukawa, which travels to Abashiri once a year to take its quota of four Baird's beaked whales.[71]

Japan compares aboriginal whaling to small-type coastal whaling and argues that the quotas set by the IWC for the resumption of whaling for the Makah tribe after 70 years of non-whaling[72] should also be a basis for setting quotas for these four towns, as small coastal-type whaling. It also argues that there is a distinction between large pelagic whaling and small-type coastal whaling given the size of these operations. Therefore, the ban should only be applicable to large pelagic whaling and small coastal whaling should be exempted. It may be said that the IWC in its Resolution recognised the negative impact on these communities caused by the cessation of minke whaling.[73] However, it declined to set quotas for this type of whaling. In the view of the present author the denial of small coastal whaling for Japanese communities is not fair and amounts to denial of their cultural identity. The IWC has never explained in detail what the difference is between indigenous whaling and Japanese small coastal whaling. The persistent denial of quotas for this type of whaling remains one of the bones of contention within the IWC.[74]

---

[67] <uk.whales.org/sites/default/files/japanese-small-type-whaling.pdf>.
[68] *Ibid.*
[69] *Ibid.*
[70] *Ibid.*
[71] *Ibid.*
[72] *See* generally Fitzmaurice (n 26).
[73] IWC Resolution 2000-1, 'Resolution on Community-Based Whaling in Japan', available at <http://www.tandfonline.com/doi/abs/10.1080/13880290009353956?journalCode=uwlp20>.
[74] Butler-Stroud noted on 18 September 2014 that 'The IWC today resoundingly rejected Japan's attempt to gain a commercial whaling quota for its commercial coastal whaling operations and passed a resolution that instructs the Commission and its Scientific Committee to implement further controls on Article VIII whaling'; C Butler-Stroud, 'IWC says no to Japanese whaling', available at <http://uk.whales.org/blog/chrisbutler-stroud/2014/09/iwc-says-no-to-japanese-whaling>.

2.3 Inter-generational Equity

Inter-generational equity directly links successive generations to environmental issues. As explained by Brown-Weiss, who introduced and elaborated this concept, the use of our natural resources raises at least three kinds of equity problems between generations: depletion of resources for future generations; degradation in the quality of resources; and discriminatory access to use and benefit from resources received from past generations.[75] There is a multitude of ways in which the depletion of natural resources may occur. The present generation may deplete a more expensive natural resource, thus making it unavailable (or available only at higher price) for future generations. Natural resources may also be exploited by the present generation in ignorance of their potential economic importance. In some cases the present generation may exhaust certain natural resources by destruction of areas of high biological diversity. Depletion of resources may reduce the diversity of resources available for adapting to climate change and the destruction of forests can result in an increase of global greenhouse emissions.[76] Degradation of the quality of the environment also poses questions of equity. The quality of the natural environment (globally and locally) has certainly deteriorated, especially in the last 50 years. Pollution, which degrades many components of the environment (such as marine areas, the atmosphere, fresh water, soil, and land), affects both the uses that future generations can make of the environment and the cost of doing so.[77]

Every generation has the right to use the environment and natural resources. The main problem is finding a balance between the respective rights of the present generation and future generations, taking into account such factors as poverty, which may prevent certain communities from securing an equitable share in their legacy.[78] Inter-generational equity is formed on the basis of two relationships: the relationship between generations and that between the human species and the natural system. Humans bear the main responsibility for maintenance of the natural system. The notion of equality is at the core of the legal framework connecting generations, in their care and use of the natural system. The concept of inter-generational equity is based on a notion of a partnership between generations themselves and between natural systems and generations. The present generation holds the natural environment in trust for future generations and is also its beneficiary with the right to use it. Each generation has an obligation to pass on the natural environment to future generations in a state no worse than it had received from the past generations.[79] As noted above, Brown-Weiss distinguished three elements of the principle of inter-generational equity: conservation of diversity of natural and cultural resources (or comparable option); conservation of environmental quality (or comparable quality); and equitable or non-discriminatory

---

[75] E Brown-Weiss, 'Implementing Intergenerational Equity' in Fitzmaurice, Ong and Merkouris (n 36), 100. *See* also, by the same author, *In Fairness to Future Generations* (United Nations University and Transnational Publishers, 1989).
[76] Brown-Weiss, 'Implementing Intergenerational Equity' (n 75), 101.
[77] *Ibid.*, 101.
[78] *Ibid.*, 102.
[79] *Ibid.*

access to the earth and its resources. The first principle (conservation of diversity) means that each and every generation has an obligation to conserve the diversity of natural and cultural resources so as not to restrict the options available to future generations to meet their own needs and satisfy their own values. Conservation of quality means that every generation must maintain the quality of the natural environment so it can be passed on to future generations in no worse condition than inherited from past generations. Equitable access means that each generation has a non-discriminatory right to access and benefit from the natural environment.[80] These principles should themselves meet three criteria. First, they should encourage equality among generations, by introducing balance in the use of natural resources, i.e., using natural resources in a manner not exclusionary to future generations, on the one hand; and not imposing undue burdens upon the present generation, on the other hand. Secondly, the present generation should not interfere with or attempt to predict the values of future generations, but rather should offer flexibility to future generations with regard to how to pursue their own goals and aims. Third, these principles should be sufficiently clear as to be applicable in different social and legal systems.[81]

Neither the rights of future generations nor the notion of keeping our planet in trust for future generations are new ideas. Even MEAs and soft law documents drafted many years ago include, at least in the Preamble, the invocation of future generations.[82] The question of inter-generational equity has also been the subject of national and international case law. The classic case concerning the application of this principle is the 1993 *Minors Oposa* claim before the Supreme Court of the Philippines,[83] where an action was filed by several minors, represented by their parents, against the Department of Environment and Natural Resources to cancel existing timber licence agreements in the country and to stop issuance of new ones. The claim alleged violation of the

---

[80] *Ibid.*, 103.

[81] *Ibid.*

[82] *See*, for example, the ICRW itself (n 1): 'Recognising the interest of nations of the world in safeguarding for future generations the great natural resources represented by whale stocks ...'; CITES (n 15): 'Recognising that wild fauna and flora in their many beautiful and varied forms are an irreplaceable part of the natural systems of the earth which must be protected for this and generations to come ...'; the 1979 Bonn Convention on the Conservation of Migratory Species of Wild Animals (1980) 19 ILM 15: 'Aware that each generation of man holds the resources of the earth for future generations and has an obligation to ensure that this legacy is conserved and, where utilised, is used wisely ...'; the 1979 Convention on the Conservation of European Wildlife and Natural Habitats, ETS 104: 'Recognising that wild flora and fauna constitute a natural heritage of aesthetic, scientific, cultural, recreational, economic and intrinsic value that needs to be preserved and handed on to future generations ...'; Principle 2 of the 1972 Stockholm Declaration on the Human Environment (1972) 11 ILM 1416: 'The natural resources of the earth including the air, water, land, and fauna and especially representative samples of natural ecosystems must be safeguarded for the benefit of present and future generations through careful planning or management, as appropriate'; Principle 3 of the 1992 Rio Declaration on Environment and Development (1992) 31 ILM 874: 'The right to development must be fulfilled so as to equitably meet developmental and environmental needs of present and future generations'.

[83] *Minors Oposa v. Secretary of the Department of Environment and Natural Resources (DENR)*, Supreme Court of the Philippines, 30 July 1993, (1994) 33 ILM 173.

constitutional rights to a balanced and healthful ecology and to health (sections 16 and 15, Article II of the Constitution) and of the concept of inter-generational equity (the petitioners represented others of their generation as well as generations yet unborn). Finding for the petitioners, the Court stated that even though the right to a balanced and healthful ecology is recognised under the Declaration of Principles and State Policies of the Constitution and not under the Bill of Rights, it does not follow that it is less important than rights noted in the latter. The Court upheld both bases for the claim and stated that the petitioners enjoyed standing on behalf both of themselves and of future generations as 'the minors' assertion of their right to a sound environment constitutes, at the same time, the performance of their obligation to ensure the protection of that right for the generations to come'.

It must be mentioned, however, that national courts have not always been willing to apply this concept. In Pakistan, for example, this concept has not been applied in any case.[84] Furthermore, in Bangladesh in the 1997 case of *M. Farooque v Bangladesh and Others*,[85] the petitioners submitted that they were representatives not only of their own generation but of generations to come, relying on the *Minor Oposa* claim. The Court, however, rejected this argument, stating that in the Philippines minors had *locus standi*, since the Constitution of the Philippines grants the fundamental right to a clean environment, but that this did not exist in Bangladesh.

The concept of inter-generational equity has also been applied in international case law. In the ICJ, Judge Weeramantry has shown himself to be a great supporter of this concept, for example in the *Nuclear Test II* case.[86] The Court itself dealt with this issue in the *Nuclear Weapons Advisory Opinion*,[87] saying:

> The use of nuclear weapons could constitute a catastrophe for the environment. The Court also recognizes that the environment is not an abstraction but represents the living space, the quality of life and the very health of human beings, including generations unborn. The existence of the general obligation of States to ensure that activities within their jurisdiction and control respect the environment of other States or of areas beyond national control is now part of the corpus of customary international law relating to the environment.

> Further: 'the use of nuclear weapons would be a serious danger to future generations. Ionizing radiation has the potential to damage the future environment, food and marine ecosystem, and to cause genetic defects and illness in future generations.'[88]

There are some other examples of the mentioning of this concept before the ICJ. In the *Gabčíkovo-Nagymaros Project* case the Court also referred to future generations:

---

[84] For discussion of the case law in certain other developing countries, *see* J Razzaque, 'Human Rights and the Environment: the National Experience in South Asia and Africa', Joint UNEP-OHCHR Expert Seminar on Human Rights and the Environment, 14–16 January 2002, Geneva; Background Paper No 4.

[85] (1997) 9 DLR (AD) 1.

[86] *Request for an Examination of the Situation in Accordance with paragraph 63 of the Court's Judgment of 20 December 1974 in the Nuclear Tests (New Zealand v. France) Case*, Order of 22 September 1995, Dissenting Opinion of Judge Weeramantry (1995) ICJ Rep 317.

[87] *Legality of the Threat or Use of Nuclear Weapons, Advisory Opinion* (1996) ICJ Rep 226, para. 29.

[88] *Ibid*., para. 36.

... Owing to new scientific insights and to a growing awareness of the risks for mankind – for present and future generations – of pursuit of such interventions at an unconsidered and unabated pace, new norms and standards have been developed, set forth in a greater number of instruments during the last two decades.[89]

It may be mentioned that several Constitutions of States contain the invocation to future generations. Some authors are of the view this is the manifestation of the concept of inter-generational equity.[90] Three types of clauses relating to inter-generational justice are distinguished: general clauses;[91] ecological generational justice;[92] and financial generational justice.[93]

---

[89] *Case Concerning the Gabčíkovo-Nagymaros Project (Hungary v. Slovakia)* (1997) ICJ Rep 7, para. 140.

[90] JC Tremmel, 'Establishing Intergenerational Justice in National Constitutions' in JC Tremmel (ed), *Handbook of Intergenerational Justice* (Edward Elgar Publishing, 2006); see also very critical analysis of this concept by V Lowe, 'Sustainable Development and Unsustainable Arguments' in AE Boyle and DAC Freestone (eds), *International Law and Sustainable Development* (OUP, 1999), 19–38; and, in the same volume, AE Boyle, 'Codification of International Environmental Law and the International Law Commission: Injurious Consequences Revisited', 61–87.

[91] Tremmel, *ibid.*, 191. There are several Constitutions with such clauses: e.g. Poland, Preamble: 'Recalling best traditions of the First and Second Republic, obliged to bequeath to future generations all that is valuable from our over thousand years' heritage'; Switzerland, Preamble (to the Federal Constitution): 'In the name of God Almighty! Whereas, we are mindful of our responsibility towards creation; ... conscious of our common achievements and our responsibility towards future generations'; Estonia, Preamble: 'Unwavering in our faith and with unwavering will to safeguard and develop a state ... which shall serve to protect internal and external peace and provide security for present and future generations, the Estonian people ... adopted the following Constitution'.

[92] Again, there are numerous such clauses, for example Argentina, Article 41, clause 1: 'All inhabitants are entitled to the right to a healthy and balanced environment fit for human development in order that productive activities shall meet present needs without endangering those of future generations; and shall have the duty to preserve it. As a first priority environmental damage shall bring about the obligation to repair according to law'; Poland, Article 74, clause 1: 'Public authorities shall pursue policies ensuring ecological security of current and future generations'; South Africa, section 24: 'Everyone has the right (1) to an environment that is not harmful to their health and well-being; and (2) to have the environment protected, for the benefit of present and future generations, through reasonable legislative and other measures that (a) prevent pollution and ecological degradation, (b) promote conservation; and (c) secure ecologically sustainable development and use of natural resources while promoting justifiable economic and social development'.

[93] Examples here include Estonia, Article 116: 'proposed amendments to the national budget or to its draft, which require a decrease in income, and increase in expenditures, as prescribed in draft national budget, must be accompanied by the necessary financial calculations, prepared by the initiators, which indicate the sources of income to cover the proposed expenditures'; Poland, Article 216(5): 'it shall be neither permissible to contract loans nor provide guarantees and financial sureties which would engender a national public debt exceeding three-fifths of the value of the annual gross domestic product. The method for calculating the value of the annual gross domestic product and national public debt shall be specified by statute'; Germany, Article 109, clause 2 of the Basic Law: 'The Federation and the Länder shall perform jointly the obligations of the Federal Republic of Germany resulting from legal acts of

One of the most important aspects of the Brown-Weiss articulation of the concept of inter-generational equity was the suggestion of the appointment of an ombudsman representing the interests of future generations. At the global level, this idea appeared to be quite far-fetched, if not completely unworkable at the time. However, certain States set up special organs whose function is the representation of such interests. In 1993, the government of France established a Council for the Rights of Future Generations, whose task was to consider issues related to nuclear power. This body lapsed, however, when France resumed nuclear testing in the Pacific.[94] Another example is Israel, where the Knesset established a Commission for Future Generations, though this body is also now defunct. Its function was to assess draft Bills that were of particular relevance for future generations within the area of the environment, natural resources, development, health, the economy, planning and construction, education, quality of life, technology and all matters which were determined by the Knesset's Constitution, Law and Justice Committee as having particularly important consequences for future generations. In practice the Commission dealt with protecting children.[95] In 2007 the Hungarian Parliament enacted legislation establishing an Ombudsman for Future Generations, vested with broad powers, including advising the Parliament on the impact of certain legislation on future generations and intervening to enjoin activities that could have a detrimental impact on future generations. As of 2012 the office of the Future Generations Ombudsman has been abolished. The functions of protecting the interests of future generations are now performed by the deputy Commissioner for Protecting the Rights of Future Generations within the office of the Commissioner for Fundamental Rights with a reduced staff of four people.[96] The above examples show that there is a certain rather limited degree of concern by States regarding future generations. It may be said, however, that there are not that many examples of special organs being established and some of these no longer operate. Furthermore, the one in Israel was limited in its operation even when in existence: Shoham and Lamay note that '[t]he Commission preferred to consider future generations as the next baby to be born tomorrow morning, a definition that relates to the immediate future generation, consisting of currently existing children'.[97]

It might be concluded from the literature that the concept of inter-generational equity features very prominently in international environmental law; it may be even said that it is a leading concept. However, its precise normative content and practical application remain ambiguous. It is unclear whether it has acquired the normative content of a

---

the European Community for the maintenance of budgetary discipline pursuant to Article 104 of the Treaty Establishing the European Community and shall, within this framework, give due regard to the requirements of overall economic equilibrium' and Article 115(1): 'The borrowing of funds and the assumption of surety obligations guarantees, or other commitments that may lead to expenditures in future fiscal years shall require authorisation by a federal law specifying or permitting computation of the amounts involved'.

[94] E Brown-Weiss, 'Implementing Intergenerational Equity' (n 75), 110.
[95] Ibid.
[96] See comment on the 2011.CXI Act of the Parliament <http://intezet.greendependent.org/documents/Converge_workshop/Converge-POSTER-JNO.pdf>.
[97] S Shoham and N Lamay, 'Commission for Future Generations' in Tremmel (ed.) (n 90), 252.

principle, or remains merely a concept, or perhaps a philosophical theory (as to which, see below). Attempts to invoke it before national courts have largely proved unsuccessful; and before international courts and tribunals inter-generational equity has not acquired a great prominence. Its ambiguous legal content also raises questions of legal standing before courts and tribunals.

2.3.1 Inter-generational equity and whaling

The next question concerns how the whaling regime of the ICRW addresses the issue of inter-generational equity. One of the necessary elements of implementing the principle of inter-generational equity is the sustainable use (or utilisation) of resources. Sustainable use is undoubtedly one of the constitutive elements of the concept of sustainable development. Of particular relevance in this respect is, therefore, the analysis of the conservation methods adopted by the IWC, particularly regarding the three types of whaling – commercial, scientific, and indigenous – in light of the preservation of whale stocks for future generations. One of the principles on which Brown-Weiss' theory of inter-generational equity is based is the conservation of options according to which future generations should be entitled to diversity comparable to that enjoyed by previous generations.[98] Therefore it can be said that the preservation of whale stocks should aim to implement this principle.

In this context, mention must be made of the 'Future of the IWC' process, which was established at the IWC's 59th Annual Meeting in 2007 and aimed at addressing the main issues faced by the IWC. The majority of this work was completed by the Commission's 62nd Annual Meeting in 2010. The main result was the tabling of a document called 'A Proposed Consensus Decision to Improve the Conservation of Whales'. This document was discussed at the 62nd Annual Meeting, but ultimately the Chair concluded that the Commission was not in a position to come to a consensus on the measures proposed.

In 2011 the Commission stated its desire to maintain the progress achieved through the Future of the IWC process and agreed to:

- Encourage continuing dialogue amongst Contracting Governments regarding the future of the International Whaling Commission;
- Continue to build trust by encouraging Contracting Governments to coordinate proposals or initiatives as widely as possible prior to their submission to the Commission;
- Encourage Contracting Governments to continue to co-operate in taking forward the work of the Commission, notwithstanding their different views regarding the conservation of whales and the management of whaling.[99]

2.3.1.1 Conservation methods under the ICRW[100] The first method used by the IWC for annual catch limits was the so-called 'blue whale unit', the aim of which was to regulate the total number of whales to be hunted every year. It was based on the amount of oil a species could produce in comparison with other species; thus, one blue whale unit equated to one blue whale, two fin whales, two and a half humpback whales

---

[98] See E Brown-Weiss, 'Implementing Intergenerational Equity' (n 75), 108.
[99] <http://iwc.int/future>.
[100] For detailed discussion, see Davies (n 3), 164–168.

or six sei whales.[101] This system did not prove to be very successful in protecting whale stocks, especially when the focus of whaling shifted from hunting them for oil to meat products.[102] During its first 15 years, the Schedule did very little to protect whales and the hunting of whales further increased during the early 1960s; the number killed in the 1960–61 season reached approximately 60,000.[103]

The blue whale unit was abandoned and in 1976 the IWC adopted a new system, the New Management Procedure (NMP), which was based on a division of each species into approximately 20 different stocks, with quotas set on a stock-by-stock basis. Each stock was classified as 'initial management stock', 'sustained management stock' or 'protection stock'. This classification depended on the relationship between the population level of the stock and the level of its maximum sustainable yield. The hunting of protection stocks was prohibited altogether and the two other categories were to be hunted in a sustainable manner.[104] The NMP was also disappointing from the point of view of protection of whale stocks, as it proved ineffective in establishing the maximum sustainable yield and stock sizes, above all because the IWC did not have accurate and reliable data on whale population levels. These were the reasons underlying the introduction of the moratorium (entailing zero quotas) on commercial whaling from the 1986 coastal and 1985–86 pelagic whaling seasons.[105] Unfortunately, due to the opting-out (tacit acceptance) procedure, States that registered objections to the decision on introducing a moratorium, such as Norway, are not bound by it. The Soviet Union (now Russia) also objected to the moratorium and has not withdrawn its objection. However, Russia has not actually conducted a commercial whale hunt since the 1986/87 season. Iceland did not initially object to the IWC moratorium, but left the IWC in 1992. It re-joined in 2002 with a reservation which declared its opposition to an extension of the moratorium beyond a reasonable time frame and stipulated that its further (unfounded) extension would result in the resumption of commercial whaling by Iceland, which happened in 2006.[106]

The direct result of the moratorium was the introduction of a 'comprehensive assessment' of whale stocks by the Scientific Committee, which included the development of a Revised Management Procedure (RMP) to replace the NMP. This was adopted by the IWC in 1994, but has not yet been implemented, as a comprehensive

---

[101] Paragraph 8(b) of the 1949 Schedule (published in 1950), 16.
[102] Davies (n 3), 164.
[103] *Ibid.*
[104] *Ibid.*,165.
[105] The moratorium was introduced by way of an amendment to the Schedule: 'Notwithstanding the other provisions of paragraph 10, catch limits for the killing for commercial purposes of whales from all stocks for the 1986 coastal and the 1985/86 pelagic season and thereafter will be zero. This provision will be kept under review, based upon scientific advice, and by 1990 at the latest the Commission will undertake a comprehensive assessment of the effects of this decision on whale stocks and consider modification of this provision and the establishment of catch limits'.
[106] <http://animalrights.about.com/od/wildlife/a/Japan-Iceland-Norway-And-Commercial-Whaling_2.htm>. On Iceland's reservation, *see* A Gillespie, 'Iceland's Reservation at the International Whaling Commission' 14(5) EJIL (2003), 977.

Revised Management Scheme (RMS) must first be adopted.[107] The RMP takes into account uncertainties as to the size of whale stocks and also imperfect data on certain issues such as environmental changes.[108] The RMP is aimed at 72 per cent recovery of the initial level of all whale stocks. If stocks fall below 54 per cent of their initial level, their exploitation will be prohibited and no catch limits will be set.[109] As Davies observes, the 'RMP offers some hope that the regulation of whaling will be more effective if commercial whaling ever resumes in the future'.[110] The re-introduction of commercial whaling may not be supportive of intra-generational equity as there is no real current need for whale products and many opportunities for non-consumptive exploitation; the IWC has noted that 'the governance regime for whales has contributed to more sustainable practices and a change in mind-sets, allowing a transition from predominantly consumptive exploitation of a natural resource (whaling) to non-consumptive use such as whale watching and related tourism'.[111]

The principal drawback of the RMP is its reliance on annual national reports, which as Davies notes, are not always reliable. Those of the Soviet Union, for example, did not reflect actual catches for the period 1947–1972, excluding altogether its catch in the southern hemisphere, which totalled over 100,000 in this period.[112] The RMS Working Group was established in 1994 to deal with the following issues: an effective inspection and observation scheme; arrangements to ensure that total catches over time are within the limits set under the RMS; incorporation into the Schedule of the specification of the RMP, and all other elements of the RMS.[113]

The development of the RMS has proved very tortuous. In July 2000, the IWC adopted Resolution 2000-3. This Resolution recognised that it is important for the future of the Commission that the process to complete the RMS proceeds expeditiously. In 2001, during the Monaco meeting, some progress was made on revising the section of the Schedule that deals with supervision and control and developing a text to incorporate the structure and elements of the RMS, including the RMP, into the Schedule. The focus of discussions at the meeting was on the development of an effective inspection and observation scheme. There are, however, numerous issues which are outstanding in this regard: the level of international observer coverage required, the type and level of tracking of whaling vessels required, the timing (e.g. daily, weekly) of reporting of whales hunted, struck and killed, maintenance and availability of a register of DNA profiles of all whales killed, procedures to monitor the origins of whale products on the market, and the funding of the scheme.[114] In the 2006

---

[107] IWC Resolution 1994-5, Report of the International Whaling Commission. 45:43-4, 1995; Davies (n 3), 166–167.
[108] *Ibid.*, 167.
[109] *Ibid.*
[110] *Ibid.*
[111] Opening Statement to the 65th Meeting of the International Whaling Commission, September 2014, available at <http://www.hsi.org/assets/pdfs/iwc_65_hsi_opening_statement.pdf>.
[112] Davies (n 3), 167.
[113] <https://iwc.int/rmp>.
[114] Davies (n 3), 168.

Annual Meeting, the IWC accepted that an impasse had been reached at the Commission level on RMS discussions.[115]

In conclusion it may be said that the conservation methods used by the IWC are not fully fostering inter-generational equity. The RMP and RMS are not yet operational and due to unbridgeable differences between the member States of the IWC it is unlikely that they will be implemented in the foreseeable future. The introduction of the moratorium on commercial whaling, without doubt, bolstered the maintenance of whaling stocks for future generations. It may also be noted, however, that the lack of a coherent policy and cooperation between the member States of the IWC is one of the factors interfering with attempts to introduce long-lasting conservation methods for preserving whale stocks for future generations that would be capable of accommodating the interests of all States.

2.3.1.2 Scientific whaling Scientific whaling is one of the most divisive issues concerning whaling, so much so that it was ultimately referred for the consideration of the ICJ. The ICRW regulates this type of whaling in Article VIII.[116] The main risk is that it may prove unsustainable and thus in breach of the principles of inter-generational equity. According to the Convention, governments of States Parties are allowed to grant special permits to their nationals for the purpose of scientific whaling on a purely unilateral basis. Therefore, such whaling falls almost outside the jurisdiction of the IWC, with the exception of the powers of the IWC Scientific Committee to review the permits and comment on them.[117] This is a very weak provision, which does not authorise the IWC to approve or ban the granting of such a permit.[118] There are several States that carry out scientific research, namely Iceland, Japan and Norway, though South Korea has abandoned its plans to commence scientific whaling.

Japan has issued scientific permits in the Antarctic and in the western North Pacific every year in recent years, following a two-year feasibility study that commenced in 1987/88. The 2004/05 Antarctic season was the final year of the 16-year JARPA programme. The objectives of JARPA were as follows: estimation of biological parameters (especially the natural mortality rate) to improve management; elucidation of stock structure to improve management; examination of the role of whales in the Antarctic ecosystem; and examination of the effect of environmental changes on cetaceans.

JARPA II commenced during the austral summer of 2005/06 and was implemented in full from 2009/10, following two seasons of feasibility studies. The objectives for

---

[115] *Ibid.*

[116] *See* A Gillespie, 'Whaling under a Scientific Auspice: The Ethics of Scientific Whaling Operations' (2000) 3 JIWLP 1.

[117] Para. 30 of the Schedule. The Scientific Committee's review concentrates on whether (1) the permit adequately specifies its aims, methodology and the samples to be taken; (2) the research is essential for rational management, the work of the Scientific Committee or other critically important research needs; (3) methodology and sample size are likely to provide reliable answers to the questions being asked; (4) the questions can be answered using non-lethal research methods; (5) the catches will have an adverse effect on the stock; (6) the potential for scientists from other nations to join the research is adequate: *see* <http://iwc.int/permits>.

[118] Davies (n 3), 175.

JARPA II differ from those for JARPA and were defined by Japan as (1) monitoring of the Antarctic ecosystems; (2) modelling competition among whale species and developing future management objectives; (3) elucidation of temporal and spatial changes in stock structure; and (4) improving the management procedure for Antarctic minke whale stocks. The permit issued was for 850 Antarctic minke whales, 50 fin whales and 50 humpback whales annually, though in practice Japan has refrained from actually taking humpback whales.[119] Japan also carried out a research permit programme known as JARPN (for 100 minke whales per year) in the western North Pacific from 1994 to 1999. The aims of this programme were to clarify questions of stock identity in order to improve the design of RMP Implementation Simulation Trials for the North Pacific and to act as a feasibility study for the development of a programme on feeding ecology. The Scientific Committee agreed that the information obtained was useful for management and will continue to be used in the refinement of Implementation Simulation Trials for North Pacific common minke whales. However, no consensus view was reached on whether the results could have been obtained using non-lethal research techniques in a suitable time frame.[120] The value of the JARPA programmes' research proved extremely contentious within both the Scientific Committee and the Commission – in particular the relevance of the proposed research to management, appropriate sample sizes and applicability of alternate (non-lethal) research methods.[121] In 2007 the IWC adopted a Resolution asking Japan to suspend indefinitely lethal methods of research under JARPA II in the Southern Ocean Whale Sanctuary.[122]

Japan is not alone in conducting such research. The scientific research programme of Iceland (2003–07) stated its overall objective to be increasing understanding of the biology and feeding ecology of important cetacean species in Icelandic waters for improved management of living marine resources based on an ecosystem approach. The intention of the programme was also strengthening the basis for conservation and sustainable use of cetaceans and at the same time making a contribution to multi-species management of living resources in Icelandic waters.[123] The original research programme had multiple specific objectives among which the order of priority differs between whale species.[124] However, in practice, the government of Iceland only issued permits for the common minke whale segment of the original proposal. A total of 200 common minke whales were caught from 2003 to 2007 as originally proposed. The proposal of Iceland raised some doubts concerning its purpose and some of the IWC members held the view that the most important research objective should be the effect of pollutants on whale stocks. The other most divisive issue was the use of lethal

---

[119] <http://iwc.int/permits>.
[120] *Ibid.*
[121] *Ibid.*
[122] IWC Resolution 2007-1.
[123] <http://iwc.int/permits>.
[124] *Ibid.* For common minke whales the primary specific objective was to increase the knowledge of the species' feeding ecology in Icelandic waters. For fin and sei whales, the primary specific objective was the study of biological parameters during the apparent increase in population size in recent decades. These objectives were the basis for the proposed sample sizes. Other research objectives include studies of population structure, pollutants, parasites and pathogens, and the applicability of non-lethal methods.

methods, since certain members of the IWC were not at all convinced that the use of such methods was absolutely necessary for Iceland to conduct its research.[125] There is also uncertainty as to the sustainability of the methods (lethal or non-lethal) of scientific whaling. As Iceland explains, first a feasibility study must be conducted in order to choose the method.[126]

Scientific whaling raises at least two legal questions. It has been Japanese practice to sell meat products obtained from scientific research for restaurants. This practice, which was noted by the IWC in its Resolution,[127] has been criticised as being merely a means for Japan to bypass the moratorium on commercial whaling. Such practice is, however, allowed under Article VIII(2) of the ICRW which states that '[a]ny whales taken under these special permits shall so far as practicable be processed'. Therefore the sale of whale products derived from scientific whaling, unlike the sale of products obtained from aboriginal (subsistence) whaling, is not per se illegal. Interestingly, however, during the hearing in the whaling case, Japan's deputy foreign affairs minister, Koji Tsuruoka, told the Court that Tokyo was conducting a 'comprehensive scientific research program with the aim of demonstrating that commercial whaling could be sustainable'.[128] Therefore, it appears that Japan added a new element to the conduct of its scientific whaling with this latter aim regarding commercial whaling.

The Judgment of the Court on whaling by Japan in the Antarctic rendered on 31 March 2014, has added an important new dimension to the issue of inter-generational equity.[129] The Court held Japan responsible for three breaches of the ICRW regarding

---

[125] Ibid.

[126] 'For the scientific results to be statistically relevant there is a need for a certain sample size. The necessary sample size depends inter alia on each particular research objective. For studies of descriptive nature, like diet composition of minke whales in Icelandic waters, it is very difficult to determine statistically a priori the required sample size as no previous studies have been conducted in this area as this depends largely on spatial and temporal variability in feeding habits. The research proposal is for a so-called feasibility or pilot study, intended to serve as a basis for future research. The nature of these future studies, including the extent to which lethal and non-lethal methods will be applied, will depend on the outcome of these feasibility studies. Iceland's plan does not suggest taking more whales than is needed for making sure the results are scientifically significant. In fact, in 2003 the take will be less than was envisaged for the first year of research in the original scientific plan, partially due to a later start of the programme': 'Iceland and whaling for scientific purposes', available at <http://uk.whales.org/issues/whaling-in-iceland>.

[127] It was observed that whales killed in the research programme by Japan provided over 3,000 tons of product sold commercially (IWC Resolution 2007-1).

[128] He said that 'The lifting of the [1982] moratorium requires that convincing scientific data be presented'; 'Japan slams anti-whaling stance as "emotional crusade"', available at <http://www.nzherald.co.nz/world/news/article.cfm?c_id=2&objectid=10894494>.

[129] *Whaling in the Antarctic (Australia v. Japan: New Zealand Intervening)*, Judgment of 31 March 2014, at <http://www.icj-cij.org/docket/files/148/18136.pdf> For discussion, see CR Payne, 'Australia v. Japan: ICJ Halts Antarctic Whaling', ASIL Insights, at <http://www.asil.org/insights/volume/18/issue/9/australia-v-japan-icj-halts-antarctic-whaling>; MR Rahman, 'Battle for Whales in The Hague: Analysis of ICJ Judgment in Australia v. Japan' at <http://papers.ssrn.com/sol3/papers.cfm?abstract_id=2418817>.

scientific whaling under JARPA II,[130] namely: the moratorium on all commercial whaling; the moratorium on use of factory ships to process whales; and (insofar as the taking of fin whales was concerned) the prohibition on whaling in the Southern Ocean Sanctuary.[131] However, Japan was not found to be in breach of the procedural requirements regarding scientific whaling, as prescribed by paragraph 30 of the Schedule. As a consequence of the Judgment, Japan is under an obligation to revoke any extant authorisation, permit or licence to kill, take or treat whales in relation to JARPA II and to refrain from granting any further permits in implementation of this programme. The Court noted a lack of transparency in the reasons for selecting particular sample sizes for individual research items (in particular in selection of Antarctic minke whales).[132] The Court came to the conclusion that although Article VIII of the ICRW exempts from the Convention the grant of special permits, scientific whaling is not outside of the Convention altogether and therefore the 'margin of appreciation' enjoyed by members of the IWC in such a type of whaling as pleaded by Japan is not unlimited and must conform with an objective standard based upon the requirement of good faith.[133] The Court was of the view that JARPA II could in principle be characterised as scientific research, based on its objectives, and it did not condemn lethal methods altogether. However, it found that important aspects of JARPA II's design and implementation were not reasonable in relation to its research objectives and put in question several issues relating to this programme. Most importantly, the

---

[130] Australia had accused Japan of a variety of breaches to the Whaling Convention:

35. In proposing and implementing JARPA II, Japan has breached and is continuing to breach its international obligations.
36. In particular, Japan has breached and is continuing to breach the following obligations under the ICRW:
(a) the obligation under paragraph 10 (e) of the Schedule to the ICRW to observe in good faith the zero catch limit in relation to the killing of whales for commercial purposes; and (b) the obligation under paragraph 7 (b) of the Schedule to the ICRW to act in good faith to refrain from undertaking commercial whaling of humpback and fin whales in the Southern Ocean Sanctuary.
37. Moreover, having regard to the scale of the JARPA II program, to the lack of any demonstrated relevance for the conservation and management of whale stocks, and to the risks presented to targeted species and stocks, the JARPA II program cannot be justified under Article VIII of the ICRW.

Further, it was alleged that Japan had breached the following: '(a) under the Convention on International Trade in Endangered Species of Wild Fauna and Flora ("CITES"), the Fundamental Principles contained in Article II in relation to "introduction from the sea" of an Annex I listed specimen 1lother than in "exceptional circumstances", and the conditions in Article III (5) in relation to the proposed taking of humpback whales under JARPA II; and (b) under the Convention on Biological Diversity, the obligations to ensure that activities within their jurisdiction or control do not cause damage to the environment of other States or of areas beyond the limits of national jurisdiction (Article 3), to co-operate with other Contracting Parties, whether directly or through a competent international organization (Article 5), and to adopt measures to avoid or minimize adverse impacts on biological diversity (Article 10(b)).'

[131] Payne (n 129).
[132] Para. 188 of the Judgment.
[133] Para. 62 of the Judgment.

Court emphasised the lack of Japan's willingness to cooperate with the IWC in the use of non-lethal scientific methods which became available in the intervening years. The Court interpreted Article VIII of the ICRW and found that Japan had not sufficiently substantiated the scale of lethal sampling. The Judgment does not, however, preclude all future scientific whaling on Japan's part; indeed, the Court noted that Japan will rely on the Judgment's findings 'as it evaluates the possibility of granting any future permits under Article VIII, paragraph 1, of the Convention'.[134]

The question now arises as to how this Judgment contributes to the furtherance of the concept of inter-generational equity. At first blush it might appear that the Judgment is a milestone in protecting the rights of future generations. The Court in painstaking manner investigated the legal and scientific aspects of the case (JARPA II). However, from the point of view of a more general analysis (which of course the regime for future generations should reflect), the Judgment is, in the view of the present author, of only limited value. The Judgment only relates to one programme of scientific whaling (i.e., to Japanese Antarctic whaling, and not to Japanese Pacific whaling or Icelandic scientific whaling), and has no relevance at all to the key and complex issue of commercial whaling. The increasing demand for whaling quota allocations by indigenous peoples is also an issue of contention between States that are parties to the IWC. Therefore a judgment concerning specific issues, which form only a part of a very complicated picture of contemporary whaling, will not influence the rights of future generations to any significant degree.

In accordance with its judicial function, the Court restricted itself to the determination of the dispute as presented to it, and as a result did not find it necessary to formulate any definition of scientific whaling or to adopt a specific set of criteria. It declined to discuss either commercial whaling or indigenous whaling. It made some general observations linking scientific whaling to the whole nexus of rights and obligations of States under the ICRW and clarified the issue of the margin of appreciation regarding the issuance of special permits; both constitute useful observations but leave much to future implementation. It also admitted the possibility of future Japanese whaling, which cannot indeed be prohibited. In the meantime, Japanese scientific whaling in the north-western Pacific is to continue, and a revised plan for the Antarctic has been proposed.[135] The Icelandic programme is also to continue. Japan announced it would resume its whaling programme in the Antarctic at the IWC meeting in 2014. Japan stated that it would amend its programme in order to make it more scientific and allow hunting to continue. That will most likely mean the quotas will be reduced from about 1,000 to a few hundred.[136] The Judgment has not resolved the underlying conflicts and has not addressed general issues and therefore its contribution

---

[134] Para. 246 of the Judgment.

[135] 'This announcement is a huge disappointment and flies in the face of the UN's International Court of Justice ruling last month' said Greenpeace Japan's Executive Director Junichi Sato: see 'Japan to redesign Antarctic whale hunt after UN court ruling' at <http://news.yahoo.com/japan-continue-scientific-whaling-pacific-reports-054921014.html>.

[136] 'Japanese whaling operations in Antarctic to resume after six-month stoppage', available at <http://www.abc.net.au/news/2014-09-18/japan-to-resume-whaling-in-antarctic-waters/5752476>.

to implementing the concept of inter-generational equity cannot be said to be of a fundamental importance.[137]

2.3.1.3 Sanctuaries By virtue of Article V(1)(c) of the ICRW, the IWC can designate certain marine areas as sanctuaries within which whaling is prohibited. Sanctuaries are designed to provide whales with a refuge allowing species to recover from the serious over-exploitation that occurred throughout much of the 20th century. According to the Australian government their aim is to benefit long-term whale conservation by:

> facilitating the recovery of seriously depleted great whale populations by protecting important areas such as feeding or breeding grounds and migratory routes; providing economic benefits to range states by providing opportunities to develop non-lethal economic uses of cetacean species (such as ecotourism and whale watching); fostering interest and cooperation in non-lethal research into the behaviour and biology of cetacean species; providing the Commission with a broad management tool to protect multiple species; setting aside a place where whales can play their important role in the ecosystem; increasing public awareness and appreciation of the value and vulnerability of marine ecosystems.[138]

It is particularly notable that this statement by the government of Australia makes reference only to scientific research being conducted in sanctuaries utilising non-lethal methods. The first international whale sanctuary was established in the Antarctic region in 1938 and was only intended to maintain unexploited stock levels for potential future whaling. Although the sanctuary had been created prior to the establishment of the IWC, the latter agreed to maintain it. The area was opened to commercial whaling in 1955 when the sanctuary was closed.[139]

There are now two IWC sanctuaries still in place: the Indian Ocean Sanctuary and the Southern Ocean Sanctuary. The Indian Ocean Sanctuary was proposed by the Seychelles and currently prohibits commercial whaling throughout the Indian Ocean as it extends north from 55 degrees southern latitude. Its status is reviewed every 10 years.[140] The Southern Ocean Sanctuary was established in 1994 and prohibits commercial whaling in Antarctic waters in the area south of 40 degrees southern latitude apart from in the Indian Ocean, where the Southern Ocean Sanctuary joins up with the Indian Ocean Sanctuary at 55 degrees southern latitude, and in the area around South America and parts of the South Pacific, where the sanctuary follows the Antarctic convergence at 60 degrees southern latitude. This sanctuary's status is also reviewed every 10 years.[141] Japan lodged an objection to the Southern Ocean Sanctuary in relation to the taking of minke whales, which means that, if the commercial whaling moratorium is lifted in the future, Japan will be able to harvest minke commercially in Antarctic waters even if the Southern Ocean Sanctuary remains in place. Japan's

---

[137] Though some of its observations shed light on this indirectly – for example, in determining whether the catch permits were issued 'for purposes of scientific research', it noted the paucity of peer-reviewed scientific papers that emerged from the JARPA II programme.
[138] <http://www.environment.gov.au/coasts/species/cetaceans/international/iwc.html>.
[139] <https://www.google.co.uk/#q=The+area+was+opened+to+commercial+whaling+in+1955+when+the+sanctuary+was+closed>.
[140] *Ibid.*
[141] *Ibid.*

justification for this lies in the fact that an amendment to the Schedule is needed to establish a sanctuary. By virtue of Article V(2)b any such amendment 'shall be based on scientific findings'. Japan argued that the scientific basis for the establishment of both current sanctuaries was lacking; a claim which has been contested by New Zealand.[142]

There have been attempts to establish further sanctuaries. Brazil, Argentina and South Africa first proposed the establishment of a South Atlantic Whale Sanctuary in 2001 for the region south of the equator extending from South America to Africa with a view to guaranteeing protection for all the large whales of the region in their breeding, calving and feeding grounds. The sanctuary was intended to stimulate coordinated research in the region and to develop the sustainable and non-lethal economic use of whales. In addition, from 2000 to 2004 Australia and New Zealand put forward to the IWC a proposal for a South Pacific Whale Sanctuary, to prohibit commercial whaling and foster cooperative research on cetaceans and increase public awareness and understanding. Neither proposal succeeded in gaining the three-quarters majority required for adoption. Since 2004, Australia and New Zealand have decided to continue to seek support for the proposal for a South Pacific sanctuary but have not put the issue to a vote. They have also continued to raise awareness of efforts in the South Pacific to improve the protection of cetaceans.[143]

There are also national whale sanctuaries, such as the Australian Whale Sanctuary, which has been established to protect all whales and dolphins found in Australian waters. Under the Environment Protection and Biodiversity Conservation Act 1999 (EPBC Act), all cetaceans (whales, dolphins and porpoises) are protected in Australian waters. Within the Sanctuary it is an offence to kill, injure or interfere with a cetacean and severe penalties apply to anyone convicted of such offences. The Australian Whale Sanctuary comprises the Commonwealth marine area, an area beyond the coastal waters of each state and the Northern Territory. It includes all of Australia's Exclusive Economic Zone (EEZ).[144]

The contentious character of whaling within the Southern Ocean Sanctuary has been evidenced by Australia's aforementioned initiation of proceedings against Japan before the ICJ in 2010 in the *Whaling in the Antarctic* case (with New Zealand intervening).[145] Australia emphasised in its Request to the Court the inter-generational emphasis of the ICRW:

> Under Article VI of the ICRW the IWC may from time to time make recommendations to any or all parties on any matters which relate to whales or whaling and to the objectives and purposes of the ICRW which include, first and foremost, 'safeguarding for future generations the great natural resources represented by the whale stocks'.[146]

---

[142] Annual Report, International Whaling Commission, 2001.
[143] <http://www.scoop.co.nz/stories/PA0107/S00369/iwc-asked-to-support-south-pacific-sanctuary.htm>.
[144] <http://www.environment.gov.au/coasts/species/cetaceans/conservation/sanctuary.html>.
[145] Application Instituting Proceedings by Australia, available at <http://www.icj-cij.org/docket/files/148/15951.pdf>.
[146] *Ibid.*, para. 17.

The establishment of whale sanctuaries by amendment of the Schedule can be said to be in line with the inter-generational element of the Whaling Convention's object and purpose as sanctuaries help to maintain whale stocks for future generations by the prohibiting of commercial whaling in these designated areas. However, an unresolved question is the legality of lethal scientific research within sanctuaries (such as any future research by Japan in the Southern Ocean under a revamped research programme). If such research is permitted, the taking of whales in significant numbers under scientific permit may contribute to the depleting of whaling stocks, and thus impact adversely on the rights of future generations. On the other hand, it may be argued that the aims of intra-generational equity would be implemented if certain whale catches were permitted. In such an event whale meat might provide much-needed food for developing countries.

2.3.1.4 Small and medium-size cetaceans and the jurisdiction of the IWC The mandate of the IWC regarding small to medium cetaceans has been a long-standing issue,[147] and arises from definitional problems relating to the classification of cetaceans. The biological order in question is commonly divided into two groups, comprising large and small cetaceans. However, a difficulty lies in the allocation of individual species to either category, as there are no accepted definitions that would combine both biological and political aspects.[148] An alternative distinction is between toothed and baleen species: the larger species are almost all baleen whales, though the sperm whale is a notable exception. The original Annex of Nomenclature attached to the Final Act of the 1946 ICRW did not explicitly adopt either distinction, merely listing certain species with their scientific and popular names. However, the Schedule to the ICRW, which has been amended over time, is based on the assumption that the toothed and baleen whales are different sub-orders.[149] In the intervening years the ongoing debate in the IWC regarding the cetacean taxa that fall within its mandate had become more of a political issue than one based on biological definitions. The debate regarding the IWC's mandate over the management of small cetaceans is certainly not settled. In the early 1970s there was a strong recommendation by the Scientific Committee that the IWC review the issues concerning the conservation and management of small cetaceans and a Special Sub-committee on small cetaceans was established. The fundamental issues regarding the IWC's jurisdiction were defined in more precise terms at the 32nd meeting of the IWC: namely, whether the ICRW bestowed competence upon the IWC to manage small cetaceans; the possible conflict with coastal States' jurisdiction; and the role of regional organisations in the administration of small cetaceans.[150] The first issue (and to an extent also the second as will be noted) relates to the interpretation of the Convention and related documents. The ICRW

---

[147] *See* A Gillespie, 'Small Cetaceans, International Law and the International Whaling Commission' (1997) 12 International Journal of Marine and Coastal Law 257, and A Gillespie, *Whaling Diplomacy: Defining Issues in International Environmental Law* (Edward Elgar Publishing, 2005).
[148] Gillespie, *Whaling Diplomacy, ibid.*, 277.
[149] *Ibid.*, 278.
[150] *Ibid.*, 282–289.

refers to a 'whale' in general terms, without a distinction between large and small cetaceans.[151] The objections of some States regarding the competence of the IWC to manage small cetaceans is based on the lack of specific reference to such small cetaceans in the Annex on Nomenclature. Some States have suggested that if the type of whale in question was not originally listed in this Annex its regulation lies outside the jurisdiction of the IWC (unless otherwise agreed by the Parties).[152] Many States, including the UK, however, have taken the view that the original Annex on Nomenclature cannot be used as a limiting factor to the jurisdiction of the IWC. It is submitted that such documents should indeed be treated only as guidance to the Parties, not as exhaustive listings. Practice can evolve over time, as has been noted by the 10th Conference of the Parties of the Convention on Biological Diversity.[153] However, a host of States, including Mexico, Norway and Russia, have strongly objected to this approach.

Gillespie argues that whether a type of whale falls within the regulatory competence of the IWC depends on whether it is placed in the Schedule by a three-quarters majority of its voting members.[154] In the view of the present author, however, even if the IWC voted in favour of the inclusion of small cetaceans in the Schedule in this way, the theoretical question remains as to whether the Convention actually allows for such inclusion and for the exercise of IWC competence. It might be argued (as, for example, Norway and Japan have done) that the current wording of the Schedule, which does not expressly include reference to small cetaceans, thereby excludes IWC jurisdiction over small cetaceans altogether.

There have been several Resolutions of the IWC (e.g. the 'Berlin Initiative') which have called on the Parties to recognise the IWC as the relevant body to govern small cetaceans, but these resolutions can only be regarded as amounting to 'soft law'. There are also other conventions, such as the Agreement on the Conservation of Small Cetaceans in the Baltic, North East Atlantic, Irish and North Seas (ASCOBANS), which are aimed specifically at the regulation of small cetaceans, and which might be regarded as providing more effective regulation than the IWC. However, their geographical scope is limited and it could certainly be argued that small cetaceans would benefit from being regulated by the one body which has international competence, namely the IWC. Despite continuing disagreement among the Parties as to whether the IWC enjoys regulatory competence over small cetaceans, it has at least been agreed by the latter that international cooperation is required to conserve them.

The second issue concerning the jurisdiction of the IWC over small and medium-sized cetaceans relates to the provisions of the 1982 United Nations Law of the Sea

---

[151] See the preamble to the Convention and its Article V where it refers to 'whales' generically rather than identifying individual species.

[152] Gillespie, *Whaling Diplomacy* (n 147), 288.

[153] Decisions of the Conference of the Parties: 'Ecosystems, ecological processes within them, species variability and genetic variation change over time whether or not they are used. Therefore, governments, resource managers and users should take into account the need to accommodate change, including stochastic events that may adversely affect bio-diversity and influence the sustainability of a use'; available at <https://www.cbd.int/doc/handbook/cbd-hb-dec-en.pdf>.

[154] Gillespie, *Whaling Diplomacy* (n 147), 289.

Convention (UNCLOS).[155] The Whaling Convention applies to 'all waters'[156] in which whaling is carried out by factory ships, land stations or by whale catchers, and therefore very arguably not only applies to the high seas, but also the territorial sea and the Economic Exclusive Zone (EEZ). However, under Article 56(1)a of UNCLOS, States enjoy sovereign rights 'for the purposes of ... exploiting, conserving and managing the natural resources, whether living or non-living' within the EEZ. The concept of the EEZ is regarded as being of customary status under international law, and it is of importance to note that small cetaceans are very likely to frequent the coastal waters of this area.[157] The question arises whether the UNCLOS regime supersedes the regime established by the ICRW on this jurisdictional point. In several resolutions adopted by member states of the IWC, States have objected to its jurisdiction within EEZs and have treated any jurisdiction in such zones as ultra vires. It is submitted that, if a specific Party could be said to have agreed to cede their sovereignty to the IWC allowing the latter to regulate the taking or to manage the conservation of whales within coastal waters, the issue of IWC general competence would certainly be resolved in its favour. The majority of Parties have indeed accepted IWC competence within their 200-mile coastal zones (though not necessarily in relation to the regulation of small cetaceans), although certain Latin American countries refuse to acknowledge the IWC's general competence in such areas.[158] Gillespie makes reference to Article 65 UNCLOS in which the wording 'competent organisation' is regarded by him as making reference to the IWC, thereby making the IWC 'the central and uppermost international authority for cetaceans'.[159] According to the same author any ambiguity as to what is the competent international organisation was resolved by Chapter 17 of Agenda 21, which not only reiterates Article 65 UNCLOS, but specifically refers to the IWC as studying larger whales as well as other cetaceans, and working with other organisations in the area of management and conservation of cetaceans and other marine mammals. In conclusion it might be said that the controversy as to whether the IWC enjoys competence to conserve and manage small cetaceans has not contributed to the promotion of inter-generational equity in that it has resulted in a lack of effective regulation and near extinction of small cetaceans, such as the narwhal and beluga whale. Indeed, the Japanese continue to take certain small cetaceans within coastal waters and this may, to a certain extent, be said to have a negative impact on the promotion of inter-generational equity in relation to the preservation of such cetaceans.

2.3.1.5 Pirate whaling There has in the past been a great amount of 'pirate whaling' or whaling outside the jurisdiction of the IWC, including by certain indigenous

---

[155] Entered into force 1994, 1836 UNTS 42.
[156] Article I(2).
[157] Gillespie, *Whaling Diplomacy* (n 147), 289–297.
[158] Davies (n 3), 155.
[159] Gillespie, *Whaling Diplomacy* (n 147), 292. Article 65 of the UNCLOS: 'Nothing in this Part restricts the right of a coastal State or the competence of an international organization, as appropriate, to prohibit, limit or regulate the exploitation of marine mammals more strictly than provided for in this Part. States shall co-operate with a view to the conservation of marine mammals and in the case of cetaceans shall in particular work through the appropriate international organizations for their conservation, management and study.'

peoples. Although pirate whaling peaked during the 1950s and 1960s,[160] there was still evidence of such activities in the 1980s.[161] The IWC adopted a Resolution in 1979 in which it called on Parties to take a number of measures in this regard: to cease buying whale products from non-Parties; not to allow vessels of such States to hunt for whales in waters under their national jurisdiction; and to ban the transfer of any whaling vessels and the export of equipment used in whaling operations.[162] The implementation of this Resolution, together with the highlighting of the 'pirate whaling' issues by NGOs, certainly significantly reduced the impact of such activities. However, a considerable amount of whale hunting is still carried on outside the purview of the IWC by indigenous people in States such as Equatorial Guinea, Indonesia, the Philippines and Canada.[163] Great difficulties exist in assessing the impact of such whaling on whale populations owing to a lack of accurate data on such operations. It should be noted that such States may well regulate aboriginal whaling unilaterally, but prefer to avoid the coordinated processes conducted within the IWC in this regard.[164]

Pirate whaling is potentially one of the most detrimental activities impacting upon the preservation of whaling stocks, and therefore undermining the promotion of inter-generational equity. As mentioned above, the lack of relevant data on catches is one of the factors which make an assessment of the true impact of such activities on whale stocks very problematic.

2.3.1.6 The 1992 North Atlantic Marine Mammal Commission (NAMMCO) NAMMCO was established on the basis of a 1992 Agreement signed in Nuuk by Norway, Iceland, Greenland and Faroe Islands.[165] Its aim is to study ecosystems as a whole and to understand better the role of marine mammals in such ecosystems. More specifically, it aims to 'contribute through regional consultation and cooperation to the conservation, rational management and study of marine mammals in the North Atlantic'.[166] The main organ of NAMMCO is its Council in which all members participate. Advice on research comes through two main committees: the Scientific and Management committees. The Council is the decision-making body of the NAMMCO. NAMMCO provides a mechanism for cooperation on conservation and management for all species of cetaceans (whales and dolphins) and pinnipeds (seals and walruses) in the region.[167] The Agreement was concluded as a reaction by dissatisfied States and other entities (Faroe Islands and Greenland) with the IWC regime, more particularly the continuation of the commercial whaling moratorium. NAMMCO advises on the setting of national quotas for commercial and aboriginal whaling, including small and medium-sized

---

[160] Endangered Species Handbook, <http://www.endangeredspecieshandbook.org/en/trade_pirate.php>.
[161] On pirate whaling in Korea in 1985, see EarthTrust, <http://earthtrust.org/korea.html>.
[162] Report of the International Whaling Commission, 30–38, 1990 (Appendix 9).
[163] See RR Reeves, 'The Origin and the Character of "Aboriginal Subsistence Whaling": A Global Review' (2002) 32 Mammal Review 88–89.
[164] Fitzmaurice (n 26), 17.
[165] The text of the agreement is available via NAMMCO's website at <http://www.nammco.no/>.
[166] Article 2.
[167] *See* further <http://www.nammco.no/about/nammco-council>.

whales such as narwhal and beluga.[168] The establishment of such an organisation can be said to lead to the fragmentation of whaling regulation, a fact which is further exacerbated by the setting of individual quotas by States which are not Party to any international or regional whaling organisation at all, such as Canada, which allows a robust programme of aboriginal whaling outside the remit of the IWC.

The lack of a coherent global policy approach regarding all whaling activities (for example, the setting of quotas and agreed methods of taking), is in clear contravention of the principles of inter-generational equity. For example, frequent inconsistencies in the IWC's policy approach on whaling (such as whether to raise quotas for indigenous whaling in Greenland) cannot be said to be forward looking. The presence of divergent and often contentious approaches does not foster the necessary long-term policy needed to secure not only the interests of future generations, but also the interests of the present generation. The lack of certainty as to the regulation of different forms of whaling, as well as the ill-defined functions of the IWC, are all contributory factors in poor management of whales stocks. As such, they have a damaging effect on the interests of all generations and undermine the principles and concepts underlying the theory of Brown-Weiss.

3. REGIONAL FISHERIES MANAGEMENT ORGANISATIONS (RFMOS)

A question might be asked whether the inter- and intra-generational dimensions of IWC policy can be compared to the approaches of RFMOs, such as the Northwest Atlantic Fisheries Organization (NAFO). There is not an easy or obvious reply to this question. On the one hand, it has been argued that recent amendments to some of the treaties establishing older RFMOs 'emphasize the long-term objective of conservation efforts'.[169] Additionally, 'recent regional [fisheries] agreements have included in their texts ... the obligations to adopt the precautionary and ecosystem approaches to fisheries management', which may be said to foster inter-generational equity,[170] and indeed intra-generational fairness. However, it has also been noted that the setting of

---

[168] The Norwegian explanation of its role is that 'NAMMCO has adopted a control and monitoring regime that covers all catches of marine mammals, including traditional coastal whaling in Norway, catches by Iceland and the Faeroe Islands and hunting by the indigenous people of Greenland': see 'Norwegian whaling – based on a balanced ecosystem', available at <http://www.fisheries.no/ecosystems-and-stocks/marine_stocks/mammals/whales/whaling/>. For example, in 2006 the Management Committee on Cetaceans of NAMMCO expressed serious concern relating to the takes of beluga and narwhal in Western Greenland: Report of the Meeting of the Management Committee for Cetaceans, 2–3 September 2008, Sisimut Greenland, available at <http://www.fisheries.no/ecosystems-and-stocks/marine_stocks/mammals/whales/whaling/#.VjzHfYuPklo>.

[169] MCE Palma, 'Allocation of Fishing Opportunities in Regional Fisheries Management Organizations: A Legal Analysis in the Light of Equity' (Dalhousie University, 2010) 143 <http://dalspace.library.dal.ca/bitstream/handle/10222/13060/Engler%20Palma,%20Maria%20Cecilia,%20LLM,%20Law,%20August%202010.pdf?sequence=1>,136.

[170] *Ibid.*

TACs (Total Available Catch) and the allocating of national quotas, which should be considered as tools of fostering inter- and intra-generational equity, have in fact produced a very adverse effect as a result of the selfish behaviour of States;[171] practice in the allocation of fisheries resources has been said to 'create incentives to disregard long-term, and therefore inter-generational, considerations'.[172]

From an institutional point of view, RFMOs tend to seek the involvement only of fishing States, with the result that consumer States may be excluded. This could suggest a greater commitment to intra-generational equity under the ICRW, where both kinds of actors are included on an equal footing. Practice of the IWC suggests that the attitude of the various factions involved are determined more by entrenched political positions rather than any commitment to equity as such.

4. CONCLUSIONS

The above contentious issues concerning the regulation and conservation of whaling do not facilitate the drawing of any immediate conclusion that the principles of intra- and inter-generational equity are fully implemented and propagated by the IWC and its Parties. Despite the invocation to the interests of future generations in the Preamble of the ICRW, the practice of States and the conflicts within the IWC do not allow for the conclusion that these principles underpin the current system regulating whales and whaling. The moratorium on commercial whaling and the establishing of sanctuaries are definitely promoting and supporting inter-generational equity. However, the opting-out of the moratorium by certain States, as well as the controversies surrounding scientific whaling and the establishment of sanctuaries, could be said to undermine to a certain degree their usefulness in accommodating the rights of future generations.

The same can be said of the promotion of intra-generational equity in so far as it relates to indigenous whaling. The accommodating of cultural diversity of indigenous peoples is without doubt a crucial role of the IWC. However, arguments within the IWC as to quotas to be allocated to indigenous peoples, including in some cases a lack of application of the principle of sustainable use, defeats to a certain extent the principles of intra-generational equity. It must also be mentioned that the States outside the IWC (for example, Canada) do not always conduct their whaling activities according to the principles of sustainability in so far as the allocation of indigenous whaling quotas is concerned. Therefore, the general conclusion must be drawn that the practice of contemporary whaling does not sufficiently follow the principles of intra- and inter-generational equity, whether within or outside the framework of regulation established by the IWC.

Clearly there is ample room for further research on these issues. Indeed, whaling in the light of legal and ethical issues is one of the main research interests of the author of this chapter. In 2015 a monograph entitled *Whaling and International Law* was published by Cambridge University Press on these issues. It deals with problems connected with all types of whaling from the points of view of law, ethics and cultural

---

[171] *Ibid.*, 150.
[172] *Ibid.*

diversity. Additionally, the author of this chapter in cooperation with the University of Kobe will edit a book on legal issues raised by the *Whaling* case in which she will write a chapter on the question of the interpretation of treaties in this case.

SELECT BIBLIOGRAPHY

Bowman, MJ, '"Normalizing" the International Convention for the Regulation of Whaling' (2008) 29 Michigan Journal of International Law 293

Bowman, MJ, 'Transcending the Fisheries Paradigm: Towards a Rational Approach to Determining the Future of the International Whaling Commission' (2009) 7 New Zealand Yearbook of International Law 85

Bowman, MJ, P Davies and C Redgwell, *Lyster's International Wildlife Law* (2nd edn., CUP, 2010)

Couzens, E, *Whales and Elephants in International Conservation Law and Politics: A Comparative Study* (Taylor and Francis Group, 2013)

Dorsey, K, *Whales and Nations: Environmental Diplomacy on the High Seas* (Washington University Press, 2014)

Fitzmaurice, M, 'Indigenous Whaling and Environmental Protection' (2012) 55 German Yearbook of International Law 419

Fitzmaurice, M, *Whaling and International Law* (Cambridge University Press, 2015)

Gillespie, A, *Whaling Diplomacy: Defining the Issues in International Environmental Law* (Edward Elgar Publishing, 2005)

Kishigami, N, H Hamaguchi and JM Savelle (eds), *Anthropological Studies of Whaling* (National Museum of Ethnology, 2013)

Nagtzaam, GJ, 'The International Whaling Commission and the Elusive Great White Whale of Preservationism', available at <http://works.bepress.com/gerry_nagtzaam/2>

Oberthür, S, 'The International Convention for the Regulation of Whaling: From Over-Exploitation to Total Prohibition' (1998/1999) Yearbook of International Cooperation on Environment and Development 29

12. Common concern, common heritage and other global(-ising) concepts: rhetorical devices, legal principles or a fundamental challenge?

Duncan French

1. INTRODUCTION

The purpose of this chapter is to consider some of the underlying justifications for international legal action in relation to the conservation of natural resources and environmental protection more generally. As will be explored, the emphasis is not at the level of political or strategic considerations per se,[1] but rather upon how States express to themselves, and to the wider international community, their motivations for action. While the relevance of why certain States act in certain ways should not be underplayed, this chapter focuses upon what conceptual justifications States have elaborated to provide the legal basis for doing so. In short, it reflects on those foundational aspects of international environmental law which are at the juncture of legal framework, sovereign discretion, collective interest and normative obligation.

The focus of this chapter is primarily – though not exclusively – on multilateral aspects of international environmental law, rather than upon bilateral or regional commitments, or transboundary obligations in customary environmental law.[2] This is not to deny the relevance or the contribution of such measures to the broader discipline,[3] but only to indicate that for the current discussion, the focus is on such concepts as common concern, common heritage and natural heritage.[4] These – and similar – ideas have had a profound effect on the reach and ambition of international environmental law, transforming international rules in the area from the specific,

---

[1] *See* S Barrett, *Environment and Statecraft: The Strategy of Environmental Treaty-Making* (OUP, 2005).

[2] *See*, for instance, the 2001 International Law Commission (ILC) Articles on Prevention of Transboundary Harm from Hazardous Activities, Yearbook of the International Law Commission, 2001, vol. II, Part Two.

[3] *See*, in particular, the relevance of agreements between range States as a key feature of the 1979 Convention on the Conservation of Migratory Species of Wild Animals (1980) 19 ILM 15.

[4] Though this chapter focuses upon nature conservation, this must be broadly considered as many of the concepts range generally across the wider discipline of environmental law, including notably climate change.

perhaps even in certain instances parochial, to the truly global.[5] However, notwithstanding their significance, much remains unclear about the rationale, consequences and normative impact of these founding global(-ising) concepts.

Indeed, as a matter of first order, it may be questioned why there is a need to reference such concepts at all as a foundation for action. In other areas of international law, the justification – typically found in a treaty's preamble or at best perhaps in an opening article – is often descriptive, implicit, confused or, alternatively, needs little explanation. Rarely, however, is the necessity of international action phrased as a principle, as seems to be the case with multilateral environmental agreements. One might, of course, point to 'human dignity' as the justification for international human rights law[6] as an antecedent to international environmental law, but in addition there is something else going on in environmental law, which this chapter explores, questioning the absolutist nature of the Westphalian model of territorial sovereignty.

Namely, it is that neither territorial control, on the one hand, nor the international regulation of areas beyond territorial control, on the other, is capable of providing an effective structure for the global regulation of environmental problems. In other words, the very structure of the international legal order is found to be wanting and consequently alternatives, however inchoate, must be considered. Moreover, though common concern, in particular, has been accepted in part out of political expediency, there is potential scope for it to have much broader impact, recognising that it justifies legal action by reference to a general consensus that States share a common responsibility to act at the global level. The chapter thus concludes by outlining briefly three such implications centring on institutional development, enforcement and normative change. Together, they present a tantalising imagining of what international environmental law might become.

2. COLLECTIVE INTEREST AND GLOBAL ENVIRONMENTAL RESPONSIBILITY

Since the emergence of international environmental law, there has always been something of a tripartite division between those matters which are transboundary in nature, those that are beyond the jurisdiction of any State and those that are primarily internal to the State itself. This division – nowhere made truly explicit and surely capable of further refinement (especially in relation to the treatment of shared natural

---

[5] As well as a spatial dimension, Judge Weeramantry in his separate opinion in *Gabčíkovo-Nagymaros Project (Hungary/Slovakia)* ICJ Reports (1997) 7, 98, reflected on the temporal development of the global nature of the discipline: 'In the context of environmental wisdom generally, there is much to be derived from ancient civilizations and traditional legal systems in Asia, the Middle East, Africa, Europe, the Americas, the Pacific, and Australia – in fact, the whole world. This is a rich source which modern environmental law has left largely untapped.'

[6] 1948 Universal Declaration of Human Rights (UNGA Res. 217A (III)), preamble: 'Whereas recognition of the inherent dignity and of the equal and inalienable rights of all members of the human family is the foundation of freedom, justice and peace in the world'.

336 *Research handbook on biodiversity and law*

resources) – has, however, been significant in the development and operation of international environmental law.

The dichotomous structuring of the international legal order around, on the one hand, units of statehood possessing territorial sovereignty[7] – emboldened still further during the 1960s and 1970s by the notion of permanent sovereignty over natural resources[8] – and, on the other, areas beyond national jurisdiction (notably the high seas, outer space and, by virtue of treaty, Antarctica) has thus had a profound impact upon the balanced development of international environmental law and its relative success.[9] In equal measure, issues that do not neatly fall within this understanding of the legal order – straddling fish stocks[10] and migratory terrestrial species (especially bird-life) – have consistently struggled over the years to be adequately protected.

In some instances, this division is explicit; on reading Principle 21 of the 1972 Stockholm Declaration/Principle 2 of the 1992 Rio Declaration on Environment and Development,[11] the distinction between the 'sovereign right' of a State within its borders 'to exploit their own resources pursuant to their own environmental [and developmental][12] policies' and its responsibilities external thereto[13] is immediately apparent. It concretises, in a very clear sense, the paradoxical nature of environmental responsibility; *in extremis*, shielding the State as regards the management of its 'own' natural resources, but subjecting its external impact to international scrutiny and responsibility.[14] Nevertheless, Principle 21 also oversimplifies international law, as no distinction is made between the responsibility for transboundary impact and the negative effects of harm caused in areas beyond national jurisdiction. It has thus justified the evolution of a single meta-norm (the singularly inappropriately-named 'no

---

[7] See D French (ed), *Statehood and Self-Determination: Reconciling Tradition and Modernity in International Law* (CUP, 2013).

[8] See UN Res 1803 (XVII) (1962), affirming that all legal measures must be based upon 'the inalienable right of all States freely to dispose of their natural wealth and resources in accordance with their national interests, and on respect for the economic independence of States'.

[9] 1972 Stockholm Declaration on the Human Environment, UN Doc. A/CONF.48/14/Rev. 1 (1972), preamble: 'A growing class of environmental problems, because they are regional or global in extent or because they affect the common international realm, will require extensive co-operation among nations and action by international organizations in the common interest.'

[10] For instance, see the 1995 Agreement for the Implementation of the United Nations Convention on the Law of the Sea Relating to the Conservation and Management of Straddling Fish Stocks and Highly Migratory Fish Stocks (1995) 34 ILM 1542 ('Straddling Fish Stocks Convention'), particularly Article 5: 'to conserve and manage straddling fish stocks and highly migratory fish stocks, *coastal States and States fishing on the high seas shall, in giving effect to their duty to cooperate* …' (emphasis added).

[11] UN Doc. A/CONF.151/26/Rev.1 (Volume 1) (1992).

[12] As inserted by the 1992 Rio Declaration.

[13] *Ibid.*: 'the responsibility to ensure that activities within their jurisdiction or control do not cause damage to the environment of other States or of areas beyond the limits of national jurisdiction'.

[14] See the preamble to the ILC Articles on Transboundary Harm (n 2): '*Bearing in mind* the principle of permanent sovereignty of States … *Bearing also in mind* that the freedom of States to carry on or permit activities in their territory or otherwise under their jurisdiction or control is not unlimited'.

harm' principle) to come into existence,[15] notwithstanding the very significant legal and practical differences that exist between these two scenarios.[16]

2.1 Environmental Regulation within the Dichotomy of the Traditional Legal Order

International environmental law operates upon one of the most significant fault-lines of the international system, whereby such matters as the management of natural resources, responsibility for environmental damage and the capacity to ensure respect for international environmental law invariably depend upon whether the matter is subject to rules applicable to territorial sovereignty or those relating to common areas. Where is the resource to be found? Where did the damage occur? Which (if any) State has jurisdiction and control? And which State is harmed? There is, of course, a risk of giving too much credence to such a demarcation, as if the problems resulting were insurmountable, or incapable of legal resolution. Clearly they are not, and international environmental law has often been singularly successful in devising both normative and institutional innovations to tackle such questions.

Nevertheless, the mere fact that such innovation has been necessary – that there are often issues to overcome – is itself testament to the obduracy of this division within the system. Moreover, though States have occasionally found the political will and legal techniques to make change, such progress as has occurred has generally been through treaty, without necessarily commensurate advancement in customary international law. The discussion later in the chapter on *actio popularis* for obligations *erga omnes* highlights a theme that developments in treaty law are not always reflected in general legal doctrine. And, of course, it should not need saying that, whatever merits a particular treaty regime may possess, there are inherent limitations of a treaty-based approach to environmental protection both in the face of a free-rider problem and in legal terms: *pacta tertiis nec nocent nec prosunt*.[17]

Within this context of a historical division between national territory and common areas, and the consequent allocation of differing rights and responsibilities between States, there has developed a range of legal rules and principles which run counter to what we would now consider necessary to achieve sustainable development and environmental protection. Mention has already been made of the primacy traditionally given to permanent sovereignty over natural resources, whereby international supervision over domestic resources is strictly curtailed, unless and until States themselves are prepared to consent to self-imposed restriction. Of potential equal detriment has been the very opposite standard, namely the non-appropriation and freedom of the high seas, which neither generates any incentive for individual restraint nor provides a basis

---

[15] *See Legality of the Threat or Use of Nuclear Weapons*, ICJ Reports (1996) 226, at 241–242: 'the existence of the general obligation of States to ensure that activities within their jurisdiction and control respect the environment of other States or of areas beyond national control is now part of the corpus of international law relating to the environment'.

[16] For instance, on standing *see Nuclear Tests Cases (Australia v France; New Zealand v France)* ICJ Reports (1974) 253.

[17] Article 34, 1969 Vienna Convention on the Law of Treaties; trans: 'a treaty binds the parties and only the parties'.

for institutional control, with the tragedy of the commons being a widely regarded consequence.[18]

Thus, despite the ecological inter-connectedness of natural resources, there is often a marked reluctance in international law to reflect the need for more progressive and overarching obligations. And though in both instances we have seen significant developments in international environmental law, the scale of this ambition is generally relatively low. In the case of domestic natural resources, it has been through the development of a range of resource-specific and habitat-specific treaties and then, in 1992, the adoption of the more comprehensive, if less normatively ambitious, Convention on Biological Diversity.[19] In the case of international marine resources, it has been through the formulation of new principles and standards that seek to restrain such absolute freedoms, mention especially being made of the 1946 International Convention for the Regulation of Whaling,[20] various provisions of the 1982 UN Convention on the Law of the Sea (UNCLOS)[21] and, more specifically, the 1995 Straddling Stocks Agreement, as well as the work of regional fisheries management organisations.[22]

Of course, this is not to suggest that the principles of sovereign control or free access cannot be used, in appropriate contexts and with sufficient safeguards, to achieve ecological outcomes. Indeed, one of the principal mechanisms to curtail the tragedy of the high seas commons has been the establishment of exclusive economic zones, thus extending sovereign rights at the expense of freedom of access. The point being made, rather, is more subtle. It is not that the existence either of territorial sovereignty or of common areas is inappropriate as a model of governance per se, but that in both instances traditional assumptions operate: namely, that one begins from the starting point of either permanent sovereignty in the case of territorial resources or freedom of access in the case of the high seas.[23] Though it is – probably – going too far to say this operates as a 'Lotus' presumption of unfettered sovereign freedom,[24] it is invariably the case that traditional principles are difficult to dislodge. As the *Bering Sea Fur Seals*

---

[18] G Hardin, 'The Tragedy of the Commons' (1968) 162 *Science* 1243–1248. *See* further the chapter by Goodwin in this handbook.

[19] (1992) 31 ILM 818.

[20] 161 UNTS 72. Most notably, the moratorium on commercial whaling since the 1985/1986 season, though the Schedule annexed to the Convention also contains, inter alia, provisions on designated sanctuaries and various limitations on catch methods.

[21] (1982) 21 ILM 1261; see in particular Articles 117–119, relating to the conservation and management of living resources on the high seas.

[22] On the success and challenges of regulating high seas fisheries, *see* K Gjerde, 'High Seas Fisheries Management under the Convention on the Law of the Sea' in D Freestone, R Barnes and DM Ong (eds), *The Law of the Sea: Progress and Prospects* (OUP, 2006) 281–306.

[23] See the First Report of the ILC Special Rapporteur, Professor Shinya Murase, on the Protection of the Atmosphere (A/CN.4/667, 14 February 2014) paragraph 86: 'Common property, or *res communis*, refers to areas such as the high seas that are open for legitimate use by all States and that may not be appropriated to the sovereignty of any individual State. The airspace above the high seas is in this sense "common property". However, like sovereign airspace, common property is fundamentally a spatial dimension and is therefore insufficient to deal with the atmosphere as a global unit.'

[24] *The Case of the SS Lotus (France v Turkey)* (1927) PCIJ, Series A, No. 10, paragraph 44.

Arbitration[25] demonstrated, the default position is generally reverted to as a means of defending the traditional legal order – in the case of that arbitration, the freedom to fish on the high seas.[26]

Moreover, unless customary international law develops alongside treaty law, innovation and best practice remain patchy, inconsistent and sporadic. It is for this reason that the emergence of concepts such as common heritage of mankind and common concern of humankind have significance in international environmental law; not because they are sufficiently developed to change the structure of international law, but because they are influencing, in a more nuanced manner, the overall narrative of the debate. Thus, though we are unlikely yet to be at a stage of normative evolution that such concepts allow us to fundamentally question key features of the legal order – features that have so far been considered immutable – the use and language of such concepts is itself telling.

2.2 'Community of Interest' – Precursor to Common Interest and Collective Responsibility

An attempt at rethinking the system is, of course, not new. As Bowman notes,[27] an earlier example of this was the Permanent Court of International Justice's attempt to find new solutions to problems surrounding cross-border and shared waterways, including through the formulation of the legal framework of a 'community of interest'.[28] This is significant, not only because of its contribution to more recent developments in international water law,[29] but also because it encapsulates something of the collective, or communal, nature of – in this case – a shared natural resource;[30] a theme that features heavily in both common heritage and common concern.

But the notion of 'community of interest' highlights two further facets of the current discussion. First, the phrase encapsulates something of the controversy, but also the resolution, of attempting to limit sovereignty when confronted with matters that are incapable of being effectively managed through a strict adherence to territorial control. Neither an absolute sovereignty approach nor any theory which gives preference to downstream States has usually been able to resolve such disputes in a manner that was mutually satisfactory to all parties. The fact that it is the 'interests' of riparian States which are the focus of concern, rather than their strict legal entitlement – however that

---

[25] (1898) 1 Moore's International Arbitration Awards 755.
[26] *See* MJ Bowman, 'Environmental Protection and the Concept of Common Concern of Humankind' in M Fitzmaurice, DM Ong and P Merkouris (eds), *Research Handbook on International Environmental Law* (Edward Elgar Publishing, 2010), 494–497.
[27] *Ibid.*, 498.
[28] *Territorial Jurisdiction of the International Commission of the River Oder* (1929) PCIJ, Series A, No. 23, p. 27: '[the] community of interest in a navigable river becomes the basis of a common legal right, the essential features of which are the perfect equality of all riparian States in the user of the whole course of the river and the exclusion of any preferential privilege of any one riparian State in relation to the others'.
[29] *Gabčíkovo-Nagymaros Project* (n 5), 56.
[30] *See* generally the 1978 UNEP Principles on Conservation and Harmonious Utilization of Natural Resources Shared by Two or More States (1978) 17 ILM 1094.

is determined – also provides an interesting lesson for current debates.[31] The move away from a direct conflict with sovereign claims is particularly relevant when exploring the differences between the controversies surrounding the common heritage ideal and the more widely accepted common concern approach, as discussed below.

Secondly, 'community of interests' reflects something of the non-hierarchical nature of how these concepts evolve and how they operate. As Brunnée has pertinently noted, '[t]o the extent that there are foundational values and interests common' to humanity, they may well find their way into international law. However, in terms of the making and enforcement of international law, states remain the key players.'[32] Thus, when reflecting on how the international community has reached the point that it has, with the emergence of these legal principles, it is significant to remember the pivotal role of States themselves in devising the way forward. It also explains the sometimes uneven and inconsistent manner of their application. These are not general principles imposed from without. They are devised through the political-cum-legal dialogue of States themselves, whenever such a fragile consensus is able to come together and then, more importantly, to be maintained.

The next section looks at some of the more specific concepts of common heritage and common concern as representations of the international community's collective response. It is, however, useful to remind oneself that these concepts reflect a perennial tension between States seeking to achieve a community response (and thus, however implicitly, to impede their own individual autonomy), and that it is the States themselves that have defined what it is they want this very community to achieve. Thus, the tension itself is important in examining the significance and normative reach that can be expected of such concepts.

3. COMMON HERITAGE, COMMON CONCERN AND NATURAL HERITAGE

In promoting global environmental responsibility, international law has developed a range of conceptual devices, most notably common concern of humankind as found in the 1992 'Rio' Climate Change Convention and Convention on Biological Diversity (CBD),[33] to justify collective international action. As will be discussed, the decision to include explicit reference to common concern within these Conventions was partially in response to the emergence of a positive willingness among the international community to contemplate global action but also, more negatively, it was a reaction against the earlier principle of common heritage of mankind. The original purpose of common heritage, as will be noted, was removed somewhat from the conservation of the global

---

[31] See, in particular, the endorsement of the principle of equitable utilisation in the 1997 Convention on the Non-Navigational Uses of International Watercourses (1997) 36 ILM 719.

[32] J Brunnée, 'Common Areas, Common Heritage, and Common Concern' in D Bodansky, J Brunnée and E Hey (eds.), *The Oxford Handbook of International Environmental Law* (OUP, 2007), 556.

[33] 1992 UN Framework Convention on Climate Change (1992) 31 ILM 851; 1992 Convention on Biological Diversity 1760 UNTS 79.

environment. Rather, it was – and ultimately, remains – fundamentally concerned with the fair allocation of global natural resources and the equitable sharing of benefits arising from their utilisation.[34] Its most significant, and truly only *operational*, incarnation has been in relation to the minerals of the deep seabed.[35]

3.1 Common Heritage of Mankind

As the Maltese Ambassador to the United Nations, Arvid Parvo, famously argued before the General Assembly in 1967, there was a strong economic and moral argument to subject these natural resources to a regime other than the freedom of the high seas, which did little other than promote a winner-takes-all mentality, allowing those more technologically developed to take full, and first, advantage. Unsurprisingly, the common heritage principle as it was subsequently elaborated in the 1970 Declaration[36] was very much of its time; a key component of the New International Economic Order promoted by developing States, which reached its high watermark with the 1974 Charter of Economic Rights and Duties of States.[37] The original version of the 1982 UNCLOS incorporated the rules and institutions considered necessary to give effect to the underlying principles of the common heritage concept, notably non-appropriation by States, institutionalised control over the natural resources, fair division of benefits, and an overriding precept of peaceful use. And though the 1994 Implementation Agreement[38] made fundamental changes to the governance and operation of the Area, the outline legal framework and principles remained the same. Significantly, though Part XI of UNCLOS contains certain references to marine pollution, these are arguably limited and, in the light of subsequent developments, seem now rather derisory.

Nevertheless, both because of further developments – including the advanced elaboration of the Mining Code by the Seabed Authority[39] and, importantly, the 2011 Advisory Opinion of the Seabed Disputes Chamber on various environmental matters[40] – as well as the innate challenge the common heritage principle poses to the traditional legal order, common heritage has both a historical and contemporaneous significance. This is not to understate the continuing controversy surrounding the common heritage principle, as well as its limited application beyond the deep seabed, but rather to reaffirm that its themes of collective oversight of natural resources, global responsibility and institutional supervision and, as the Seabed Disputes Chamber makes clear in

---

[34] *Responsibilities and Obligations of States Sponsoring Persons and Entities with respect to Activities in the Area*, Advisory Opinion of 1 February 2011, (2011) ITLOS Reports 10, paragraph 151.

[35] For a general discussion of the Area, as the deep seabed is known, *see* S Nandan, 'Administering the Mineral Resources of the Deep Seabed' in Freestone, Barnes and Ong (n 22), 75–92.

[36] 1970 Declaration of Principles Governing the Sea-Bed and the Ocean Floor, and the Subsoil Thereof, Beyond the Limits of National Jurisdiction, UNGA Res 2749 (XXV) (1970).

[37] UN Doc A/RES/29/3281 (1974).

[38] Agreement on the Implementation of Part XI of the 1982 Law of the Sea Convention (1994) 33 ILM 1309.

[39] For the text, *see* <https://www.isa.org.jm/mining-code>.

[40] *See* (n 34).

the Advisory Opinion, the normative impact – in contrast to simply normative rhetoric – of its general principles,[41] are all relevant to the subsequent discussion on common concern. Of course, common heritage has failed to find traction beyond the deep seabed, as the common heritage regime for the moon and other celestial bodies never took full effect,[42] and the initial conceptualisation of plant genetic resources as common heritage was almost immediately retracted.[43] Moreover, a similar argument for climate change and biodiversity to be considered common heritage was never sensibly going to find favour in the final drafts of those Conventions.

However, perhaps two further comments are useful at this stage. First, the customary status of common heritage remains unsettled, though perhaps it is not as much a matter of division as it has been previously. Nevertheless, with the United States remaining outside UNCLOS, there is still scope for uncertainty and some concern on this point. Secondly, regardless of the question of customary status, its elaboration and subsequent implementation by treaty regime is a *sine qua non* of its success. Both of these aspects are relevant to understanding the impact of other cognate concepts, notably common concern, to which this chapter now turns.

3.2 Common Concern of Humankind

As noted above, common concern was the preferred justification for the 1992 Rio Conventions, reflecting as it did a less radical alternative to the common heritage principle.[44] Nevertheless, it is important to consider the concept as more than just a more acceptable form of justification, though this was undoubtedly the case. In particular, while common concern may have proved more palatable to States, as it avoided direct conflict with either territorial sovereignty or sovereign rights and privileges (in the case of the freedom of the high seas), this should not be taken to mean that common concern was perceived as being devoid of legal consequence altogether, and thus politically uncontroversial. Of course, there is a risk of aligning how we are now beginning to understand the concept, relating it to distinct but mutually supportive ideas based especially around *erga omnes* and *actio popularis*, with how it was understood in 1992. It is perhaps a moot point as to what the international community *en masse* understood by common concern in the run up to

---

[41] See D French, 'From the Depths: Rich Pickings of Principles of Sustainable Development and General International Law on the Ocean Floor – the Seabed Disputes Chamber's 2011 Advisory Opinion' (2011) 26 International Journal of Marine and Coastal Law 525, 544: 'Much of the stringency of the tone of the Advisory Opinion is affected by the importance the Chamber gives to the Area as a common heritage of mankind, which elevates these particular natural resources, and this particular geographical region, to a special position in international law'.

[42] 1979 Agreement Governing the Activities of States on the Moon and Other Celestial Bodies (1979) 18 ILM 1434.

[43] See Bowman (n 26), 501. More generally, see F Biermann, '"Common Concern of Humankind": The Emergence of a New Concept of International Environmental Law' (1996) 34 Archiv des Völkerrechts 426–481.

[44] MJ Bowman, PGG Davies and CJ Redgwell, *Lyster's International Wildlife Law* (2nd edn, CUP, 2010), 52.

those negotiations;[45] though invariably the inclusion of 'concern' instead of 'heritage' reflected both a general political awakening of global environmental issues, while at the same time adopting an incremental and cautious attitude in their resolution.

Nevertheless, certain points can be made. First, it is of fundamental significance that, as regards the matters that were addressed in the 1992 Conventions, that which was of common concern related to resources both *within* States as well as those *external* to them. This is particularly so if one is prepared to view the global climate for these purposes as a natural resource. Indeed, if one reviews the CBD, its provision on jurisdictional scope is particularly interesting. Article 4 provides that its scope extends:

> (a) In the case of components of biological diversity, in areas within the limits of its national jurisdiction; and (b) In the case of processes and activities, regardless of where their effects occur, carried out under its jurisdiction or control, within the area of its national jurisdiction or beyond the limits of national jurisdiction.

For the sake of clarity, the CBD also notes that 'Contracting Parties shall implement this Convention with respect to the marine environment consistently with the rights and obligations of States under the law of the sea'.[46] Thus, unlike the common heritage principle, the resources in question are not exclusively found in common areas subject to (without treaty intervention) almost unfettered exploitation. What this might mean in terms of the standard of behaviour now expected of States in the conservation of resources of common concern – and whether any particular generalised rule can thus be imagined which encompasses utilisation in all contexts – is considered further below.

Of course, it is perhaps not surprising that matters affecting a common 'area', such as the climate, can be classified as common concern, since one of the classical features of international law is that it is able to provide agreed rules and principles beyond the boundaries of States. The regulation of domestic biodiversity – however weakly – by international treaty is more challenging,[47] and for that to occur prompts a broader challenge to the legal order. What flows from categorising resources as of common concern in such a way? It is not to question the status quo – sovereignty remains intact.[48] Nor does it set out what level of intervention is permitted; that is left to further normative detail and subsequent implementation. Indeed, intervention is perhaps the wrong term – common concern was chosen precisely to avoid raising the spectre of (direct) international interference. Rather, by seeing domestic resources as being of common concern it generates global interest, and thus removes the conceit of an

---

[45] PW Birnie, AE Boyle and CJ Redgwell, *International Law and the Environment* (3rd edn, OUP, 2009), 198.
[46] CBD, Article 22(2).
[47] Birnie, Boyle and Redgwell (n 45), 600: 'The doctrine of permanent sovereignty over natural resources has also encouraged over-exploitation in the absence of clearly established and implemented international conservatory obligations.'
[48] Bowman, Davies and Redgwell (n 44), 52: 'The essential implications are that, while existing conceptions of sovereignty are left formally undisturbed, all such matters are to be considered legitimate topics for international debate.'

exclusive domestic domain.[49] Just as the Universal Declaration of Human Rights removed both a legal and an almost psychological obstacle – if not necessarily the practical implications – of human rights being considered as a matter for an individual State alone, common concern arguably achieves much the same effect for global environmental responsibility.

However, that in itself cannot be a sufficient explanation for identifying biodiversity as a natural resource of common concern. What of oil, gas and other 'exhaustible' natural resources? These patently fall outside the definition of components of biological diversity, and thereby the regime of the CBD. Consequently they are not thought of as being of common concern. Legally, this is inevitable, but why should this be the case? As fundamental resources for the global economy, certainly a hypothetical argument could be made for their inclusion as matters of common concern. Indeed, from the perspective of global development, as drivers of the global economy, they are very much of global interest.[50] But as matters of common concern in the *environmental* context, they remain outside of the debate. Part of the answer may lie in the preamble of the CBD:

> Conscious of the intrinsic value of biological diversity and of the ecological, genetic, social, economic, scientific, educational, cultural, recreational and aesthetic values of biological diversity and its components,
>
> Conscious also of the importance of biological diversity for evolution and for maintaining life sustaining systems of the biosphere,
>
> Affirming that the conservation of biological diversity is a common concern of humankind,

These recitals provide enormous insight into the rationale of seeing biological diversity, whether one likes that term or not, as being of common concern. But what we are looking for here is whether one can identify reasons that support its inclusion in the panoply of issues that justify the nomenclature of common concern. If one cannot do this, and it is therefore not possible to provide normative coherence as to why certain issues are of common concern and some are not, there is a real risk that the discussion descends into little more than retrospective realpolitik.

In reviewing these recitals, several things become apparent. The reference to 'social, economic, scientific … values' might seem a relevant consideration, but ultimately is inconsequential in this regard as almost any issue of international concern raises some or all of these values to some extent. Of more importance is the reference to the 'intrinsic value' of biodiversity, as well as its 'ecological, genetic' significance.

---

[49] Article 2(7), UN Charter. Note also the changing nature of the domestic domain, as indicated in *Nationality Decrees Issued in Tunis and Morocco* (Advisory Opinion) (1923) PCIJ Series B, No. 4, paragraph 40: 'The question whether a certain matter is or is not solely within the jurisdiction of a State is an essentially relative question; it depends upon the development of international relations.'

[50] Rarely mentioned, see in particular Principle 5 of the 1972 Stockholm Declaration: 'The non-renewable resources of the Earth must be employed in such a way as to guard against the danger of their future exhaustion and to ensure that benefits from such employment *are shared by all mankind*' (emphasis added).

Combined with the 'importance of biological diversity for evolution and for maintaining life sustaining systems of the biosphere', this would seem to begin to identify what makes biodiversity truly of common concern. What gives the conservation of biological diversity its importance, and what links it to climate change, is its significance to the health of the global biosphere for the benefit of humankind. And as the preamble of the Climate Change Convention notes:

> Recalling the provisions of General Assembly resolution 44/228 of 22 December 1989 on the United Nations Conference on Environment and Development, and resolutions 43/53 of 6 December 1988, 44/207 of 22 December 1989, 45/212 of 21 December 1990 and 46/169 of 19 December 1991 *on protection of global climate for present and future generations of mankind*.[51]

Such matters are of common concern not simply because they impact upon the environment per se, but because the scale and extent of the impact potentially affects all of humanity, both of present and future generations. Of course, if this were too strictly applied, or the requirement of anthropocentric impact too rigidly interpreted, one would risk being too exclusive and ignoring the application of common concern to other, particularly, habitat- or species-specific Conventions, e.g., wetlands or whales. For instance, as Bowman notes, the 'tangible acknowledgement of this common interest [in the Whaling Convention] was apparent in the (evidently deliberate) omission from the text of any provision limiting participation to those states that were actively engaged in whaling activities'.[52] Of course, the CBD has done much to bring such habitat and species-specific agreements into one overarching conceptual frame, even if normative and institutional fragmentation remains problematic.[53] But, equally, one might argue that it is precisely because of the more remote human impact of some of these threats that it is more difficult to justify the issues as matters of common concern and, commensurately, why there has been more controversy within the international community surrounding conservatory efforts relating to them. If the consensus surrounding them is weaker, it is perhaps because the rationale is not as universally accepted.

On this point, there is no better example than the 1994 Desertification Convention,[54] which, though adopted only two years after the Rio Conventions, does not use the formula of 'common concern'. Though it expresses several related ideas in several similar ways, significantly it never uses the words together themselves. An extensive quotation from the preamble is useful here.

---

[51] Emphasis added.
[52] Bowman (n 26), 502.
[53] K Scott, 'Managing Fragmentation through Governance: International Environmental Law in a Globalised World' in A Byrnes, M Hayashi and C Michaelson (eds), *International Law in the New Age of Globalization* (Martinus Nijhoff/Brill, 2013) 207–238.
[54] 1994 Convention to Combat Desertification in Those Countries Experiencing Drought and/or Desertification, Particularly in Africa (1994) 33 ILM 1016.

346 *Research handbook on biodiversity and law*

> The Parties to this Convention,
>
> Affirming that human beings in affected or threatened areas are at the centre of concerns to combat desertification and mitigate the effects of drought,
>
> Reflecting the urgent concern of the international community, including States and international organizations, about the adverse impacts of desertification and drought,
>
> Aware that arid, semi-arid and dry sub-humid areas together account for a significant proportion of the Earth's land area and are the habitat and source of livelihood for a large segment of its population,
>
> Acknowledging that desertification and drought are problems of global dimension in that they affect all regions of the world and that joint action of the international community is needed to combat desertification and/or mitigate the effects of drought,

This would seem to be no drafting error, or diplomatic oversight; indeed the full title of the Convention is instructive: 'to Combat Desertification *in Those Countries* Experiencing Drought and/or Desertification'.[55] True, there are such phrases as 'human beings … are at the centre of concerns', 'urgent concern' and 'problems of global dimension … affect[ing] all regions of the world', which in layman's terms may come close to saying the same thing – but, importantly, nowhere is combating desertification itself said to be of common concern. And if one looks more closely at these preambular paragraphs, one can see the global nature of the problem is tempered with a decidedly regional perspective. Only those human beings in 'affected or threatened areas' are of concern; the 'urgent concern' is intergovernmental – of States and international organisations – not of humankind; and the areas under threat affect a 'large segment', but not the whole, of the world's population. Of course, the final paragraph cited does recognise that 'joint action of the international community is needed' and one might validly ask how this differs from common concern.

Well, at one level it does not; even without using the formula of 'common concern' the international community was more than able to negotiate and adopt a treaty on the matter; a treaty that has universal acceptance.[56] Significantly, later preambular paragraphs also give recognition to the importance of combating desertification for present and future generations, as well as 'Bearing in mind the relationship between desertification and other environmental problems of global dimension facing the international and national communities'. Nevertheless, the absence of common concern may not be without meaning.

Later in the chapter, there is consideration of current, emerging and future implications of common concern; will they apply to desertification as easily as climate change and the conservation of biological diversity? Or will they apply only to the extent that there is a 'relationship between desertification and other environmental problems of global dimension'? However, as noted in the next paragraph, in light of the fact that many older treaties do not mention common concern, but may well be now included within its dimensions, it would be churlish to refuse to countenance this. But equally, only two years after 1992, what the Desertification Convention does is to

---

[55] Emphasis added.
[56] As at 1 February 2014, the Desertification Convention has 195 Parties, which is the same number of parties as the Climate Change Convention and two more than the CBD.

caution us against over-extending its scope. Similarly, the 1995 Straddling Fish Stocks Convention does not utilise the concept either, but rather restricts its justification to a more descriptive account of the issue; namely that the parties are 'Conscious of the need to avoid adverse impacts on the marine environment, preserve biodiversity, maintain the integrity of marine ecosystems and minimize the risk of long-term or irreversible effects of fishing operations'. Part of the issue here may, of course, be that States did not wish to introduce new ideas into UNCLOS by means of an implementation agreement. Nevertheless, as a natural resource still very much governed in the absence of treaties by general international law under the rubric of *res communis*, similar challenges may arise in the future in the over-reliance on a concept not expressly included.

Thus before considering other aspects of common concern, it is perhaps only right to pause and ponder whether common concern is a necessary nomenclature for international action. The arguments have been made that paradoxically it may well matter that the term was not included in certain treaties while at the same time acknowledging that there may be treaties, especially earlier ones, which did not use the term but have in essence the same effect. The preamble of the 1946 International Convention for the Regulation of Whaling explicitly refers to 'the interests of nations of the world in safeguarding for future generations the great natural resources represented by the whale stocks' and affirms that 'it is in the common interest to achieve the optimum level of whale stocks'. The 1971 Ramsar Convention on Wetlands of International Importance talks about the 'fundamental ecological functions of wetlands' as habitats and adds that 'waterfowl in their seasonal migrations may transcend frontiers and so should be regarded as an international resource'.[57] The 1973 Convention on International Trade in Endangered Species recognises that 'wild fauna and flora in their many beautiful and varied forms are an irreplaceable part of the natural systems of the earth'.[58] The 1972 World Heritage Convention conceptualises matters ever so slightly differently[59] – formulating the notion that 'parts of the cultural or natural heritage are of outstanding interest and therefore need to be preserved as part of the world heritage of mankind as a whole'[60] – but in tone and consequence, this feels more like 'common concern' than 'common heritage',[61] certainly as understood within the institutional context of UNCLOS.

The 1991 Madrid Protocol categorises Antarctica 'and dependent and associated ecosystems' as 'a natural reserve, devoted to peace and science'.[62] Though it is unclear whether this is anything more than a descriptor of the cumulative effect of the various provisions of, and measures taken under, the 1959 Antarctic Treaty and this Protocol,

---

[57] 996 UNTS 245; preamble.
[58] 993 UNTS 243; preamble.
[59] *See* C Redgwell, 'Article 2: Definition of Natural Heritage' in F Francioni, with F Lenzerini (eds), *The 1972 World Heritage Convention: A Commentary* (OUP, 2008) 63–84, who highlights the significant work of Russell E Train in the development of the 'world heritage' concept; *see* further the contribution of Train himself in *World Heritage 2002: Shared Legacy, Common Responsibility* (UNESCO, 2003) 36–37.
[60] (1972) 11 ILM 1358; preamble.
[61] Birnie, Boyle and Redgwell (n 45), 198.
[62] (1991) 30 ILM 1461, Article 2.

rather than having any form of substantive effect, the unique role of Antarctica in maintaining balance in the global environment, as well as more broadly its special status in international law, gives this reference particular significance. Moreover, though there is not universal membership, the fundamental obligations of peaceful use and environmental protection are almost invariably objective in nature, and thus opposable to all.

A further recent instance – recognising that these are but examples – is the 2013 Minamata Convention on mercury pollution.[63] Though again not referring to the matter as one of common concern, the preamble to the Convention views mercury as 'a chemical of global concern'. This is a treaty which identifies and seeks to tackle a particular threat, and to that extent it follows the 2001 Stockholm Convention on Persistent Organic Pollutants, which took a more descriptive approach to the matter, noting that 'persistent organic pollutants possess toxic properties, resist degradation, bioaccumulate and are transported, through air, water and migratory species, *across international boundaries and deposited far from their place of release*, where they accumulate in terrestrial and aquatic ecosystems'.[64] In both cases, however, the necessity for a collective response is predicated on the global nature of the environmental risk.

A final point is that it has been argued[65] that the Rio Conventions do not describe the climate and biological diversity per se as being of common concern, but rather 'change in the Earth's climate and its adverse effects' and 'the conservation of biological diversity' respectively. Thus, what is of concern is not the resources in and of themselves: this sets to one side questions both of proprietorship and sovereignty (as well as in the case of the climate the complex issue of quite how to classify it)[66] while transferring the focus onto the necessity for international cooperation. Nevertheless, regardless of what precisely is the subject-matter of the concern, the raison d'être of common concern is the *collective* responsibility to act.

As an aside – if an important one – there might appear to be a difference of emphasis between the references in the two treaties, whereby it is the threat of 'change' in the case of climate change and the more positive obligation of conservation in the case of biodiversity that is of common concern. This would seem to be purely a matter of drafting. It is certainly not clear that had the forms of reference been reversed – that is, the conservation of the climate and the prevention of damage to biodiversity – this would in any way change the nature of the obligations. Of course, it might be pointed out that in fact there is a notable difference between seeking to prevent environmental damage and responding to the damage; not only in terms of relative cost and the nature of commitment undertaken but also whether responding to damage includes adaptation and restoration, as well as more traditional forms of reparation. These have certainly

---

[63] For the text of the convention, *see* <http://www.mercuryconvention.org/>.
[64] (2001) 40 ILM 532; preamble, emphasis added.
[65] *See* Brunnée (n 32), 564–565.
[66] For a recent development, *see* the 2014 International Law Association Declaration on Legal Principles relating to Climate Change, Article 3(1): 'States shall protect the climate system as a common natural resource for the benefit of present and future generations, within the broader context of the international community's commitment to sustainable development' (<http://www.ila-hq.org/en/committees/index.cfm/cid/1029>).

been issues for the climate change regime,[67] as well as raising interesting questions of liability under the CBD.[68] Nevertheless, as a matter of overarching principle, it is not apparent that the wording itself on this point matters, and certainly little seems to have flowed from this distinction, at least so far.

3.3 Commonality and Differentiation

Finally, it also cannot be ignored that at the Rio Conference where the conventions opened for signature, the Rio Declaration was agreed, which specified, in Principle 7, the importance of differentiation of obligations in the face of global environmental responsibility. It is worth quoting Principle 7 in full, not because this chapter wishes to devote excessive attention to differentiation, in contrast to commonality, but because Principle 7 is 'the flipside of the concept of common concern'.[69]

> States shall cooperate in a spirit of global partnership to conserve, protect and restore the health and integrity of the Earth's ecosystem. In view of the different contributions to global environmental degradation, States have common but differentiated responsibilities. The developed countries acknowledge the responsibility that they bear in the international pursuit of sustainable development in view of the pressures their societies place on the global environment and of the technologies and financial resources they command.

Differentiation between countries, especially but not exclusively between North and South, is of course not reserved in international law to environmental matters,[70] but its significance in this context has been a key feature of international environmental law at least since 1990,[71] though in light of shifts in approach, this may now be beginning to wane, at least in the context of the climate change regime.[72] Differentiation, though important in itself, is however as relevant for what it also says about commonality. Ultimately, it is differentiation from a common responsibility. It is thus within the context of global environmental responsibility that differentiation springs. Differentiation does not repudiate common concern; rather, it substantiates it.

Although Principle 7 is most routinely remembered for its second and third sentences, which incorporated differentiation as a guiding theme of international environmental policy, the first sentence has been somewhat overlooked, which is unfortunate because on reflection it provides a much stronger link with the principle of

---

[67] See the recent discussions on liability and redress within the context of the climate change regime, with particular regard to the establishment in 2013 of the 'Warsaw International Mechanism for Loss and Damage associated with Climate Change Impacts', viewable on the UNFCCC website at <http://unfccc.int/adaptation/workstreams/loss_and_damage/items/6056.php>.

[68] See the 2010 Nagoya–Kuala Lumpur Supplementary Protocol on Liability and Redress to the Cartagena Protocol on Biosafety, text at <http://bch.cbd.int/protocol/NKL_text.shtml>.

[69] Brunnée (n 32), 566.

[70] Preferential trading relations are permitted, for instance, under strict conditions within the disciplines of the World Trade Organization.

[71] L Rajamani, *Differential Treatment in International Environmental Law* (OUP, 2006).

[72] D French and L Rajamani, 'Climate Change and International Environmental Law: Musings on a Journey to Somewhere' (2013) 25 JEL 437, 440–441.

350   *Research handbook on biodiversity and law*

common concern as found in the Rio treaties. Indeed, the sentence provides clear evidence that common concern is not purely issue-specific to those matters that have simply incorporated the terminology within their treaties: as it says, 'States shall cooperate in a spirit of global partnership to conserve, protect and restore the health and integrity of the Earth's ecosystem'. What the first sentence of Principle 7 arguably also does is to provide support for a more expansive interpretation of that commonality; that common concern is not a principle that applies in a segmented way to specific matters of global environmental responsibility – be it climate change, wetlands, marine mammals – but, rather, that the principle is of general application, of which these are merely examples.

This is much more than a semantic difference. If taken to mean what the text certainly implies, it might help situate common concern as a general principle, detached and not reliant upon treaty law for its authority. If it is true that common concern provides a conceptual 'umbrella' for many multilateral environmental agreements, including those agreed in previous decades, sentence one of Principle 7 is as good as any other statement of jurisprudence or soft law in encapsulating that expression.[73] In short, common concern would be viewed as a putative customary norm, on which there is further discussion below.

4. NORMATIVE ASPECTS OF COMMON CONCERN: CURRENT, EMERGING AND FUTURE IMPLICATIONS

The final part of this chapter seeks to sketch some of the potential normative implications of common concern. These include, first, the elaboration of institutional processes to support the attainment of the common concern objective; secondly, the use of *actio popularis* as a mode of enforcement in international environmental law by individual States acting on behalf of the international community as a whole; and, thirdly, the argument that common concern may over time generate common standards of conduct of universal application – be it prevention, a precautionary approach, due diligence, or sustainable use, inter alia. As will become clear, these implications are increasingly accepted, to a greater or lesser degree, in international (environmental) law. To that extent, they are not necessarily innovative per se; rather, the argument is made that individually – and more importantly, collectively – they reveal something significant about the potential normative consequences of common concern:[74] namely, that through the elaboration of such developments there will materialise a more coherent understanding of what common concern may require. Thus, in considering the future direction of common concern, one should be wary of being too conservative in considering future effects. To that end, reference is particularly made to the 2011

---

[73] A similar justification can be found in the 1982 World Charter for Nature (UNGA Res 37/7), Principle 3: 'All areas of the earth, both land and sea, shall be subject to [these] principles of conservation.'

[74] Moreover, as these are general developments in international environmental law, the comments in this chapter are limited to what they may reveal about the future of common concern.

Advisory Opinion of the Seabed Disputes Chamber to show how what was previously considered a primarily non-operational, non-justiciable, legal concept – namely the common heritage principle – has proved itself open to reinterpretation and revitalisation by an independent tribunal to achieve a more dynamic understanding.

4.1 Institutional Coordination for Common Concern

One of the most established implications of common concern has been the creation of a plethora of institutional regimes for the more effective, and collective, performance of environmental treaty commitments. Such a framework is now considered a functional necessity of multilateral environmental agreements. As has been noted,

> Assuming that conservation and management principles can be agreed, the basic legal requirements for the institution of an effective conservation and management regime which provides for conservation of biodiversity are … establishment of the source of jurisdiction over the resource or resources concerned and their habitats; obligations to conduct scientific research and take account of scientific advice, subject now to the need, as appropriate, to adopt a 'precautionary approach'; prescription of regulations; establishment of permanent international institutions to provide a forum for discussion, evaluation, coordination, and adoption of required measures, inter alia; compliance and enforcement mechanisms; and dispute settlement arrangements.[75]

It is beyond the scope of this chapter to consider the features of these regimes in any detail, but three points are worth highlighting. First, the very formation of bodies such as Conferences (CoP) or Meetings (MoP) of the Parties as quasi-institutional forums,[76] or the use[77] (and, very occasionally, the establishment[78]) of international institutions proper for this purpose reinforces the global nature of the collective response, justifying corporate oversight and international action. Secondly, the practice of these institutional forums over the last few decades in taking their respective regimes forward, not only in terms of treaty development, but also in terms of promulgating a myriad of policy initiatives and 'soft law' decisions, reveals an ongoing process of international activity. In short, there is a perpetual dynamism through the regularity of institutional deliberations.[79] And while such negotiations are inherently political in nature, they operate within a legal framework which generates an inevitable regime-pull, as well as placing constraint at some level upon impermissible action. Thirdly, many of these regimes have established non-compliance procedures to provide a non-contentious process by which matters of implementation can be addressed without recourse to third-party

---

[75] Birnie, Boyle and Redgwell (n 45), 602.
[76] R Churchill and G Ulfstein, 'Autonomous Institutional Arrangements in Multilateral Environmental Agreements: A Little-Noticed Phenomenon in International Law' (2000) 94 AJIL 623–659.
[77] *See*, for instance, the International Maritime Organisation in relation to various marine pollution and related treaties.
[78] *See*, for instance, the International Whaling Commission.
[79] For a wide-ranging discussion, *see* MJ Bowman, 'Beyond the "Keystone" CoPs: The Ecology of Institutional Governance in Conservation Treaty Regimes' (2013) 15 International Community Law Review 5–43.

352 *Research handbook on biodiversity and law*

adjudication.[80] These procedures exemplify, perhaps as strongly as the development of the institutions themselves, the collective interest of States Parties in achieving the effective implementation of a Convention.

Thus, the first substantial aspect of common concern is treaty-institutionalism as a proxy for true global governance.[81] Moreover, in the light of changing State priorities and interests and the very real possibility of stalled normative progress in the future – despite the putative opportunities within the ideal of common concern, including those considered below – current treaty regimes will remain the foci of much international attention. This is particularly the case if States refuse to move to the next level of global environmental governance and ambition, reflecting a more realist view that 'such references ... camouflag[e] thin cooperation among states as high community aspiration'.[82] To that extent, the challenge will be merely to protect and to implement better the array of treaty arrangements that are already in place.

4.2 Common Concern and Environmental Obligations *Erga Omnes*

A second, tantalising, aspect of common concern is to consider whether, and how far, it provides support to the argument that certain environmental commitments are within that category of obligations that are of interest to the international community more generally: that is, that they are obligations *erga omnes*.[83] As the International Court made plain in the *Barcelona Traction Case*, these are obligations that '[b]y their very nature ... are the concern of all States. In view of the importance of the rights involved, all States can be held to have a legal interest in their protection.'[84] Significantly the Court went on to note these rights have acquired this status either because they have 'entered into the body of general international law ... [or] are conferred by international instruments of a universal or quasi-universal character'.[85] Though environmental matters were not specifically mentioned by the International Court in *Barcelona Traction*, both subsequent jurisprudence[86] and, indeed, the implication of the commentary of the International Law Commission (ILC) draft Articles on State Responsibility[87] leave little doubt that rules relating to the protection of the global environment, including those that seek to curtail marine damage, to conserve marine mammals, and to protect internationally significant habitats and global spaces – in shorthand, issues that are accepted as being of common concern – fall also within the rubric of obligations *erga omnes*.

---

[80] *See* the chapter by Scott in this handbook.
[81] *See* D French, 'Finding Autonomy in International Environmental Law and Governance' (2009) 21 JEL 255, 288.
[82] Brunnée (n 32), 554.
[83] *See* M Ragazzi, *The Concept of International Obligations Erga Omnes* (OUP, 1997).
[84] *Barcelona Traction, Light and Power Company, Limited, Second Phase (Belgium v Spain)* ICJ Reports (1970) 3, 33.
[85] Ibid.
[86] *Nuclear Weapons Advisory Opinion* (n 15), paragraph 241: 'The Court also recognizes that the environment is not an abstraction but represents the living space, the quality of life and the very health of human beings, including generations unborn.'
[87] Yearbook of the International Law Commission, 2001, vol. II (Part Two).

Nevertheless, the point is well made that the International Court 'did not find it necessary to create a means of enforcing *erga omnes* obligations ... and, twenty-five years later, [in *East Timor*] confirmed their view that recognition of a right *erga omnes* does not automatically permit any State standing to vindicate that right'.[88] Notwithstanding this, present discussions are returning to earlier debates as to whether there is a right of *actio popularis* in international law as a necessary supplement to the enforcement of obligations *erga omnes*. Certainly, the inclusion of Article 48 of the 2001 ILC Articles on State Responsibility, providing for non-injured States to invoke the responsibility of States in respect of obligations owed to the international community, has prompted renewed interest,[89] though, importantly, neither in the express wording of the draft articles nor in the commentary is reference made to the related, but doctrinally separate, issue of standing. Thus it is perhaps not surprising that, more cautiously, standing has first been considered within the tighter framework of enforcing obligations *erga omnes partes* between contracting parties to multilateral agreements, as recently upheld by the International Court in the 2012 *Belgium v Senegal* judgment concerning alleged breaches of the 1984 Torture Convention,[90] and much more pertinently argued by Australia against Japan in *Whaling in the Antarctic*.[91]

Moreover, as the Seabed Disputes Chamber in its 2011 Advisory Opinion noted, 'Each State Party may also be entitled to claim compensation in light of the *erga omnes* character of the obligations relating to preservation of the environment of the high seas and in the Area.'[92] Again, though not explicitly tackling the issue of standing, this is a very clear endorsement of the principle, especially as it goes on to rely on Article 48 expressly.[93] Indeed, this would seem to be an interesting development of the ILC Draft Articles, where the extent of a claim by non-injured States is limited to a request for cessation of the wrongful act, assurances of non-repetition and 'performance of the obligation of reparation ... in the interest of ... the beneficiaries of the obligation

---

[88] D French and K Scott, 'International Environmental Treaty Law' in MJ Bowman and D Kritsiotis (eds), *Conceptual and Contextual Perspectives on the Modern Law of Treaties* (CUP, in press) relying on *Case Concerning East Timor (Portugal v Australia)* ICJ Reports (1995) 90, at 102. The Court in *East Timor* explained that though the right in question (namely, self-determination) 'irreproachabl[y]' 'has an *erga omnes* character', nevertheless 'the *erga omnes* character of a norm and the rule of consent to jurisdiction are two different things'.

[89] J Peel, 'New State Responsibility Rules and Compliance with Multilateral Environmental Obligations – Some Case Studies of How the New Rules Might Apply in the International Environmental Context' (2001) 10 RECIEL 82–97.

[90] *Questions relating to the Obligation to Prosecute or Extradite (Belgium v Senegal)* ICJ Reports (2012) paragraphs 68–69.

[91] *Whaling in the Antarctic (Australia v Japan; New Zealand intervening)* Judgment of 31 March 2014. Interestingly, the Court in its judgment does not take the opportunity of clarifying further the law on standing in this area.

[92] Seabed Advisory Opinion (n 34), paragraph 180.

[93] Significantly, the Chamber also found, *ibid.*, that the Seabed Authority itself would also have a right to bring a claim: 'No provision of the Convention can be read as explicitly entitling the Authority to make such a claim. It may, however, be argued that such entitlement is implicit in Article 137, paragraph 2, of the Convention, which states that the Authority shall act "on behalf" of mankind'.

breached'.[94] By referencing Article 48, however, the Chamber risks conflating compensation for direct injury and the remedies appropriate to the invocation of *erga omnes* obligations. If its reference to compensation was only to the extent that it must be used for the benefit of the collective interest, the Opinion was less than clear on this point.

But leaving this to one side, there is no doubt that obligations *erga omnes* (*partes*), in particular, and common concern have real potential to be mutually reinforcing concepts, the former providing a general framework for the environmental-specificity of the latter. Thus, perhaps there is a very obvious link between the concepts – which, in some senses, it would seem almost trite to remark upon – namely, that international law already has the conceptual tools by which to develop and promote matters of common concern, especially where they are accepted to be of general application. Though there remain genuine questions over the consequences of obligations *erga omnes* and the extent of standing to secure their enforcement, especially beyond *inter partes* treaty commitments, in the context of the global environment, the common interest in the subject-matter and the apparent universality of the obligation will become increasingly difficult to disentangle.

4.3 Normative Development of Common Concern

The third normative aspect of common concern is to consider whether it might prompt substantive rule change. It is perhaps the most far-reaching, and from that perspective currently least likely, implication. It is premised upon the view that once a threat to a natural resource is recognised as being of common concern, certain common standards may begin to emerge, building upon particular treaty rules but eventually being of wider application. In short, common concern has the potential to create a nascent customary framework of conservation principles and standards.[95] The initial evolution of such ideas may already be beginning to emerge, prompted by recent jurisprudence. The International Court's 2010 judgment in *Pulp Mills*[96] and the Seabed Disputes Chamber's 2011 Advisory Opinion both reveal an increasing willingness to develop customary rules around the conservation of natural resources. Indeed, the Advisory Opinion expressly built upon the International Court's ruling in several important respects. For instance, the Chamber developed the Court's finding that there was now a customary obligation to undertake an environmental impact assessment in the case of

---

[94] Article 48(2), ILC Articles on State Responsibility.
[95] Alternatively, might common concern fall within the category of a general principle of international law? See P-M Dupuy, 'Formation of Customary Law and General Principles' in Bodansky, Brunnée and Hey (n 32), at 461: 'It might be argued that "general principles" of international environmental law differ from customary norms based only on the level of generality of their formulation. Nevertheless, both kinds of norms proceed from the same progressive sedimentation of general statements, together with more or less coherent state practice and sometimes assisted by judicial consolidation.'
[96] See *Pulp Mills on the River Uruguay (Argentina v Uruguay)* ICJ Reports (2010) 14, at 74, where the Court acknowledges 'not only the need to reconcile the varied interests of riparian States in a transboundary context and in particular in the use of a shared natural resource, but also the need to strike a balance between the use of the waters and the protection of the river consistent with the objective of sustainable development'.

the risk of transboundary harm, adding: 'The Court's reasoning in a transboundary context may also apply to activities with an impact on the environment in an area beyond the limits of national jurisdiction; and the Court's references to "shared resources" may also apply to resources that are the common heritage of mankind.'[97]

Thus, can one begin to see the emergence of expected standards of conduct in relation to natural resources that are of common concern, regardless of geographical location? Certainly, the International Law Association comes close to suggesting this in its 2002 New Delhi Declaration of Principles of International Law relating to Sustainable Development, where it states that: 'The protection, preservation and enhancement of the natural environment, particularly the proper management of climate system, biological diversity and fauna and flora of the Earth [sic], are the common concern of humankind'.[98] Equally, the Seabed Disputes Chamber in particular makes a number of important and general statements on such key issues as prevention, precaution and due diligence.[99] Together they provide an interesting insight into the next stage of international environmental law, on which other chapters in this handbook are able to provide further insight.

Of course, bestowing a general status upon common concern would not in and of itself transform the legal obligations of States. As noted previously, it would remain inchoate without further substantiation through the adoption of more detailed international rules and institutional processes. Nevertheless, to assert that conserving the global environment is of common concern would ultimately raise the potential spectre of broader, and more general, legal implications. Of course, one is reminded not only of the inherent risks of translating treaty rules too readily into custom,[100] but also of the warning in *Dispute Concerning Access to Information under Article 9 of the OSPAR Convention (Ireland v United Kingdom)* (2003), in which it was noted that 'A treaty is a solemn undertaking and States Parties are entitled to have applied to them and to their peoples that to which they have agreed and not things to which they have not agreed'.[101] Thus, it would be wrong to suggest that such rule development is inevitable, much less relate it directly to the notion of common concern. However, it cannot escape attention that the International Court in *Pulp Mills* expressly references sustainable development to justify what is, in effect, progressive development, and the Seabed Chamber rightly premises its opinion on common heritage. Of particular interest is the Chamber's re-envisioning of the common heritage principle. It has taken an inherently controversial, somewhat fractured, principle and revealed its continuing relevance in the functioning of the deep seabed mining regime, especially as it now begins the next phase in its operational development. Certainly, the Chamber makes the point very clearly that the express wording of the treaty must be considered as determinative and

---

[97] Seabed Advisory Opinion (n 34), paragraph 148.
[98] UN Doc. A/57/329 (1992), annex.
[99] *See* generally French (n 41).
[100] *See* the *North Sea Continental Shelf Cases (Federal Republic of Germany v Denmark; Federal Republic of Germany v Netherlands)* ICJ Reports (1969) 3, at 42, where it is noted that 'this result is not lightly to be regarded as having been attained'.
[101] Award of 2 July 2003, 23 RIAA 59, paragraph 102.

cannot be displaced by external values, however important.[102] Yet on reading the Advisory Opinion, one cannot ignore how the Chamber relies upon the common heritage principle in its decision-making. This is no dead provision but rather one that retains significant, and ongoing, relevance. Mention has already been made, for instance, of the Chamber's reliance on common heritage in justifying the broadening of invocation of responsibility for environmental harm.

This final section has been in part speculative; not necessarily in terms of the future direction of the law in this area, but in terms of its relationship to common concern. The proposition has been made that common concern might itself generate legal implications – institutional, enforcement and substantive – that over time will become of general application. Even if this were not to occur, as the Seabed Disputes Chamber has highlighted, concepts are rarely so inutile that they cannot provide ongoing relevance directly within treaty regimes. Common concern and other global(-ising) principles have enormous future potential. Brunnée is surely correct to give pre-eminence to the extant treaty regimes, acknowledging the advantage they have in transforming 'pragmatic cooperation into genuine normative communities',[103] but equally this chapter has also reflected on the aspiration that common concern contains and that it 'is no longer being applied for purely narrative purposes, but rather that it is intended to import some reasonably specific legal consequences'.[104] To that extent, common concern is much more than a political tool; it has both the content and normative longevity of a meaningful legal principle.

5. CONCLUSION

The chapter has reviewed the concept of common concern, noting its particular – though not exclusive – relevance to matters of global environmental responsibility. There is no doubt that common concern was in part selected for the Rio Conventions on a negative basis, as it had fewer of the difficulties associated with the common heritage principle. Nevertheless, seeing it only in this light would undermine its importance, both then and, increasingly, now. Common concern reflects, in a rather summary manner, a willingness by the international community to act collectively to conserve the integrity of the biosphere. And be it for reasons of functionality, legal necessity or global aspiration, common concern encapsulates a need to respond holistically to this ecological imperative regardless of traditional doctrines of either permanent sovereignty over natural resources, on the one hand, or *res communis*, on the other. Thus, though the scope of the obligation appears relatively weak, its potential to raise challenges to the established legal order is highly significant.

Of course, the emerging and future normative implications of common concern remain unclear, certainly in contrast to the relatively concrete institutional and, to some extent, substantive treaty measures taken up to this point. Nevertheless, the direction of movement is reasonably ascertainable – an increasing level of engagement by the

---

[102] Seabed Advisory Opinion (n 34), paragraph 156.
[103] Brunnée (n 32), 572.
[104] Bowman (n 26), 503.

international community, either collectively or through the active agency of individually concerned States. Moreover, as the Seabed Disputes Chamber has shown, in the right context, normative meta-principles, such as the common heritage of mankind, are capable of surprising application.

To that extent, we may not yet have seen the most interesting implications of what common concern will come to require of States in conserving the global environment. It may have been a concept borne out of expediency and political negotiation, but common concern may in future reveal a willingness on the part of the international community collectively to engage in a broader, if incremental, project of refashioning – in however minor a way – the legal order for the benefit of future generations and their natural environment.[105]

As indicated above, future research questions include how far common concern will be strengthened in matters of institutional development, enforceability and normative development. Beyond this, there is the related issue of how far common concern – and concepts like it – will become part of, and influence, the mainstream of general international law. In the discussion of the fragmentation of international law, discerning ideas that reach beyond a sub-disciplinary focus to help us understand the wider system will become ever more important. And finally, how far will common concern be able to develop in challenging the Westphalian distinctions between territorial sovereignty, on the one hand, and common areas, on the other?

What is certain is that, as with all such autonomous and quasi-autonomous concepts, once agreed by States, their future direction is never wholly predictable, be it in the jurisprudence of international courts and tribunals, or through the collective dynamism of Conferences of the Parties. On that basis alone, common concern is so much more than a mere rhetorical device.

SELECT BIBLIOGRAPHY

Biermann, F, '"Common Concern of Humankind": The Emergence of a New Concept of International Environmental Law' (1996) 34 Archiv des Völkerrechts 426

Bowman, MJ, 'Environmental Protection and the Concept of Common Concern of Humankind' in M Fitzmaurice, D Ong and P Merkouris (eds), *Research Handbook on International Environmental Law* (Edward Elgar Publishing, 2010) 493

Bowman, MJ, PGG Davies and CJ Redgwell, *Lyster's International Wildlife Law* (2nd edn, CUP, 2010)

---

[105] Nevertheless. there remains a reluctance within general international law to endorse the principle too readily. See Report of the International Law Commission (Sixty-seventh session) (4 May–5 June and 6 July–7 August 2015) ILC Report, A/70/10, 2015, chap. V (Protection of the Atmosphere) paragraph 54: 'While a number of treaties and literature demonstrate some support for the concept of "common concern of humankind", the Commission decided not to adopt this language for the characterization of the problem, as the legal consequences of the concept of common concern of humankind remain unclear at the present stage of development of international law relating to the atmosphere. It was considered appropriate to express the concern of the international community as a matter of a factual statement, and not as a normative statement, as such, of the gravity of the atmospheric problems'.

Boyle, AE, 'International Law and the Protection of the Global Atmosphere: Concepts, Categories and Principles' in RR Churchill and DAC Freestone (eds), *International Law and Global Climate Change* (Graham and Trotman, 1991) 7

Brunnée, J, 'Common Areas, Common Heritage, and Common Concern' in D Bodansky, J Brunnée and E Hey (eds), *The Oxford Handbook of International Environmental Law* (OUP, 2007) 550

Churchill, R and G Ulfstein, 'Autonomous Institutional Arrangements in Multilateral Environmental Agreements: A Little-Noticed Phenomenon in International Law' (2000) 94 AJIL 623

French, D, 'Finding Autonomy in International Environmental Law and Governance' (2009) 21 JEL 255

French, D, 'From the Depths: Rich Pickings of Principles of Sustainable Development and General International Law on the Ocean Floor – the Seabed Disputes Chamber's 2011 Advisory Opinion' (2011) 26 International Journal of Marine and Coastal Law 525

French, D (ed), *Statehood and Self-Determination: Reconciling Tradition and Modernity in International Law* (CUP, 2013)

French, D and K Scott, 'International Environmental Treaty Law' in MJ Bowman and D Kritsiotis (eds), *Conceptual and Contextual Perspectives on the Modern Law of Treaties* (CUP, in press)

2014 International Law Association Declaration on Legal Principles relating to Climate Change, at <http://www.ila-hq.org/en/committees/index.cfm/cid/1029>

Murase, S, First Report of the ILC Special Rapporteur on the Protection of the Atmosphere (A/CN.4/667, 14 February 2014)

Peel, J, 'New State Responsibility Rules and Compliance with Multilateral Environmental Obligations – Some Case Studies of How the New Rules Might Apply in the International Environmental Context' (2001) 10 RECIEL 82

Ragazzi, M, *The Concept of International Obligations Erga Omnes* (OUP, 1997)

Rajamani, L, *Differential Treatment in International Environmental Law* (OUP, 2006)

PART IV

REGULATORY CHALLENGES AND RESPONSES

13. Biodiversity, knowledge and the making of rights: reviewing the debates on bioprospecting and ownership

Emilie Cloatre

1. INTRODUCTION

This chapter reviews some of the debates surrounding access to genetic resources and the ethics of bioprospecting, and unpacks some of their complexity. It questions what the debates and legal answers provided to these debates indicate with regard to the status of biodiversity in global discourses, and in turn for our understanding of the role that law may play in controversies around biological resources.

The term bioprospecting covers practices of systematic search through natural resources, with a view to pursuing further scientific development. In many cases, though not always, such developments also have a commercial aim – this constitutes the main focus of much of the controversy in this domain, and the main focus of the analysis below. Since the late 1990s, bioprospecting has become a highly contentious practice and politically loaded issue, in globalised as well as local settings.[1] Since it brings together questions of ownership, access to natural resources, rights over nature and knowledge, the problem is a very 'messy' one, which seems to evade straightforward solutions.[2] The 1992 Convention on Biological Diversity (CBD), without being the only forum in which these issues were debated, has been a focal point for some of the concerns raised by communities, (some) governments and NGOs over the years, but the solutions proposed still leave many core questions unanswered.[3] Key paradigms such as 'benefit-sharing' and 'prior informed consent' have been developed that seek to ensure that the uses of biodiversity are compatible with aspirations for both environmental conservation and global justice. While these have certainly provided important new tools for those seeking to engage with the ethics of bioprospecting, many maintain that legal solutions, within and outside the CBD, do not fully address the political,

---

[1] DF Robinson, 'Locating Biopiracy: Geographically And Culturally Situated Knowledges' (2010) 42(1) Environment and Planning A 38; K Moran, SR King and TJ Carlson, 'Biodiversity Prospecting: Lessons and Prospects' (2001) 30 Annual Review of Anthropology 505; G Dutfield, 'Sharing the Benefits of Biodiversity: Is There a Role for the Patent System?' (2002) 5(6) Journal of World Intellectual Property 899.

[2] C Hayden, *When Nature Goes Public: The Making and Unmaking of Bioprospecting in Mexico* (Princeton University Press, 2003); R Coombe, *The Cultural Life of Intellectual Properties: Authorship, Appropriation and the Law* (Duke University Press, 1998); M Brown, *Who Owns Native Culture?* (Harvard University Press, 2003).

[3] D Schroeder and T Pogge, 'Justice and the Convention on Biodiversity' (2009) 23(3) Ethics and International Affairs 267.

social and ethical issues that bioprospecting has raised, and as a result are rarely able to solve all the dilemmas it generates.[4]

The complexity of these debates is largely due to the fact that their meanings are deeply rooted in politically shaped understandings of the value of natural resources, and of the knowledge that relates to them. As a result, discourses surrounding bioprospecting fold into much larger questions of cultural and social valuing. In turn, this makes the problem difficult to address without engaging those deep-running issues of social justice and sociological constitutions. In order to fully grapple with the issues at play, it is therefore useful to engage not only with legal commentaries on the issues, but also with critical insights; in particular the extensive analyses that have emerged from Science and Technology Studies (STS) or critical intellectual property scholarship. In doing so, it becomes possible to pay attention to the complex ways in which genetic resources are not only natural products, but also social objects. This in turn helps with problematising the ways in which biodiversity (in its materiality and sociality) is conceived[5] and with seizing better the complexity of the debate, and the ability of law to provide workable answers.

This chapter starts by reviewing the history of the debates on bioprospecting, and the legal problems these have raised (Section 2). It then explores how environmental law, in the shape of the CBD and subsequent documents issued by the Conference of the Parties to the CBD, has sought to answer the debates at stake (Section 3). In Section 4, the translation of these principles into practice is explored, and some of the limitations of the legal paradigms are interrogated. Section 5 explores why the issue of biopiracy is as much a conflict of knowledge as it is a conflict over resources, and the links between bioprospecting and intellectual property are considered. The chapter then turns towards a more critical exploration of the problems at play, and argues that the legal solutions proposed to what are defined as the key 'problems' of bioprospecting fail to fully seize and respond to the complex patterns of power and inequality that are visible in its practice (Section 6). In turn, this means that the issues raised by bioprospecting are inherently representative of the problematique of global justice, access and distribution that have riddled much of the life sciences, and in particular biomedical sciences, in recent years. Importantly, debates on bioprospecting become illustrative and symbolic of conversations in law and politics that are significant for the positioning of biodiversity in contrasting understandings of modernity.

2. BIOPROSPECTING, ETHICS AND POLITICS

This section briefly introduces the issues surrounding practices of bioprospecting, and the shifts in some political discourses towards its denunciation and relabelling as

---

[4] C Hayden, 'Bioprospecting's Representational Dilemma' (2005) 14(2) Science as Culture 185; C Hayden, 'Taking as Giving: Bioscience, Exchange, and the Rise of an Ethic of Benefit-sharing' (2007) 5 Social Studies of Science 729; A Osseo-Asare, *Bitter Roots: The Search for Healing Plants in Africa* (University of Chicago Press Books, 2014).
[5] *See* Hayden (n 2); A Pottage, 'Too Much Ownership: Bioprospecting in the Age of Synthetic Biology' (2006) 1(2) Biosocieties 137.

'biopiracy'. I explore aspects of its problematique and some of the ways in which the debate has been framed in academic commentaries and policy conversations.

The use by industrial researchers of natural resources and associated traditional knowledge has a long history. The development of many well-established pharmaceutical substances has been traced back to plants and traditional knowledge, in particular since the colonial era.[6] For example, the discovery of quinine is known to have been derived from the cinchona tree collected by researchers in the Amazonian forest[7] – an example commonly used as one of the earliest cases of biopiracy.[8] Other examples of medicines derived from plants used in traditional medicine include aspirin and the birth control pill.[9] Similarly, studies have explored how poisonous arrows seized by colonisers in West Africa provided the basis for the development of potent medicines, providing a particularly poignant example of the seizing of knowledge and deployment of conflicting power.[10] In these processes, materials and knowledge are transformed, from plants to medicines, from traditional usages to industrial application. At the same time, the form of intellectual property systems means that these products often shift from 'freely accessible' status (legally if not morally) to subjects of particular forms of individual rights and monopolies – a point I return to below. The use of natural resources, and of indigenous knowledge that had helped identify potentially powerful plants and substances, therefore has well-established roots which are entangled[11] in processes of appropriation characteristic of the colonial era.

Over the years, bioprospecting also became a common and organised practice for many researchers and companies that developed missions, partnerships or strategies to systematically explore the value of these resources.[12] Progressively, it emerged as an organised, often systematic, way of collecting knowledge through biodiversity. In many

---

[6] L Schiebinger, *Plants and Empire: Colonial Bioprospecting in the Atlantic World* (Harvard University Press, 2004).

[7] MC Torri and TM Hermann, *Bridges Between Tradition and Innovation in Biomedicine* (Springer, 2011).

[8] For a legal discussion of the events, see *Merrell Dow Pharmaceuticals Inc v Norton & Co* [1996] RPC 76, 88–89.

[9] Hayden (n 2); MJ Balick and PA Cocks, *Plants, People and Culture: the Science of Ethnobotany* (WH Freeman, 1996).

[10] A Osseo-Asare, 'Bioprospecting and Resistance: Transforming Poisoned Arrows into Strophantin Pills in Colonial Gold Coast, 1885–1922' (2008) 21(2) Social Science and Medicine 269.

[11] A term that, following others in STS, I use to refer to the inherent intermingling of objects, actors or fields of activity. The long history of the word in sociology of science and technology links to the key claims of the discipline that all key divides that we have imagined (nature/society, humans/non-humans) are in fact more blurred and complex than assumed. When studying the role of any entity, their complex interrelations with others therefore become essential to understanding their mechanics. In this chapter I use this term more generally when referring to actors that bear tight co-relationship with other actors or processes. For further conversations, see B Latour, *We Have Never Been Modern* (Harvard University Press, 1993); B Latour, *Reassembling the Social: An Introduction to Actor-Network Theory* (OUP, 2005).

[12] S Green, 'Indigenous People Incorporated? Culture as Politics, Culture as Property in Pharmaceutical Bioprospecting' (2004) 45(2) Current Anthropology 211.

ways, bioprospecting therefore became a specific enterprise to be regulated, raising its own set of legal, ethical and regulatory questions.

It is important that many of the stories that have emerged of plant-led pharmaceutical research have taken place in the so-called Global South. The rich biodiversity of many southern nations has made them particularly abundant sources of genetic materials and many stories of plants being turned into drugs have emerged in developing nations. Indeed, the rich biodiversity of the South has often been put forward as the main reason why research on plants was particularly effective there. At the same time, this vision overlooks other important factors, both for the historical explanation of events, and for their social significance. In particular, the colonial history of these places facilitated the emergence of bioprospecting and of cases of biopiracy. The dispossession, power imbalances, strategies of dominance and, later on, maintenance of economic dominance, in particular by global corporations, have created in such places the conditions that enabled some of the most problematic cases of ethically fragile practices and 'biopiracy'.[13]

The historical legal context is relevant here too. In the 1890s, the Bering Sea Fur Seals Arbitration established the principle that natural resources were subject to standard principles of sovereignty, and that states should therefore have control of natural resources within their own territory. The principle survived throughout the remainder of the colonial era, giving rise to many of the early cases of seizing of natural resources for purposes that would rarely benefit local populations, including those mentioned above. In 1983, the Food and Agriculture Organization (FAO) International Undertaking on Plant Genetic Resources created a turning point, at least in the agricultural field, by making genetic resources the 'heritage' of mankind. This emerging principle, then of clear benefit to developed countries, was later contested again by developing countries, who saw themselves becoming deprived of the benefits that the principle of sovereignty had granted to colonial powers for many decades.[14] Both the history of bioprospecting, and the legal developments that are explored below, should therefore be positioned within this particular historical perspective.

For some, the importance of plants in pharmaceutical development has been overstated, particularly in recent years, as research has increasingly become based on synthetic chemistry. They read this as meaning that research had been diverting and will continue to move away from the need to initiate discovery from plants.[15] Indeed, both the budgets invested in research on plants, and the share of drugs that are ultimately derived from them, are relatively small compared to the size of pharmaceutical budgets and profits as a whole.[16] Nonetheless, evidence demonstrates that

---

[13] C Hayden, 'From Market to Market: Bioprospecting's Idioms of Inclusion' (2003) 30(3) American Ethnologist 359.

[14] MJ Bowman and C Redgwell, *International Law and the Conservation of Biological Diversity* (Kluwer, 1996), chapter 8; P Sands and J Peel, *Principles of International Environmental Law* (CUP, 2012), 507–510.

[15] AL Harvey and N Gericke, 'Bioprospecting: Creating a Value for Biodiversity' in I Pavlinov (ed), *Research in Biodiversity – Models and Applications* (InTech, 2011), available at <http://www.intechopen.com/books/research-in-biodiversity-models-and-applications>.

[16] A Artuso, 'Capturing the Chemical Value of Biodiversity: Economic Perspectives and Policy Prescriptions' in F Grifo and J Rosenthal (eds), *Biodiversity and Human Health* (Island

plants remain one important, albeit not exclusive, source that the pharmaceutical industry is keen not to neglect in the constant pressure they experience for competitive innovation. Similarly, reminders of the importance of plants and knowledge about plants in drug discovery are regularly published in medical journals.[17] The importance of plant-based research, in addition, should not be considered exclusively in terms of its significance from the industry's point of view, but also in relation to its ethical, financial and environmental meaning from the point of view of source nations and communities.

2.1 Emerging Controversies and the Framing of 'Biopiracy'

From the 1990s, however, the approach adopted by bioprospecting, and the persistence of its practice, started being brought to the forefront of global debates, both in environmental politics, and in broader debates on cultural rights and global justice.[18] Alongside debates within policy circles and broader civil society, scholars in law and social sciences started taking an interest in what these debates meant for contemporary understandings of nature, and of the interrelationship between knowledge, culture and (relative) power.[19] A series of particularly controversial cases galvanised a new form of resistance to the way in which local resources were used for private interests. These debates have been extensively discussed in the legal literature,[20] and I will here just try to summarise some of the key aspects. In particular, this section will start by briefly recalling the contestations that were voiced from the 1990s, and that illustrate the complex ways in which practices and resources became problematised.

In the 1990s, several controversial cases, which became labelled as cases of 'biopiracy', emerged in relation to the patenting of products derived from natural resources or traditional knowledge. These surfaced as applications were made for inventions that came to be contested, by representatives of indigenous communities or interested NGOs, as being a misappropriation of knowledge that had been held for many years by various communities. A few of these examples are recounted to illustrate the controversies, before moving on to exploring their legal complexity. Although the chapter returns in more detail to the question of intellectual property (IP)

Press, 1997), 184–204; B Aylward, 'The Role of Plant Screening and Plant Supply in Biodiversity Conservation, Drug Development, and Health Care' in T Swanson (ed), *Intellectual Property Rights and Biodiversity Conservation* (CUP, 1995), 93–126.

[17] EJ Buenz and others, 'Searching Historical Texts for Potential New Drugs' (2006) 333 British Medical Journal 1314.

[18] Brown (n 2); ME DeGeer, 'Biopiracy: The Appropriation of Indigenous Peoples' Cultural Knowledge' (2003) 9 New England Journal of International and Comparative Law 179.

[19] Coombe (n 2); R Coombe, 'Legal Claims to Culture in and Against the Market: Neoliberalism and the Global Proliferation of Meaningful Difference' (2005) 1(1) Law, Culture and the Humanities 35; Hayden (n 13); Hayden (n 2); Hayden, 'Taking as Giving' (n 4); Pottage (n 5).

[20] *See*, for example, Robinson (n 1); DeGeer (n 18); I Mgbeoji, 'Patents and Traditional Knowledge of the Uses of Plants: Is a Communal Patent Regime Part of the Solution to the Scourge of Biopiracy?' (2001) 9 Indiana Journal of Global Legal Studies 163; V Shiva, *Biopiracy: The Plunder of Nature and Knowledge* (South End Press, 1997).

below, a brief introductory point is useful here. Patent law assumes that, in order to be patentable, inventions need to be new, to involve an inventive step and to be capable of industrial application.[21] Both the notions of novelty and industrial application have long been recognised to be problematic as far as the protection of traditional knowledge is concerned. Similarly, most of the cases of biopiracy have revolved around the question of novelty, and it has been for courts to determine when a product developed on the basis of traditional knowledge differs sufficiently from traditional usage to be deemed 'new'. While the global IP system, in the form of the WTO Agreement on Trade-Related Aspects of Intellectual Property Rights (TRIPS), provides the general principles of patentability, it is then for national courts of the country within which a patent application is filed to make decisions and judgments as to how the notion of novelty should be applied. For the purpose of this chapter, a description of some of the best-known cases of patents, challenged on the basis of the lack of novelty of the invention they sought to protect, and following claims of biopiracy, is sufficient to understand some of the key issues at stake.[22]

The revocation of a US patent on the use of turmeric powder for the treatment of wounds granted to the University of Mississippi Medical Center in 1995[23] is a useful example of the early cases that generated worldwide opposition, and has become symbolic of the problem of biopiracy.[24] Turmeric powder has a long history of use in Ayurvedic medicine in India, and its potential uses have been explored and identified by traditional practitioners for many years. The patenting of its use for wound healing in 1995 therefore, unsurprisingly, generated opposition and a formal challenge by the Indian government. As patents are in principle rights granted on an invention, to a named inventor, they should not be provided on pre-existing knowledge. However, what is considered as pre-existing knowledge is not necessarily straightforward, nor commonsensical: in US law, therefore, what is to be considered as 'state of the art', or in other words the state of pre-existing knowledge, needs to have been documented in writing if this knowledge is based outside of the US. In the case of the turmeric, the Indian government therefore managed to contest the patent because it was able to produce documentary evidence of the fact that this particular medicinal use was indeed already known, and therefore what the applicant was seeking to protect was not an invention for the purpose of patent law.

The turmeric case itself built on earlier US patenting practices that had proved controversial in relation to Indian knowledge, most symbolically in relation to the Neem tree. The Neem tree has played a significant part in medicinal practices in India for centuries. Each of its parts, from leaves to seeds, oil and wood has been used in some way in the daily health practices of millions of families. The tree also holds various properties as an insecticide, and, aside from its medicinal functions, farmers

---

[21] 1994 WTO Agreement on Trade-Related Aspects of Intellectual Property Rights, Article 27(1) available at <https://www.wto.org/english/docs_e/legal_e/27-trips.pdf>.
[22] For more detailed engagement with specific IP systems, *see,* for example, G Dutfield, *Intellectual Property Rights, Trade and Biodiversity* (Earthscan, 2000); or for a general overview of IP law *see* L Bently and B Sherman, *Intellectual Property Law* (OUP, 2014).
[23] U.S. Patent No. 5, 401, 504.
[24] For further details *see,* for example, Dutfield (n 22); Shiva (n 20).

have developed various agricultural uses. In addition, and importantly, the Neem tree is considered as sacred in some parts of India and is therefore an important cultural symbol.[25] Western researchers have been aware of these properties since at least the 1960s, and it became a subject of interest for research. In the 1990s, WR Grace and Co took a particular interest in the agricultural practices that had emerged from the Neem tree, and produced a Neem-based insecticide product that could be developed and sold for industrial purposes. The product was largely based on furthering the potentialities of the extracts traditionally used by Indian farmers, by turning it into a product that was long-lasting enough to be used on an industrial scale. In 1992, the company were granted a patent for this particular product, and for particular industrial versions of the Neem extracts.[26] Soon, activists in India and elsewhere organised to challenge (unsuccessfully) the granting of this patent, and expose it as an illegitimate use of pre-existing, collective knowledge, for private and economic interests.[27] Controversies around the Neem tree patent have continued, with various other patents being granted in the US and through the European Patent Office. The revocation of these after challenges by activists in 2005 became held as one of the symbolic victories against biopiracy.[28]

Such cases emerged across the globe in the 1990s, arguably not so much as a result of changes in the law that sought to address such issues, but largely as a result of an increasing awareness and mobilisation against the issue by various NGOs and community organisations. If opposition was particularly vocal and organised in India, other controversies emerged in Central and South America or in Africa. On the former, the example of the Enola bean – a bean species understood and known to Peruvian farmers before being patented by an agricultural company and its patent subsequently being challenged – is an interesting example. One well-documented example of the latter is the patenting of various substances isolated from the Hoodia plant, a resource used by Bushmen in South Africa as part of various medicinal practices.[29]

Before exploring in the next section some of the solutions that have been proposed to the problem of biopiracy, it is useful to reflect further on the nature of those conflicts, both in legal terms (how does the law allow practices that may seem deeply unethical?) and in more conceptual or social terms (what are the various social concerns, and concerns for justice, that arise from those conflicts?).

One first level of complexity of the various cases, taken as a whole, is that they do not necessarily provide a uniform picture in terms of the links between traditional knowledge, natural resources and final product for which patents are sought. For example, the final invention may be more or less closely related to information provided by indigenous groups, it may be closer or more distant from the traditional

---

[25] See, for example, Shiva (n 20).
[26] See US Patent No. 5, 124, 349.
[27] E Marden, 'The Neem Tree Patent: International Conflict over the Commodification of Life' (1999) Boston College International and Comparative Law Review 279.
[28] EPO Boards of Appeal, Decision of 8 March 2005, Case No. T0416/01-3.3.2. See also the Decision of 23 July 2008, Case No. T0096/06-3.3.01.
[29] Dutfield (n 1); Harvey and Gericke (n 15).

remedy, or it may be used for the same or a different purpose. As a result, both ethical and legal dilemmas might vary.

In addition, configurations of the actors involved are by no means uniform – for example, some cases involve transnational movements while others involve primary actors (providers of resources and applicants) based in one jurisdiction. In turn, this means that the sets of interest that states in provider countries may have might vary – we can think, for example, of the interests of the Indian government in the turmeric or Neem cases, and that of the South African government in relation to the Hoodia case.

To return to the issue of the links between plants, knowledge and final products, the processes at play can be summarised as follows. Bioprospectors, following research activities that adopt a variety of forms, from exploring natural settings per se to researching local markets, having identified the potential value of a particular plant, undertake further research on it to establish the way in which it works, and how it can be used for industrial purposes (one of the requirements of patent law for a product to be deemed patentable). Once this has been done, a particular form of the extracts from the plant, or a particular type of use of the plant, is proposed in a patent application. The new product derived from the plant, is argued to be 'novel' and of 'industrial application' by the researchers, which therefore potentially enables them to obtain a patent, and the financial rewards that this brings through market exclusivity.[30]

The amount of distance and transformation evident between plant discovery and industrial product development varies substantially in different cases, as do investments and research. At one level, therefore, debates on biopiracy surround essentially the use of biodiversity for research, and complex questions in the early stages of the debate arise in terms of how much transformation would be needed for a 'product' to be more than a 'biological organism' – on which patents, in theory, cannot be allocated. Legal doctrine can, to some extent, answer these questions, although not always consistently across jurisdictions. At another level, however, the debates are further complicated by the potential role played by indigenous/traditional/local knowledge (qualifications that are themselves subject to complex processes of social making, as I return to below) in the process of the substance's discovery.

3. ENVIRONMENTAL LAW AND BIOPROSPECTING: FROM THE CBD TO THE NAGOYA PROTOCOL

In the face of the increasing tensions and debates surrounding bioprospecting, legal responses have been deployed since the 1990s, cutting across arenas in which the various key constituents of the problem were considered. While environmental treaties and organisations became a privileged forum in which the significance of bioprospecting for the valuing and meaning of nature was considered, these negotiations were taking place against a backdrop of increasing attention to indigenous peoples' rights.[31]

---

[30] I Mgbeoji, *Global Biopiracy: Patents, Plants and Indigenous Knowledge* (Cornell University Press, 2006); Shiva (n 20).

[31] Including, for example, the early negotiations of the Declaration on the Rights of the Indigenous People, ultimately adopted by the UN General Assembly in 2007, UNGA Res.

In addition, debates and conversations surrounding biopiracy, and the solutions proposed, took place at a range of levels, from global organisations to local conversations, and international treaties to contractual arrangements.

In spite of this, the legal responses to biopiracy, and the legal framing of bioprospecting, came to be defined around a few key events. One such defining and well-known event came in the shape of the FAO's 2001 International Treaty on Plant Genetic Resources for Food and Agriculture.[32] This text has been significant in shifting understandings of plant genetic resources away from the notion of common heritage and asserting farmers' rights as they relate to plant genetic resources for food and agriculture (Article 9). In turn, the text reaffirms the need to protect traditional knowledge, and the right to benefit-sharing for farmers, and restricts the possibility for IP claims to be made on plant genetic resources (Article 10). It therefore forms an important background to some of the discussion below. However, the scope of the treaty itself, being restricted to plants used for food and agriculture (Article 3), meant that vast resources were excluded from its reach. Notably, plants used in pharmaceutical research, which this chapter is most concerned with, are not affected by the treaty. Therefore, while it constitutes an important element of background to the issues explored below, it will not be a key focus of this analysis.

The other key event, of more specific import here, and of particular relevance to this handbook, is the CBD. The negotiation of the CBD was the result of long, and at times, tense negotiations – a history that has been told elsewhere in more detail.[33] Within these negotiations, the issue of the formal assertion of the rights of countries and indigenous people over their own resources was one of the central points of tension, as conflicting interests between providers and potential users were played out. Importantly, this aspect of the debates became essential to the participation of biodiversity-rich developing countries to the negotiations of the CBD. The significant financial and institutional efforts that were requested of them in order to protect and sustain biodiversity was seen to be at least in part compensated if the Convention was to also pay attention to their relative position of power in negotiated access to their resources.

The CBD provided one of the first forums in which biodiversity-rich developing countries could assert the notion that access to natural resources should be negotiated, and raised questions of rights and sovereignty of both people and states. This was important in a context where, previously, bioprospectors' access had not been mediated by national law, with the result that genetic resources were often considered to be unregulated and freely accessible. In turn, this was symbolic of a broader set of discursive and political changes. For some, the simple fact of engaging with the issue of access to genetic resources was an important step, as it meant that the patterns of inequality and global (in)justice that had come to be rewarded by bioprospecting practices and associated IP opportunities came to be recognised by the international

---

61/295. Note also the 1989 Convention Concerning Indigenous and Tribal Peoples in Independent Countries (ILO 169), 28 ILM 1382.

[32] Text of the treaty available at <http://www.planttreaty.org>.

[33] *See* C McManis (ed), *Biodiversity and the Law: Intellectual Property, Biodiversity and Traditional Knowledge* (Earthscan, 2007); F McConnell, *The Biodiversity Convention – A Negotiating History* (Springer, 1996).

community (or sections of it) as being problematic and needing to be addressed.[34] However, in order to explore the significance of the legal contributions of the CBD, it is useful to question the agreement and its consequences more closely.

The CBD sets itself three objectives in Article 1. While the first two relate respectively to the conservation and sustainable use of biodiversity (what could be described as the original aims pursued by those who triggered the negotiations), its third objective is the fair and equitable sharing of the benefits arising out of the utilisation of genetic resources. In itself, this is of significance in that it illustrates the success of biodiversity-rich negotiators in putting this issue on the agenda of the CBD. References to the third objective continue in the text of the CBD itself. Article 8, which sets the core principles that have been developed in further decisions, is worth quoting here:

> Each contracting Party shall, as far as possible and as appropriate: ...
>
> (j) Subject to national legislation, respect, preserve and maintain knowledge, innovations and practices of indigenous and local communities embodying traditional lifestyles relevant for the conservation and sustainable use of biological diversity and promote their wider application with the approval and involvement of the holders of such knowledge, innovations and practices and encourage the equitable sharing of the benefits arising from the utilization of such knowledge innovations and practices.

This article has two main effects. First, and most notably, it recognises the principle of benefit-sharing in the context of traditional knowledge over biological resources.[35] Second, it formally recognises the significance of traditional knowledge in the preservation and valuation of biodiversity. The entanglement of knowledge and nature, society and biodiversity, is recognised, with the effect of acknowledging the importance of knowledge and practices that are not geared towards either science or industry. While the possibilities for traditional knowledge to be effectively protected in other legal domains (including through IP) often remain uncertain, as will be developed below, this is nonetheless an important statement in this major environmental Convention.[36]

Article 15, in turn, sets the principle that access to genetic resources should be on mutually agreed terms[37] and subject to prior informed consent.[38] Again, in its letter, the CBD therefore seems to engage some of the concerns put forward by provider parties. Shortly after the CBD, various regional and national regimes for access and benefit-sharing (ABS) were put into place that have been analysed in various sections of the

---

[34] Schroeder and Pogge (n 3).
[35] P Miller, 'Impact of the Convention on Biological Diversity: The Lessons of Ten Years of Experience with Models for Equitable Sharing of Benefits' in C McManis (ed), *Biodiversity and the Law: Intellectual Property, Biodiversity and Traditional Knowledge* (Earthscan, 2007).
[36] Schroeder and Pogge (n 3); Hayden (n 2).
[37] CBD Article 15(4).
[38] *Ibid.*, Article 15(5).

literature.[39] At the same time, beyond the political and symbolic shifts, the principles established by the CBD leave a lot of questions unanswered, and to be addressed in future meetings of the parties. Indeed, one of the key contributions of the CBD in that respect is that it also sets an extensive institutional system that enables further and long-term discussions. The refining of the implications of recognising some rights for local communities was further pursued by subsequent work of the parties and relevant working groups, producing more detailed articulations, such as the 2002 Bonn Guidelines on 'Access to Genetic Resources and Fair and Equitable Sharing of the Benefits Arising out of their Utilization', the 2004 Addis Ababa 'Principles and Guidelines for Sustainable Use of Biodiversity' and more recently the 2010 'Nagoya Protocol on Access to Genetic Resources and the Fair and Equitable Sharing of Benefits Arising from their Utilization to the Convention on Biological Diversity' (Nagoya Protocol).[40] The new principles that emerged also shaped policy discussions, working groups and guidelines produced by international and regional institutions.[41] Alongside these developments, some of the ideas introduced by the CBD, such as the concept of benefit-sharing, shaped conversations in a variety of local contexts, as developed below.[42]

3.1 The Nagoya Protocol and its Contribution to Access and Benefit-sharing Conversations

The Nagoya Protocol aims to provide a more detailed set of guidelines for the implementation of the general concept of benefit-sharing and other ABS requirements put forward in the CBD and developed during further COP discussions. Before exploring the various critiques that followed, a brief summary of its substance is useful.

---

[39] *See*, for example, C Correa, 'The Access Regime and the Implementation of the FAO International Treaty on Plant Genetic Resources for Food and Agriculture in the Andean Group Countries' (2003) 6(6) Journal of World Intellectual Property 795.

[40] For a brief recap of the events leading to the Nagoya Protocol, *see*, for example, K Bavikatte and DF Robinson, 'Towards a People's History of the Law: Biocultural Jurisprudence and the Nagoya Protocol on Access and Benefit-Sharing' (2011) 7(1) Law, Environment and Development Journal 35. The text of the Protocol is available at <www.cbd/int/abs/text>. For discussion *see* T Greiber and others, *An Exploratory Guide to the Nagoya Protocol on Access and Benefit-Sharing* (IUCN Envtl. Policy and Law Paper No. 83, 2012); E Morgera, M Buck and E Tsioumani (eds), *The 2010 Nagoya Protocol on Access and Benefit-Sharing in Perspective* (Brill/Nijhoff, 2012).

[41] *See*, for examples, S Bhatti and others (eds), *Contracting for ABS: The Legal and Scientific Implications of Bioprospecting Contracts* (IUCN Envtl Policy & Law paper No. 67/4) available at <https://portals.iucn.org/library/efiles/documents/EPLP-067-4.pdf>; PIIPA (Public Interest Intellectual Property Advisors), *Bioprospecting Resource Guide* (2013), available at <http://www.piipa.org/images/PDFs/PIIPA_Bioprospecting_Resource_Guide_2013_Final.pdf>; C Chiarolla and others, *Biodiversity Conservation: How Can the Regime of Bioprospecting under the Nagoya Protocol Make A Difference?*, IIDRI Studies No. 06/13, available at <http://www.iddri.org/Publications/Collections/Analyses/Study0613_CC%20RL%20RP_bioprospecting.pdf>.

[42] Hayden (n 13); Hayden, 'Taking as Giving' (n 4); Green (n 12).

The Nagoya Protocol states as its core aim the third objective of the CBD – fair and equitable access to genetic resources. However this is linked back to the first two objectives of the CBD, by stating that fair and equitable access shall contribute to 'the conservation of biodiversity and the sustainable use of its components'. The links here may be perceived as direct,[43] in the sense that controlling bioprospecting may be seen to limit some practices that directly deplete natural resources. However, given the relatively small scale of many bioprospecting operations, it is mostly constructed as indirect, and flows from the protection of, and respect for, traditional knowledge that more ethical bioprospecting practices would bring.[44] The text of the Nagoya Protocol, in addition, suggests in Article 9 that benefits arising from the utilisation of biodiversity should be redirected towards the conservation of biodiversity.

In order to achieve its main objective, the Protocol sets detailed guidelines for parties. Parties are to take 'legislative, administrative or policy measures' to ensure that the principles of equal and fair sharing of benefits are turned into practice[45] and that such benefits are shared fairly with indigenous and local communities.[46]

Article 6 sets out specific guidelines for negotiating access to genetic resources. Access should be only after prior informed consent from the country of origin;[47] and in accordance with national laws, with the prior informed consent of indigenous and local communities 'where they have the established right to grant access to such resources'.[48] Parties are required to take necessary legislative, administrative and policy steps to improve the transparency and reliability of the decision-making process, by providing inter alia for: 'legal certainty, clarity and transparency of their domestic access and benefit-sharing legislation or regulatory requirements';[49] 'fair and non-arbitrary rules and procedures on accessing genetic resources';[50] 'information on how to apply for prior informed consent';[51] and a 'clear and transparent written decision by a competent national authority, in a cost-effective manner and within a reasonable period of time'.[52] They should also 'Where applicable, and subject to domestic legislation, set out criteria and/or processes for obtaining prior informed consent or approval and involvement of indigenous and local communities for access to genetic resources'.[53]

---

[43] For a more detailed exploration of the links between biodiversity and the protection of indigenous cultures, *see* J Woodliffe, 'Biodiversity and Indigenous People' in MJ Bowman and C Redgwell (eds), *International Law and the Conservation of Biological Diversity* (Kluwer, 1996), chapter 13.

[44] *See*, for example, Woodliffe, *ibid.*; European Parliament, *Report on Development Aspects of Intellectual Property Rights on Genetic Resources: The Impact on Poverty Reduction in Developing Countries* (2012/2135(INI)), 14.

[45] Nagoya Protocol Article 5(1) and (3).
[46] *Ibid.*, Article 5(2) and (5).
[47] *Ibid.*, Article 6(1).
[48] *Ibid.*, Article 6(2).
[49] *Ibid.*, Article 6(3)(a).
[50] *Ibid.*, Article 6(3)(b).
[51] *Ibid.*, Article 6(3)(c).
[52] *Ibid.*, Article 6(3)(d).
[53] *Ibid.*, Article 6(3)(f).

Access to traditional knowledge associated with genetic resources should be with prior informed consent from local communities, and on mutually agreed terms.[54] This is developed further in Article 12. This states that parties shall pay attention to indigenous and traditional communities' customary laws and procedures in implementing their obligations,[55] and support the development by these communities of 'protocols in relation to access to traditional knowledge associated with genetic resources and the fair and equitable sharing of benefits arising out of the utilization of such knowledge';[56] 'Minimum requirements for mutually agreed terms to secure the fair and equitable sharing of benefits arising from the utilization of traditional knowledge associated with genetic resources';[57] and 'Model contractual clauses for benefit-sharing arising from the utilization of traditional knowledge associated with genetic resources'.[58]

The various procedures and decisions are to be put into practice by a designated 'national focal point' on ABS, and relevant national authorities – which may in practice be the same entity.[59]

Finally, the Protocol contains provisions on compliance and monitoring. Compliance is mostly left to states parties themselves and remains relatively vague: states parties are expected to 'take appropriate, effective and proportionate measures to address situations of non-compliance' with procedures on ABS both of genetic resources and of traditional knowledge.[60] Monitoring is to be organised through 'checkpoints', to be designated by each party, and the transfer of information from these to the CBD's Clearing House Mechanism.

Overall, the Protocol therefore reiterates, in further detail and with more specific articulation, the principles set out in the CBD in relation to access to and utilisation of genetic resources, and associated traditional knowledge. Its significance, to a great extent, is to be assessed in relation to that of the CBD itself in solving complex and multiple situations that have arisen in the context of bioprospecting. Commentators of the Protocol have been quick to point to some of its limitations, both with regard to compliance procedures,[61] and because of the uncertainties that remain after its adoption. In particular, two debated issues were carefully sidestepped by the Protocol. First, the precise scope of application of measures of ABS, namely whether only genetic resources per se, or their derivatives, should be covered, was not addressed directly. Secondly, the question of whether the provisions should apply retrospectively to bioprospecting activities that had already started was also sidestepped.[62] A second

---

[54] Ibid., Article 7.
[55] Ibid., Article 12(1).
[56] Ibid., Article 12(3)(a).
[57] Ibid., Article 12(3)(b).
[58] Ibid., Article 12(3)(c).
[59] Ibid., Article 13.
[60] Ibid., Articles 15 and 16.
[61] CC Kamau, B Fedder and G Winter, 'Nagoya Protocol on Access to Genetic Resources and Benefit-Sharing: What is New and What Are the Implications for Provider and User Countries and for the Scientific Community?' (2010) 6(3) Law, Environment and Development Journal 248.
[62] C Aubertin and G Filoche, 'The Nagoya Protocol on the Use of Genetic Resources: One Embodiment of an Endless Discussion' (2011) 2(1) Sustentabilidade em Debate – Brasília 51.

significant – though expected – limitation of the Protocol is that it does not directly engage with IP, and the question of how patent law should relate to questions of access. There are reasons for this, the clearest one being that IP debates are dominated by conversations taking place in the context of the WTO TRIPS agreement, which has remained the main forum for engaging with IP. At the same time, commentators have welcomed the fact that a step forward has been taken in providing clearer expectations on states parties and setting clearer guidelines, as well as for acknowledging the need to recognise and reward contributions from indigenous communities through traditional knowledge.[63]

However, other issues can also be pointed out as significant but on which legal commentators have been less vocal. These affect the potential impact of the CBD and agreements that emerged from it in a rather fundamental way. The CBD, and the Nagoya Protocol in turn, are based on a number of categorisations of the actors of bioprospecting, but also of its practice. These are themselves based on assumptions as to what type of problem it is, and in turn how it may be solved. Paying further attention to the detailed explorations that critiques have provided of the way bioprospecting works in practice allows us, however, to problematise some of these assumptions, and in turn to engage more deeply with the question of whether the law as it is conceived in the CBD can ever provide a meaningful and unified answer to the problems at stake. In other words, it is not necessarily obvious that further refinements of and guidelines on the principles of the CBD can provide a solution to a problem of such messiness, complexity and variability in its power relations as that of biopiracy. I now turn to the further problematisation of the law that research on the practice of benefit-sharing has demonstrated.

4. THE COMPLEX PRACTICE OF BENEFIT-SHARING CONTRACTS

Once the principle of benefit-sharing is established, its articulation in practice still needs to be organised, and has been left mostly to the local level. As indicated by the CBD and the Nagoya Protocol, states have mostly been left to develop their own legal framework to ensure the fair distribution of benefits. The standard template for its specific allocation became the benefit-sharing contract i.e. multi-party agreements between private bioprospectors, local collaborators and the government of the host country. These set the general lines of how benefits would be shared should lucrative discoveries emerge from the collaboration. Both due to their legal nature, and to the complexity of the political and social problem of biopiracy, bioprospecting contracts have remained riddled with difficulties, and have generally been seen as only a partial solution to the problems with which communities are confronted. This complexity and uncertainty plays out at several levels, and some in particular have been widely explored. I will here briefly return to, and summarise, some of the practical considerations that they have posed, before reflecting on the conceptual and political debates

---

[63] *Ibid.*

with which they are entangled. It is worth noting, however, that in spite of the wide set of commentaries that have been produced on benefit-sharing contracts in general, in-depth empirical engagement with their practice remains relatively limited. The few examples provided and that I rely upon below, however, provide very subtle and detailed explorations of the complexity of these instruments, and raise key questions for legal scholars.

A first obvious difficulty that those seeking to develop benefit-sharing contracts have met has been that of establishing standards of fairness in profit allocation. Here, the question has been in evaluating what would be a 'fair' share of benefits to be repaid to local providers of resources, and/or of knowledge. This is a question inevitably complicated by the uncertainty of the activities at stake (from bioprospecting to research and development), but also by the conflicting perspectives involved in determining the relative share and importance of various types of input in the process. Indeed, establishing how the sharing of benefits might be 'fair' is necessarily dependent on how the relative importance of different forms of techniques and knowledge are rewarded in global societies. It is also dependent on the relative importance that is allocated to the provision of 'bare' resources in the process of pharmaceutical or biotechnological development.[64] In addition, the very concept of 'bare' or 'untouched' resources has been actively challenged by anthropologists, for example, who remind us of the inevitable entanglement of nature with social practices and traditions.[65] The reliance on ad hoc solutions that are expected with bioprospecting contracts cannot be extracted from the larger political culture in which the valuation of knowledge is often conditional on its commercial application. Here, much of the determination of what may be a fair share of benefits is left to ad hoc negotiations. These are inherently dependent on the relative bargaining power of different groups. In turn, this relative bargaining power is itself entangled with broader systems of rewards and value, including through intellectual property, as I explore below.

In addition to the issue of value and fairness of distribution, bioprospecting contracts also raise complex questions in relation to defining the most appropriate parties to the contract. The type of resources that those contracts are concerned with – hybrid mixtures of genetic resources, per se, and of associated or entangled knowledge – are often spread across vast spaces and national boundaries, and used by a variety of communities and groups. In practice, each bioprospecting mission will have identified particular partners, concerned communities or groups, as well as state representatives, all of whom become party to the agreement.[66]

There, resources come to be associated with a predefined set of actors and representatives of groups and communities. Quite obviously, the delimitation of those that are 'concerned' by the matter, or have any form of claim or right over a particular resource and its derived knowledge, is a contested element, and the solutions provided by bioprospecting contracts might create as many ethical questions as they answer. Whether we conceive of natural resources as being attached to a particular sample, or as a sample being only symbolic of shared resources to which a broader set of

---

[64] Schroeder and Pogge (n 3).
[65] Hayden (n 2).
[66] Green (n 12); Hayden (n 13).

376 *Research handbook on biodiversity and law*

populations are attached, impacts on how we conceive of the fair allocation of rights. A similar argument can be developed in relation to shared knowledge. A risk here is that some groups, in areas where governance, politics, or landscapes make access for bioprospectors more complex, will be systematically excluded from benefit-sharing schemes. Importantly, the underlying notion of 'indigenous', 'local', or 'national' resources and knowledge are the product of significant social constructions, in which histories and politics are deeply entangled. Much of the ethnographic work that has sought to engage with issues of bioprospecting has demonstrated the complex processes that participate in shaping contemporary forms of knowledge over nature. For example, Cori Hayden, in one of the most in-depth early inquiries of the practice of bioprospecting contracts, demonstrated how ethnobotanical knowledge has been defined and redefined in relation to its traditional, indigenous or national characteristics as the state of Mexico came to engage with bioprospecting contracts. Cases in which clear distinctions between categories and origins of knowledge can be clearly drawn, in turn, have often been recognised as the more successful, but also the more exceptional, experiences of bioprospecting contracts. For example, although the benefit-sharing agreement surrounding the Hoodia plant in South Africa has often been considered as particularly positive, Abena Osseo-Asare also points to the exceptional historical context that facilitated this process.[67] While international law and its implementation seem to be resting on the notion that the origin of knowledge and the relationship of biodiversity with communities may be determined in advance, it is also important to acknowledge the multiple movements that affect the definition and perceived nature of traditional knowledge.[68]

When looking more closely at any of these agreements, such contestations and tensions exist at a variety of levels, to the extent that most aspects of the contract raise ethical dilemmas that are symbolic of the broader social and political conflicts at play. From the selection of legitimate spokespersons, to the identification of adequate parties, to the definition of what benefits should be shared, how, and under what circumstances, bioprospecting contracts become multi-layered sites in which the ethical conflicts surrounding the use of biodiversity become permanently displaced. Although legal discourses have tended to view these as problems that can be solved as conditions become clarified and consensus is reached, the terrain is inherently fluid and changing, to the extent that any answer is itself productive of the categories it comes to rest upon, and participates in displacing rather than limiting their complexity. Overall, one of the criticisms that can, therefore, be made of the CBD is that it seems to rest on a slightly simplified view of the bioprospecting mission. In particular, it does not fully reflect the messiness of the reality of knowledge practices and knowledge exchange that are involved in those processes. The problem is that, if the law rests on a flawed understanding of the processes it sets out to regulate, or an incomplete understanding, the solutions it offers will remain riddled with uncertainties that will almost inevitably play against the most vulnerable actors: in other words, the key notions that the law

---

[67] Osseo-Asare (n 4), chapter 5.
[68] Hayden (n 2).

rests upon, from 'state provider', to 'indigenous communities', to 'traditional knowledge', are all problematic in ways that the legal system does not explicitly engage with.

Of course, the CBD also offers flexibility in its implementation that governments can seize when implementing national law, therefore benefiting from opportunities to iron out some of these uncertainties. However, and this is not only the problem of the CBD but the problem of law more generally, whom such flexibilities will benefit, how, and to what extent, inevitably rests on fragile contextual lines of power. As one example of such issues and shifts, the relative novelty of the legal systems put in place around the CBD in many places has also meant that, in practice, NGOs have often become arbitrators of some of the issues that follow from the legal uncertainty or flexibility. However, their own involvement has itself been problematised in terms of the political questions this creates, and how it impacts on the relative role of the state.[69] In other words, much uncertainty remains as to the effects and possibilities of the law, on which legal scholars seeking to engage with the area should maintain a critical gaze.

I return below to some of the conceptual issues that this raises in relation to visions and understandings of biodiversity, but first I turn to the other key legal area that has been engaged in debates on biopiracy: IP. However, rather than understanding benefit-sharing paradigms as being separate from those on IP, it is preferable to see them as raising interrelated issues. This is important because the conversations held in the context of bioprospecting, and of the CBD, for example, parallel those taking place in the IP system. These arguably provide the predominant system of relative value-allocation for the products of 'knowledge'. In IP, as in bioprospecting contracts, however, ideas which are taken for granted, about nature/society, about the networks surrounding genetic resources, or about their entanglement in particular social processes, seem to obscure some of the complexity of the events at play.

5. BIOPROSPECTING, IP AND COMPETING KNOWLEDGE SYSTEMS

While benefit-sharing contracts have been posited as one of the key solutions to the ethical dilemmas that bioprospecting poses – or at least to the dilemmas as framed and defined in international politics – the questions at stake are part of broader issues in IP. The hybrid nature of genetic resources used for research purposes (which ranges between biodiversity and its imaginary of shared belonging, historically complex forms of shared and co-generated knowledge, and industrial development) places them in an uncertain position in the face of IP.

IP is built on the joint ideas of rewards and incentives to produce particular forms of knowledge deemed to be of social value. In the context of bioprospecting, however, ambivalence exists both in terms of what is considered as significant knowledge in particular activities, and in terms of what is considered as a valuable final product. The process of research from bioprospecting to final product involves a series of

---

[69] K Peterson, 'Benefit-Sharing for All: Bioprospecting NGOs, Intellectual Property Rights, New Governmentalities' (2001) 24(1) Political and Legal Anthropology Review 78.

knowledge-exchanges and transformations (of plants, of extracts, of products, and of ideas into products). Further definitions of what constitutes communal, traditional, national or individual contributions are uncertain, while what are deemed to be 'natural' or 'created' goods is itself contested. Indeed, bioprospecting is useful in exploring the underlying assumptions of IP, as much as IP is a helpful starting point to understanding the limits of law in dealing with bioprospecting/biopiracy. In spite of efforts to try to fit some of the dilemmas surrounding biopiracy into the IP framework, or the deployment of a parallel system of reward for traditional resources,[70] more radical questions remain as to whether the foundations of IP are themselves inherently inadequate when engaging with either the hybridity of nature/society assemblages or the fluid forms of co-generated knowledge.[71]

Underlying much of the debate in IP is the question of when, and under what conditions, knowledge and the natural resources to which it relates are deemed to be 'valuable'. Inevitably, IP can only provide a limited and densely political answer to this question, and has been pointed out as a symbolic site in which the limits of 'global justice' and the possible role of law in approaching it could be found.[72]

In effect, plants collected through bioprospecting often find themselves at the crossroads of two competing knowledge systems, whose valuation differs significantly in law and in social practice. First, localised, long-established, 'traditional'[73] forms of knowledge over plants are often a starting point in the identification of particular genetic material. Traditional medicine (typically, though not exclusively) provides a useful entry point to identifying plants that may have effects on human bodies. While such knowledge may be recorded in writing, the primary way of accessing it is often through direct interactions with some of its holders, who may or may not be representatives of a clearly defined broader community. Here, the questions that arise in relation to rights and ownership are both particularly complex and particularly sensitive.

Second, plants collected through bioprospecting missions also become almost immediately part of a very different knowledge system – one based on scientific research and oriented towards industrial application – which could turn a natural product into a manufactured one, and in turn transform and generalise its usage.[74] An immediate effect of being associated with this second knowledge system is the possibility for plants to become transformed and redefined within the predominant system of IP – which in turn enables them to perpetuate their status in an industrial complex and its expectations for particular forms of reward and valuations. Biopiracy therefore becomes symptomatic of deeper questions about the valuation of knowledge by law, and in turn about the empowerment that arises from holding competing forms

---

[70] DA Posey and G Dutfield, *Beyond Intellectual Property: Towards Traditional Resource Rights for Indigenous People and Local Communities* (International Development Research Centre, 1996).

[71] C Oguamanam, 'Local Knowledge as Trapped Knowledge: Intellectual Property, Culture, Power and Politics' (2008) 11(1) Journal of World Intellectual Property 29.

[72] Schroeder and Pogge (n 3).

[73] Although, as we have seen, the definition of tradition is itself an ambivalent one.

[74] V Adams, 'Randomized Controlled Crime: Postcolonial Sciences in Alternative Medicine Research' (2002) 32(5–6) Social Studies of Science 659.

of knowledge (i.e. traditional vs science-based; communal vs individual; written vs verbal; industrially relevant and generalisable vs locally grounded and small scale). In turn, questions emerge here in terms of the various ways in which biodiversity itself comes to translate as holding value – economic or other.[75]

The problematique of bioprospecting is therefore positioned at the crossroads of a series of complex and politically loaded debates: on the position and value of nature in scientific development, in law and in society more broadly; on the value attached to contrasting forms of knowledge and knowledge systems; on the question of global inequalities in biotechnological development; on the ethics of IP law. Each of these also runs deeply into post-colonial tensions.

As far as IP is concerned, the legal problematic at play here can be recalled briefly: contemporary systems of IP, and patents in particular, are largely geared towards modes of work that are directed at industrial applications, forms of reproduction, and practical uses and distributions that are mostly maladapted to the ways in which many forms of knowledge, including traditional (or defined as such), can be regarded. In addition, the single-owner model that patent law in particular privileges is unhelpful when seeking to allocate rights to collective forms of knowledge.[76] As far as natural resources are concerned, the relevance of IP has been even more centrally criticised: naturally occurring organisms are deemed not to be patentable, but the relevance of this to the protection of nature is largely one-sided – how do we then imagine possibilities for reward for those who protect, maintain and foster natural resources? While benefit-sharing contracts provide a partial answer both in terms of incentives and in terms of rewards, without a more radical shift in the dominant system of valuing and rewarding, they do not have the potential to offer the more significant changes that many would hope for.

Commentators, and policy-makers, have therefore struggled to adapt or rethink IP frameworks to make them speak to biopiracy. Two approaches have been considered. First, the possibility of 'rewriting' IP from the inside, and adapting the existing IP systems to the needs of traditional knowledge holders has been explored. One of the most common proposals to be put forward has been to inscribe a requirement of disclosure of origin within patent law. Several versions of this have been considered, from more light-touch versions in which such disclosure would essentially be a formal requirement (or for the weakest forms of protection a simple encouragement to disclose) to stricter ones where both disclosure, and evidence of compliance with the CBD's ABS provisions would be conditions of patentability and failure would be a potential ground for revocation. The proposed systems, however, all have limitations. Those at the weaker end of the spectrum have been questioned in terms of efficacy, and whether they have any chance of improving practices. The stricter requirements in turn have raised questions in terms of their compatibility with international law on IP, in the form of the TRIPS agreement. Indeed, some have argued that this could result in the

---

[75] For an overview of the various sets of values attached to nature by environmental law, *see*, for example, A Gillespie, *International Environmental Law, Policy and Ethics* (2nd edn., OUP, 2014).

[76] *See*, for example, Dutfield (n 1).

addition of a substantive condition to patent requirements, which would be incompatible with TRIPS in its current form. Commentators such as Graham Dutfield,[77] however, have highlighted how some of the requirements associated with such an approach could nonetheless be argued to be compatible with TRIPS, on the ground that the burdens they would create would be primarily formal and administrative, while the substantive requirement of disclosing the use of traditional knowledge would not add to the basic expectations to reflect the state of the art in patent applications. He points, however, to a more practical difficulty with any such provision: that patent officers may not be able to straightforwardly identify applications that simply make no mention of resources they have used. The remaining value of disclosure requirements would be in facilitating revocation if and when local groups are sufficiently mobilised and aware to challenge a particular patent – which, again, presupposes a lot.[78]

A second set of proposals commonly made in order better to adapt IP practices to avoid more cases of biopiracy has been centred on the creation of traditional knowledge databases.[79] Here, the idea is to facilitate the ability of patent officers to compare patent applications to the state of the art, and also to accommodate the provisions of particular states, notably the United States, that will only recognise knowledge held outside of their territory as being part of the state of the art if it is in written form. However, whether such a solution is able to solve any of the difficulties at play is again uncertain. Some of this is due to the mechanics of IP law itself. For example the definition of what constitutes prior knowledge, and what constitutes novelty or an inventive step, varies across countries. A full overview of these different understandings is impossible here for reasons of space although others have covered this in great detail.[80] For now, it is useful to simply recall that one of the complexities of the debates lies in the fact that, in most cases of biopiracy, the product on which the patent is sought will not be a direct replication of traditional knowledge, and therefore the deviations it proposes could be sufficient for patent law to see it as 'something different', and therefore a new, patentable, product. Here, while some of the most obvious cases of biopiracy may be avoided, the question of ownership in the more subtle, and more common cases, is not solved. In addition, the idea of databases itself raises difficulties: as more knowledge becomes formalised in this form, the forms of knowledge that remain excluded in turn become, arguably, more vulnerable. Typically, this is likely to be the case for those that are most clearly on the fringes of society in the first place, or those groups who are most distant from governmental support or practices. Other authors, who more enthusiastically embrace current frameworks of IP, have suggested that the existing system could be more effectively used to protect traditional knowledge, be it by engaging more actively with patents, geographical indications, or most prominently, plants breeders' rights.[81] The latest, which constitute a *sui generis* system of protection

---

[77] Ibid.
[78] For a more detailed analysis of the links between patents, bioprospecting and indigenous knowledge, *see* Mgbeoji (n 30).
[79] Dutfield (n 1); European Parliament (n 44).
[80] *See*, for example, Dutfield (n 22); Mgbeoji (n 20).
[81] *See*, for example, D Downes, 'How Intellectual Property Could Be a Tool to Protect Traditional Knowledge' (2000) 25 Columbia Journal of Environmental Law 253; Mgbeoji

for new varieties of plants, have in practice not always been used in a way that is most effectively protective of indigenous people's rights, but have been a focal point of discussion in the agricultural context.[82]

Alongside these various proposals that seek to fit within the current IP system, proposals for a new understanding of rights, specifically aimed at the protection of traditional knowledge, were developed and centred around the protection of traditional resources.[83] Here, Traditional Resource Rights – a concept that emerged 'to define the many "bundles of rights" that can be used for protection, compensation and conservation'[84] – provided the basis for early reflections on how resources could be protected, aside from IP per se. Here, the persistent difficulty, however, is that the de facto power of IP, supported by significant economic actors, will remain hard to challenge. More recently, some have advocated the protection of Intangible Biological Resources through mechanisms designed to support and reinforce the prior consent and benefit-sharing requirements of the CBD. Again, their practice has shown the limitations of the prospects they offer.[85]

On the basis of these various ideas, working groups across international organisations have pursued attempts at reflecting on how rights systems could be designed better – within the strict limitations for flexibility left by the TRIPS Agreement – so as to open various avenues but without any settled solution yet being reached.

One of the problems here, as has been pointed out repeatedly by critical IP scholarship, is that the issues at stake may run too deep for solutions to be found without a radical challenge to the underlying assumptions about knowledge, value and power that justify the current IP system. Rather than simply a problem of law being ill-adjusted to specific ethical or social conflicts in its details or in its forms, the issue here is one of global imbalances of power having materialised in a particular system of rights and rewards that is fundamental to the shape of global capitalism.[86] The complex historical and political context of IP, through its colonial travels, and contested political pressures for recent harmonisation, have all become part of a long-term set of mechanisms through which particular activities are perceived as more valuable for markets than others.[87] In this quest, conceptions of 'modernity' and 'development' have become an inherent part of the law, and contributed to the progressive side-lining and

---

(n 20); I Walden, 'Intellectual Property and Biodiversity' in MJ Bowman and C Redgwell (eds), *International Law and the Conservation of Biological Diversity* (Kluwer, 1996), chapter 9.

[82] Dutfield (n 22); GP Nabhan and others, 'Sharing the Benefits of Plant Resources and Indigenous Scientific Knowledge' in S Brush and D Stabinsky (eds), *Valuing Local Knowledge: Indigenous People and Intellectual Property Rights* (Island Press, 1996), chapter 9.

[83] DA Posey, 'The Relation between Cultural Diversity and Biodiversity' in S Bilderbeek (ed), *Biodiversity and International Law* (IOS Press, 1992), 44–47; Posey and Dutfield (n 70).

[84] Posey and Dutfield (n 70).

[85] EB Rodrigues, 'Property Rights, Biological Resources and Two Tragedies: Some Lessons from Brazil' in T Bubela and ER Gold (eds), *Genetic Resources and Traditional Knowledge* (Edward Elgar Publishing, 2012), chapter 5.

[86] Coombe (n 2).

[87] Coombe (n 2); Coombe (n 19).

exclusion of particular groups.[88] Rather than side-effects of the IP system, the ethical issues that arise are therefore inherent parts of it. In effect, the problems associated with bioprospecting are symbolic of a deep-running logic of which IP is an inherent part. This illustrates the co-constitutive nature of law, or in other words the fact that law participates in generating social relations and events in its attempt to solve them. The crisis of values, or of valuing, that cases of 'biopiracy' are representative of, cannot be dissociated from either the emergence of IP as a tool in shaping and allocating ownership, nor from the movements that animate communities whose resources are being used.[89] IP is a significant part both of global capitalism and of the post-coloniality of many of the places where bioprospecting takes place, and its underlying logic is therefore in direct contradiction with alternative ways of imagining nature, culture and values that opponents of biopiracy have long put forward.[90]

The debates at play here therefore also raise two sets of fundamental questions to which this chapter will now turn. First, the debates on biopiracy/bioprospecting are illuminating in relation to what they reveal about biodiversity itself, as an object or a site of global tensions and conflicting claims. Second, the tensions that surround discussions of the links between law, global justice and environmental protection that surround biopiracy have significance in relation to the positionality of law in creating either problems or solutions to the issue.

6. CONCEPTUALISING (LOSS OF) BIODIVERSITY IN BIOPROSPECTING DEBATES

The problematique of bioprospecting is obviously entangled with conceptualisations of biodiversity at a number of levels. At the same time, the role and position of biodiversity in discourses on bioprospecting is also ambivalent. In particular, the raising of awareness regarding biopiracy needs to be understood as one part of broader strategies of political resistance that have allowed other causes to gain support – including the rallying of environmental organisations in the progressive resistance to IP rights in other contexts. In addition, resistance to bioprospecting is almost inevitably part of a broader resistance to neoliberal world orders.[91] It is useful to briefly unpack how various dimensions of biodiversity as a concept (as well as a material actor) emerge from the debates.

First, narratives of destruction, appropriation and 'plundering' of nature are prominent in conversations on bioprospecting. Here, the activities of researchers are considered as being a potential threat to nature itself, but also to the knowledge and care that has participated in its preservation over centuries – either because this knowledge is under direct threat through its exclusion from reward systems, or because nature

---

[88] SB Banerjee, 'Who Sustains Whose Development? Sustainable Development and the Reinvention of Nature' (2003) 24(1) Organization Studies 143.
[89] C Oguamanam, 'Protecting Indigenous Knowledge in International Law: Solidarity Beyond the Nation-State' (2004) 8 Law, Text, Culture 191; Bavikatte and Robinson (n 40).
[90] Coombe (n 19).
[91] Ibid.

becomes part of a neoliberal system in which raw resources are given little value. These conversations are also inherently mixed with broader discussions about how resources, held mostly in the so-called Global South, should be used by private interests. All of this is reminiscent of broader concerns about the exploitation of natural resources in those countries since the colonial era. Therefore the 'plunder of nature' denounced by the activist Vandana Shiva[92] is best understood in the context of decades of exploitation of minerals and other natural resources from localities with an inadequate level of compensation. The immediate aims of protecting nature for conservation and environmental purposes are therefore almost impossible to separate from the larger developmental and global justice concerns of which they are reminiscent.[93] The central importance of protecting and maintaining natural resources is in turn enhanced by ethical and political contexts. As a consequence, proposed solutions such as bioprospecting contracts often have as one of their key dimensions the fostering of activities that aim to guarantee the availability of resources.[94] Biodiversity in those contracts becomes positioned at the crossroad between environmental, economic and scientific aims, and though partly instrumentalised also becomes defined as a set of resources to be maintained and protected.

Second, biodiversity (and its protection) has been largely deployed as a powerful discursive tool to engage identities and the relative importance of various forms of knowledge. While this does not mean that the risks posed by the instrumentalisation of nature are not significant, it is also important to recognise that biodiversity has been part of complex processes of community, identity and nation-building. In many places, the question of what biodiversity meant for particular identities only emerged as bioprospecting started becoming a visible practice, as identities and resources became co-produced and mutually reshaped. Visions of nature as existing as a distinct sphere from society, social identities and conflicts, which have become central in shaping the specific debates on bioprospecting, and broader environmental discourses run the risk of losing some of the inherent links between nature and social practices.[95] As one example, Cori Hayden points to the narrative of biodiversity as unknown as a popular metaphor: 'the metaphor of natural diversity as unknown information is central to many popular and popular scientific accountings of the loss of "value" posed by species extinction'.[96] Indeed, the vision of nature as untouched, unsocialised and clearly distinct from culture is a powerful one, and one that has been criticised famously, among others by Bruno Latour, as being the weak basis for the myth of 'modernity'.[97]

Third, and most importantly maybe, the debates on bioprospecting can be read as an opportunity to reflect on the shifting value of nature in the face of global capitalism, and the ways in which economics and nature are permanently co-productive. Here, while at one level one may describe the global economic and legal system in its current

---

[92] Shiva (n 20).
[93] Schroeder and Pogge (n 3); Banerjee (n 88).
[94] Hayden (n 2).
[95] Latour (n 11).
[96] Hayden (n 2), 57.
[97] Latour (n 11).

form as being inadequate in rewarding the work on nature produced by some, certain more radical and in-depth critiques also emerge. For Rosemary Coombe:

> Perhaps the emergence of a widespread rejection of absolute dependence on consumerist consumption and the desire for greater local autonomy to reconstruct sustainable livelihoods that are not totally subsumed by markets expressed by numerous interlinked peoples' movements constitutes the most radical of the new countercultural currents in the anti-corporate movement.[98]

Following from this, bioprospecting and its debates become symbolic of why conceptualisations of the value of nature are inherently entangled with political visions, and interlinked with the social production of identities, and of new forms of economic and social power.[99] While an idealised metaphor of nature as unknown, untouched and separable from society serves important political purposes, and participates in sustaining social movements, it does not capture the complexity of the history of nature as inherently part of cultural processes and social paths that are loaded with effects and consequences we never object to on 'purely' environmental or social grounds, but also on political grounds. The objections to the way in which biodiversity is used and rewarded in bioprospecting missions are part of much broader opposition to visions of modernity as necessarily and inevitably based on capitalist and industrial processes. Similarly, market discourses and the production of a 'productive' nature are inherently entangled and responsive to each other. As Cori Hayden notes: 'nowhere is this faith in the market more evident than in the ways nature, as biodiversity, is being framed as a storehouse of valuable genetic resources and as a resource to be managed as an explicitly economic enterprise'.[100]

In recent years, a number of symbolic paradigm changes have taken place at the global level, whereby the international community has reasserted the inherent value of nature. One such example is the UN Harmony with Nature Project.[101] The CBD inscribes itself in this general movement, and indeed it has sought to embed a pluralistic approach to the question of value. At the same time, the deployment of one of its mains tools of translation from principles to practice (i.e. bioprospecting contracts) has remained based around conversations on the economic value of plants, knowledge and contributions to particular products. In this context, the question of whether individual missions are themselves damaging to biodiversity, and their defence as being potentially supportive of conservation (for example through carefully designed bioprospecting contracts) does little to appease concerns about how we approach, value and reward various forms of resources, including genetic resources, that some may choose to describe as raw but are always inevitably the product of a long social history, and loaded with social futures.

---

[98] Coombe (n 19), 49.
[99] B Parry, 'Hunting the Gene-hunters: The Role of Hybrid Networks, Status, and Chance in Conceptualising and Accessing "Corporate Elites"' (1998) 30(12) Environment and Planning A 2147.
[100] Hayden (n 2), 49.
[101] For further details *see* <http://www.harmonywithnatureun.org>.

To sum up, the question of (bio)prospecting is therefore immediately reminiscent of a much longer history of complex – and unbalanced – relationships. At the same time, it adopts a particular shape, whose projection in global discourses is also constitutive of a particular vision of biodiversity, and how it is to be understood and valued in neoliberal markets. As we develop solutions to address the immediate concerns of bioprospecting, biodiversity is predominantly viewed as a resource to be protected in ways that remain fully compatible with the tenets of neoliberal thoughts.[102] In fact, progress towards solutions so far has remained limited because a true solution would require a radical political shift that needs to happen beyond this particular case. While an attention to biodiversity within existing market-oriented frameworks appeases some concerns, the limits of the framework itself for nature are also apparent, or as phrased by Shane Green: 'A greener, moral politics requires a greener, more moral capitalism. Yet, there are inherent contradictions in the idea of "green capitalism" when it seems more profitable to maintain the capitalist status quo.'[103]

7. RETHINKING THE PLACE OF LAW IN BIOPIRACY DEBATES

Some of the central difficulties in engaging with the debates around bioprospecting are rooted in the conceptual complexity of rethinking the role of law. In many respects, these debates are illustrative of some deep-running issues of global equality, but also of chosen paths to conflicted modernities that have become part of the global legal landscape. While some victories and milestones have been achieved through agreements such as the CBD and the Nagoya Protocol, the persistence and economic influence of other frameworks, in particular IP, is reminiscent of the fact that these issues are not discrete problems to solve, but are symbolic of much broader ethical and political challenges in the context of an uncertain 'global development'. The role of law itself, and the function of legal scholarship, is at stake in reflecting on where we may be taking these issues. Legal answers, as is illustrated both by benefit-sharing contracts and by adjustments to IP, remain contested and insufficient, and reveal new difficulties as they unfold. At the same time, given the broader picture within which bioprospecting is taking place, legal adjustments can probably never quite suffice to address the problem itself, and even less its deeper political and social significance. Importantly, too, legal solutions can never be seen as purely external to the problem that they are seeking to address, but instead become generative of new identities, mobilisation, imaginaries and claims as they unfold – each loaded with its own consequences, which may be either productive or reductive of new inclusions, rights and equalities. At the same time, these transformations pose complex issues for legal scholarship, with which some of the recent critical and interdisciplinary scholarship on biopiracy has most positively engaged. The complexity of each of the jointly productive processes at play here, and the messiness of the issues that unfold, are part of the story to be told by legal scholars, and for critical environmental scholarship in search of a concrete solution to

---

[102] K McAfee, 'Selling Nature to Save It? Biodiversity and Green Developmentalism' (1999) 17 Environment and Planning D: Society and Space 133.
[103] Green (n 12), 12.

this fluid, changing and messy problem. Ultimately, the questions raised by bioprospecting are entangled with those linked to the position of nature, and biodiversity, in the global political economy and contemporary societies. The vision of nature/culture as being separable, and of law as being potentially external to the problems it seeks to regulate, are deeply challenged by an area in which nature, cultures, identity, modernity and politics are interlinked in the constantly shifting terrains of negotiations over knowledge, natural resources and the values and ownership of each. This leaves the question of how to address conflicts surrounding biopiracy largely unanswered, and research agendas in this field largely open. For these to be productive, however, and to progress in new directions, it seems essential to engage with the multiplicity of conflicts and perspectives that these raise. This will be best achieved through genuine interdisciplinary engagements, where legal problems are considered and imagined with the critical tools that enable them to progress in a way that is consistent with the messiness of the worlds they are seeking to regulate, and to be reassessed with thorough attention to the political contexts that have made them appear in their particular form as social problems.

SELECT BIBLIOGRAPHY

Aubertin, C and G Filoche, 'The Nagoya Protocol on the Use of Genetic Resources: One Embodiment of an Endless Discussion' (2011) 2(1) Sustentabilidade em Debate – Brasília 51
Brush, S and D Stabinsky, *Valuing Local Knowledge: Indigenous People and Intellectual Property Rights* (Island Press, 1996)
Coombe, R, 'Legal Claims to Culture in and Against the Market: Neoliberalism and the Global Proliferation of Meaningful Difference' (2005) 1(1) Law, Culture and the Humanities 35
Dutfield, G, *Intellectual Property Rights, Trade and Biodiversity* (Earthscan, 2000)
Dutfield, G, 'Sharing the Benefits of Biodiversity: Is There a Role for the Patent System?' (2002) 5(6) Journal of World Intellectual Property 899
European Parliament, *Report on Development Aspects of Intellectual Property Rights on Genetic Resources: The Impact on Poverty Reduction in Developing Countries* (2012/2135(INI))
Green, S, 'Indigenous People Incorporated? Culture as Politics, Culture as Property in Pharmaceutical Bioprospecting' (2004) 45(2) Current Anthropology 211
Hayden, C, 'From Market to Market: Bioprospecting's Idioms of Inclusion' (2003) 30(3) American Ethnologist 359
Hayden, C, *When Nature Goes Public: The Making and Unmaking of Bioprospecting in Mexico* (Princeton University Press, 2003)
McManis, C (ed), *Biodiversity and the Law: Intellectual Property, Biodiversity and Traditional Knowledge* (Earthscan, 2007)
Mgbeoji, I, *Global Biopiracy: Patents, Plants and Indigenous Knowledge* (Cornell University Press, 2006)
Osseo-Asare, A, *Bitter Roots: The Search for Healing Plants in Africa* (University of Chicago Press Books, 2014)
Posey, DA and G Dutfield, *Beyond Intellectual Property: Towards Traditional Resource Rights for Indigenous People and Local Communities* (International Development Research Centre, 1996)

14. Ecological restoration in international biodiversity law: a promising strategy to address our failure to prevent?

Kees Bastmeijer

1. INTRODUCTION[1]

Humankind has proved to be very successful in using the natural resources of the earth to ensure sufficient food and material wealth for at least a large part of the population. As explained by Crispin Tickell 'unlike other animals, we made a jump from being successful to being a runaway success ... because of our ability to adapt environments for our own uses in ways that no other animal can match'.[2] Over the last 50 years in particular, we have also become increasingly aware of the darker side of this success: it is clear that not all people have an equal share in this success and we have learned that our activities in a market economy have substantial 'externalities'. These externalities include the over-exploitation of various natural resources, significant disturbance of ecosystems and severe negative effects on many other species on our planet. As long ago as 1969, the American National Research Council emphasized that humankind is in an extreme time period, characterized by imbalance between development and available natural resources: 'It now appears that the period of rapid population and industrial growth that has prevailed during the last few centuries, instead of being the normal order of things and capable of continuance into the indefinite future, is actually one of the most abnormal phases of human history.'[3]

A few years later, on the occasion of the United Nations Stockholm Conference on the Human Environment (1972), the international community explicitly stressed the need for fundamental change,[4] an acknowledgement that has resulted in several decades of law-making in respect of many environmental concerns. In relation to the natural

---

[1] Parts of this introduction build on K Bastmeijer, 'Addressing Weak Legal Protection of Wilderness: Deliberate Choices and Drawing Lines on the Map' in S Carver and S Fritz (eds), *Mapping Wilderness: Concepts, Techniques and Applications* (Springer, 2016), 117–136.
[2] C Tickell, 'The Human Species: A Suicidal Success?' (1993) 159(2) The Geographical Journal 219, 219.
[3] M King Hubbert, 'Energy Resources' in National Research Council – Division of Earth Sciences – Committee of Resources and Man, *Resources and Man* (WH Freeman & Co., 1969) 238.
[4] Stockholm Declaration, adopted at the 1972 UN Stockholm Conference on the Human Environment, Preamble, para. 6; *see* also Preamble, para 1: 'A point has been reached in history when we must shape our actions throughout the world with a more prudent care for their environmental consequences. Through ignorance or indifference we can do massive and irreversible harm to the earthly environment on which our life and wellbeing depend.'

world, many international conventions, regional binding instruments and domestic laws have been adopted, particularly since the early 1970s, to protect the variety of life forms (species of plants and animals), habitats and ecosystems; in other words to protect biological diversity (hereinafter, biodiversity).[5]

At all governance levels success stories may be identified on the role of law in preventing the extinction[6] and recovery of certain species.[7] However, the overall effectiveness of nature conservation law is limited. The above noted ability to change our environment for our own benefit has resulted in the current situation in which only about 30 per cent of the earth's land surface can still be qualified as relatively untouched by humans ('wilderness').[8] Fragmentation of nature reserves, environmental pollution and large-scale hunting and fishing have had significant negative impacts on the earth's biodiversity.[9] According to the 2005 Millennium Ecosystem Assessment, '[i]t is *well established* that losses in biodiversity are occurring globally at all levels, from ecosystems through species, population, and genes'.[10] The loss of biodiversity should have been significantly reduced by 2010, however worldwide governments have failed to reach this objective:[11]

---

[5] Convention on Biological Diversity (1992) 31 ILM 818 defines 'Biological diversity' as: 'the variability among living organisms from all sources including, inter alia, terrestrial, marine and other aquatic ecosystems and the ecological complexes of which they are part; this includes diversity within species, between species and of ecosystems.' See also the definitions of 'ecosystem' and 'habitat' in this provision. For the purpose of this chapter the term biodiversity is meant to include the variety of genes, species, habitats and ecosystems. See also, for this approach, the World Conservation Monitoring Centre of the United Nations Environmental Programme: 'It has become a widespread practice to define biodiversity in terms of genes, species and ecosystems, corresponding to three fundamental and hierarchically-related levels of biological organisation' available at <http://old.unep-wcmc.org/what-is-biodiversity_50.html>. For a different interpretation, see P Birnie, A Boyle and C Redgwell, *International Law & the Environment* (OUP, 2009), 588.

[6] *See*, for example, the 'Bald Eagle Protection Act' of the USA, 8 June 1940, Ch. 278, 54 Stat. 250 (codified as amended at 16 U.S.C. §§ 668–668d), available at <http://www.animallaw.info/statutes/stus16usc668.htm>.

[7] *See*, for example, S Deinet and others, *Wildlife Comeback in Europe: The Recovery of Selected Mammal and Bird Species*, Final Report to Rewilding Europe by ZSL, BirdLife International and the European Bird Census Council (2013), available at <http://static.zsl.org/files/wildlife-comeback-in-europe-the-recovery-of-selected-mammal-and-bird-species-2576.pdf>.

[8] CF Kormos and H Locke, 'Introduction' in CF Kormos (ed), *A Handbook on International Wilderness Law and Policy* (Fulcrum Publishing, 2008).

[9] *See*, for example, the *Millennium Ecosystem Assessment: Synthesis* (World Resources Institute, 2005) 834.

[10] *Ibid.*, 834, emphasis in original text.

[11] Secretariat of the CBD, *Global Biodiversity Outlook 3* (Montreal, 2010), available at <http://www.cbd.int/doc/publications/gbo/gbo3-final-en.pdf>. *See* also E Morgera and E Tsioumani, 'Yesterday, Today and Tomorrow: Looking Afresh at the Convention on Biological Diversity', University of Edinburgh School of Law Working Paper No. 2011/21, 2011, available at <http://ssrn.com/abstract=1914378> in which it is observed 'Lack of effective implementation of the CBD was clearly demonstrated in the international community's failure to meet the global target of reducing significantly the rate of biodiversity loss by 2010.' *See* also SR Harrop, 'Living in Harmony with Nature? Outcomes of the 2010 Nagoya Conference of the Convention on Biological Diversity' (2011) 23(1) Journal of Environmental Law 117, 120. Harrop refers to

Ecological restoration in international biodiversity law 389

> There are multiple indications of continuing decline in biodiversity in all three of its main components – genes, species and ecosystems ... Species which have been assessed for extinction risk are on average moving closer to extinction ... Natural habitats in most parts of the world continue to decline in extent and integrity ... and [e]xtensive fragmentation and degradation of forests, rivers and other ecosystems have also led to loss of biodiversity and ecosystem services.[12]

Each year, this ineffectiveness is painfully illustrated by the 'Red List' of the International Union for the Conservation of Nature (IUCN).[13] The general view is that these problems will only further increase due to climate change, the proliferation of non-native species and the intensifying strain of human activity.

These and many other monitoring studies on the status of biodiversity show that the dominant approach in nature conservation law, the approach of preventing nature degradation through the prohibition or regulation of human acts and activities that may cause negative impacts on natural values (hereinafter referred to as the 'preventive approach'), is important but not sufficient to reach the conservation objectives that have been agreed in conventions and regional instruments for nature protection (e.g. EU Directives). These insights have stimulated debates on the role of law in adopting a variety of strategies for biodiversity protection. One strategy that has been receiving increasing attention in policy making and in various research disciplines is active investment in nature to ensure the recovery, creation or expansion of natural values, such as ecosystems and species (hereinafter referred to as the 'restoration approach').[14] This chapter focuses more closely on both approaches and their interrelationship under existing international biodiversity law.

The discussion starts with a general introduction on both approaches, their recognition in existing international nature conservation conventions, and their mutual interrelationship as considered by the Conferences of the Parties (COP) of these conventions (Section 2). Next, some general weaknesses of the prevention approach in existing nature conservation law are discussed, with the aim to get a better understanding of why the effectiveness of this approach is limited (Section 3). In part based on this understanding, the growing attention to ecological restoration in the literature, and particularly COP decisions (including strategic plans and policy targets) under the selected conventions, is discussed (Section 4). The discussions will be concluded with a general discussion on the question whether we may expect the restoration approach to be successful, taking note of the limitations of the prevention approach (Section 5).

The scope of the discussions is limited to a number of selected international nature protection conventions: the Convention on Wetlands of International Importance

---

M Walpole and others, 'Tracking Progress Toward the 2010 Biodiversity Target and Beyond' (2009) 325 Science 1503.

[12] *Global Biodiversity Outlook 3*, ibid., 9.
[13] *IUCN Red List of Threatened Species*, available at <www.iucnredlist.org/>.
[14] For a more comprehensive discussion on the role of law in ecological restoration in various parts of the world, *see* RC Gardner, 'Rehabilitating Nature: A Comparative Review of Legal Mechanisms that Encourage Wetland Restoration Efforts' (2003) 52 Catholic University Law Review 573.

(Ramsar Convention),[15] the Convention on the Conservation of Migratory Species of Wild Animals (Bonn Convention),[16] the Convention on the Conservation of European Wildlife and Natural Habitats (Bern Convention),[17] the World Heritage Convention (WHC),[18] and the Convention on Biological Diversity (CBD).[19] For the purpose of this discussion, the terms 'restoration' or 'restoration approach' do not refer to 'normal' nature management activities that are regularly conducted by people to ensure good conditions for already existing natural values. It is about taking active measures to restore lost natural values (species, habitats, ecosystems) or to create new or extra values compared to the current situation, in order to reach the objectives that have been set by, and under, the selected conventions (e.g., reaching and maintaining a favourable conservation status of species, habitats and ecosystems). Consequently, 'restoration' as discussed in this contribution includes the concept of ecosystem restoration as defined by the Society of Ecological Restoration (SER),[20] but it has a broader scope as it also relates to active measures to support the recovery of species populations.[21] It also includes the creation of 'new' natural values that may never have been available at the particular location ('new nature'), but still may be considered as 'restoration' at the level of the overall conservation status of species, habitats or ecosystems. Furthermore, the restoration approach as discussed in this chapter is not based on the idea that nature should in all circumstances be brought back in 'pre-disturbed conditions',[22] but rather on the idea that humankind, while acknowledging ecological dynamics and climate

---

[15] 1971 Convention on Wetlands of International Importance (Ramsar Convention or Wetland Convention), available at <www.ramsar.org>.
[16] 1979 Convention on the Conservation of Migratory Species of Wild Animals (Bonn Convention), available at <http://www.cms.int/en/node/3916>.
[17] 1979 Convention on the Conservation of European Wildlife and Natural Habitats (Bern Convention), available at <http://conventions.coe.int/Treaty/en/Treaties/Html/104.htm>.
[18] 1972 World Heritage Convention, available at <http://whc.unesco.org/archive/convention-en.pdf>.
[19] CBD (n 5).
[20] *See* the Society for Ecological Restoration International Science & Policy Working Group, *The SER International Primer on Ecological Restoration* (2004) 3, available at <www.ser.org>: 'Ecological restoration is the process of assisting the recovery of an ecosystem that has been degraded, damaged, or destroyed.'
[21] Recovery of species is sometimes excluded from restoration definitions. *See*, for example, CBD, COP 11, 8–19 October 2012, Hyderabad, India, Information Note XI/1, 'Available Guidance and Guidelines on Ecosystem Restoration. Annotated Compilation of Publically Available Documents on Ecosystem Restoration Guidance and Guidelines', 1: '… the object being restored or recovered is an ecosystem, not an individual species nor the habitat of any one species'.
[22] *Ibid.*, 1: 'It refers not only to activities aimed at returning an ecosystem to its pre-disturbance conditions, insofar as possible, but also to rehabilitation and other activities focused on the recovery of biodiversity, ecosystem functioning, or other indicators of ecological health.' *See* also L Roberts, R Stone and A Sugden, 'The Rise of Restoration Ecology' (2009) 325 Science 555: 'The goal of restoration ecology is not necessarily to restore an ecosystem to a pristine, pre-human ideal, but a long-term view is still important.'

change influences,[23] can actively support species, habitats and ecosystems to recover and reach a favourable status of conservation.

2. RESTORATION IN INTERNATIONAL BIODIVERSITY CONVENTIONS AND THE PRIORITY OF PREVENTION

Most legal 'tools' embedded in international nature protection conventions and domestic legislation fall within the prevention approach. Human activities that may cause adverse impacts on important natural areas or sites, or that may have negative impacts on species or habitat types, are subjected to certain requirements, prohibitions or restrictions. In implementing these instruments, Contracting Parties have a certain discretionary space to ensure that the provisions and instruments fit well in the domestic legal system.[24] Often such instruments take the shape of a governmental authorization system (e.g. permit requirements) or consist of a combination of prohibitions and (procedures for) exceptions. The main aim of such instruments is to prevent or limit the adverse impacts on the relevant natural values. To support the positive functioning of these instruments, they are often linked with requirements on prior environmental impact assessments (EIAs), and monitoring and reporting obligations. Often the instruments leave the relevant governmental authorities a certain degree of discretionary power to balance the various interests involved, for instance nature conservation interests and socio-economic interests.

At the other end of the spectrum of 'tools' in the relevant conventions are instruments that fall within the restoration approach, as described in the introduction. Instead of a passive approach of assessing, authorizing and conditioning activities, the relevant instruments require an active approach of taking restoration measures. Many experts consider 'restoration ecology' as a separate branch of ecological science, and it is the central theme of a number of special academic journals, books and societies.[25] In legal academic debates, ecological restoration has received less attention; however, also from a legal perspective the approach is certainly not new. In fact the discussion below shows that restoration receives explicit attention in most nature conservation treaties. Furthermore, restoration measures may well form part of the efforts that are needed to comply with more general requirements under the conventions, particularly requirements related to a certain conservation status for species, habitats and/or ecosystems that must be achieved and enforced. Before looking more closely at the relationship between prevention and restoration on the basis of COP decisions (Subsection 2.2), the

---

[23] See J van Andel and J Aronson (eds), *Restoration Ecology. The New Frontier* (Wiley-Blackwell, 2012) 2: 'ecological restoration requires a dynamic, adaptive approach to problem solving and resource management, especially in this era of rapid and irreversible change in climate, land use and species assemblages'.

[24] MJ Bowman, PGG Davies and C Redgwell, *Lyster's International Wildlife Law* (2nd edn., CUP, 2010), 109–110.

[25] See, in particular, the Society of Ecological Restoration (SER), founded in 1987. A selection of SER journals and books is available at <http://www.ser.org/resources>.

most important provisions in international nature protection conventions that explicitly or implicitly relate to restoration are briefly identified below.

2.1 Restoration in International Nature Protection Law: A General Inventory

Article 8(f) of the CBD states that 'Each Contracting Party shall, as far as possible and as appropriate ... [r]ehabilitate and restore degraded ecosystems and promote the recovery of threatened species, *inter alia*, through the development and implementation of plans or other management strategies'.[26] Furthermore, Article 10(d) of this convention states that '[e]ach Contracting Party shall, as far as possible and as appropriate ... (d) [s]upport local populations to develop and implement remedial action in degraded areas where biological diversity has been reduced'. As explained by Gardner, '[r]emedial actions, with respect to degraded wetlands, would obviously include restoring the site to its previous condition'[27] and this would certainly also apply for other types of ecosystems. As will be discussed in Section 4 in more detail, more recently restoration has received a more prominent position under the CBD system.

Ecological restoration is less explicitly worked out in the text of the Ramsar Convention. However, the strategy of restoration has received comprehensive attention in COP decisions, guidance documents and implementation practices. The main obligation under the convention is the obligation 'to promote the conservation of the wetlands included in the List, and as far as possible the wise use of wetlands in their territory' (Article 3(1)).[28] The COP of the Ramsar Convention has defined 'wise use' of wetlands as 'the maintenance of their ecological character'.[29] The concrete meaning of this obligation depends on the specific wetland, but – as explained in the 2013 edition of the Ramsar Convention Manual – it may also imply restoration obligations:

> To achieve the wise use of a wetland so that present and future generations may enjoy its benefits, a balance must be attained that ensures maintenance of the wetland type. Activities may vary between strict protection with no resource exploitation; a small amount of resource exploitation; large-scale sustainable resource exploitation; or active intervention in the wetland, including restoration.[30]

---

[26] CBD (n 5), Article 8(f).
[27] Gardner (n 14), 582.
[28] In the literature, it has been explained that the concept of 'conservation' for listed wetlands does not exclude human use of such wetlands and therefore does not imply a stricter protection regime compared to the 'wise use' concept for all wetlands; *see* Bowman, Davies and Redgwell (n 24), 415–416.
[29] *Ibid.*, 417. The authors refer to the revised 'Conceptual Framework for the Wise Use of Wetlands and the Maintenance of Their Ecological Character', included in Ramsar Resolution IX.1, Annex A.
[30] Ramsar Convention Secretariat, 'The Ramsar Convention Manual. A Guide to the Convention on Wetlands' (6th edn., Gland, 2013), 49, available at <http://www.ramsar.org/sites/default/files/documents/library/manual6-2013-e.pdf>. *See*, in similar wording, Bowman, Davies and Redgwell (n 24), 417: '... since wise use demands maintenance of the ecological character of the wetland in question, each case must be treated on its merits with a view to achieving that goal. Thus small or large-scale exploitation may be permissible (or, indeed, necessary) in appropriate cases, whereas in others strict protection may be required, or even active intervention

Relevant for the issue of restoration is also Article 4(2) of the Ramsar Convention, which states:

> Where a Contracting Party in its urgent national interest, deletes or restricts the boundaries of a wetland included in the List, it should as far as possible compensate for any loss of wetland resources, and in particular it should create additional nature reserves for waterfowl and for the protection, either in the same area or elsewhere, of an adequate portion of the original habitat.

Article III(4)(a) of the Bonn Convention states that 'Parties that are Range States of a migratory species listed in Appendix I shall endeavour ... to conserve and, where feasible and appropriate, restore those habitats of the species which are of importance in removing the species from danger of extinction'.[31] The guidelines in Article V for regional agreements under Article IV(3) of the convention to protect Annex II species,[32] also states that '[w]here appropriate and feasible, each agreement should provide for but not be limited to ... (e) conservation and, where required and feasible, restoration of the habitats of importance in maintaining a favourable conservation status ...'. While the above provisions of the Bonn Convention refer to 'conservation' and 'restoration', the titles of most Article IV(3) agreements refer only to 'conservation';[33] however, this also applies to the Bonn Convention itself, and, as noted by Gardner,[34] a closer study of the contents of these regional agreements shows that many of these agreements also include provisions on restoration. Gardner refers to provisions in the African-Eurasian Waterbirds Agreements (AEWA) of 1995.[35] Also the more recent regional agreements include provisions on restoration. For instance, Article III(1)(a) of the Agreement on the Conservation of Albatrosses and Petrels (2001)[36] states:

> In furtherance of their obligation to take measures to achieve and maintain a favourable conservation status for albatrosses and petrels, the Parties, having regard to Article XIII, shall (a) conserve and, where feasible and appropriate, restore those habitats which are of importance to albatrosses and petrels.

---

aimed at restoration.' *See* also Gardner (n 14), 579: 'The concept of wise use, which applies to both listed and other wetlands, also contemplates restoration actions.'

[31] For a discussion on the far-reaching scope but also the weak legal language of this provision, *see* S Lyster, 'The Convention on the Conservation of Migratory Species of Wild Animals (The "Bonn Convention")' (1989) 29 Natural Resources Journal 979, 985–987.

[32] *See* A Gillespie, *Conservation, Biodiversity and International Law* (Edward Elgar Publishing, 2011), 91: 'If the circumstances so warrant, agreements may also encompass Appendix I species such as with gorillas.'

[33] The term restoration has been included in the title of 'a non-binding administrative agreement' under Article IV(4) of the Bonn Convention: *see* the 'Memorandum of Understanding concerning Conservation and Restoration of the Bukhara Deer (*Cervus elaphus bactrianus*)', 16 May 2002, available at <http://www.cms.int/species/bukhara_deer/pdf/mou_e.pdf>.

[34] *See* Gardner (n 14), 584–585.

[35] *Ibid.*

[36] Agreement on the Conservation of Albatrosses and Petrels, as amended by the Fourth Session of the Meeting of the Parties, Lima, Peru, 23–27 April 2012, available at <http://www.acap.aq/index.php/acap-agreement>.

Article III(2)(b) of the CMS Agreement on the Conservation of Gorillas and Their Habitats (2007)[37] states that Parties shall 'identify sites and habitats for gorillas occurring within their territory and ensure the protection, management, rehabilitation and restoration of these sites ... ,' while paragraph (c) of this provision requires the Parties to 'coordinate their efforts to ensure that a network of suitable habitats is maintained or re-established throughout the entire range of all species and sub-species, in particular where habitats extend over the area of more than one Party to this Agreement'.

These agreements, as well as other binding regional agreements under Article IV(3) and Article IV(4) of the CSM, also provide for the instrument of an Action Plan. The Action Plan is attached to the agreement, constitutes an integrated part of it, and is therefore legally binding for the Contracting Parties. The agreements state that 'habitat conservation' should be one of the components of the Action Plan,[38] and the texts of the existing Action Plans make clear that this includes rehabilitation and/or restoration. For instance, Article 3.3 of the Action Plan under the AEWA states that 'Parties shall endeavour to rehabilitate or restore, where feasible and appropriate, areas which were previously important for the populations listed in Table 1 ...'. The Action Plan under the Agreement on the Conservation of Albatrosses and Petrels states in paragraph 2.1 (*General Principles*) that '[s]o far as is appropriate and necessary, the Parties shall take such management action, and introduce such legislative and other controls, as will maintain populations of albatrosses and petrels at, or restore them to, favourable conservation status, and prevent the degradation of habitats'. Article 2.1 of the Action Plan under the Agreement on the Conservation of Gorillas and Their Habitats uses similar wording.

In the WHC, the terms restoration or reparation are not to be found, but 'rehabilitation' is one of the components of the general obligations of Article 5(d):

> To ensure that effective and active measures are taken for the protection, conservation and presentation of the cultural and natural heritage situated on its territory, each State Party to this Convention shall endeavour, in so far as possible, and as appropriate for each country: ... (d) to take the appropriate legal, scientific, technical, administrative and financial measures necessary for the identification, protection, conservation, presentation and rehabilitation of this heritage.

Such rehabilitation measures may include 'corrective measures' to restore the 'Outstanding Universal Value' of heritage.[39] Restoration is also of direct relevance for implementing other components of the convention, particularly the arrangements in

---

[37] Agreement on the Conservation of Gorillas and Their Habitats, Paris, 26 October 2007, available at <http://www.cms.int/>.

[38] Agreement on the Conservation of Albatrosses and Petrels (n 36), Article VI(2)(b) explicitly refers to actions on 'habitat conservation and restoration'.

[39] World Heritage Centre, *Operational Guidelines for the Implementation of the World Heritage Convention*, (WHC 13/1 July 2013), para 16, available at <http://whc.unesco.org/en/guidelines>.

relation to heritage that has been placed on the 'List of World Heritage in Danger'.[40] As explained on the convention's secretariat website:

> Inscription of a site on the List of World Heritage in Danger requires the World Heritage Committee to develop and adopt, in consultation with the State Party concerned, a programme for corrective measures, and subsequently to monitor the situation of the site. All efforts must be made to restore the site's values in order to enable its removal from the List of World Heritage in Danger as soon as possible.[41]

Removal of heritage from the List of World Heritage in Danger is only possible after the 'Desired State of Conservation' has been achieved through so-called 'corrective measures', and these corrective measures will often include restoration measures as well as measures to end the threats.[42] So, while rehabilitation may be required for any heritage site, restoration is of particular importance for the removal of a site from the List of World Heritage in Danger.

A special feature of the WHC is that its provisions apply to natural as well as cultural heritage sites, and that the restoration concept in relation to cultural heritage has a long and rich history under the convention. Although this has not been studied for the purpose of this chapter, it might well be that the experiences regarding restoration of cultural heritage are of value for discussions on restoration of natural heritage and vice versa. For instance, discussions on the meaning of 'authenticity' and 'integrity' for the restoration of cultural heritage[43] may be of value for discussions on restoration of natural heritage sites.[44]

Similar to the WHC, the Bern Convention pays little explicit attention to the restoration approach. Most of the more specific provisions strongly focus on the avoidance of negative impacts and the need to prohibit certain activities (the prevention approach), without explicit references to restoration measures. The provision that reflects more explicitly the restoration approach is Article 11:

> Each Contracting Party undertakes: (a) to encourage the reintroduction of native species of wild flora and fauna when this would contribute to the conservation of an endangered species,

---

[40] World Heritage Convention, Article 11(4).

[41] Website of the World Heritage Convention, available at <http://whc.unesco.org/en/158/>.

[42] World Heritage Centre, *Operational Guidelines* (n 39), para. 183. See also the World Heritage Centre, Guidance Note 'Desired State of Conservation for the Removal of a Property from the List of World Heritage in Danger'.

[43] For an explanation of the criteria for determining authenticity and integrity (as requirements for nominating heritage), *see* World Heritage Centre, *Operational Guidelines* (n 39) para II.E. *See* also H Stovel, 'Effective Use of Authenticity and Integrity as World Heritage Qualifying Conditions' (2007) 2(3) City & Time 21.

[44] Restoration without attention for authenticity of the heritage may even be one of the reasons for listing heritage as World Heritage in Danger. This was, for instance, the case for the listing of the Royal Palaces of Abomey (cultural heritage in Benin) in 2003: 'The In-Danger listing was also due to the observation that restoration was carried out without respect for the authenticity of materials, volumes or colours.' See ICOMOS, *World Heritage in Danger Compendium II. A compendium of key decisions on the conservation of cultural heritage properties on the UNESCO List of World Heritage in Danger* (ICOMOS April 2009), 32.

provided that a study is first made in the light of the experiences of other Contracting Parties to establish that such reintroduction would be effective and acceptable.

However, restoration measures are also of great importance to fulfil other obligations of the convention. For instance, Article 2 of the convention requires the Contracting Parties to 'take requisite measures to maintain the population of wild flora and fauna at, or adapt it to, a level which corresponds in particular to ecological, scientific and cultural requirements ...'. This may well require restoration measures. As will be discussed in Section 4 in more detail, the restoration approach has become increasingly important in implementing the Bern Convention, particularly in the form of actively promoting the development of so-called Species Action Plans and Species Recovery Plans.

2.2 The Interrelationship between Prevention and Restoration

The interrelationship between both approaches under international biodiversity conventions has received little explicit attention in the legal literature. Most often terms like prevention, avoidance and conservation are used in parallel to terms such as restoration, recovery and repair as different approaches to reach the conservation objectives. This is also a common approach in documentation under certain conventions, particularly the Bonn Convention. Often restoration appears to be considered a component of conservation. However, a closer look at the treaty systems makes clear that various decisions, resolutions, recommendations and guidance documents that have been adopted by the COPs of some of the selected conventions clearly indicate that the preventive approach should receive priority.

This prioritization is particularly emphasized in the documents on restoration that have been developed in the framework of the Ramsar Convention. As early as 1990, with the adoption of Recommendation 4.1, the COP recommended that 'all Parties examine the possibility of establishing appropriate wetland restoration projects', however the preamble explicitly stresses that the Parties were '[c]onvinced that maintenance and conservation of existing wetlands is always preferable and more economical than their subsequent restoration'.[45] Probably one of the reasons for this priority of prevention is that it is often uncertain whether restoration activities will actually be undertaken and, if such activities take place, whether restoration efforts will be successful in ecological terms. Restoration may also take much time and is often expensive. Moreover, a strong emphasis on restoration options may limit the willingness of stakeholders to take the preventive approach: why block economic development if we can compensate or restore damage to natural values at a later moment in time and/or at a location that fits better in our spatial planning? The awareness of this risk was already reflected in the 1990 Recommendation which recalls '... that restoration schemes must not weaken efforts to conserve existing natural systems'.[46]

---

[45] Ramsar Convention, Recommendation 4.1 ('Wetland restoration'), available at <http://www.ramsar.org/sites/default/files/documents/library/key_rec_4.01e.pdf>.
[46] *Ibid.*

Also, on more recent occasions the priority of prevention has been emphasized within the Ramsar Convention framework.[47] For instance, the preamble of Resolution XI.9, adopted in 2012, explains:

> Resolutions adopted by the Parties consistently urge that a three-step approach should be taken to responding to current or likely changes in the ecological character of wetlands, whether or not such wetlands are included in the Ramsar List, namely:
>
> (a) avoiding impacts (e.g. systematic assessment of projected negative changes to ecological character of potentially impacted wetlands through strategic planning to systematically identify potential areas for conservation);
> (b) mitigating on-site for unavoidable impacts (e.g. through minimizing project impacts and restoring area after the project); and
> (c) compensating for, or offsetting, any remaining impacts (e.g. off-site restoration).[48]

This prioritization of approaches is also part of the main text of the Resolution:

> ... reaffirms the Contracting Parties' commitment to avoiding negative impacts on the ecological character of Ramsar Sites and other wetlands as the primary step in strategies for stemming the loss of wetlands, and where such avoidance is not feasible, to applying appropriate mitigation and/or compensation/offset actions, including through wetland restoration.[49]

This policy and its history in the Ramsar Convention system have been explained in more detail in the Annex to the Resolution: the 'Integrated Framework and guidelines for avoiding, mitigating and compensating for wetland losses'. This Annex also lists all 'Resolutions and Recommendations which recognize the three-stage sequence of avoiding, mitigating (or minimizing), and compensating for wetland losses'.[50]

This prioritization of prevention over restoration is also highlighted in soft-law documents adopted by COPs of other nature conservation conventions. For instance,

---

[47] See, for example, Scientific and Technical Review Panel of the Ramsar Convention, Briefing Note 4 (May 2012), available at <http://www.ramsar.org/sites/default/files/documents/library/bn4-en.pdf>: 'Restoration is not a substitute for protecting and ensuring the wise use of wetlands, i.e., the potential to restore a wetland is not a justification or suitable trade-off for the continued degradation of wetlands. Furthermore, while restoration can play an important role in enhancing wetland benefits, experience shows that a "restored" wetland rarely provides the full range and magnitude of services delivered by a wetland that has not been degraded.'

[48] Ramsar Convention, Resolution XI.9 (An Integrated Framework and guidelines for avoiding, mitigating and compensating for wetland losses), para 10, available at <http://www.ramsar.org/sites/default/files/documents/library/cop11-res09-e.pdf>.

[49] Ibid., para. 14.

[50] See ibid., 6 (Box 1). See also the Scientific and Technical Review Panel, Briefing Note 3 (April 2012), available at <http://www.ramsar.org/sites/default/files/documents/library/bn3.pdf>, Recommendation 2.3 (1984) and Resolutions VII.24 (1999), X.12 (2008), X.17 (Annex) (2008), X.19 (Annex) (2008), X.25 (2008), and X.26 (2008). In the same Briefing Note (at 8), the STRP explained that a review 'of environmental laws and policies demonstrates that an avoid-mitigate-compensate approach is common throughout all Ramsar regions' and that this 'approach is often applied to all ecosystems, not just wetlands'; however, it explicitly notes that 'the extent to which these laws and policies are applied in a manner that results in effective avoidance, mitigation, and compensation requires further study'.

Decision XI/16 on 'Ecosystem Restoration', adopted at the 11th COP meeting of the CBD in 2012, notes in the very first paragraph of the preamble 'that ecosystem restoration is not a substitute for conservation, nor is it a conduit for allowing intentional destruction or unsustainable use'.[51] This may have been based on one of the conclusions of what has been called 'the flagship publication of the Convention on Biological Diversity',[52] the Global Biodiversity Outlook 3 (2010): '[T]he biodiversity and associated services of restored ecosystems usually remain below the levels of natural ecosystems. This reinforces the argument that, where possible, avoiding degradation through conservation is preferable (and even more cost-effective) than restoration after the event.'[53]

Discussions within the selected treaty systems and in the literature also show that the relationship between prevention and restoration is not just a matter of prioritization but is more complex. For instance, also after restoration measures have been taken, prevention of damage to the restored values is important to ensure long-term effectiveness. For instance, the removal of non-native predators from islands with ground-breeding birds has little effect without efforts to prevent new introductions of such non-native species. And the reintroduction of a species that has been hunted to extinction does not make sense if the reintroduction is not paralleled by the adoption and enforcement of strict hunting regulations. This notion of 'prevention after restoration' is, for instance, clearly reflected in the Guidance Note on the de-listing of heritage from the World Heritage in Danger List:

> The decision to remove a property from the List of World Heritage in Danger should therefore be based on demonstrating the reduction of threats, the restoration of deteriorated attributes, and the capacity of the property's protection and management system to prevent the threats from recurring.[54]

3. THE WEAK SIDE OF PREVENTION[55]

As stressed by Veit Koester, '[m]ost likely the global situation state of wetlands, properties of outstanding universal value, endangered species of wild fauna and flora, migratory species of wild animals and biodiversity as such would have been considerably worse without the existence of the conventions.'[56] Indeed, the preventive approach has resulted in numerous success stories. Nonetheless, the great number of monitoring reports and effectiveness studies obliges us to acknowledge the limited effectiveness of the efforts to prevent damage to biodiversity. The overall picture is that many human

---

[51] CBD, Decision XI/16 (Ecosystem Restoration), Preamble para. 1 available at <https://www.cbd.int/doc/decisions/cop-11/cop-11-dec-16-en.pdf>.
[52] See the webpage of the CBD Secretariat on the *Global Biodiversity Outlook 3* available at <https://www.cbd.int/gbo3/>.
[53] *Global Biodiversity Outlook 3* (n 11), 13. *See also* n 50.
[54] World Heritage Centre (n 42).
[55] Parts of this section build on Bastmeijer (n 1).
[56] *See* V Koester, 'Book Review of Karin Baakman, Testing Times: The Effectiveness of Five International Biodiversity-related Conventions' (2012) 21(1) RECIEL 67, 70.

activities that are adversely impacting biodiversity are continuing or even increasing in scale and intensity, and that the objectives of international nature conservation conventions are not achieved. 'The five principal pressures directly driving biodiversity loss (habitat change, overexploitation, pollution, invasive alien species and climate change) are either constant or increasing in intensity.'[57]

This limited effectiveness of international nature conservation conventions has been the subject of many reports and academic publications. One might argue that a comparison of objectives and results simply shows limited effectiveness, but the explanation for this is complex and involves many different factors. Certain factors may have little to do with law (e.g. lack of financial means for implementation), some factors will apply more generally to international law (e.g. vague terminology and weak enforcement mechanisms), while other factors may be characteristic of one particular treaty system. For instance, based on an analysis of the Bonn Convention's 25 years of experience, Richard Caddell explains that 'the treaty regime itself faces a number of obstacles' that are 'common afflictions that plague international environmental law generally' (e.g. 'lack of initial activity' within the treaty system and 'limited compliance with the measures adopted by the CMS and its subsidiary bodies'), while effectiveness is also limited by aspects that are specific to the Bonn Convention ('lack of scientific knowledge generally on migratory wildlife').[58] In addition to weaknesses in the international systems, the domestic systems to implement the conventions have certain weaknesses as well.

A comprehensive discussion of the reasons for ineffectiveness of nature conservation law is far beyond the scope of this chapter.[59] However, underneath the many concrete legal and practical problems, there appear to be at least four more fundamental characteristics of nature protection conventions, as described in the following subsections, that contribute to the limited effectiveness of the preventive approach.

3.1 Procedural and Vague Obligations Leave Space for Prioritizing Short-term Interests

Many provisions and obligations in nature conservation law have a procedural character. Examples include obligations to develop policy plans, to notify certain activities, to cooperate with other Parties, to assess environmental impacts of plans and projects and to monitor change. These obligations have several advantages, but do not contain clear standards of what activities and related influences on natural values are to be considered acceptable. Those provisions in nature conservation law that do include more substantial standards are often characterized by vague formulations. This

---

[57] *Global Biodiversity Outlook 3* (n 11), 9.
[58] R Caddell, 'International Law and the Protection of Migratory Wildlife: An Appraisal of Twenty-Five Years of the Bonn Convention' (2005) 16 Colorado Journal of International Environmental Law & Policy 113, 140–141.
[59] For a comprehensive study on this theme, *see* K Baakman, *Testing Times: The Effectiveness of Five International Biodiversity-related Conventions* (Wolf, 2011).

approach makes it possible to reach consensus among a large group of state governments[60] and it supports the 'living instrument' idea behind many conventions in the sense that the interpretation of the provisions may be adjusted to new challenges and circumstances. However, directly connected to these advantages are the disadvantages from the perspective of effectiveness of nature conservation law: the legal obligations and prohibitions leave so much space for interpretation that in implementing practice short-term economic interests are often prioritized over natural values. In other words, in balancing interests, governments may decide to sacrifice natural values to economic plans and projects without clear violations of the relevant legal instruments. This is, in fact, also the weak side of the ideal of sustainable development. Balancing interests is at the heart of this ideal, however in practice it also leaves space for prioritization, and, often, safeguarding natural values is in a weak position compared to short-term economic interests. Often this results in weak sustainability approaches in which the limitation of adverse impacts on nature ('doing less bad') is considered sufficient for labelling the plan or project as 'sustainable'.

3.2 The Difficulty of Denying Authorization for Economic Plans and Projects

The previous characteristic is even more problematic because, particularly in many Western states, the process of modernization and liberalization of the last 200 years has resulted in a great emphasis on the right of the individual to 'develop' and to accumulate wealth through continued appropriation of private property. This development is complex and has its roots in many legal, philosophical and economic theories. For instance, Locke has stated that a person may acquire components of nature as his private property by mixing it with his labour 'as much as any one can make use of to any advantage of life before it spoils'.[61] However, as explained by MacPherson, Locke considered this 'spoil-limitation' for acquisition irrelevant after the introduction of money: 'Gold and Silver do not spoil; a man may therefore rightfully accumulate unlimited amounts of it, "the exceeding of the bounds of his just Property not lying in the largeness of his Possession, but the perishing of anything uselessly in it".'[62]

This line of reasoning has been strengthened by economic theories, such as the 'invisible hand' theory of Adam Smith.[63] If individuals in a society act to the benefit of their own interests, this will also be best for society: 'By pursuing his own interest he

---

[60] *See*, for example, Birnie, Boyle and Redgwell (n 5), 617.

[61] J Locke, 'Two Treatises of Government, The Second Treatise' (1690), para 31. *See* T Brooks, *Locke and Law* (Ashgate, 2007), xvi: 'We can only have as much property as we can enjoy.'

[62] CB MacPherson, *The Political Theory of Possessive Individualism – Hobbes to Locke* (OUP, 1962), 204. MacPherson refers to Locke, *The Second Treatise*, *ibid.*, s. 46. *See* also SA Bell, JF Henry and L Randall Wray, 'A Chartalist Critique of John Locke's Theory of Property, Accumulation, and Money: or, is it Moral to Trade Your Nuts for Gold?' (2004) LXII(1) Review of Social Economy 51.

[63] A Smith, *The Theory of Moral Sentiments* (1759), Glasgow Edition of the Works and Correspondence Vol. 1, 203, available at <http://oll.libertyfund.org/titles/2620>. *See* also B van Heerikhuizen, 'An academic lecture on Adam Smith and Bernard de Mandeville', available at <http://www.youtube.com/watch?v=A8wMy1JvSsY>.

[a person] frequently promotes that of the society more effectually than when he really intends to promote it.'[64] Such legal and economic theories are further strengthened by socio-psychological theories. For instance, Veblen explained at the end of the 19th century that the concept of private property results in a competition within society that is based on comparison and imitation: 'The motive that lies at the root of ownership is emulation. ... The possession of wealth confers honour; it is an invidious distinction.'[65] Over time, private property has increasingly been viewed 'as evidence of the prepotence of the possessor of these goods over other individuals within the community. The invidious comparison now becomes primarily a comparison of the owner with the other members of the group.'[66] Veblen explains that, as a consequence, in Western societies accumulation of wealth even becomes a necessity:

> With the growth of settled industry, therefore, the possession of wealth gains in relative importance and effectiveness as a customary basis of repute and esteem. ... It therefore becomes the conventional basis of esteem. Its possession in some amount becomes necessary in order to any reputable standing in the community. It becomes indispensable to accumulate, to acquire property, in order to retain one's good name.[67]

This theory relates very well with recent socio-psychological research that shows that selfish behaviour of individuals in society is not so much motivated by the wish to be selfish for its own sake, but rather by the desire of the individual to prevent a weak position in society.[68]

Although this discussion is of course far from complete, such historic theories, the strong belief in a liberal market economy, and the related views on a limited role of government will at least in part explain why in the 21st century many governments appear to have difficulties in saying 'no' to plans and projects for reasons of nature conservation. In discussions on the acceptability of human plans and projects there appears to be a general starting point that any person may conduct any activity at any time and place, without firm requirements to prove the importance of the initiative for society. The government often has the burden of proof to demonstrate why a private initiative should not take place and it appears that nature conservation is often too weak to overcome that burden. Moreover, also within governments, for instance in processes of developing policy and taking decisions on permit applications, short-term interests are often prioritized. Certainly, in 2014, the intensifying environmental concerns have made it acceptable that activities are subjected to procedural requirements (e.g. EIA) and to certain conditions to limit adverse impacts on the environment, but denying authorization still appears to be a taboo.

---

[64] A Smith, *An Inquiry into the Nature and Causes of the Wealth of Nations* (ed by S. M. Soares) (MetaLibri Digital Library, 2007), 562.
[65] T Veblen, *The Theory of the Leisure Class. An Economic Study of Institutions* (Macmillan Company, 1899), chapter 2 ('Pecuniary Emulation') (1909 edition, available at <http://www.gutenberg.org/files/833/833-h/833-h.htm>).
[66] *Ibid.*
[67] *Ibid.*
[68] *See,* for example, C de Dreu, 'Verbinden als sociaal dilemma' (April 2010) Idee, 10, 11.

3.3 Inability to Address Cumulative Adverse Impacts

The previous characteristics result in situations in which many human activities are considered to be acceptable or which are explicitly authorized, while they still have a certain adverse impact on nature. Certainly, due to environmental legislation, such impacts are subjected to prior assessments, and may be minimized by permit conditions and regulations; however, it often is the accumulation of all these smaller impacts that causes the greatest concerns. Vöneky refers to Francioni, who has stated that 'most environmental damage is caused by lawful acts that have had adverse effects on the environment'[69] and this problem has become even more apparent over the last two decades. Most of the serious concerns for biodiversity are caused by accumulative impacts of 'lawful' activities; activities that also grow in number, intensity and geographical scope. At the global scale we may refer to climate change and the over-exploitation of certain minerals, and at the regional level examples include over-exploitation of fish and fresh water stress. These examples may also be relevant at the domestic or even local level, in parallel to many other examples, such as the accumulation of nitrate deposition or even the scarcity of space.

Probably the fact that most causes of biodiversity loss have an accumulative character makes the prevention approach so difficult. Most of the prevention instruments (e.g. EIA, permit requirements, prohibitions) relate to individual activities that are generally considered acceptable. The international conventions appear to leave much space for this practice, while at the end the accumulative impacts are serious hurdles for reaching the conservation objectives. Experiences with environmental law in the European Union indicate that the prevention approach is likely to be more effective if the source-related instruments (e.g. permitting) are linked with clear and strict environmental quality standards. In such systems, decisions on the allowance of individual activities may not simply be based on a discretionary power of a governmental authority to balance interests, as permits can only be issued as long as the overall environmental quality standard is respected.[70] However, this approach is missing in most international biodiversity conventions.

---

[69] S Vöneky, 'The Liability Annex to the Protocol on Environmental Protection to the Antarctic Treaty' in D König and others (eds), *International Law Today: New Challenges and the Need for Reform? Beiträge zum ausländischen öffentlichen Recht und Völkerrecht* (Springer, 2008) 165, 176–177; F Francioni, 'Liability for Damage to the Common Environment: The Case of Antarctica' (1994) 3 RECIEL 223.

[70] Examples include the air quality standard for particulate matter (PM10) and – although of a different nature – conservation objectives for Natura 2000 sites. The functioning and consequences of such systems depend on how these systems of quality standards have been shaped and implemented; *see* C Backes and M van Rijswick, 'Effective environmental protection: towards a better understanding of environmental quality standards in environmental legislation' in L Gipperth and C Zetterberg (eds), *Miljörättsliga perspektiv och tankevändor. Vänbok till Jan Darpö & Gabriel Michanek* (Iustus Förlag, 2013), 19–50, available at <https://www.researchgate.net/publication/260136848>.

3.4 Nature Conservation Law's Signal to the Public to Only Protect 'Special' Nature?

Furthermore, the above discussed space for balancing interests and prioritizing economic interests is enlarged because the more common natural values receive little or no protection. Many legal instruments to protect nature focus on certain values that are endangered or for other reasons considered 'special'. For example, history indicates legal protection in treaties, as well as domestic laws, was attributed to those species that were almost extinct.[71] In the literature this type of nature conservation is referred to as 'deathbed conservation'.[72] The conventions discussed in this chapter clearly have a broader purpose and scope, however many of the more clear and strict prohibitions and requirements apply only to species or sites that have been listed (Bonn Convention, Bern Convention, Ramsar Convention).

On the one hand this approach is logical as particularly threatened natural values require protection and there are quite a number of success stories of such focused approaches. A recent report in Europe provides an overview of the 'comeback' of a significant number of large mammals and iconic bird species, and nature conservation law is indicated as one of the important explanations.[73] On the other hand, however, the approach of protecting only 'the special' values has several weaknesses. There is a risk that protection simply comes too late. It also provides the remarkable signal to the public that nature should only receive care if it is scarce, endangered or for other reasons special. As Doremus explains in an article with the beautiful title 'The Special Importance of Ordinary Places':[74] 'The rhetoric of specialness sets up a dichotomy between special places, which are worth saving, and non-special ones, which by definition are not. If we truly want to protect nature, that distinction is ultimately untenable.'[75]

4. ADDRESSING WEAK PREVENTION: THE NEED TO RESTORE AS REFLECTED IN THE LITERATURE AND COP DECISIONS

A logical response to problematic prevention is the adoption of approaches to deal with the negative consequences. This is clearly illustrated by the climate change debate. The UN Framework Convention on Climate Change (UNFCCC) includes both approaches, prevention (mitigation by emission reduction) and restoration (as part of the broader concept of adaptation), and while for a long time all efforts were focused on prevention, it has become increasingly clear that restoration and other adaptation

---

[71] Birnie, Boyle and Redgwell (n 5).
[72] *See*, inter alia, A Trouwborst, 'Seabird Bycatch – Deathbed Conservation or a Precautionary and Holistic Approach?' (2008) 11(4) Journal of International Wildlife Law & Policy 293.
[73] *See*, for example, Deinet and others (n 7).
[74] H Doremus, 'The Special Importance of Ordinary Places' (1999) 23 Environmental Law & Policy Journal 3, available at <http://scholarship.law.berkeley.edu/facpubs/511>.
[75] *Ibid.*, 4.

approaches (e.g. relocation) are a necessary second pillar of climate change policy.[76] Similar to this discussion, ecological restoration has received increased attention in the literature, and in monitoring studies, as a necessary approach to meet the nature conservation objectives. 'Ecological restoration, rooted in the early developments and visionary work of a few individuals and programmes in the nineteenth and twentieth centuries ... has grown to a respectable "size" and volume only in the last few decades ... but is now gaining momentum and attention as never before.'[77] Also the *Global Biodiversity Outlook 3* states that '[i]ncreasingly, restoration of terrestrial, inland water and marine ecosystems will be needed to re-establish ecosystem functioning and the provision of valuable services'.[78]

As in the convention systems discussed above, so also in the literature restoration is not presented as an alternative for prevention but rather as an additional approach. For instance, Lars Brudvig states:

> Our abilities to recreate ecosystems are simply not – and may never be – sufficient to warrant habitat destruction; however, passive protection of habitat remnants alone will not suffice in many landscapes. ... Certainly, we should preserve what remains, but the present century must additionally usher in an era of restoration, in which lands that have been transformed by human land use are modified to better support desired biodiversity and functions.[79]

Thus, while prevention continues to be essential, '[i]t is ecological restoration that most proximally will reverse biodiversity declines'.[80] Hobbs and Harris use similar wording:

> In terms of nature conservation, there is no substitute for preserving good quality habitat, and the maintenance and management of this is a number one priority. However, in many parts of the world, this is either no longer an option because few areas of unaltered habitat remain, or it is no longer sufficient since the remaining habitat on its own cannot sustain the biota, and hence needs to be improved or expanded.[81]

Francisco Comin has emphasized that all this is even more true due to negative influences on nature conservation relating to climate change:

> In addition, biodiversity loss and ecosystem degradation will continue at high rates under any current development scenario, due to the momentum of global climate change impacts and

---

[76] J Verschuuren, 'Climate Change Adaption under the United Nations Framework Convention on Climate Change and Related Documents' in J Verschuuren (ed.), *Research Handbook on Climate Change Adaptation Law* (Edward Elgar Publishing, 2013), 22: '[i]t was only in 2010 that the COP [of the UNFCCC] placed adaptation high on its agenda'.

[77] Van Andel and Aronson (n 23).

[78] *Global Biodiversity Outlook 3* (n 11), 13.

[79] LA Brudvig, 'The Restoration of Biodiversity: Where Has Research Been and Where Does it Need to Go?' (2011) 98(3) American Journal of Botany 549, 549.

[80] *Ibid.*, 556.

[81] RJ Hobbs and JA Harris, 'Restoration Ecology: Repairing the Earth's Ecosystems in the New Millennium' (2001) 9 Restoration Ecology 239, 240, available at <http://onlinelibrary.wiley.com/doi/10.1046/j.1526-100x.2001.009002239.x/pdf>.

continued extraction and consumption of natural resources. So, the ecosystem approach to restoration is critical in any global strategy leading to a desirable habitat for all species.[82]

Furthermore, the Contracting Parties to the nature conservation conventions have become increasingly aware of their limited ability to prevent harm in practice and that the preventive approach and related legal provisions have limited grip on the loss of biodiversity. While restoration was already part of the main text of most international nature protection conventions (see Section 2.1 above), the COPs have assigned an increasingly important role to the restoration approach for achieving conservation objectives. A particularly important development is that restoration has become an important component of the CBD Strategic Plan for Biodiversity 2011–2020, adopted as Decision X/2 at the 10th COP:[83] The vision of this Strategic Plan is a world of 'Living in harmony with nature' where '[b]y 2050, biodiversity is valued, conserved, restored and wisely used, maintaining ecosystem services, sustaining a healthy planet and delivering benefits essential for all people'. Restoration is also part of the mission of this plan and the central topic of two separate targets of the Aichi Biodiversity Targets: 'By 2020, ecosystem resilience and the contribution of biodiversity to carbon stocks has been enhanced, through conservation and restoration, including restoration of at least 15 per cent of degraded ecosystems, thereby contributing to climate change mitigation and adaptation and to combating desertification.'[84]

The COP decision clearly reflects the importance of restoration in parallel to the prevention approach: 'While longer-term actions to reduce the underlying causes of biodiversity are taking effect, immediate action can help conserve biodiversity, including critical ecosystems, by means of protected areas, habitat restoration, species recovery programmes and other targeted conservation interventions.'[85] Restoration is also an important component for other Aichi targets. For instance, target 11 states that:

> By 2020, at least 17 per cent of terrestrial and inland water areas, and 10 per cent of coastal and marine areas, especially areas of particular importance for biodiversity and ecosystem

---

[82] FA Comin, 'The Challenges of Humanity in the Twenty-first Century and the Role of Ecological Restoration' in FA Comin (ed), *Ecological Restoration – A Global Challenge* (CUP, 2010) 3, 13. *See* also Comin at 7: 'Ecological restoration at a global scale will be critical, since the Earth's capacity to sustain us has been diminished significantly by degradation of the environment.'

[83] CBD, Decision X/2 (The Strategic Plan for Biodiversity 2011–2020 and the Aichi Biodiversity Targets), para. 10(c), available at <http://www.cbd.int/doc/?meeting=cop-10>.

[84] *See* target 15 of the CBD Aichi Targets in Secretariat of the CBD, 'Plan for Biodiversity 2011–2020 and the Aichi Targets "Living in Harmony with Nature"' (Montreal, 2011), available at <http://www.cbd.int/doc/strategic-plan/2011-2020/Aichi-Targets-EN.pdf>. *See* also Aichi target 14. *See* also J Aronson and S Alexander, 'Ecosystem Restoration is Now a Global Priority: Time to Roll up our Sleeves' (2013) 1 Restoration Ecology 1, available at <http://www.landscapes.org/wp-content/uploads/2013/10/Aronson-2013.pdf>: 'Twenty years after the adoption of the CBD Convention Text and Article 8(f) … , the Parties have now fully recognized the critical role of restoration in the implementation of the Convention and the Aichi Biodiversity Targets for 2020.'

[85] CBD Decision X/2 (n 83).

services, are conserved through effectively and equitably managed, ecologically representative and well connected systems of protected areas and other effective area-based conservation measures, and integrated into the wider landscapes and seascapes.[86]

As one of the measures to meet this target, Decision X/31 on 'Protected areas' includes a separate paragraph on '[r]estoration of ecosystems and habitats of protected areas', in which the Parties are urged to:

(a) Increase the effectiveness of protected area systems in biodiversity conservation and enhance their resilience to climate change and other stressors, through increased efforts in restoration of ecosystems and habitats and including, as appropriate, connectivity tools such as ecological corridors and/or conservation measures in and between protected areas and adjacent landscapes and seascapes;
(b) Include restoration activities in the action plans of the programme of work on protected areas and national biodiversity strategies.[87]

A similar target had already been included in the 'Programme of Work on Protected Areas', to which subparagraph (b) refers: it is suggested that the Parties '[e]stablish and implement measures for the rehabilitation and restoration of the ecological integrity of protected areas'.[88] CBD Decision X/31 also

[*i*]*nvites* Parties to: (a) Achieve target 1.2 of the programme of work[89] on protected areas by 2015, through concerted efforts to integrate protected areas into wider landscapes and seascapes and sectors, including through the use of connectivity measures such as the development of ecological networks and ecological corridors, and the restoration of degraded habitats and landscapes in order to address climate change impacts and increase resilience to climate change.[90]

The important role of restoration was even more explicitly emphasized at the COP in 2012 (Hyderabad, India), where the Contracting Parties, with the adoption of Decision XI/16, explicitly noted 'that ecosystem restoration will play a critical role in achieving the Strategic Plan for Biodiversity 2011–2020, including the conservation of habitats and species'.[91] On the margins of this COP, the 'Ecosystem Restoration Day' was held on 17 October 2012, and concluded with the adoption of the 'Hyderabad Call for a

---

[86] *Ibid.*, 9.
[87] CBD Decision X/31 (Protected Areas), para. 7, available at <http://www.cbd.int/decision/cop/?id=12297>.
[88] CBD Programme of Work on Protected Areas, S. 1.5.3, available at <http://www.cbd.int/programmes/pa/pow-goals-alone.pdf>. *See* also IUCN/WCPA Ecological Restoration Taskforce – K Keenleyside and others (eds), *Ecological Restoration for Protected Areas, Principles, Guidelines and Best Practices*, Best Practice Protected Area Guidelines Series No. 18 (IUCN, 2012) available at <https://portals.iucn.org/library/efiles/edocs/PAG-018.pdf>.
[89] *See* the CBD Programme of Work on Protected Areas (n 88), 1.2.5: 'Rehabilitate and restore habitats and degraded ecosystems, as appropriate, as a contribution to building ecological networks, ecological corridors and/or buffer zones.'
[90] CBD Decision X/31 (n 87).
[91] CBD Decision XI/16 (Ecosystem Restoration), Preamble, para. 2.

Concerted Effort on Ecosystem Restoration'.[92] This document, agreed upon by a number of governments, the secretariats of the CBD, the UNFCCC, the Ramsar Convention, and several other organisations, also underlined that prevention is clearly not enough:

> *Acknowledging* that there is an emerging consensus that the restoration and rehabilitation of degraded lands, ecosystems and landscapes is increasingly important as conservation alone is no longer sufficient and the destruction of natural habitat remains the largest driver in the loss of biodiversity and ecosystem services.[93]

The involvement of the Ramsar secretariat in this 'call' is not surprising, as the COP to this convention has also increasingly emphasized the importance of restoration.[94] For instance, the Ramsar Strategic Plan 2009–2015, adopted in 2008 by the 10th COP as Resolution X.1, includes as one of the key objectives to establish 'wise use', that by 2015 all Parties 'have identified priority sites for restoration' and that 'restoration projects [are] underway or completed in at least half the Parties'.[95] At this COP the SER was also added to the list of observers.[96] In January 2012, the secretariat of the Ramsar Convention and the SER signed a Memorandum of Cooperation.[97] In this memorandum, '[t]he Convention on Wetlands and the Society for Ecological Restoration recognize the fundamental role that wetland restoration and rehabilitation plays in protecting biodiversity values, enhancing the delivery of ecosystem services, fostering sustainable development, and mitigating and adapting to climate change'.[98] Concrete joint activities have been identified in an Annex to the memorandum.

---

[92] 'Hyderabad Call for a Concerted Effort on Ecosystem Restoration', adopted on the margins of COP XI of the CBD (India 2012), available at <https://www.cbd.int/doc/restoration/Hyderabad-call-restoration-en.pdf >.

[93] *Ibid.*

[94] For a relatively early example, *see* Ramsar Convention, Resolution VII.17 (Restoration as an element of national planning for wetland conservation and wise use), para. 10, available at <http://www.ramsar.org/sites/default/files/documents/library/key_res_vii.17e.pdf>: 'CALLS UPON all Contracting Parties to recognise that although restoration or creation of wetlands cannot replace the loss of natural wetlands, and that avoiding such loss must be a first priority, a national programme of wetland restoration, pursued in parallel with wetland protection, can provide significant additional benefits for both people and wildlife, when the restoration is ecologically, economically and socially sustainable.'

[95] Ramsar Convention Secretariat, 'The Ramsar Strategic Plan 2009–2015: Goals, Strategies, and Expectations for the Ramsar Convention's Implementation for the Period 2009 to 2015' (Ramsar Convention Secretariat, 2010), key-result area 1.1.i and Strategy 1.8 <http://www.ramsar.org/sites/default/files/documents/pdf/strat-plan-2009-e-adj.pdf>.

[96] Ramsar Convention, Resolution X.9 (Refinements to the modus operandi of the Scientific & Technical Review Panel (STRP)), available at <http://www.ramsar.org/sites/default/files/documents/pdf/res/key_res_x_09_e.pdf>.

[97] 'Memorandum of Cooperation between The Secretariat of the Convention on Wetlands (Ramsar, Iran, 1971) and The Society for Ecological Restoration (SER)', January 2012, available at <http://www.ramsar.org/sites/default/files/documents/pdf/moc/01_31_2012-Ramsar-SER_MoC.pdf>.

[98] *Ibid.*

Although the COP of the Ramsar Convention continuous to reiterate that effective wetland protection begins with avoidance of adverse wetland impacts (see Section 2.2), the 11th COP in 2012 explicitly acknowledged that the objectives of the convention will not be met with the current approach and implementation efforts. In Resolution XI.9 of 2012, the Contracting Parties state that they are 'concerned that the total area and condition of natural wetlands in many countries, and the species they support, are still declining'.[99] In the resolution the Contracting Parties note 'that these wetland losses are occurring despite the provisions of the Ramsar Convention on Wetlands and the existence of wetland protection laws and practices in many countries that require that adverse wetland impacts be avoided, and where this is not possible, mitigated or compensated by offsets such as wetland restoration'.[100]

Additionally, under the other international nature protection conventions the importance of the restoration approach to reach the relevant targets has been emphasized.[101] While the Bern Convention itself does not include many references to restoration, under the convention many efforts have been made to promote the development of Species Action Plans and Species Recovery Plans. According to Eladio Fernández-Galiano, Head of Natural Heritage and Biological Diversity Division of the Council of Europe, 'Species Action Plans and Species Recovery Plans are one of the most important tools in conservation of biological diversity'.[102] Based on Article 14 of the convention, the 'Standing Committee of the Convention on the Conservation of European Wildlife and Natural Habitats' adopted in 1997 'Recommendation No. 59 (1997) on the drafting and implementation of Action Plans of Wild Fauna Species'.[103] This recommendation expresses an awareness 'that in many instances wild species which have an unfavourable conservation status (particularly those listed in Appendix II of the Convention) may require special conservation efforts to acquire a population level which corresponds to their ecological requirements …' and 'that Species Action Plans (of which Species Recovery Plans are a particular case) may be appropriate

---

[99] Ramsar Convention, Resolution XI.9 (An Integrated Framework and Guidelines for Avoiding, Mitigating and Compensating for Wetland Losses), para. 4, available at <http://www.ramsar.org/sites/default/files/documents/library/cop11-res09-e.pdf>.

[100] *Ibid.*, para. 5.

[101] In comparison to other conventions, there is little attention to restoration under the Bonn Convention. However, also in relation to migratory species, the importance of restoration has been highlighted. *See*, for instance, the UNEP report 'Migratory Species and Climate Change' (UNEP, 2010) available at <http://www.unep.org/chinese/iyb/pdf/CCnMigratorySpecies.pdf>: 'Investing in ecological restoration should not be forgotten. A deliberate effort to reintroduce extirpated species, and to build wetlands where they have been wiped out will be required.'

[102] E Fernández-Galiano, Foreword to the Action Plan for the Great Snipe, available at <http://www.unep-aewa.org/publications/technical_series/ts5_great_snipe.pdf>.

[103] Standing Committee of the Convention on the Conservation of European Wildlife and Natural Habitats, 'Recommendation No. 59 (1997) on the drafting and implementation of Action Plans of Wild Fauna Species', available at <http://www.eko-g.cz/wp-content/uploads/2015/03/bernska_umluva.pdf>. *See also* 'Recommendation no. 48 (1996) concerning the conservation of European globally threatened birds' (Standing Committee 22 January 1996), available at <https://wcd.coe.int/>: 'Aware that the design and implementation of Recovery Plans may be a useful tool to redress the situation of European globally threatened birds.'

conservation tools to restore threatened populations in some circumstances'.[104] The recommendation included 'Guidelines on the Drafting and Implementation of Action Plans of Wild Fauna Species' and according to paragraph 4.2 and 4.2.4, it must be ensured that

> the plan takes into consideration ... : Habitat conservation and habitat restoration in the natural range of the species (including present sites and those in which the species was present in recent times); while designing areas for conservation, corridor areas permitting genetic flow among neighbouring populations should to be taken into account.

The guidelines have constituted the basis for the development of numerous Species Action Plans, of which many include concrete restoration measures. Just one of the many examples is the Action Plan[105] for 'one of the most threatened fish species in Europe', the European sturgeon (*Acipenser sturio*), which plan includes objectives for, for example, the re-opening and reconstruction of sturgeon migration routes, improvement of water quality and reintroduction 'to re-establish self-sustaining populations in as many areas of its natural range as possible'.[106] The guidelines as well as the concrete examples of Species Action Plans and Recovery Plans make clear that it is not 'just' about a species. Eladio Fernández-Galiano, explains:

> By focussing the attention on the fate and problems of a particular threatened species, many other issues come to light: the effects that agriculture, urban development or pollution are having on nature, the interconnection of species, habitats and management, and the complexities of ecological processes. In a time where most conservation efforts in Europe are faithfully devoted to habitat protection in the hopeful wish that it will automatically yield species conservation, looking at the precise case of some species is highly revealing and can tell us where to address in priority scarce conservation resources.[107]

5. CONCLUDING REMARKS: RESTORATION AS A PROMISING STRATEGY TO ADDRESS OUR FAILURE TO PREVENT?

The prevention approach, the dominant approach in nature conservation conventions, has resulted in numerous successes in nature conservation. However, in view of the great number of monitoring reports and effectiveness studies, we must acknowledge the limited effectiveness of the efforts to prevent damage to biodiversity. As stated by Katharine N. Suding: 'Although a key aim of environmental management should be the avoidance of degradation in the first place, an unfortunate truth is that humans are

---

[104] *Ibid.*
[105] For a recent overview of all Recovery Plans for European birds, *see* Doc. T-PVS/Inf(2013)14E for the 33rd meeting of the Standing Committee (Strasbourg 3–6 December 2013) 'Overview of the Species Action Plans endorsed by the Bern Convention and need for update' (July 2013), available at <https://wcd.coe.int/ViewDoc.jsp?id=2063389>.
[106] 'Action Plan for the Conservation and Restoration of the European Sturgeon (Acipenser Sturio)', Bern Convention, Nature and environment, No. 152 (Council of Europe Publishing, July 2010), paras 6.2.2 and 6.2.3, available at <https://wcd.coe.int/ViewDoc.jsp?id=1653821>.
[107] Fernández-Galiano (n 102).

impacting most ecosystems around the globe.'[108] Weaknesses of prevention go beyond legal deficiencies that could easily be addressed by small adjustments of the legal system. Most likely the limits of prevention are based in weaknesses of humankind itself, such as the difficulty to accept limitations to our social and economic ambitions and our inability to deal with cumulative impacts. These problems are also clearly reflected in some other global environmental problems, such as marine litter and climate change.

As a response to the limited effectiveness of prevention, the restoration approach is receiving increasing recognition in the literature as well as within the systems of various international nature protection conventions as a necessary approach to achieve the convention's objectives.[109] While the priority of prevention is continuously highlighted, in particular in the systems of the Ramsar Convention and the CBD, restoration may ensure that the results of human efforts to protect nature go 'beyond damage control'. Limitation of negative impacts – which is often the maximum result of legal tools that fall within the prevention approach – will not be enough to stop biodiversity loss, particularly because the current conservation status of many species and habitats is already unfavourable and pressures on many natural values are likely to increase due to climate change.

However, restoration has some important weaknesses as well. For instance, more emphasis on restoration may further weaken prevention. Short-term interests may result in approaches that may be characterized as 'no prevention now because we will restore later'. Resolution VIII.16 of the Ramsar Convention emphasizes that 'trading high-quality habitat or ecosystems for promises of restoration should be avoided except in the case of overriding national interests'.[110] However, the problem is that governments are often very creative in showing the general public that new plans and projects are within the scope of this category of interests. This may be particularly concerning because guarantees for successful restoration do not exist. Restoration is complex, simply because nature itself is complex, which means that there are good reasons to be careful in making predictions about what restoration will bring.[111] 'Both quantitative data and subjective assessments clearly show that currently available restoration techniques almost never lead to conditions that match those of pristine natural

---

[108] KN Suding, 'Toward an Era of Restoration in Ecology: Successes, Failures, and Opportunities Ahead' (2011) 42 Annual Review of Ecology, Evolution, and Systematics 465, 465.

[109] See also Aronson and Alexander (n 84), 1–4: 'Many … global agreements, commitments, and initiatives have established the imperative for ecosystem restoration, which is now recognized as a global priority for countries and communities alike.' See also D Jørgensen, 'Ecological Restoration in the Convention on Biological Diversity Targets' (2013) 22(12) Biodiversity and Conservation 2977, 2978: '[t]he focus on ecological restoration in international conservation policymaking has reached an almost feverish pitch'.

[110] Ramsar, Resolution VIII.16 (Principles and Guidelines for Wetland Restoration), available at <http://www.ramsar.org/sites/default/files/documents/pdf/res/key_res_viii_16_e.pdf>.

[111] See, for example, IUCN/WCPA Ecological Restoration Taskforce (n 88), V: 'As we increase our efforts to restore protected area values, however, we must also act with caution and humility, recognizing that ecological restoration is a complex and challenging process and that our interventions can have unforeseen consequences.'

ecosystems.'[112] This is particularly true for ecosystems that develop over (and are only 'complete' after) time periods of several hundred years.

Possibly an even more crucial concern is that it is uncertain whether the 'political will' exists among the Contracting Parties of the relevant conventions to make the restoration approach – parallel to prevention efforts – a really successful strategy. This political will is essential as the provisions on restoration in the international nature protection conventions are even weaker than those relating to prevention. As explained by Gardner in respect of wetland restoration, 'countries retain a great deal of discretion in how, or whether, to implement and support restoration projects'.[113] Caddell states in relation to Article III(4) of the Bonn Convention that 'these requirements are subject to the wide qualification that such habitat conservation and restoration is to be conducted where it is "feasible and appropriate," allowing a Range State considerable leeway in electing whether or not to undertake such measures'. This space for interpretation, and related lack of adequate monitoring options, is also highlighted by Jørgensen, who states that '[a]lthough restoration is incorporated as a key conservation measure in the MEAs, none of the texts actually define what the term means in the context of the MEA'.[114] He argues 'that the failure to define ecological restoration before writing it into its targets will pose future problems when evaluating progress toward the 2020 biodiversity goals'.[115]

This dependency on politics raises the crucial question why we should expect the restoration approach to be more successful in comparison to the prevention approach, particularly taking into account the high costs of restoration.[116] Or, in other words, why would governments be willing to restore while often the political will to prevent is lacking? Answering this question would require more research, but there are several options for a positive response. For instance, one could argue that 'saying no' to human activities with adverse impacts on natural values is more difficult than 'saying yes' to human ambition to restore nature. Restoration may provide governments, and particularly politicians, with the chance to be acknowledged as facilitators of positive action, instead of a negative power that is 'blocking' private initiative. This may even be truer when restoration projects are beneficial for short-term socio-economic interests (in fact, the same interests that often make strict prevention difficult): 'Efforts to restore natural

---

[112] Resolution VIII.16 (n 110) para 12. *See* also Section 2.2 above. For a meta-analysis that confirms this statement, *see* JMR Benayas and others, 'Enhancement of Biodiversity and Ecosystem Services by Ecological Restoration: A Meta-analysis' (2009) 325 Science 1121.

[113] Gardner (n 14), 578. *See* in relation to the Bonn Convention (Art. III(4)) Caddell (n 58), 117: 'In addition, the strength of the obligation to perform such activities is also open to question. Article III of the Bonn Convention simply requires the parties to "endeavour" to undertake these tasks. What, in practice, does the term "endeavour" involve in the context of the Bonn Convention?'

[114] See also Jørgensen (n 109), 2978.

[115] *Ibid.*

[116] *See* also the 'Joint Statement by the Representatives of the World's Youth to the Tenth Meeting of the Conference of the Parties to the Convention on Biological Diversity', CBD, Report of COP 10 (UNEP/CBD/COP/10/27), 15, para. 42, available at <http://www.cbd.int/doc/?meeting=cop-10>: 'Mr. Matsui … said that the more extensive the loss of biodiversity, the more extensive would be the need for its restoration. That would create greater financial demands than the costs of prevention of the loss of biodiversity in the first place.'

capital may require massive programs initiated by governments and international institutions. Implementation of such national programs can bring about positive socio-economic consequences (e.g. job creation and training, and social fabric weaving) and possibly reduce social estrangement and political unrest.'[117]

A possible other reason for an optimistic view in relation to restoration may be the increased acknowledgement that the degradation of ecosystems and related biodiversity also affects the services that nature provides to humans (so-called 'ecosystem services'). Many of the COP documents discussed above emphasize these services and the importance of ecological restoration to restore and safeguard these services.[118] Although this approach has a severe risk of an overly narrow approach as the obligations under the discussed treaties relate to biodiversity and not only to biological resources, the acknowledgement of the relationship between ecosystem services and restoration may have advantages. It may, for instance, increase public and political support for restoration and could constitute the foundations for new financial arrangements that could make nature protection and restoration less dependent on the availability of governmental budgets. New mechanisms and instruments are being developed to ensure that those who enjoy the benefits of nature also ensure a solid protection and restoration of these benefits.

Probably we need such positive thinking, as well as the development of new, creative approaches, to get beyond damage control to reach the biodiversity protection targets. In the words of Crispin Tickell: 'We are supremely adaptable and ingenious mammals. If our brains get us into a mess, our brains should at least try to get us out of it.'[119] Restoration appears to be a promising path to follow, but it should not be considered as a simple alternative to prevention. As reflected in the literature and treaty systems discussed in this chapter, it is clear that the conservation objectives may only be achieved through strengthening prevention, paralleled by effective restoration. This requires, for instance, political will of governments and a responsible attitude and involvement of the private sector, but also additional good research by experts from various research disciplines. In view of the limited size of existing legal literature on restoration, and on the interrelationship between prevention and restoration, it would be of high value if lawyers were also to intensify their contribution to this debate. Such research could, for instance, focus on the question of how the weaknesses of prevention could be addressed (within the treaty systems or at the implementation level), how the priority of prevention over restoration could be legally ensured, and how the private sector could be stimulated or required to take restoration measures (e.g. to create space for economic development within a system of strict quality standards, or as a legal requirement to pay for ecosystem services).

---

[117] AF Clewell and J Aronson, 'Motivations for the Restoration of Ecosystems' (2006) 20(2) Conservation Biology 420, 425.

[118] *See*, for example, Aichi Targets (n 84), Target 14: 'By 2020, ecosystems that provide essential services, including services related to water, and contribute to health, livelihoods and well-being, are restored and safeguarded, taking into account the needs of women, indigenous and local communities, and the poor and vulnerable.'

[119] Tickell (n 2), 222.

SELECT BIBLIOGRAPHY

Allison, SK, *Ecological Restoration and Environmental Change: Renewing Damaged Ecosystems* (Routledge, 2012)

Benayas, JMR and others, 'Enhancement of Biodiversity and Ecosystem Services by Ecological Restoration: A Meta-analysis' (2009) 325 Science 1121

Bowman, MJ, PGG Davies and C Redgwell, *Lyster's International Wildlife Law* (2nd edn., CUP, 2010)

Caddell, R, 'International Law and the Protection of Migratory Wildlife: An Appraisal of Twenty-Five Years of the Bonn Convention' (2005) 16 Colorado Journal of International Environmental Law & Policy 113

Clewell, AF and J Aronson, 'Motivations for the Restoration of Ecosystems' (2006) 20(2) Conservation Biology 420

Comin, FA (ed), *Ecological Restoration – A Global Challenge* (CUP, 2010)

Gardner, RC, 'Rehabilitating Nature: A Comparative Review of Legal Mechanisms that Encourage Wetland Restoration Efforts' (2003) 52 Catholic University Law Review 573

IUCN/WCPA Ecological Restoration Taskforce (K Keenleyside and others), 'Ecological Restoration for Protected Areas, Principles, Guidelines and Best Practices', Best Practice Protected Area Guidelines Series No. 18 (IUCN, 2012)

Jørgensen, D, 'Ecological Restoration in the Convention on Biological Diversity Targets' (2013) 22(12) Biodiversity and Conservation 2977

Suding, KN, 'Toward an Era of Restoration in Ecology: Successes, Failures, and Opportunities Ahead' (2011) 42 Annual Review of Ecology, Evolution, and Systematics 465

Van Andel, J and J Aronson (eds), *Restoration Ecology. The New Frontier* (Wiley-Blackwell, 2012)

15. Non-compliance procedures and the implementation of commitments under wildlife treaties
Karen N. Scott

1. INTRODUCTION

Following the proliferation of multilateral environmental instruments (MEAs), which characterised the development of international environmental law during the latter part of the twentieth century, the focus of the first part of the twenty-first century has been the consolidation and implementation of those instruments. In particular, priority has been given to issues of compliance and the development of novel and effective mechanisms internal to treaty regimes designed to support the adherence of states to their treaty obligations.

Biodiversity-related and other MEAs are typically characterised as law making or standard setting as opposed to contractual or synallagmatic, and their obligations are *erga omnes* in the sense of being owed to all parties or even to the international community as a whole. The relative lack of reciprocity between contracting parties and the notion of community interest, 'over and above any interests of the contracting parties individually'[1] creates both conceptual and practical challenges when applying the rules of treaties and state responsibility to the operation and termination of MEAs.[2] Nowhere is this more acute than in the area of responding to breach of treaties. Whilst Article 60 of the 1969 Vienna Convention on the Law of Treaties (VCLT)[3] constitutes one of the few provisions in the Convention to distinguish between law making and contractual treaties, it nevertheless requires a party to have been especially affected by the breach in order to suspend the operation of a treaty against the party in breach.[4] Combined with the high threshold associated with the definition of 'material breach'[5] and the crude consequences of ultimately relieving the party in breach of future

---

[1] J Pauwelyn, 'A Typology of Multilateral Treaty Obligations: Are WTO Obligations Bilateral or Collective in Nature?' (2003) 14 EJIL 907, 908.

[2] *See* further Duncan French and Karen N. Scott, 'International Environmental Treaty Law' in M Bowman and D Kritsiotis (eds), *Conceptual and Contextual Perspectives on the Modern Law of Treaties* (CUP, *forthcoming*).

[3] 1969 Convention on the Law of Treaties (Vienna), (1969) 8 ILM 689 (VCLT).

[4] 1969 VCLT, Article 60(2)(b) and (c). States parties to an agreement may choose to suspend or terminate a treaty against a party in breach by unanimous agreement (Article 60(2)(a)).

[5] 1969 VCLT, Article 60(3) defines a material breach as a repudiation of the treaty not sanctioned by that treaty or a violation of a provision essential to the accomplishment of the object or purpose of that treaty.

obligations under the treaty, the application of Article 60 of the VCLT to the typical MEA is undeniably inapposite. Furthermore, for breaches less than 'material' little guidance is provided for under treaty law and contracting parties are forced to rely on general rules found in the laws of state responsibility and countermeasures.

Consequently, it is unsurprising that one of the most significant developments in the evolution of international environmental treaty law over the last couple of decades has been the creation of mechanisms and procedures designed to manage compliance, or, more accurately, non-compliance within the parameters of the treaty regime itself. Non-compliance procedures (NCPs) typically adopt a positive or supportive approach to compliance, and avoid the problems associated with reciprocity by decoupling the initiation of proceedings from the consequences (or harm) resulting from the breach. Moreover, by permitting and encouraging participation in proceedings by a wide range of actors, including treaty institutions, NGOs and even individuals, NCPs comprise an undoubtedly multilateral approach to addressing issues of non-compliance and, in doing so, respond to the *erga omnes* or community interest in the regime in question.

Over 20 MEAs have established NCPs to date,[6] and the inclusion of a compliance mechanism over and above the provision made for dispute resolution is now almost ubiquitous within modern environmental instruments.[7] Whilst academic and indeed public attention has largely focused on compliance mechanisms developed within regimes concerned with pollution prevention and control such as climate change[8] or ozone depletion[9] or within the field of access to environmental information,[10] the earliest NCPs were in fact developed largely within wildlife and biodiversity-focused

---

[6] See KN Scott, 'Non-compliance Procedures and Dispute Resolution Mechanisms under International Environmental Agreements' in D French, M Saul and ND White (eds), *International Law and Dispute Settlement: New Problems and Techniques* (Hart Publishing, 2010), 225.

[7] In circumstances where older instruments are revised it is similarly common for the revised version to include provision for a NCP. For example, the 2003 Revised African Convention on the Conservation of Nature and Natural Resources requires parties to develop a NCP once it enters into force (Article XXII). The text is available at: <http://au.int/en/content/african-convention-conservation-nature-and-natural-resources-revised-version>.

[8] See F Romanin Jacur, 'The Non-Compliance Procedure of the 1987 Montreal Protocol to the 1985 Vienna Convention on Substances that Deplete the Ozone Layer' in T Treves and others (eds), *Non-Compliance Procedures and Mechanisms and the Effectiveness of International Environmental Agreements* (Asser Press, 2009), 11; M Koskenniemi, 'Breach of Treaty or Non-Compliance? Reflections on the Enforcement of the Montreal Protocol' (1992) 3 Yearbook of International Environmental Law 123.

[9] See R Lefeber, 'From the Hague to Bonn to Marrakech and Beyond: A Negotiating History of the Compliance Regime under the Kyoto Protocol' (2001) 14 Hague Yearbook of International Law 25; Sabrina Urbinati, 'Procedures and Mechanisms Relating to Compliance under the 1997 Kyoto Protocol to the 1992 United Nations Framework Convention on Climate Change' in Treves and others (eds) (n 8), 63.

[10] One of the most active NCPs operates under the auspices of the 1998 Aarhus Convention on Access to Information, Public Participation in Decision-making and Access to Justice in Environmental Matters (1999) 38 ILM 517 ('Aarhus Convention'). *See* C Pitea, 'Procedures and Mechanisms for Review of Compliance under the 1998 Aarhus Convention on Access to Information, Public Participation and Access to Justice in Environmental Matters' in Treves and others (eds) (n 8), 221.

instruments such as the 1972 World Heritage Convention, the 1973 Convention on International Trade in Endangered Species and the 1979 Bern Convention. The early evolution of NCPs within biodiversity instruments has influenced their development within other MEAs and has undoubtedly enriched academic and practical responses to non-compliance in international environmental law more generally.

This chapter will begin with a brief introduction to the notion of 'compliance' in international law before going on to examine the evolution of the NCP under MEAs and highlighting selected core common features. It will focus on four case study instruments in order to illustrate the strengths and weaknesses of the application of NCPs within the sector of biodiversity conservation. The chapter will conclude with some brief observations on the relationship between NCPs and dispute resolution more generally, and offer some final remarks with respect to the future of NCPs within biodiversity instruments.

2. COMPLIANCE

The term 'compliance'[11] describes state behaviour that conforms to a specified treaty or customary norm[12] or, to use the definition adopted by the United Nations Environment Programme (UNEP) in its 2002 Guidelines on Compliance with and Enforcement of Multilateral Environmental Agreements (Compliance Guidelines), constitutes 'the fulfilment by the contracting parties of their obligations under a multilateral environmental agreement'.[13] It is related to, but conceptually distinct from, the notions of implementation – defined in the 2002 UNEP Compliance Guidelines as the adoption of laws, regulations, policies and other measures and initiatives by states designed to meet their obligations under environmental treaties[14] – and effectiveness, which focus on changes in state behaviour[15] or the success that an individual treaty has in addressing the problem it was intended to solve.[16]

Scholars across a range of disciplines – law, political science and international relations – have long explored the paradoxical question as to *why* nations comply with

---

[11] *See* generally B Kingsbury, 'The Concept of Compliance as a Function of Competing Conceptions of International Law' (1997–1998) 19 Michigan Journal of International Law 345.
[12] *See* especially: R Mitchell, *International Oil Pollution at Sea: Environmental Policy and Treaty Compliance* (MIT Press, 1994), 30; K Raustiala, 'Compliance & Effectiveness in International Regulatory Cooperation' (2000) 32 Case Western Reserve Journal of International Law 387, 391; E Weiss and H Jacobson, *Engaging Compliance with International Environmental Accords* (MIT Press, 1998), 40.
[13] Governing Council of UNEP Decision SS.VII/4 *Compliance with and enforcement of multilateral environmental agreements*, para. 9(a).
[14] *Ibid.*, para. 9(b).
[15] Raustiala (n 12), 394.
[16] M Ehrmann, 'Procedures of Compliance Control in International Environmental Treaties' (2002) 13 Colorado Journal of International Environmental Law and Policy 377, 378.

their international commitments.[17] Supporters of the realist or rationalist school of international law, for example, argue that it is self-interest that underpins a state's approach to compliance.[18] Where the costs of compliance are low – economically or politically – or the benefits of compliance substantial, a state is much more likely to comply with its treaty commitments than in situations where the costs of compliance are high or the perceived benefits are low.[19] Related to self-interest is a theory of compliance which focuses on the regard states have for their international reputations. Good global citizens comply with their commitments and, consequently, states are motivated to comply with their international law obligations in order to maintain and enhance that reputation.[20] Liberal theorists, on the other hand, focus on the domestic context, and argue that a strong democratic tradition contributes to how positively states view compliance at the international level.[21] Scholars subscribing to a transnational theory of compliance similarly concentrate on the importance of the national level, and emphasise the importance of the relationship between international commitments and the domestic implementation of those commitments as key to supporting compliance at the international level.[22] Yet other scholars accentuate the nature of the international norms themselves, and the perception that those norms are fair or legitimate[23] as influencing the extent to which nations are prepared to comply with those norms.

---

[17] *See* especially: J Cameron, J Werksman and P Roderick, *Improving Compliance with International Environmental Law* (Earthscan Publications, 1996) chapter 1; T Crossen, 'Multilateral Environmental Agreements and the Compliance Continuum' (2004) 16 Georgetown International Environmental Law Review 473; H Koh, 'Why Do Nations Obey International Law?' (1996–1997) 106 Yale Law Journal 2599; R Mitchell, 'Compliance Theory: Compliance Effectiveness, and Behaviour Change in International Environmental Law' in D Bodansky, J Brunnée and E Hey (eds), *The Oxford Handbook of International Environmental Law* (OUP, 2007) 893.

[18] Raustiala (n 12), 400; G Vigneron, 'Compliance and International Environmental Agreements: A Case Study of the 1995 United Nations Straddling Fish Stocks Agreement' (1998) 10 Georgetown International Environmental Law Review 581.

[19] See especially: S Barrett, *Environment and Statecraft: The Strategy of Environmental Treaty Making* (OUP, 2003), chapter 7.

[20] See A Guzman, 'A Compliance-Based Theory of International Law' (2002) 90 California Law Review 1823.

[21] E Neumayer, 'Do Democracies Exhibit Stronger International Environmental Commitment? A Cross-Country Analysis' (2002) 39 Journal of Peace Research 139. Edith Brown Weiss has notably challenged liberal theory supporters perspicuously noting that whilst democracies may be more responsive to public opinion, it cannot be assumed that public opinion is always sympathetic to environmental commitments. The ambivalent attitude of (arguably) a large proportion of the general public in the West towards making personal changes in their lives with the aim of minimising greenhouse gas emissions would appear to support Weiss's conclusion. See E Weiss, 'Understanding Compliance with International Environmental Agreements: The Baker's Dozen Myths' (1999) 32 University of Richmond Law Review 1555, 1579.

[22] See especially: C Kelly, 'Enmeshment as a Theory of Compliance' (2005) 37 New York University Journal of International Law and Politics 303; H Koh, 'Transnational Legal Process' (1996) 75 Nebraska Law Review 181.

[23] See especially: T Franck, *The Power of Legitimacy Among Nations* (OUP, 1990); T Franck, *Fairness in International Law and Institutions* (Clarendon Press, 1995).

In contrast to domestic legal systems, the role played by coercion and sanction in motivating compliance at the international level is much diminished although it is nevertheless extant.[24] Within the field of international environmental law the threat of sanctions to motivate compliance is particularly limited,[25] although punitive measures have been developed within some regimes, notably the climate change regime.[26] Scholars such as Abram Chayes and Antonia Handler Chayes argue that 'coercive enforcement is as misguided as it is costly'[27] on the basis that non-compliance with environmental commitments generally results from a lack of capacity, priority or information rather than wilful disobedience. Rather than focusing on external factors to motivate compliance, Chayes and Handler Chayes and other scholars within the so-called 'managerial school' concentrate on the internal design of a treaty, and argue that compliance is more likely to be achieved within regimes where the obligations are legitimate, clear and precise and the implementation of the treaty is supported by dynamic treaty institutions and transparent procedures and where mechanisms are provided for to facilitate reporting, verification, monitoring, capacity building and technical assistance, dispute resolution, and the regular review and adaptation of treaty commitments.[28]

NCPs, which are the focus of this chapter, perform an increasingly important function in managing and facilitating the compliance of states with their commitments under biodiversity instruments.[29] With a strong emphasis on managing compliance within designated treaty institutions on the basis of measures designed to proactively

---

[24] See especially: G Downs, D Rocke and P Barsoom, 'Is the Good News about Compliance Good News about Cooperation?' (1996) 50 International Organisation 379; T Yang, 'International Treaty Enforcement as a Public Good: Institutional Deterrent Sanctions in International Environmental Agreements' (2006) 27 Michigan Journal of International Law 1131.

[25] See A Chayes and A Handler Chayes, *The New Sovereignty: Compliance with International Regulatory Agreements* (Harvard University Press, 1995), 29–108.

[26] See C Hagem and H Westskog, 'Effective Enforcement and Double-edged Deterrents: How the Impacts of Sanctions also Affect Complying Parties' in O Schram Stokke, J Hovi and G Ulfstein (eds), *Implementing the Climate Regime: International Compliance* (Earthscan, 2005), 107.

[27] Chayes and Handler Chayes (n 25), 22.

[28] Ibid.

[29] Other mechanisms which support compliance with biodiversity commitments but which are beyond the scope of this chapter include reporting requirements, monitoring, fact-finding, inspection/observation schemes, verification processes, funding mechanisms and action supporting compliance taken outside of the treaty regime. See especially: Bodansky and others (eds) (n 17), chapters 41, 42 and 43; Michael Bowman, Peter Davies and Catherine Redgwell, *Lyster's International Wildlife Law* (2nd edition, CUP, 2010) chapter 4; D Victor, K Raustiala and E Skolnikoff (eds), *The Implementation and Effectiveness of International Environmental Commitments: Theory and Practice* (MIT Press, 1998); E Weiss and H Jacobson, *Engaging Compliance with International Environmental Accords* (MIT Press, 1998); R Wolfrum, 'Means of Ensuring Compliance with and Enforcement of International Environmental Law' (1998) 272 Recueil des Cours 9. See also the *Guidelines for Strengthening Compliance with and Implementation of Multilateral Environmental Agreements (MEAs) in the ECE Region* (2003) ECE/CEP/107 (20 March 2003); Decision SS VII/4 of the UNEP Governing Council (Special Session) (2002), *Guidelines on Compliance with and Enforcement of Multilateral Environmental Agreements*.

facilitate compliance, NCPs fit naturally within the managerial approach to compliance theory. However, incentives designed to support compliance such as technical and financial assistance also reinforce rational or self-interest explanations of compliance. Moreover, most NCPs provide for significant public participation within the compliance process through permitting NGO observers or even allowing non-state actors to initiate complaint proceedings, and thus have the potential to support a reputational theory of compliance. The role played by civil society in NCPs as well as their primary focus on facilitating compliance may also underpin compliance theories based on democracy, legitimacy or transnational process. Finally, a small number of NCPs do provide for more coercive or punitive measures such as trade sanctions, and it is likely that the desire to avoid the practical and political implications of these sanctions also serves as an incentive to comply with international treaty commitments.

3. NON-COMPLIANCE PROCEDURES

Whilst no two NCPs are exactly the same it is nevertheless possible to identify a number of features common to almost all compliance mechanisms. This section of the chapter will seek to identify the core components common to all biodiversity NCPs[30] before going on to focus on four case study conventions, which will be used to illustrate many of the NCP innovations developed within the context of biodiversity instruments.

The first feature common to most if not all NCPs is the creation of a designated institutional mechanism central to the implementation of the non-compliance procedure. Variously described as a Compliance, Implementation or Standing Committee, Commission or Bureau, these bodies normally comprise representatives drawn from across the states parties although some bodies, as in the case of the Bern Convention for example, comprise all contracting parties. Within a number of regimes, such as the Convention on International Trade in Endangered Species (CITES), NGOs and international organisations may sit on the compliance body as observers. Typically, compliance bodies perform a range of functions in addition to dealing with individual cases of non-compliance. These often include receiving reports, providing advice on general compliance-related issues, information sharing with other compliance and technical bodies and monitoring the overall implementation of and compliance with treaty provisions. A strong institutional framework to manage compliance is crucial to its effectiveness, and is an increasingly standard component of what has been

---

[30] For a more detailed overview of these core features, from which some of the information in this part of the chapter is drawn, see Scott (n 6). On NCPs more generally see M Fitzmaurice and C Redgwell, 'Environmental Non-Compliance Procedures and International Law' (2000) 31 Netherlands Yearbook of International Law 35; J Klabbers, 'Compliance Procedures' in Bodansky and others (eds) (n 17) 995–1009; Treves and others (eds) (n 8), and PGG Davies, 'Non-compliance – a Pivotal or Secondary Function of CoP Governance?' (2013) 15 International Community Law Review 77–101.

memorably described as the 'autonomous institutional arrangements' that underpin most modern MEAs.[31]

The second feature common to most NCPs relates to the initiation of NCP proceedings. Almost all biodiversity NCPs permit self-referral whereby a contracting party may pre-emptively initiate proceedings in circumstances where it is failing or about to fail to comply. Self-referral supports a precautionary approach to compliance central to most MEAs and the importance of addressing issues of non-compliance in its early stages.[32] It also reflects the priority given to facilitative as opposed to coercive compliance measures and the fact that initiating compliance proceedings within most MEAs can precipitate the grant of technical or even financial assistance. Non-compliance proceedings may also be initiated by a treaty institution such as a Secretariat or the compliance body itself under most MEAs. Some treaties – such as the 1973 CITES – provide the Secretariat with a substantial investigative role before passing the matter on to the compliance body, in the case of CITES, the Standing Committee.[33] Finally, in a minority of cases, proceedings may be initiated by NGOs or even individuals. This highly innovative and effective initiative in fact began within biodiversity instruments such as the 1979 Bern Convention and 1971 Ramsar Convention, although today is a particularly important feature of regimes dealing with access to environmental information and environmental impact assessment.[34] Given that states are naturally reluctant to initiate proceedings for non-compliance against one another, the creation of additional options for referral – by treaty institutions or NGOs/individuals – is of great practical importance and non-state actors in fact initiate by far the overwhelming majority of non-compliance proceedings under all MEAs, including biodiversity instruments. Moreover, the participation of NGOs and other non-state actors in compliance proceedings is more generally beneficial in terms of the advice and information they can provide in relation to particular incidents or compliance generally – the relationship between CITES and TRAFFIC is notable in this regard – and their capacity to bring compliance proceedings and other matters to a wider public audience.

The final feature of NCPs relates to the nature and scope of non-compliance measures. Typically, such measures tend to be facilitative and are designed to support a state in achieving compliance with its treaty commitments rather than punishing that state for non-compliance. The language used in describing NCPs demonstrates this approach: 'non-compliance' rather than 'breach'; 'compliance committee' or even

---

[31] R Churchill and G Ulfstein, 'Autonomous Institutional Arrangements in Multilateral Environmental Agreements: A Little-Noticed Phenomenon in International Law' (2000) 94 American Journal of International Law 623–659. See also, M Bowman, 'Beyond the "Keystone" CoPs: The Ecology of Institutional Governance in Conservation Treaty Regimes' (2013) 15 International Community Law Review 5, 21–43.

[32] O Yoshida, 'Soft Enforcement of Treaties: The Montreal Protocol's Noncompliance Procedure and the Functions of Internal International Institutions' (1999) 10 Colorado Journal of International Law and Policy 95, 121.

[33] See further below, Section 4.1.

[34] See the NCPs established under the 1998 Aarhus Convention and 1991 Convention on Environmental Impact Assessment in a Transboundary Context (1991) 30 ILM 735 ('Espoo Convention').

'implementation/standing committee' rather than non-compliance committee. Compliance measures consequently commonly comprise the provision of technical assistance, often through an on-the-ground mission to the contracting party concerned, training or even financial aid, which may be provided by a financial mechanism or private donation. More coercive measures are available within some MEAs – such as trade sanctions under CITES or de-listing a site under the 1972 World Heritage Convention – but these are less well developed in biodiversity instruments in contrast to other MEAs such as the 1997 Kyoto Protocol, which has established a designated enforcement branch of its NCP.[35] The reluctance among biodiversity, and indeed other, MEAs to develop more punitive measures in respect of compliance may – in part – lie in the ambivalent status of the measures themselves. The decisions of most compliance bodies, particularly those within the biodiversity sector, are either explicitly not binding or, at most, not explicitly binding. This is the case for even the most complex and sophisticated of NCPs, established under the 1987 Montreal Protocol and 1997 Kyoto Protocol in respect of ozone depletion and climate change respectively.[36] In the context of biodiversity instruments it is the non-mandatory status of the NCP and associated compliance measures rather than a lack of innovation that ultimately limits the effectiveness of these proceedings in promoting compliance with instruments in this sector.

4. SELECTED CASE STUDIES

4.1 The 1973 Convention on International Trade in Endangered Species[37]

CITES constitutes one of the most sophisticated wildlife instruments with respect to its provisions on compliance. The Convention itself, as might be expected in an instrument over 40 years old, is relatively spare when it comes to setting out compliance procedures. The Secretariat is given an oversight function, and is permitted to request information from parties relating to the implementation of their obligations under the Convention.[38] Where it receives information that a party is not implementing its obligations under CITES, the Secretariat must communicate its concerns to the party in question[39] and that party must provide the information requested by the Secretariat and, where appropriate, carry out an enquiry into the allegation of non-compliance.[40] The information provided or the results of the enquiry must then be reviewed by the Conference of the Parties (COP), which may make whatever recommendations it deems

---

[35] Decision 27/CMP.1 *Procedures and mechanisms relating to compliance under the Kyoto Protocol*.
[36] See Scott (n 6), 247–249.
[37] 1973 Convention on International Trade in Endangered Species of Wild Fauna and Flora (CITES) 12 ILM 2085 (1973). For an overview of CITES see Bowman and others (n 29), chapter 15. See also R Reeve, *Policing International Trade in Endangered Species. The CITES Treaty and Compliance* (Earthscan Publications, 2002).
[38] 1973 CITES, Article XII(2)(d).
[39] 1973 CITES, Article XIII(1).
[40] 1973 CITES, Article XIII(2).

appropriate.[41] In the event of a dispute between parties over the interpretation or implementation of its obligations, CITES requires that such disputes be resolved by negotiation with the possibility that it may be referred to arbitration if a peaceful resolution cannot be achieved.[42]

Today, the Secretariat still plays a key role in implementing the CITES NCP by initiating proceedings and engaging in early communication with the party alleged to be in non-compliance and providing assistance where appropriate.[43] However, proceedings may also be initiated by other parties to the Convention or through self-referral by the state itself in situations where it is concerned that it is not able to fully implement its obligations under CITES.[44] Where remedial action is not taken within a reasonable period of time, the Secretariat may refer the matter to the Standing Committee, which comprises 18 members representing a regional distribution of the parties.[45] The Standing Committee is given significant powers of investigation into compliance matters, and may seek an invitation from the party concerned to undertake information gathering and verification activities within their territory.[46] Like other NCPs, CITES endorses 'a supportive and non-adversarial approach ... towards compliance matters, with the aim of ensuring long-term compliance'.[47] In order to support a party in the implementation of its obligations the Standing Committee may, for example, provide advice, technical assistance and other forms of support as well as requiring that party to develop and submit a compliance action plan.[48] However, the CITES NCP also provides for more punitive options, including, for example, the issue of a written warning to the party in non-compliance, public notification to all parties that a compliance issue has not been addressed and, ultimately, a recommendation that all trade or commercial trade in one or more CITES species with the party in non-compliance be suspended.[49] CITES is notable among MEAs in that the authority to issue recommendations associated with compliance has been delegated by the COP to the Standing Committee[50] although the COP has the power to review those decisions.[51]

The progression from the carrot to the stick was recently illustrated by action taken by CITES against Guinea in respect of a range of compliance issues relating to the adequacy of its implementing legislation, the commercial trading of Appendix I species

---

[41] 1973 CITES, Article XIII(3).
[42] 1973 CITES, Article XVIII.
[43] CITES Conf. 14.3 *CITES compliance procedures*, paras. 16, 17 and 20.
[44] *Ibid.*, paras. 18 and 19.
[45] *Ibid.*, paras. 21–25. For further information on the Standing Committee see CITES Conf. 11.1 (Rev. CoP16) *Establishment of Committees* and the *Rules of Procedure of the Standing Committee* (as amended at the 62nd Meeting, Geneva, July 2012).
[46] CITES Conf. 14.3 *CITES compliance procedures*, para. 26.
[47] *Ibid.*, para. 4.
[48] *Ibid.*, para. 29.
[49] *Ibid.*, paras. 29 and 30.
[50] *Ibid.*, para. 10(d).
[51] *Ibid.*, para. 10(c). The CITES COP may also decide to carry out those tasks it has delegated to the Standing Committee. See *ibid*, para. 11. For a discussion on the authority of the Standing Committee see EJ Goodwin, 'The World Heritage Convention, the Environment, and Compliance' (2009) 20 Colorado Journal of International Environmental Law and Policy 157, 187–188.

and the issue of permits for so-called captive-bred species even though Guinea does not list captive-breeding facilities for the purposes of CITES.[52] The matter was referred to the Standing Committee by the Secretariat in 2011, which authorised the Secretariat to carry out a mission to Guinea later that year.[53] Following the Secretariat's report in 2012, the Standing Committee issued a written warning to Guinea, and urged it to take action to fully implement its commitments under the Convention.[54] In 2013, however, following Guinea's failure to adequately address the compliance concerns raised, the COP recommended, on the advice of the Standing Committee, that no party engage in commercial trade in CITES listed species with Guinea.[55]

More generally, as of 2014, 31 parties to CITES are subject to full or partial trade restrictions against them resulting from non-compliance proceedings. The types of infractions range from: consistent failure to submit annual reports (Afghanistan and Lesotho); failure to adopt appropriate domestic legislation implementing CITES (Djibouti, Mauritania and Somalia);[56] failure to monitor ranching operations appropriately (Madagascar in respect of the Nile crocodile); and permitting significant trade in CITES species contrary to the provisions of the Convention (e.g. Cameroon in respect of the Hippopotamus, Kazakhstan in respect of the European Sturgeon, Vietnam in respect of the common seahorse).[57]

The Standing Committee also has a role in providing advice more generally to parties or to CITES institutions with respect to compliance and the implementation of CITES commitments.[58] This advice or, increasingly, direction is normally issued through the COP. For example, at the Sixteenth COP held in Bangkok in 2013, decisions were adopted directing selected parties which had not otherwise done so to report on their implementing legislation,[59] other parties to provide technical assistance to categories of parties in need[60] and for all range states to report on their efforts to address the illegal trade in the Asian pangolin (or scaly anteater).[61]

Finally, it is worth noting the Review of Significant Trade Procedure, adopted by the parties in 1992.[62] Under the procedure trade in Appendix II specimens is monitored by

---

[52] CITES CoP16 Doc.29 (Rev.1), *Compliance and Enforcement: Enforcement Matters* (Sixteenth Meeting of the Conference of the Parties, Bangkok (Thailand), 3–14 March 2013) paras. 57–62.
[53] *Ibid.*, para. 60.
[54] *Ibid.*
[55] CITES Notification to the Parties No. 2013/017 Concerning Guinea: Recommendation to Suspend Trade (Geneva, 16 May 2013).
[56] See further CITES CoP16 Doc. 28, *Compliance and Enforcement: National Laws for Implementation of the Convention* (Sixteenth Meeting of the Conference of the Parties, Bangkok (Thailand), 3–14 March 2013).
[57] See <http://www.cites.org/eng/resources/ref/suspend.php>.
[58] CITES Conf. 14.3 *CITES compliance procedures* paras. 10–12.
[59] CITES Decision 16.33 *National laws for implementation of the Convention* adopted at COP 16 (Bangkok, 2013).
[60] CITES Decision 16.35 *National laws for implementation of the Convention* adopted at COP 16 (Bangkok, 2013).
[61] CITES Decision 16.41 *Pangolins (Manis spp.)* adopted at COP 16 (Bangkok, 2013).
[62] CITES Conf. 12.8 (Rev. CoP13) *Review of Significant Trade in specimens in Appendix II species.*

the Secretariat on the basis of data provided by states and NGOs in order to identify those species subject to significant levels of trade with the potential to impact on the survival of those species. Species identified to be of urgent or possible concern may then be subject to appropriate recommendations including quotas by the Animal or Plants Committee after consultation with the relevant range states. In the event that those recommendations are not implemented by range states the Standing Committee may take action and, ultimately, may suspend trade in respect of the species with the range state in question.[63]

The nature of CITES – which is not a typical MEA in that it generally requires parties to prohibit the private activities of individuals involved in the illegal wildlife trade rather than imposing positive obligations on states to protect and preserve the environment – has meant that it has developed a comparatively dynamic NCP. The detailed rules relating to compliance[64] and more particularly the NCP,[65] draw on a range of administrative and quasi-judicial principles. For example, parties subject to non-compliance proceedings are entitled to respond to the allegations against them and to participate in the proceedings before the Standing Committee[66] and may even be entitled to access financial assistance in order to attend the relevant meeting.[67] In contrast to other NCPs, only states and the Secretariat are entitled to refer compliance matters to the Standing Committee. However, in practice NGOs such as TRAFFIC and the International Union for the Conservation of Nature (IUCN) play an important role in the implementation of CITES, and information provided by NGOs will regularly form the basis of a referral by the Secretariat.

Nevertheless, the NCP, like comparable procedures provided for within other MEAs, is not binding[68] and compliance measures, including trade restrictions, are expressed in recommendatory as opposed to mandatory form. Moreover, despite CITES' efforts, illegal trade in wildlife is rising and is estimated to be worth between US$8 and US$10 billion per year (excluding fisheries and timber).[69] Trade in illegal ivory alone doubled between 2007 and 2014.[70] The scale of the problem has led to the development of initiatives supportive of but outside the scope of CITES such as the 2014 London Conference on the Illegal Wildlife Trade.[71] Furthermore, the increasing links between illegal trade in wildlife and other transnational crimes[72] and, more significantly, its capacity to fuel and exacerbate conflict situations has led to recent and unprecedented action by the United Nations Security Council (UNSC). In January 2014, the UNSC

---

[63] See further Bowman and others (n 29), 522–524.
[64] CITES Conf. 11.3 (Rev. CoP16) *Compliance and Enforcement*.
[65] CITES Conf. 14.3 *CITES Compliance procedures*.
[66] *Ibid.*, para. 27.
[67] *Ibid.*, para. 28.
[68] CITES Conf. 14.3 *CITES compliance procedures* para. 1.
[69] K Lawson and A Vines, *Global Impacts of the Illegal Wildlife Trade: The Costs of Crime, Insecurity and Institutional Erosion* (Royal Institute of International Affairs, 2014), viii.
[70] *Ibid.*
[71] See <https://www.gov.uk/government/topical-events/illegal-wildlife-trade-2014>.
[72] See for example UN ECOSOC Resolution 2011/36 *Crime prevention and criminal justice responses against illicit trafficking in endangered species of wild fauna and flora*.

adopted resolutions in respect of the Central African Republic[73] and the Democratic Republic of the Congo,[74] and highlighted the illegal wildlife trade as an exacerbating factor in both conflicts, extending travel and financial sanctions to individuals or entities involved in financing armed groups through illegal trade in wildlife.

4.2 The 1979 Bern Convention on the Conservation of European Wildlife and Natural Habitats[75]

The NCP developed to support the implementation of the 1979 Bern Convention comprises one of the oldest and most innovative of compliance mechanisms within the field of biodiversity instruments. The so-called 'case files procedure' was initiated by the Standing Committee of the Convention[76] on an ad hoc basis almost immediately upon entry into force of the Convention in 1982.[77] Any party, individual or NGO may refer a complaint to the Standing Committee in respect of a contracting party's failure to comply with its obligations under the Bern Convention, and the Secretariat must decide – after seeking further information from the party concerned – whether there are grounds for placing the complaint as a 'file' on the agenda of the next meeting of the Standing Committee.[78] If the Standing Committee chooses to open a file it may adopt specific recommendations designed to bring the state into compliance with its obligations or authorise an on-the-spot expert enquiry to seek further information.[79] The case file procedure was formalised in 1993[80] but nevertheless applied only provisionally until 2008[81] owing to initial problems associated with applying it to certain EU Member States.[82] As of 2014, 150 compliance matters have been referred to the Standing Committee with six files open at the time of writing.[83] The longest-standing open complaint, referred to the Standing Committee in 1995, relates to the threats to

---

[73] S/RES/2134 (2014) *Central African Republic*.
[74] S/RES/2136 (2014) *Democratic Republic of Congo*.
[75] 1979 Convention on the Conservation of European Wildlife and Natural Habitats (Bern Convention) (1982) UKTS 56. For an overview of the Bern Convention see Bowman and others (n 29), chapter 10. See also S. Jen, 'The Convention on the Conservation of European Wildlife and Natural Habitats (Bern, 1979): Procedures of Application in Practice' 2 (1999) Journal of International Wildlife Law and Policy 224.
[76] The Standing Committee has responsibility for the application and implementation of the Bern Convention, including a dispute settlement role. See the 1979 Bern Convention, Articles 13, 14, 15 and 18. See also Doc. T-PVS/Inf(2013) 6 *Rules of Procedure of the Standing Committee* (33rd Meeting, Strasbourg 2013). In contrast to compliance bodies established by other biodiversity instruments, the Bern Standing Committee is composed of all parties to the Convention (Article 13(2)).
[77] Bowman and others (n 29), 338.
[78] Doc. T-PVS(93) 22 *Implementation of the Bern Convention. Opening and closing of files and follow-up to recommendations* (13th Meeting, Strasbourg 1993), paras. 1–5.
[79] *Ibid.*, para. 8.
[80] See generally Doc. T-PVS(93) 22.
[81] Doc. T-PVS (2008) 7 *Application of the Convention. Summary of Case files and complaints – Reminder on the processing of complaints and new on-line form.*
[82] Bowman, Davies and Redgwell (n 29), 338.
[83] T-PVS/Inf(2014) 2 *Register of Bern Complaints 2014.*

sea turtle nesting beaches and other habitats on the Akamas Peninsula in Cyprus.[84] Other open complaints comprise concerns relating to the Bystroe Estuary canal in Ukraine (opened in 2004), wind farms in Balchik and Kaliakra in Bulgaria (opened in 2004), the eradication and trade of the American Grey squirrel in Italy (opened in 2007), threats to marine turtles in Thines Kiparissias, Greece (opened in 2010) and degradation of turtle nesting beaches in Fethiye and Patara, Turkey (opened in 2012).[85] The Standing Committee has a broad mandate to make recommendations to individual parties and these recommendations may be site or activity specific such as the removal of buildings on a nesting beach or re-routing a road likely to impact on a critical habitat.

The evolution of the Bern NCP is underpinned by a bold approach to treaty interpretation. In contrast to more modern biodiversity instruments there is no explicit provision for a compliance-focused procedure within the Bern Convention. However, the parties determined that Article 18(1) of the Convention, which stipulates that '[t]he Standing Committee shall use its best endeavours to facilitate a friendly settlement of any difficulty to which the execution of this Convention may give rise' provided a sufficient basis for the development of the case file procedure.[86] Article 14 of the Convention provides a clear mandate to the Standing Committee to make recommendations to states parties for the purposes of the Convention as well as to arrange for meetings of groups of experts in order to discharge its functions. The latter right was interpreted as providing the basis for establishing on-the-spot appraisals[87] and the parameters of such appraisals or enquiries were further developed in the Rules of Procedure of the Standing Committee.[88]

The ability to gather information from the site in question rather than simply relying on reports from the contracting party is an important function, which allows the Standing Committee to ultimately make practical and pertinent recommendations that are site specific. The on-the-spot enquiry process has operated as a precedent for other MEAs such as the 1991 Espoo Convention on Environmental Impact Assessment.[89] More significantly, the Bern Convention has recently collaborated on joint site inspections with other MEAs or international organisations with a view to developing a range of supportive compliance measures across multiple instruments. The most high-profile of such collaborative exercises to date has taken place in respect of the Danube–Black Sea Shipping Canal in the Bystroe Estuary of the Danube delta

---

[84] See T-VPS (2013) Misc *33rd Meeting of the Standing Committee*, Strasbourg 3–6 December 2013 at 6.1.
[85] Doc. T-PVS/Inf (2014) 2 *Register of Bern Complaints 2014*. These cases and several other possible files were discussed at the latest meeting of the Standing Committee. See *ibid* paras. 6.1–6.4.
[86] Doc. T-PVS(93) 22 (n 78), introduction.
[87] Doc. T-PVS (2008) 7 (n 81), para. 5.
[88] Doc. T-PVS/Inf(2013) 6, Rule 11 and Appendix I.
[89] 1991 Convention on Environmental Impact Assessment in a Transboundary Context ('Espoo Convention') (1991) 30 ILM 735.

proposed by Ukraine, which has been the subject of concern[90] within four MEAs[91] and two international organisations[92] in addition to the Bern Convention. Joint fact finding missions were carried out in 2004, 2008 and 2009[93] but despite the efforts of these institutions – and the creation of a Trilateral Joint Commission, comprising Romania, Moldova and Ukraine[94] – compliance issues have yet to be resolved and the Danube canal project remains an open case file. Nevertheless, the collaborative approach taken to compliance by the Bern Convention is unusual and undoubtedly serves as a precedent for other biodiversity instruments.[95]

Finally, in establishing the Standing Committee, the Bern Convention originally provided for a generous degree of participation by international organisations and NGOs.[96] The extent to which NGOs and even individuals are able to participate in the Bern NCP is significant – reflecting the prominent role played by NGOs in developing the procedure in the first place.[97] Not only can NGOs and individuals refer complaints to the Secretariat, but they may also – with the permission of the Chairman or a contracting party – speak to the matter at the meeting of the Standing Committee and suggest recommendations.[98] Observers may even put recommendations to the vote provided they are sponsored by a contracting party.[99]

Nevertheless, recommendations issued by the Bern Standing Committee are, as their appellation suggests, non-binding. There are no punitive options available to the Standing Committee to address non-compliance other than publically raising the matter at regular meetings. Even where matters are raised the language used remains within the confines of polite diplomacy. For example, in respect of the protection of the Akamas Peninsula by Cyprus, which comprises a case file opened almost 20 years ago, the Standing Committee at its 2013 meeting reported that it 'regretted the absence of delegates of Cyprus, as well as the lack of comprehensive information on the concrete

---

[90] For background to this case see AS Rieu-Clarke, 'An Overview of Stakeholder Participation: What Current Practice and Future Challenges? Case Study of the Danube Delta' (2007) 18 Colorado Journal of International Environmental Law and Policy 611.

[91] 1998 Aarhus Convention on Access to Information, Public Participation in Decision-making and Access to Justice in Environmental Matters; 1979 Convention on Migratory Species; 1991 Espoo Convention; 1971 Ramsar Convention on Wetlands of International Importance.

[92] European Commission and International Commission for the Protection of the Danube River.

[93] KN Scott, 'International Environmental Governance: Managing Fragmentation through Institutional Connection' (2011) 12 Melbourne Journal of International Law 177, 210.

[94] The most recent meeting of the Commission took place in November 2013. See Doc. T-PVS/Files (2014) 6 *Specific Site – File open. Proposed Navigable Waterway in the Bystroe Estuary (Danube Delta) (Ukraine). Report of the second meeting of the Romania – Republic of Moldova – Ukraine Joint Commission, Tulcea, Romania, November 28th, 2013.*

[95] On the collaboration between compliance mechanisms more generally see Scott (n 93), 207–211.

[96] 1979 Bern Convention, Article 13(3).

[97] Initial drafts of the case file procedure were developed by three NGOs – WWF, RSPB and SHE – in 1992 before being amended and finally adopted in 1993. See T-PVS (93) 22 (n 78), introduction.

[98] *Ibid.*, para. 7.

[99] T-PVS/Inf (2013) 6, Rule 9.

measures undertaken by the Party to address the matters related to the complainant'.[100] Where a state party refuses to cooperate or adopt the relevant recommendations, the Standing Committee has, in the past, simply closed the file without the matter being resolved. The most notorious example of such a case concerned a complaint against Greece in respect of turtle nesting sites at Laganas Bay, Zakynthos, which was eventually closed after 14 years without resolution.[101] Moreover, it is notable that of the open case files currently before the Standing Committee, half came before the Committee over a decade ago. The strength of the Bern NCP would therefore appear to lie in its flexible and innovative provisions permitting proceedings to be initiated and investigated rather than in forcing determinedly recalcitrant states to comply with their treaty commitments.

4.3 The 1971 Ramsar Convention on Wetlands of International Importance[102]

As an instrument originally modest in respect of the institutions it established and vague in terms of the obligations it created, it is unsurprising that the 1971 Ramsar Convention has developed a fairly unambitious non-compliance mechanism. Relying almost exclusively on the carrot rather than the stick, the Ramsar NCP is designed principally to support states in their efforts to implement their obligations under the Convention. In contrast to other MEAs, the focus of the NCP under Ramsar is on only one key obligation under the Convention: the requirement to submit suitable wetlands to the Ramsar List of Wetlands of International Importance[103] and, more particularly, the obligation to inform the Bureau (Secretariat) of any changes in the ecological character of those listed wetlands as a result of human interference.[104] The NCP comprises two distinct but mutually supporting components.

First, where the Bureau receives information indicating that the ecological character of a listed wetland is changing or is likely to change as a result of technological development, pollution or other human interference it can provide special assistance to the affected party through the Ramsar Advisory Mission, formally known as the monitoring procedure.[105] The Mission will generally visit the site in question –

---

[100] T-VPS (2013) (n 84) para. 6.1.
[101] Doc. T-PVS (2000) 35 *Description of the 'Specific sites' case-file system under the Bern Convention* at 7. However, the matter was subsequently referred to the European Court of Justice by the European Commission as a breach of Article 12 of the Habitats Directive in *Commission v. Hellenic Republic* (2002) ECR I-01147. The Court found that Greece was in breach of the Directive but had taken sufficient remedial action to avoid further sanction under EU law.
[102] 1971 Convention on Wetlands of International Importance especially as Waterfowl Habitat (Ramsar) (1972) 11 ILM 963. For an overview of the Ramsar Convention see Bowman, Davies and Redgwell (n 29), chapter 13. See also O Ferrajolo, 'State Obligations and Non-Compliance in the Ramsar System' 14 (2011) Journal of International Wildlife Law & Policy 243.
[103] 1971 Ramsar Convention, Article 2.
[104] 1971 Ramsar Convention, Article 3(2).
[105] Recommendation 4.7 (1990) *Mechanisms for improved application of the Ramsar Convention*.

increasingly, in collaboration with other MEA institutions – and will make recommendations designed to improve the ecological condition of the site. In the period 1988–2013 76 Missions took place,[106] the latest focusing on the Mývatn-Laxá region in Iceland in 2013.[107] A key incentive for accepting a Ramsar Advisory Mission is that in addition to technical support, financial assistance may be made available to affected states from the Ramsar Small Grants Fund.[108] Finally, a Mission may recommend that the wetland be listed on the Montreux Record, a list of Ramsar sites that have undergone or are threatened by a significant change in their ecological character.

The Montreux Record itself constitutes the second limb of the Ramsar NCP. Formally created in 1990,[109] the Record is designed to implement Article 3.2 of the Convention, and constitutes those Ramsar listed wetland sites 'where an adverse change in ecological character has occurred, is occurring or is likely to occur, and which are therefore in need of priority conservation attention'.[110] It is the affected party that has the primary responsibility for requesting the inclusion of a site on the list[111] although the Bureau can act on information received from other sources such as NGOs, and contact the affected party in order to enquire whether a Ramsar site should be included on the Montreux Record.[112] Notably, in the triennium between the 10th and 11th meetings of the parties in 2008 and 2012 the Bureau received reports from 11 contracting parties highlighting 17 Ramsar sites at risk compared with reports from other sources in respect of 68 Ramsar sites at risk in 40 states.[113] Importantly however, a site can only be included on the Montreux Record with the approval of the contracting party concerned.[114] Moreover, a site can also be removed at the request of that contracting party.[115] Once so listed, in principle, the affected contracting party should identify remedial measures to address the ecological challenges to the site, and technical and financial assistance may be made available through the Ramsar Advisory

---

[106] See <http://www.ramsar.org/activity/ramsar-advisory-missions>.

[107] A copy of the Mission's report is available online at: <http://www.ramsar.org/sites/default/files/documents/library/ram76e_iceland_2013.pdf>.

[108] Established in 1990 by Resolution 4.3 *A Wetland Conservation Fund* (and subsequently developed in later meetings).

[109] Recommendation 4.8: Change in ecological character of Ramsar sites.

[110] Resolution VI.1 *Working definitions of ecological character, guidelines for describing and maintaining the ecological character of listed sites, and guidelines for operation of the Montreux Record*, para. 3.1. These guidelines, adopted in 1996, develop those adopted in 1993 (Resolution 5.4: *The Record of Ramsar sites where changes in ecological character have occurred, are occurring, or are likely to occur (Montreux Record)*. See also Resolution X.16: *A Framework for processes of detecting, reporting and responding to change in wetland ecological character*.

[111] Resolution VI.1, *ibid.*, para. 3.2.1.

[112] *Ibid.*

[113] Resolution XI.4 *The status of sites on the Lists of International Importance*, Annexes 2a and 2b.

[114] Resolution VI.1 (n 110), para. 3.2.1.

[115] *Ibid*, para. 3.3.5. Similarly, parties can remove wetland sites located within their territories from the Ramsar List because of urgent national interests (1971 Ramsar Convention, Article 2.5).

430 *Research handbook on biodiversity and law*

Mission process. There are currently 48 sites listed on the Montreux Record,[116] which is a relatively small proportion of the 2,181 Ramsar List sites.[117] Although 32 sites listed on the Montreux Record have been declared out of danger and removed from the Record over the last 20 years[118] and this arguably attests to a level of success of the procedure, against the well-known threats to wetlands – particularly in light of the impacts of climate change[119] – the Ramsar NCP appears to be significantly under-utilised.

Finally, it is worth noting that the Ramsar COP adopts a regular resolution within which it identifies more general issues associated with compliance. For example, at the 11th COP, which took place in 2013, the parties adopted Resolution XI.4,[120] which noted that, contrary to Article 2 of the Convention and Resolution VI.13,[121] no adequate map or Ramsar Information Sheet has been submitted in respect of 1,385 Ramsar sites in 149 countries (68 per cent of all Ramsar sites) and, in Annexes attached to the Resolution, listed those states in non-compliance with these obligations.[122] The Resolution also highlighted that contrary to Resolution VIII.8 (2002), many states have failed to put in place mechanisms to facilitate the local reporting of wetland sites at risk[123] and that relatively few parties have reported instances of change or likely change in the ecological character of sites pursuant to Article 3.2 of the Convention.[124] Nevertheless, whilst the Ramsar COP can recommend that parties take action to comply with their obligations, options beyond publicising instances of non-compliance – naming and shaming – are not currently available under the Convention.

Juxtaposed with other MEAs, the NCP under the Ramsar Convention compares poorly. Although in practice NGOs and even individuals can provide information to the Bureau in respect of a wetland in danger it is the contracting party that subsequently controls the process, and the deployment of a mission to their territory or the listing of the site on the Montreux Record depends upon the consent of that contracting party. Moreover, the compliance mechanisms within the Ramsar toolbox are entirely facilitative in nature, which, whilst consistent with the managerial approach to compliance, leaves little scope for alternative solutions where a party refuses to cooperate. Most significantly, compliance action focuses almost exclusively on activities associated with sites included on the Ramsar List rather than the more general obligations under the

---

[116] The list is available at <http://ramsar.rgis.ch/cda/en/ramsar-documents-montreux/main/ramsar/1-31-118_4000_0__>.
[117] The list is available at <http://ramsar.rgis.ch/cda/en/ramsar-documents-list/main/ramsar/1-31-218_4000_0__>.
[118] See <http://ramsar.rgis.ch/cda/en/ramsar-documents-montreux-montreux-record/main/ramsar/1-31-118%5E20972_4000_0__#remove>.
[119] See the IPCC Fifth Assessment Report, *Climate Change 2014: Impacts, Adaptation, and Vulnerability*, chapters 3 and 4 available at <http://ipcc-wg2.gov/AR5/report/final-drafts/>.
[120] Resolution XI.4 *The status of sites on the List of Wetlands of International Importance*.
[121] Resolution VI.13 *Submission of information on sites designated for the Ramsar List of Wetlands of International Importance*.
[122] *Ibid.*, para. 5 and Annexes 1a and 1b.
[123] *Ibid.*, para. 8.
[124] *Ibid.*, para. 10.

Convention in respect of the wise use of all wetlands and procedural obligations such as regular reporting.

4.4 The 1972 World Heritage Convention[125]

The 1972 World Heritage Convention (WHC) largely follows the Ramsar model of biodiversity conservation, creating a list of cultural and natural heritage sites deserving special protection, supplemented by broader obligations to support the aims and objectives of the Convention with respect to the protection of cultural and natural heritage more generally.[126] The NCP established under the Convention is naturally also similar to the Ramsar NCP, with a tight focus on compliance matters associated with listed sites only and a strong emphasis on facilitating and supporting compliance. Article 13 of the WHC permits contracting parties to request international assistance to support their compliance with obligations under the Convention, in particular with the identification of natural and cultural sites for potential inclusion in the World Heritage List.[127] One of the most innovative features of the WHC was the creation of the World Heritage Fund supported by regular compulsory and voluntary contributions from states parties as well as donations from international organisations and private bodies or individuals.[128] There is no doubt that the existence of the Fund and the ability of the Convention to provide technical and other support to states parties has contributed to what has been described as the 'compliance pull' of the WHC.[129]

As with the Ramsar Convention, compliance is most visibly addressed at the international level through the List of World Heritage in Danger. The List is maintained by the World Heritage Committee[130] and comprises properties appearing 'in the World Heritage List for the conservation of which major operations are necessary and for which assistance has been requested under this Convention'.[131] There are currently 44 sites on the World Heritage in Danger List.[132] The emphasis on international assistance emphasises that the List is not designed to operate as a sanction,[133] and the Convention notably makes no distinction between the various factors which may contribute to listing a site, some of which result from deliberate action on the part of the contracting party (such as the development of large-scale projects or changes in the use or

---

[125] 1972 UNESCO Convention Concerning the Protection of the World Cultural and Natural Heritage (1972) 11 ILM 1358. For an overview of the WHC see Bowman and others (n 29), chapter 14. See also Goodwin (n 51).

[126] 1972 WHC, Articles 4–6 and 11(1).

[127] Further details relating to the nature of assistance and manner of its distribution are set out in Articles 19–26 of the WHC.

[128] 1972 WHC, Articles 15–18.

[129] *See* Goodwin (n 51).

[130] The World Heritage Committee is established under Articles 8–10 of the WHC.

[131] 1972 WHC, Article 11(4). See also WHC.13/01 (July 2013) *Operational Guidelines for the Implementation of the World Heritage Convention* paras. 177–199 (hereinafter, WHC Operational Guidelines).

[132] *See* <http://whc.unesco.org/en/danger/>.

[133] This is also emphasised in the public description of the operation of the Danger List. *See* <http://whc.unesco.org/en/158/>.

ownership of land) and others which lie outside the control of the party (such as earthquakes, floods and tidal waves).[134] Significantly, the WHC Guidelines permit sites to be listed not only in the case of an ascertained danger to the property, but also in situations where the site is at risk from a *future* specific and serious threat.[135] This permits the Committee to take preventive action in order to preclude or mitigate the risk. For example, in 2012, the Liverpool Mercantile City in the UK was placed on the Heritage in Danger List in response to a proposal to undertake wide-scale redevelopment in the historic docklands area.[136]

Typically, the state within which the site is located will refer the matter to the Committee with a view to its being included on the List of Heritage in Danger. However, the Committee can also consider information received from any source about a World Heritage Property, and relies significantly on a range of NGOs such as the IUCN, ICCROM (International Centre for the Study of the Preservation and Restoration of Cultural Property) and ICOMOS (International Council on Monuments and Sites) for expert advice.[137] On receipt of a referral the Committee will normally direct the WHC Secretariat, working collaboratively with the party in question, to ascertain the current condition of the site, the dangers threatening it and the feasibility of undertaking corrective measures.[138] After considering all the information available, the Committee will make a decision on whether to include the property on the List of World Heritage in Danger and, where appropriate, make recommendations in respect of the course of action to be taken by the contracting party to avoid the danger or improve the conservation status of the site.[139] Like the recommendations made by the Standing Committee under the Bern Convention, WHC recommendations may be similarly specific to a site or activity. For example, in conjunction with the listing of Liverpool in the UK on the List of World Heritage in Danger in 2012, the Committee recommended the UK 'reconsider the proposed development to ensure that the architectural and town-planning coherence, and the conditions of authenticity and integrity of the property are sustained'.[140] Multiple recommendations were made by the Committee when listing the Belize Barrier Reef System in 2009, including a direction to better manage development rights on private or leased land, to establish a co-management system allocating responsibility between the government and conservation NGOs and to better manage the threat of invasive species.[141] Contracting parties can apply for international assistance and for grants from the World Heritage Fund in respect of implementing Committee recommendations.[142] Progress in restoring sites to a more

---

[134] 1972 WHC, Article 11(4).
[135] WHC Operational Guidelines, paras. 179 and 180.
[136] Decision 36 COM 7B.93 adopted at the 36th Meeting of the World Heritage Committee (St Petersburg, 2012).
[137] 1972 WHC, Article 8(2) and WHC Operational Guidelines, paras. 30–38.
[138] WHC Operational Guidelines, para. 184.
[139] *Ibid.*, paras. 185–187.
[140] Decision 36 COM 7B.93 adopted at the 36th Meeting of the World Heritage Committee (St Petersburg, 2012), para. 6.
[141] Decision 33 COM 7B.33 adopted at the 33rd Meeting of the World Heritage Committee (Seville, 2009).
[142] 1972 WHC, Articles 19–26 and WHC Operational Guidelines, paras. 223–257.

stable conservation status is monitored by the Committee through regular reports, reactive monitoring and on-the-spot missions.[143]

One significant limitation on the potential effectiveness of the Heritage in Danger List procedure is the uncertainty over whether the consent of the contracting party is a prerequisite to the inclusion of a site on the List.[144] In contrast to Article 11(3) of the Convention, which unequivocally requires the consent of a party in respect of the inclusion of a site in the World Heritage List, Article 11(4), which provides for the World Heritage in Danger List, is silent on the matter. The WHC Operational Guidelines provide no additional elucidation and merely stipulate that the decision on whether to add a site to the Heritage in Danger List shall be taken by a majority of two-thirds of the Committee members present and voting and that the decision must be communicated to the state party in question.[145] This issue arose in 1999 in connection with the proposed listing of Kakadu National Park in Australia.[146] In the event, the site was not so listed but Australia nevertheless maintained that the site could not have been included on the List without its consent.[147] Bowman, Davies and Redgwell note that the preparatory work of the WHC would appear to support the argument that state consent operates as a prerequisite for listing sites on *both* the Heritage List and the Heritage in Danger List.[148] However, a good-faith interpretation of the text of Article 11(4), which, in contrast to Article 11(3), omits any reference to state consent operating as a prerequisite to listing a site on the Heritage in Danger List, would suggest that consent is unnecessary. This interpretation is also consistent with the underlying ethos of the Convention that *all* states parties have an interest in the conservation and preservation of heritage listed sites.

The only compliance measure available to the WHC that could be categorised as punitive, is the de-listing procedure. Not explicitly provided for in the WHC itself, the WHC Operational Guidelines permit the Committee to delete sites from the World Heritage List in cases where the property 'has deteriorated to the extent that it has lost those characteristics which determined its inclusion' in the List in the first place or where 'the intrinsic qualities of [the site] were already threatened at the time of its nomination by action of man and where the necessary corrective measures as outlined by the State Party at the time have not been taken'.[149] The de-listing procedure may be initiated on the basis of information provided by a source other than the party in question, such as an NGO,[150] and whilst the Committee must verify all information it receives with the affected party[151] and, more generally, engage in consultations with

---

[143] 1972 WHC, Articles 4 and 29 and WHC Operational Guidelines paras. 190–191, 169–176.
[144] See EJ Goodwin, 'The Consequences of Deleting World Heritage Sites' (2010) 21 King's Law Journal 283, 290–297 and 304–308.
[145] WHC Operational Guidelines, paras. 186 and 187.
[146] Bowman (and others) (n 29), 461.
[147] Ibid.
[148] Ibid.
[149] WHC Operational Guidelines, para. 192.
[150] Ibid., para. 194.
[151] Ibid., para. 194.

that party,[152] its consent is not a prerequisite to the ultimate decision to remove the site from the Heritage List. Only two sites have been de-listed to date: the Arabian Oryx Sanctuary in Oman (2007) and the Dresden Elbe Valley in Germany (2009).[153] Although de-listing ultimately represents a failure on the part of both contracting party and the compliance mechanism, the existence of the procedure arguably provides a genuine incentive for states to comply with their commitments. Inclusion of a site on the World Heritage List carries a level of prestige and potentially has economic advantages associated with tourism. Moreover, states may be able to access the Heritage Fund and other forms of international assistance in respect of listed sites. The de-listing of a site therefore has the potential for significant economic and political detriment in terms of reduced tourist income and other financial resources, and, most significantly, a loss of prestige both nationally and internationally.[154]

The relative success of the WHC NCP can, in large part, be attributed to the institutional infrastructure supporting the Convention, which comprises a particularly dynamic Committee of just 15 member states strongly supported by formal and informal participation of international organisations and NGOs.[155] In contrast to other MEAs, the WHC compliance mechanisms are embedded within the Convention itself and the decisions of the Committee with respect to the listing and de-listing of sites are binding. Moreover, the particular emphasis on supporting contracting parties to achieve compliance with their obligations, and the robust provisions within the Convention relating to the Heritage Fund and the provision of international assistance, provide a clear incentive for states to cooperate with the Committee. However, the recommendations associated with the inclusion of a site on the Heritage in Danger List are not binding and apart from the ultimate sanction of de-listing the site altogether, the Committee lacks the 'stick' to support the 'carrot'. Moreover, as with the case of the Ramsar Convention, the WHC NCP is confined to matters associated with listed sites, and does not extend to obligations under the Convention more generally such as those associated with regular reporting or contributions to the Heritage Fund.

---

[152] *Ibid.*, para. 196.
[153] Bowman and others (n 29), 463.
[154] The 'compliance pull' of prestige and the tourist dollar has recently been demonstrated in New Zealand in the context of the proposed Dart Passage Tunnel through the Te Wāhipounamu World Heritage site, a landscape park covering 10 per cent of New Zealand in the south-west of the South Island (see <http://whc.unesco.org/en/list/551>). In response to information received from a third party, the World Heritage Committee requested details of this project from New Zealand in August 2012. Following significant local media coverage of UNESCO's interest in the project, and the possibility that a referral could be made to list Te Wāhipounamu on the Heritage in Danger List (see <http://www.nzherald.co.nz/nz/news/article.cfm?c_id=1&objectid=10825928>), the tunnel proposal was ultimately rejected by the Minister of Conservation in July 2013 (see <http://www.beehive.govt.nz/release/minister-declines-milford-dart-tunnel-proposal/>). A proposal for a monorail through the park nevertheless remains on the table.
[155] For a discussion of the importance of the Committee, as well as of reservations with respect to its legitimacy see Goodwin (n 51) generally, and at 195–196 in particular.

5. NON-COMPLIANCE PROCEDURES AND DISPUTE RESOLUTION

Most MEAs are either silent on the relationship between NCP proceedings and their provisions relating to dispute settlement or stipulate that the former are non-prejudicial to the latter. Although Article 60(4) of the 1969 VCLT provides for the priority of specialised rules relating to breach over the default provisions of the Convention, as noted above, most NCP measures are not formally binding. Moreover, many NCPs are not comprehensive in scope and those developed under biodiversity instruments, in particular, are often restricted to specified treaty obligations and therefore do not apply to non-compliance more generally. Therefore, non-compliance proceedings are best viewed as supplementing rather than replacing rules relating to breach, dispute resolution and countermeasures.[156] This pragmatic conclusion nevertheless does give rise to the potential for overlapping or even conflicting proceedings, particularly in light of the ability of multiple actors to initiate the NCP process within many biodiversity regimes.[157]

6. CONCLUDING REMARKS

The development of NCPs within the early operational stages of a number of biodiversity instruments demonstrates the dynamic and innovative approach taken to the creation of international environmental law in the 1970s and 1980s. These NCPs were undoubtedly precedential, and provided much inspiration for the development of the higher-profile compliance mechanisms developed within subsequent MEAs concerned with ozone depletion, climate change, access to environmental information and environmental impact assessment. However, initiatives that were rightly regarded as pioneering 30 years ago risk being perceived as anachronistic today. Compared to NCPs developed more recently, compliance mechanisms under a number of biodiversity instruments are too narrowly focused and, ultimately, weak when it comes to the enforcement of associated measures. Nevertheless, the importance of NCPs cannot be underestimated. States remain reluctant to hold one another to account under MEAs using orthodox principles of treaty law and state responsibility – the recent proceedings brought by Australia against Japan (with New Zealand intervening)[158] in respect of so-called scientific whaling under the 1946 Whaling Convention,[159] being an unusual and high-profile exception to the rule – and NCPs thus constitute the primary mechanism for holding states to account for their environmental obligations. That being so, biodiversity regimes should look to the next stage in the evolution of these

---

[156] See Goodwin (n 144), 295–297.
[157] For a discussion of the potential conflicts and options for resolution see Scott (n 6), 254–258.
[158] *Whaling in the Antarctic* (Australia v. Japan, New Zealand Intervening), ICJ Judgment 31 March 2014.
[159] 1946 International Convention on the Regulation of Whaling UKTS 5 (1949), in force 10 November 1948.

compliance mechanisms in order to broaden and strengthen their application to all contracting parties in respect of all potential compliance issues.

SELECT BIBLIOGRAPHY

Bowman, M, P Davies and C Redgwell, *Lyster's International Wildlife Law* (2nd edn., CUP, 2010)
Chayes, A and A Handler Chayes, *The New Sovereignty: Compliance with International Regulatory Agreements* (Harvard University Press, 1995)
Davies, PGG, 'Non-compliance – a Pivotal or Secondary Function of CoP Governance?' (2013) 15 International Community Law Review 77
Ferrajolo, O, 'State Obligations and Non-Compliance in the Ramsar System' (2011) 14 Journal of International Wildlife Law & Policy 243
Fitzmaurice, M and C Redgwell, 'Environmental Non-Compliance Procedures and International Law' (2000) 31 Netherlands Yearbook of International Law 35
Goodwin, EJ, 'The World Heritage Convention, the Environment, and Compliance' (2009) 20 Colorado Journal of International Environmental Law and Policy 157
Goodwin, EJ, 'The Consequences of Deleting World Heritage Sites' (2010) 21 King's Law Journal 283
Jen, S, 'The Convention on the Conservation of European Wildlife and Natural Habitats (Bern, 1979): Procedures of Application in Practice' (1999) 2 Journal of International Wildlife Law and Policy 224
Klabbers, J, 'Compliance Procedures' in Daniel Bodansky, Jutta Brunnée and Ellen Hey (eds), *The Oxford Handbook of International Environmental Law* (OUP, 2007) 995
Reeve, R, *Policing International Trade in Endangered Species: The CITES Treaty and Compliance* (Earthscan Publications, 2002)
Scott, KN, 'Non-compliance Procedures and Dispute Resolution Mechanisms under International Environmental Agreements' in D French, M Saul and ND White (eds), *International Law and Dispute Settlement: New Problems and Techniques* (Hart Publishing, 2010) 225
Treves, T and others, *Non-Compliance Procedures and Mechanisms and the Effectiveness of International Environmental Agreements* (Asser Press, 2009)
Victor, D, K Raustiala and E Skolnikoff (eds), *The Implementation and Effectiveness of International Environmental Commitments: Theory and Practice* (MIT Press, 1998)
Weiss, E and H Jacobson, *Engaging Compliance with International Environmental Accords* (MIT Press, 1998)
Wolfrum, R, 'Means of Ensuring Compliance with and Enforcement of International Environmental Law' (1998) 272 Recueil des Cours 9

16. 'Only connect'? Regime interaction and global biodiversity conservation[1]

Richard Caddell

> 'Only connect! ... Live in fragments no longer.'
>
> E. M. Forster, *Howards End*

1. INTRODUCTION

Since the watershed UN Conference on the Human Environment in 1972 called upon humankind to recognise its 'special responsibility to safeguard and wisely manage the heritage of wildlife and its habitat',[2] there has been a steady proliferation of treaties and management bodies responsible for the protection of biological diversity. While this institutional bounty represents an admirable alignment of political will towards the protection of threatened species and ecosystems, significant concerns have nonetheless remained over the practical cohesion and coherence of these regimes, with the problem of so-called 'treaty congestion' considered to be especially acute.[3] Treaty congestion essentially connotes the administrative and regulatory inefficiencies created by the inherent nature of international environmental law-making,[4] which often generates multilateral environmental agreements (MEAs) in an ad hoc and haphazard manner, potentially facilitating conflict, duplication and inefficiency between allied regimes. Accordingly, individual species can find themselves subject to the regulatory attentions of a wide range of disparate treaties, each purporting to advance individual management policies for the population in question. Such a situation presents clear challenges in coordinating and consolidating conservation strategies and priorities at both an international level and, more importantly, for contracting parties seeking to implement these commitments on the ground. The development of policies to address these

---

[1] This chapter seeks to outline the legal and practical position as of 30 April 2015; given that the Biodiversity-Related Conventions welcomed the International Plant Protection Convention to its group only on 16 August 2014 and has addressed no subsequent meetings or formal initiatives to advance synergies with that new treaty, the role of this latter instrument has not been substantively examined in this contribution.

[2] *Declaration of the United Nations Conference on the Human Environment*, UN Doc. A/Conf.48/14/Rev. 1 (1973), Principle 4.

[3] See BL Hicks, 'Treaty Congestion in International Environmental Law: The Need for Greater International Coordination' (1999) 32 University of Richmond Law Review 1643.

[4] For an early analysis of this problem, which remains equally pertinent in the current legal landscape, see E Brown Weiss, 'International Environmental Law: Contemporary Issues and the Emergence of a New World Order' (1993) 81 Georgetown Law Journal 675.

problems has therefore constituted a significant – yet often unheralded – concern for many such treaties.

One largely self-evident strategy towards countering the impacts of treaty congestion in a biodiversity context is to delineate the responsibilities of the various international actors in the field and to foster closer working relationships between regimes in respect of species and operational priorities of common concern. Indeed, since the inception of the 1992 Convention on Biological Diversity,[5] a clear emphasis towards the need to operate on a more collegiate and collaborative basis has been apparent within the constituent organisations of the various MEAs concerned with the conservation of global biodiversity. Prior to 1992 such treaties had largely existed in splendid isolation, addressing the conservation needs of particular species almost exclusively within the confines of their individual mandates. However, it has become increasingly clear that international conservation efforts in respect of a number of species will require the effective coordination of the individual resources and expertise of these regimes. The pursuit of 'synergies' between conventions has thereby steadily become the practical rule, rather than the tentative exception. The Rio +20 Conference, convened in 2012, further reinforced the commitment of the international community towards a greater degree of interaction between allied regimes for the conservation of biodiversity.[6] Moreover, aside from overarching political support for global initiatives to improve the coherence of multilateral environmental governance, powerful practical arguments have also been advanced in support of a greater degree of convergence between biodiversity MEAs. Financial pressures have had a dramatic impact upon the funding and resources available to such bodies, which are increasingly required to achieve substantially more with considerably less. Accordingly, the possibility of joint activities – and thereby spreading the cost of conservation initiatives – has become a highly attractive managerial pursuit, especially from the perspective of national governments that both underwrite and judiciously audit the performance of such bodies.

The leading treaties charged with addressing elements of the management of international biodiversity – which tend to focus either on particular ecosystems, specific species or certain anthropogenic influences – are commonly referred to as a collectivised bloc of seven 'Biodiversity-Related Conventions' (BRCs). In this respect, in addition to the CBD, the BRCs comprise the 1971 Convention on Wetlands of International Importance, Especially as Waterfowl Habitat,[7] the 1972 UNESCO Convention Concerning the Protection of the World Cultural and Natural Heritage,[8] the 1973 Convention on International Trade in Endangered Species of Wild Fauna and Flora,[9] the 1979 Convention on the Conservation of Migratory Species of Wild

---

[5] 1790 UNTS 79 ('CBD').

[6] Explicitly encouraging 'parties to multilateral environmental agreements to consider further measures ... to promote policy coherence at all relevant levels, improve efficiency, reduce unnecessary overlap and duplication, and enhance coordination and cooperation among the multilateral environmental agreements': United Nations General Assembly, *The Future We Want*; UN Doc. A/Res/66/288; paragraph 89.

[7] 996 UNTS 245 ('Ramsar Convention' or 'Wetlands Convention').

[8] 1037 UNTS 151 ('WHC' or 'World Heritage Convention').

[9] 993 UNTS 243 ('CITES').

Animals[10] (a regime that encompasses not only the parent convention but also its unique and expansive framework of independent subsidiary instruments),[11] the 2001 International Treaty on Plant Genetic Resources for Food and Agriculture[12] and, since August 2014, the 1951 International Plant Protection Convention (as revised).[13] While this collective does not encapsulate the full range of institutions and actors with a potentially significant conservationist role to play – a great many conventions will relate to biodiversity but have not been explicitly categorised as 'biodiversity-related' – these treaties nonetheless represent the leading actors from a global standpoint and, in many respects, those through which international efforts towards collaborative practices in the field of biodiversity conservation have been most pronounced.

This chapter considers the attempts undertaken by the BRCs to address these problems through the elaboration of linkage initiatives and cooperative arrangements. A vast array of Resolutions and Recommendations have been adopted by their various executive and management bodies, alongside a swathe of Memoranda of Understanding (MOUs) pledging support and cooperation between regimes. Rarely does a Conference of the Parties (COP) of any such body conclude without the articulation of further commitments to promote cooperation and collaboration with other pertinent bodies,[14] while significant outlays of financial and human resources are consistently made on an annual basis to this end. However, despite these now ubiquitous features of MEA

---

[10] 1651 UNTS 333 ('CMS' or 'Bonn Convention').

[11] 1990 Agreement on the Conservation of Seals in the Wadden Sea, reproduced at <http://www.waddensea-secretariat.org>; 1991 Agreement on the Conservation of Populations of European Bats 1863 UNTS 101 ('EUROBATS'); 1991 Agreement on the Conservation of Small Cetaceans of the Baltic, North-East Atlantic, Irish and North Seas, 1772 UNTS 217 ('ASCOBANS'); 1995 African-Eurasian Waterbird Agreement, 2365 UNTS I-42632 ('AEWA'); 1996 Agreement on the Conservation of Cetaceans of the Black Sea, Mediterranean Sea and Contiguous Atlantic Area 2183 UNTS 303 ('ACCOBAMS'); 2001 Agreement on the Conservation of Albatrosses and Petrels 2258 UNTS 257 ('ACAP'); 2007 Agreement on the Conservation of Gorillas and their Habitats 2544 UNTS I-45400 ('Gorilla Agreement'); African Turtles MOU; Aquatic Warbler MOU; Birds of Prey MOU; Bukhara Deer MOU; Dugong MOU; Grassland Birds MOU; Great Bustard MOU; High Andean Flamingos MOU; IOSEA; Mediterranean Monk Seal MOU; Migratory Sharks MOU; Pacific Islands Cetaceans MOU; Ruddy Headed Goose MOU; Siberian Crane MOU; Slender-Billed Curlew MOU; South Andean Huemul MOU; West African Elephants MOU; West African Marine Mammals MOU; the subsidiary instruments are reproduced at <http://www.cms.int>. On subsidiary arrangements under the CMS *see* R Caddell, 'International Law and the Protection of Migratory Wildlife: An Appraisal of Twenty-Five Years of the Bonn Convention' (2005) 16 Colorado Journal of International Environmental Law and Policy 113, 118–123 and C Shine, 'Selected Agreements Concluded Pursuant to the Convention on the Conservation of Migratory Species of Wild Animals' in D Shelton (ed.), *Commitment and Compliance: The Role of Non-Binding Norms in the International Legal System* (Oxford University Press, 2003), 196–223. On the operation of the CMS generally *see* MJ Bowman, PGG Davies and C Redgwell, *Lyster's International Wildlife Law* (2nd edn., Cambridge University Press, 2010), 535–584.

[12] 2004 UNTS 303 ('ITPGRFA').

[13] 150 UNTS I-1963; a new revised text of the Convention was adopted on 17 November 1997; 2367 UNTS A-1963 ('IPPC').

[14] See most recently the Eleventh Conference of the Parties to the CMS, convened in November 2014, at which the parties acknowledged 'the importance of cooperation and

practice, the ultimate achievements of inter-treaty cooperation have been subject to relatively little scrutiny to date.[15] Accordingly, this chapter will critically appraise the current state of cooperation between the BRCs, outlining through a series of case studies the potentially positive impacts that synergy arrangements can have for the regimes in question, as well as highlighting areas in which particular difficulties have been encountered. This chapter will argue that while a degree of benefit has been forthcoming, alongside certain discernible impacts upon MEA practices and policies, some caution ought to be exercised in the current (over-)reliance upon synergies as a means of addressing the challenges facing these bodies. Accordingly, this chapter will first provide a concise overview of institutional practices between MEAs. Subsequently there will be an examination of the role of executive synergies, before evaluating emerging trends in individual treaty interaction. Finally, this chapter will identify key challenges to treaty interactions, before advancing a series of broad conclusions.

2. LEGAL ASPECTS OF INSTITUTIONAL INTERACTIONS

As MEAs, the BRCs are representative of a distinct model of international governance. In contrast to the traditional form of multilateral regime-building through the creation of a specific intergovernmental organisation (IGO), the international community has, particularly since 1972, increasingly favoured the negotiation of 'autonomous institutional arrangements'[16] as a more flexible and organic mechanism to address environmental issues. These agreements typically embody a decision-making organ, most commonly in the form of a COP, which provides a regular forum for the parties to elaborate institutional policies. The COP ultimately shapes the overarching policies of the MEA and further elaborates the obligations incumbent upon the parties.[17] The COP is often supported by an ancillary body, providing a platform for more regular policy

---

synergies with other bodies, including multilateral environmental agreements (MEAs) and non-governmental organizations, as well as the private sector': Resolution 11.10: Synergies and Partnerships.

[15] On the travails of the biodiversity accords to improve inter-treaty governance: *see* R Caddell, 'The Integration of Multilateral Environmental Agreements: Lessons from the Biodiversity-Related Conventions' (2012) 22 Yearbook of International Environmental Law 37; *see also* JO Velázquez Gomar, 'Environmental Policy Integration Among Multilateral Environmental Agreements: The Case of Biodiversity' (2015) 15 International Environmental Agreements: Politics, Law and Economics (in press), A Jóhannsdóttir, I Cresswell and P Bridgewater, 'The Current Framework for International Governance of Biodiversity: Is It Doing More Harm Than Good?' (2010) 19 Review of European Community and International Environmental Law 139 and KN Scott, 'International Environmental Governance: Managing Fragmentation through Institutional Connection' (2011) 6 Melbourne Journal of International Law 177.

[16] For the definitive account of this process *see* RR Churchill and G Ulfstein, 'Autonomous Institutional Arrangements in Multilateral Environmental Agreements: A Little-Noticed Phenomenon in International Law' (2000) 94 American Journal of International Law 623.

[17] *See* further A Wiersema, 'The New International Law-Makers? Conferences of the Parties to Multilateral Environmental Agreements' (2009) 31 Michigan Journal of International Law 231, 237–245.

management between its meetings, often termed a Standing Committee. These arrangements are supported by a designated administrative forum, usually known as a Secretariat. This body usually serves in an administrative and ambassadorial capacity for the MEA and generally has limited powers that are derived from the constituent treaty and shaped by the COP.[18] This institutional model is typically completed by a scientific or technical body charged with providing specialist advice to develop further operational policies and priorities.[19] The remit and powers of these bodies are usually designated under the express terms of the treaty; these institutional arrangements are then brought to fruition at a preliminary stage in the life of the MEA.[20] This broad template has subsequently become a standard feature of many global and regional instruments regulating environmental concerns.[21]

These modern arrangements nonetheless raise two key legal issues pertinent to the development of institutional synergies. First, there is a need to identify the legal basis under which MEAs may conclude collaborative agreements with each other, which has had a considerable bearing on their institutional practices to date. Secondly, the precise inter-relationship between such treaties also merits consideration, not least where questions of leadership and policy direction are concerned.

In the first instance, operational difficulties have arisen regarding the precise scope of powers conferred on MEAs – and the extent to which particular organs can wield them.[22] As will be demonstrated below, MEAs have tended to elaborate instruments of cooperation outlining their respective roles, responsibilities and commitments as far as inter-treaty collaboration is concerned. For orthodox IGOs this is a relatively uncomplicated affair: cooperation with other bodies may be pursued through the express terms of their constituent treaties, or by attribution through the doctrine of implied powers.

---

[18] For a full outline of this issue *see* BH Desai, *Multilateral Environmental Agreements: Legal Status of the Secretariats* (Cambridge University Press, 2010), especially 101–169. A tight leash was initially placed on some of these scientific bodies to ensure that they did not inadvertently generate practical difficulties for the treaty by straying into areas that could be scientifically interesting but politically delicate. *See*, for example, the elaboration of the CMS Scientific Council: S Lyster, 'The Convention on the Conservation of Migratory Species of Wild Animals (The "Bonn Convention")' (1989) 29 Natural Resources Journal 979, 995. A more relaxed view of the scientific arrangements of the CMS and the inherent impartiality of its members has recently been articulated by this body: Resolution 11.04: Restructuring of the Scientific Council.

[19] Of the treaties in question only the ITPGRFA eschews a permanent advisory forum, which it considers 'premature' and instead favours the establishment of ad hoc technical bodies as the need arises: *Report of the Second Session of the Governing Body of the International Treaty on Plant Genetic Resources for Food and Agriculture* (FAO, 2007), 15.

[20] Earlier MEAs did not envisage particular institutions within their constituent treaties and have instead created them unilaterally as the regime has evolved. Notably, the Ramsar Convention did not initially provide for a designated technical body; its celebrated Scientific and Technical Review Panel was established by its COP at comparatively advanced stage in the Convention's tenure in the light of positive experiences within other MEAs: Resolution 5.5: Establishment of a Scientific and Technical Review Panel.

[21] *See* V Röben, 'Institutional Developments under Modern International Environmental Agreements' (2000) 4 Max Planck Yearbook of United Nations Law 363.

[22] For an illuminating account of these difficulties *see* Desai (n 18), 133–170.

The latter principle endows such bodies with the flexibility to act within the spirit of their mandates and undertake tasks which, while not explicitly articulated within the treaty, are nonetheless fundamental to their effective functioning.[23] Although the sustained use of implied powers can be legally problematic,[24] the development of cooperative agreements nonetheless remains a generally accepted aspect of such practices.[25]

MEAs, on the other hand, are clearly – and, in fact, deliberately[26] – not IGOs in the traditional sense. Nevertheless, it has been convincingly argued that they 'fulfil the necessary criteria for an IGO and would thus aspire to the application of international institutional law',[27] a contention that has remained essentially unchallenged by subsequent practice. Thus MEAs would seem able to pursue collaborative arrangements to at least some degree. To date, the instruments of cooperation adopted between them have taken the form of MOUs and it is clear that such arrangements are generally intended to be non-binding – even if they have ultimately proved to be highly influential in framing institutional activities and policies.

Collaborative MOUs are generally concluded between the secretariats, at the request of the COP, which has raised uncertainty over the precise legal basis for such arrangements.[28] In many MEAs it remains unclear as to whether legal personality has been vested exclusively in the COP (which in turn formally instructs and empowers the secretariat to facilitate collaborative dialogue) or whether such secretariats have an inherent degree of independent legal personality to address synergy matters. In some cases, notably the CBD, legal personality has been unequivocally bestowed upon the Secretariat by the constituent treaty.[29] Other BRCs have favoured the alternative model of conferring a limited degree of juridical personality through the COP to the secretariat to address particular issues, as exemplified in the context of the CMS[30] and, prospectively, the Ramsar Convention.[31] Otherwise, legal capacity to conclude MOUs

---

[23] See *Reparation for Injuries Suffered in the Service of the United Nations* ICJ Reports (1949) 174, at 182–183 (noting that '[u]nder international law, the Organisation must be deemed to have those powers which, though not expressly provided in the Charter, are conferred upon it by necessary implication as being essential to the performance of its duties': *ibid.*, at 182); *Effects of Awards Made by the U.N. Administrative Tribunal* ICJ Reports (1954) 47, 56–60; *Certain Expenses of the United Nations (Article 17, Paragraph 2 of the Charter)* ICJ Reports (1962) 151, 166–170.

[24] See J Klabbers, *An Introduction to International Institutional Law* (Cambridge University Press, 2002), 67–81.

[25] MA Young, *Trading Fish, Saving Fish: The Interaction between Regimes in International Law* (Cambridge University Press, 2011), 155.

[26] Churchill and Ulfstein (n 16) 629–631.

[27] *Ibid.*, 633.

[28] Young (n 25), 157.

[29] Article 24(d).

[30] Resolution 6.9: Juridical Personality of, and Headquarters Agreement for, the Convention Secretariat.

[31] Resolution X.5: Facilitating the work of the Ramsar Convention and its Secretariat; see also *Report on the Legal Personality of the Ramsar Secretariat*; Ramsar COP10 Doc. 35.

has simply been 'assumed' by particular secretariats, notably that of CITES.[32] The legitimacy of this approach has not been subject to meaningful dissent, suggesting that considerable tolerance is afforded to the BRCs in forging connections in this manner.

It is less clear, however, whether the secretariats may ultimately conclude formal *treaties* to address their relationships rather than non-binding MOUs. Some degree of treaty-making power has been demonstrated in the specific context of headquarters arrangements with the host state or organisation.[33] However, these examples do not appear to constitute conclusive evidence of the scope of competences exercised by secretariats. Although this question has considerable implications for the day-to-day work of MEA institutions, it may be considered a non-problem in the specific context of synergy arrangements. As suggested below, synergies have been cultivated through MOUs and liaison between secretariats, without the need for a formal treaty. Indeed, it is questionable whether more legalistic arrangements in the form of a treaty would ultimately deliver any meaningful improvement to current practices, hence in this context treaty-making powers of MEA institutions remains rather a moot point.

Secondly, questions arise over the formal inter-relationship of the treaties themselves. Numerous MEAs overlap on key issues of scope, coverage and application. This can create practical difficulties for parties in seeking to implement multilateral commitments, not least where there may be divergent obligations or policies established for similar conservation problems under multiple regimes. This has exposed a key shortcoming in the legal arrangements governing the relationship between treaties, which has traditionally focused upon conflict resolution, the practical expunging of defunct instruments through the position of successive treaties or the elaboration of normative hierarchies. Yet it is becoming increasingly clear, especially in the context of treaties governing natural resources and environmental protection, that such a model is too blunt to address the subtleties of the current governance arrangements between regimes that essentially seek to support and supplement rather than supplant each other.[34]

Although a number of MEAs contain formal conflict clauses, such provisions have ultimately proved to be of limited value in framing interactions with other regimes. Certain BRCs have either omitted formal conflict clauses[35] or applied them only to specific areas.[36] In the latter case, such clauses have generated controversy on issues of

---

[32] *Legal Personality of the Convention and the Secretariat*; Document SC54 Doc.8. Nevertheless, the precise root of this legal personality remains uncertain: Young (n 25), 157.
[33] Desai (n 18), 164–170.
[34] *See* especially Young (n 25), 298–306.
[35] Notably the Ramsar Convention. A similar position arises in the case of the 1992 UN Framework Convention on Climate Change 1771 UNTS 107 ('UNFCCC'), which may complicate interactions concerning the effects and mitigation of and adaptation to global climate change, which has steadily become a significant issue on the agendas of most of the BRCs.
[36] For example, the WHC incorporated a limited conflict clause addressing the position regarding future treaties on world heritage adopted specifically by UNESCO: Article 37(2). The effect of this clause remains somewhat hypothetical given that no such alternative regime has yet emerged under these auspices.

political significance. A particular example is the position of CITES,[37] which contains a rather infamous clause addressing marine species that have been listed in Appendix II to the Convention.[38] Under Article XIV(4) a CITES party

> which is also a party to any other treaty, convention or international agreement which is in force at the time of the coming into force of the present Convention and under the provisions of which protection is afforded to marine species included in Appendix II, shall be relieved of the obligations imposed on it under the provisions of the present Convention with respect to trade in specimens of species included in Appendix II that are taken by ships registered in that State and in accordance with the provisions of such other treaty, convention or international agreement.

This formulation remains highly problematic. The notion of the 'protection' of marine species remains fundamentally undefined, raising confusion over which regimes might ultimately be caught by this construction. A highly literal interpretation might, for instance, preclude consideration of instruments that are predominantly concerned with the exploitation of marine species – or those, such as the International Whaling Commission (IWC), for which protection is only expressly envisaged as a partial element of its regulatory mandate.[39] Moreover, as Young observes, the application of this clause would appear to be restricted solely to pre-1975 bodies, a problematic interpretation in the context of synergy initiatives given the widespread development of marine-orientated regimes subsequent to the entry into force of CITES.[40]

Likewise, in other instances conflict clauses have been drafted in an idiosyncratic and ambiguous manner, which has created scope for unintended friction with alternative regimes. The most striking example is Article 22 of the CBD, which addresses interactions with other international treaties generally and those addressing marine environmental considerations specifically. Article 22 was a contentious addition to the convention and a number of the negotiating delegations were unconvinced as to either

---

[37] For a full appraisal of the interactions between CITES and other pertinent instruments *see* R Caddell, 'Inter-Treaty Cooperation, Biodiversity Conservation and the Trade in Endangered Species' (2013) 22 Review of European, Comparative and International Environmental Law 264.

[38] Appendix II addresses species that 'although not necessarily threatened with extinction may become so unless trade in specimens of such species is subject to strict regulation in order to avoid utilisation incompatible with their survival': Article II(2). On this issue generally *see* E Franckx, 'The Protection of Biodiversity and Fisheries Management: Issues Raised by the Relationship between CITES and LOSC' in D Freestone, R Barnes and D Ong (eds), *The Law of the Sea: Progress and Prospects* (Oxford University Press, 2006), 210–232, 225.

[39] Indeed, the whaling regime is intended 'to provide for the proper conservation of whale stocks and thus make possible the orderly development of the whaling industry'; preamble to the 1946 International Convention for the Regulation of Whaling 161 UNTS 72. It remains a point of contention as to whether this mandate is one predicated upon the 'protection' of whales. If this is the case, it remains so largely due to the evolution of the treaty in the years subsequent to the entry into force of CITES; for a rigorous appraisal of this process *see* MJ Bowman '"Normalizing" the International Convention for the Regulation of Whaling' (2008) 29 Michigan Journal of International Law 293.

[40] Young (n 25), 66.

its necessity or the projected effect of this formulation.[41] Nevertheless, under Article 22(1) of the final version of the treaty, the CBD pledges general compatibility with other instruments 'except where the exercise of those rights and obligations would cause a serious damage or threat to biological diversity'. It has been suggested that this position offers little of practical merit in the context of inter-treaty relationships, since its application 'would require conflict so extensive that it is difficult to envisage in the context of the CBD'.[42] Indeed, discord of this nature has thus far been ostensibly avoided, largely because subsequent biodiversity-pertinent agreements have been drafted with a degree of reference (and, often, deference) to the CBD. Accordingly, the threshold of Article 22(1) has not been substantively tested during the tenure of the convention to date. On a more regional scale, however, similar wording employed by ACCOBAMS has led to sporadic difficulties with technical aspects of EU fisheries legislation.[43] In this context, practical accommodations have been reached between the respective regulators, thereby negating the need to formally invoke this provision. The scope and application of the tenets of Article 22(1), and similar provisions in other treaties, accordingly remain essentially untested.

Nevertheless, Article 22 provides opportunities for conflict with significant prior treaties and is considered 'unusual and can lead to a *de facto* precedence of the Convention on Biological Diversity in respect to other instruments'.[44] Potential interpretive complications are most clearly apparent in the context of the 1982 UN Convention on the Law of the Sea.[45] In this respect, Article 22(2) mandates that parties are to apply their CBD obligations 'with respect to the marine environment consistently with the rights and obligations of States under the law of the sea'. The CBD does not explicitly refer to the LOSC in its consideration of marine issues, a treaty that had yet to enter into force at the material time and remains controversial in certain quarters. This provision may accordingly be considered to apply to instruments and regimes beyond the specific confines of the LOSC, notwithstanding the predominant codifying position of the 1982 Convention. The precise inter-relationship between the two conventions remains somewhat uncertain and a number of areas of potential conflict remain, notably in the context of fisheries management, the establishment of protected

---

[41] *See* further M Chandler, 'The Biodiversity Convention: Selected Issues of Interest to the International Lawyer' (1993) 4 Colorado Journal of International Environmental Law and Policy 141, 149.

[42] C Mackenzie, 'A Comparison of the Habitats Directive with the 1992 Convention on Biological Diversity' in G Jones QC (ed.), *The Habitats Directive: A Developer's Obstacle Course?* (Hart, 2012) 25, 35.

[43] *See* R Caddell, 'Biodiversity Loss and the Prospects for International Cooperation: EU Law and the Conservation of Migratory Species of Wild Animals' (2008) 8 Yearbook of European Environmental Law 219, 251–254 (noting potential legal difficulties in the specific context of the EU's endorsement of – and the resistance of ACCOBAMS to – the widespread use of acoustic deterrent devices as a by-catch mitigation tool).

[44] R Wolfrum and N Matz, 'The Interplay of the United Nations Convention on the Law of the Sea and the Convention on Biological Diversity' (2000) 4 Max Planck Yearbook of United Nations Law 445, 475.

[45] 1883 UNTS 396 ('LOSC').

areas and navigational requirements.[46] In January 2015, the UN General Assembly advanced a formal mandate towards the development of a legally binding instrument under the auspices of the CBD to address the protection of biodiversity in areas beyond the limits of national jurisdiction.[47] This process remains nascent, but will require concerted coordination between the two regimes in a manner that has not yet been applied to date.

Little guidance has been prescribed under the 1969 Vienna Convention on the Law of Treaties[48] to address the specific question of synergies between conventions. Aside from the Charter of the United Nations,[49] treaties do not generally form natural hierarchies.[50] The BRCs have not in practice sought to impose hierarchical arrangements upon each other – although it has been mooted in some quarters, given the political pre-eminence of and near universal participation in the CBD regime.[51] Nevertheless, calls for relationships of hierarchy between certain treaties in relation to particular high-profile species[52] have not generally produced harmonious outcomes, as the experience of CITES and the IWC attests.[53] Here, a series of disputes over responsibility for aspects of the conservation of great whales – largely fuelled by mutual parties seeking to exploit the terms of one convention to gain leverage over the policies pursued in the other[54] – eventually ended in 2007 with a series of statements

---

[46] On this issue generally *see* A Boyle, 'Further Development of the Law of the Sea Convention: Mechanisms for Change' (2005) 54 International and Comparative Law Quarterly 563.

[47] *Outcome of the Ad Hoc Open-ended Informal Working Group to study issues relating to the conservation and sustainable use of marine biological diversity beyond areas of national jurisdiction*, Document A/69/780.

[48] 1155 UNTS 331 ('VCLT').

[49] 1 UNTS XVI.

[50] Article 103 of the UN Charter, as reinforced by Article 30 VCLT.

[51] *See* especially Jóhannsdóttir, Cresswell and Bridgewater (n 15), 147–149. *See* also R Churchill, 'Sustaining Small Cetaceans: A Preliminary Examination of the Ascobans and Accobams Agreements' in A Boyle and D Freestone (eds.), *International Law and Sustainable Development: Past Achievements and Future Challenges* (Oxford University Press, 2001), 244 (arguing that particular advantages might be forthcoming from incorporating the CMS subsidiary instruments into larger regimes).

[52] *See* most notably A Gillespie, 'Forum Shopping in International Environmental Law: The IWC, CITES and the Management of Cetaceans' (2002) 33 Ocean Development and International Law 17.

[53] On the problematic relationship between CITES and the IWC – and the uneasy accommodations that have been reached between these regimes – *see* Caddell (n 37), 267–269.

[54] *Ibid*; *see* also K Eldridge, 'Whale for Sale? New Developments in the Convention on International Trade in Endangered Species of Wild Fauna and Flora' (1995) 24 Georgia Journal of International and Comparative Law 549. These difficulties became so acute that the then Secretary-General of CITES formally protested to the IWC: *see* A Gillespie, *Whaling Diplomacy: Defining Issues in International Environmental Law* (Edward Elgar Publishing, 2005), 340–341.

endorsing mutual respect for the scientific organs of their respective institutional structures and pointedly omitting questions of normative hierarchy.[55]

The VCLT does establish a clear position concerning successive treaties relating to the same subject matter in which the later instrument prevails, through the *lex posterior derogat legi priori* rule.[56] Nevertheless, it is highly questionable whether the rules on successive treaties are applicable to purported synergies between the BRCs. As a case in point, Article 30 VCLT was explicitly rejected as a sufficient basis by which to moderate the relationship of the CBD with other bodies, on the express basis that the nascent treaty was not intended to supersede existing regimes governing natural resources.[57] Nor has the long-standing principle of *lex specialis derogate lege generali* proved to have been a helpful interpretive tool in the context of addressing the relationship between specific biodiversity treaties, notwithstanding its value in the adjudication of particular disputes with an environmental theme. The notion that a specialised branch of international law will provide a more nuanced regime than a more generalised treaty has a particular attraction in instances involving the interpretation of obligations concerning specific natural resources.[58] However, it has yet to be invoked in any meaningful way to address conflicts between specialist biodiversity agreements, which have ultimately sought more collaborative arrangements in recent years to pre-empt such difficulties. This has been exemplified by the practice of the CBD, which has established 'lead partners' from the ranks of the other BRCs rather than seeking to assert normative dominance. Indeed, the BRCs (and, to a considerable extent, MEAs more generally) have conspicuously avoided the language of successive treaties and normative hierarchies in framing their interactions. Instead, the broad modus operandi applied by the BRCs to date has been to establish strategic alignments between their various institutions to develop unified policies in areas of mutual interest and thus endeavour to avoid confusion, duplication and wastage of efforts. The precise mechanics of such arrangements, and their scope for creating clear and effective synergies, is therefore primarily determined at the executive and administrative levels of the treaties in question, either on a collective 'cluster' basis or through individual relationships between allied regimes.

3. MODELS OF COOPERATION BETWEEN THE BRCS

The general approach adopted by the BRCs to facilitate further inter-treaty cooperation may be broadly summarised as follows. In general, linkages have primarily been

---

[55] Resolution 2007-4: Resolution on CITES (adopted by the IWC) and Resolution Conf. 11.4: Conservation of Cetaceans, Trade in Cetacean Specimens and the Relationship with the International Whaling Commission (adopted by CITES).
[56] Article 30.
[57] Chandler (n 41), 149.
[58] As a case in point, the *lex specialis* principle was invoked in the *Southern Bluefin Tuna case*, in an attempt to exclude pertinent provisions of the UN Convention on the Law of the Sea; the International Tribunal for the Law of the Sea ultimately considered that both the specialist and the generalist treaties were nonetheless applicable to the dispute: *Southern Bluefin Tuna case (Australia and New Zealand v. Japan)*, Award of 4 August 2000, 23.

pursued at an institutional level between the various executive and administrative bodies as well as through individual working groups established under the auspices of these regimes. On a central level, there has been a series of initiatives towards developing collective priorities and work programmes between particular treaties. This has, however, proved to be a challenging task in practice, since there is a deceptively limited degree of common overlap between the seven constituent regimes that comprise the BRCs. Instead, individual regimes have tended to work more extensively with fellow members of the group on issues that are most closely aligned with their specific mandates, areas and species of concern. These arrangements have been facilitated by a series of MOUs and successive generations of targeted work programmes, which have yielded varying degrees of success. Notwithstanding these initiatives, however, it should be noted at this preliminary stage that such interactions have occurred predominantly at an executive level. Legitimate concerns can be raised that while on a managerial level the BRCs have formed a relatively strong – if occasionally dysfunctional – family, this broad sense of unity has not generally been transposed into clear and coordinated policies on the ground for national agencies charged with implementing multilateral commitments, synergised or otherwise.

3.1 Collective Synergies and 'Clustering' Arrangements

The development of collective synergies, in the sense that treaties with a similar purpose, scope and subject matter seek to align themselves on a more official administrative basis, has long been championed as an attractive institutional policy.[59] To this end, the 'clustering' of allied regimes has been considered a potentially effective means of addressing some of the more pervasive problems associated with treaty congestion.[60] There is no single institutional model for institutional clustering. Such arrangements can therefore comprise varying degrees of administrative interconnection, which may include joint meetings, co-located institutions and attempts to facilitate a more streamlined series of administrative practices. Throughout the past decade, a number of initiatives have been developed to establish collective synergies within various treaty groupings. However, the results of these practices and their potential to effect meaningful administrative benefits may generally be considered – and in the specific context of the BRCs – to be decidedly mixed.

Thus far, two broad approaches have emerged in relation to clustering arrangements for allied treaties. In the first instance, and in many ways perhaps the clearest example of inter-treaty collaboration, a formal mechanism for a combined COP and administrative institutions has been piloted by the so-called 'chemical cluster' of treaties.[61] In March 2010, following the elaboration of a series of internal commitments towards

---

[59] BH Desai, 'Mapping the Future of International Environmental Governance' (2000) 13 *Yearbook of International Environmental Law* 43.

[60] S Oberthür, 'Clustering of Multilateral Environmental Agreements: Potentials and Limitations' (2002) 2 *International Environmental Agreements: Politics, Law and Economics* 317, 321–328.

[61] The 'chemical cluster' comprises the 1989 Convention on the Control of Transboundary Movements of Hazardous Wastes and their Disposal 1673 UNTS 57 ('Basel Convention'), the 1998 Convention on the Prior Informed Consent Procedure for Certain Hazardous Chemicals

improving synergies, the COPs of the three conventions convened a pioneering Simultaneous Extraordinary Meeting (ExCOP). The Meeting culminated in the adoption of an Omnibus Decision pledging to establish joint services for financial and administrative support, legal issues, information technology, information and resource mobilisation, alongside full synchronisation of budget cycles, joint activities and a process for review of these arrangements.[62] This model represents the first – and, in many ways, the clearest – example of a group of autonomous MEAs formally sharing various administrative and managerial functions and seeking to align their individual policies within a broader group ethic.

Since this landmark meeting, there has been a steady process towards further administrative and executive integration. ExCOP1 mandated the creation of a new post of Executive Secretary of the Basel, Rotterdam and Stockholm Conventions, which was subsequently filled in April 2011. This approach remains firmly in its infancy and there is a limited pattern of treaty practice to illuminate its ultimate prospects for success. Nevertheless, a series of detailed proposals have been established for future organisational practices, which would see the creation of streamlined teams for administrative services, conventions operations, technical assistance and scientific support.[63] Such proposals also envisage considerable fluidity between the personnel of the conventions and are ultimately intended to deliver a more simple and consistent structure in the long term. However, they also acknowledge increased short-term costs and inevitable teething problems in grafting a new administrative culture to three individual conventions with their own long-standing working practices and processes.

A second ExCOP was convened in 2013, which noted that the collective endeavours of the triumvirate of treaties had been focused primarily on institutional development, with particular reference to strengthening administrative capacity and establishing a work programme.[64] ExCOP2 also saw the establishment of a targeted Contact Group on Synergies and Budget Matters to assist with the further elaboration of these arrangements.[65] The further aggregation of executive arrangements by the chemical cluster remains at a relatively preliminary stage, hence drawing firm and measured conclusions as to its effectiveness is a somewhat premature task. Nevertheless, it should be observed that, notwithstanding the logistical challenges presented by further

---

and Pesticides in International Trade 2244 UNTS 337 ('Rotterdam Convention'), and the 2001 Convention on Persistent Organic Pollutants 2256 UNTS 119 ('Stockholm Convention').

[62] *Omnibus Decision on Enhancing Cooperation and Coordination among the Basel, Rotterdam and Stockholm Conventions*, adopted by the Conference of the Parties to the Basel Convention as Decision BC.Ex-2/1, by the Conference of the Parties to the Rotterdam Convention as Decision RC.Ex-2/1 and by the Conference of the Parties to the Stockholm Convention as Decision SC.Ex-2/1 ('Omnibus Decision').

[63] *Joint Managerial Functions: Proposal for the Executive Secretary of the Basel, Rotterdam and Stockholm Conventions for the Organization of the Secretariats of the Three Conventions* (UNEP, 2011).

[64] *Report of the Second Simultaneous Extraordinary Meetings of the Conferences of the Parties to the Basel, Rotterdam and Stockholm Conventions*; Document UNEP/FAO/CHW/RC/POPS/EXCOPS.2/4, 2.

[65] *Ibid.*, 10.

integration, stakeholders across the three conventions have accorded these administrative proposals a generally warm reception – provided that the underlying institutional identity of each MEA is not lost[66] – and there remains genuine enthusiasm within these bodies, as well as the international community at large,[67] for a more unified working structure in this context.

However further synergy arrangements emerge within the chemical cluster, it nevertheless appears clear that there is little scope for this model to be replicated in a biodiversity context. The chemical treaties exhibit a high degree of overlap in mandate, structure, ethos and application, in a manner that is not obviously forthcoming among the various BRCs. The chemical treaties, which are considered to be 'an excellent example of thematically linked, yet apparently for no good reason, separate entities within international environmental law',[68] have generally compatible texts and machinery, with no significant amendments necessary to facilitate alignment with a common central objective. If these operational advantages represent the clearest template for the effective clustering of agreements, they are not present to such an extent within the biodiversity context. As noted in Section 4 below, as a collective of treaties the BRCs are perhaps characterised more by their differences than their similarities: beyond a general ethos towards the responsible husbandry of natural resources, there is limited commonality between these regimes in terms of conservation priorities, species and thematic range and unifying regulatory objectives. Attempts to streamline administrative processes have largely foundered, while there would appear to be unsurmountable practical difficulties in adjusting reporting timescales and requirements, aligning advisory and executive bodies and in sourcing the sustained investment of financial and human resources necessary to underwrite such a process. Indeed, even facilitating the closer alignment of an already highly compatible cluster of chemical treaties was a highly taxing undertaking and one for which 'many delegations expressed their suspicion that the synergies process was driven by the three convention secretariats rather than their COPs'.[69] As noted below, there remain significant differences in ethos, objective and outlook between the BRCs, combined with a strong desire to maintain

---

[66] *Joint Managerial Functions: Proposal for the Executive Secretary of the Basel, Rotterdam and Stockholm Conventions for the Organization of the Secretariats of the Three Conventions* (n 63), 23–26. This echoes similar language expressed clearly within the Omnibus Decision, which was '[m]indful of the legal autonomy of each of the Basel Convention on the Control of Transboundary Movements of Hazardous Wastes and Their Disposal, the Rotterdam Convention on the Prior Informed Consent Procedure for Certain Hazardous Chemicals and Pesticides in International Trade and the Stockholm Convention on Persistent Organic Pollutants'.

[67] Indeed the international community formally 'acknowledge[d] the work already undertaken to enhance synergies among the three conventions in the chemicals and waste cluster' at the Rio +20 Conference: *The Future We Want* (n 6), para 89.

[68] N Goeteyn and F Maes, 'The Clustering of Multilateral Environmental Agreements: Can the Clustering of the Chemicals-Related Conventions be Applied to the Biodiversity and Climate Change Conventions?' in F Maes, A Cliquet, W du Plessis and H McLeod-Kilmurray (eds), *Biodiversity and Climate Change: Linkages at International, National and Local Levels* (Edward Elgar Publishing, 2013), 147–177, 165.

[69] *Ibid.*, 166.

their individual identities and mandates, which largely militates against a concerted process of further collectivisation.

Instead, the BRCs have favoured a series of looser executive alignments, whereby regular meetings of senior personnel set the tone for synergy arrangements that can then be advanced at the individual treaty level. To this end – and perhaps unsurprisingly given the central role of the CBD in both initiatives – the BRCs have opted to follow a model pioneered by the so-called 'Rio cluster' of treaties, comprising the regimes established at the 1992 UN Conference on Environment and Development.[70] In 2001, the Rio cluster established a Joint Liaison Group (JLG) to improve coordination between these treaties on issues of mutual concern. This was followed in 2004 by the inauguration of the Biodiversity Liaison Group (BLG) between the CBD, CITES, the Ramsar Convention, the CMS and the WHC 'to enhance coherence and cooperation in their implementation',[71] which was subsequently expanded in 2006 to include the ITPGRFA[72] and, in August 2014, the IPPC.[73] Although both fora have largely operated in isolation from each other,[74] they have nevertheless tended to function on a broadly similar basis. Both groups provide a platform for executive staff to elaborate overarching policy priorities and identify further grounds for institutional synergies. Group meetings rotate between the various conventions and, where possible, dovetail with COPs or other major events.

Thus far, however, in and of themselves executive synergies between the BRCs have produced rather modest outcomes. Indeed, coordination through both the BLG and JLG has perhaps generated rather more in the way of rhetoric than results in improving cross-institutional practices. Ultimately, the experience of the JLG has perhaps revealed that the unifying themes between the three Rio Conventions are relatively limited. Beyond a general focus on deforestation and climate change adaptation, the JLG has, from a preliminary stage, recognised that its role is less focused on developing streamlined multilateral practices and is instead more concerned with reviewing progress towards national implementation of treaty commitments. On the other hand, given the overarching priorities at the material time, the main focus of the BLG has unsurprisingly concerned the pursuit of the 2010 biodiversity loss target,[75] with future

---

[70] In addition to the CBD, this cluster comprises the UNFCCC and the UN Convention to Combat Desertification in Countries Experiencing Serious Drought and/or Desertification, Particularly in Africa 1992 1954 UNTS 3 ('UNCCD').

[71] Decision VII/26 Strategic Plan for the Convention on Biological Diversity.

[72] Decision VIII/16: Cooperation with other conventions and international organizations and initiatives.

[73] A formal invitation for membership of the BLG was extended at the most recent meeting of this forum, with the IPPC now considered to be the seventh member of the 'biodiversity cluster': *Report of the Sixth Meeting of the Liaison Group of Biodiversity-related Conventions*, at 2.

[74] Nevertheless, the Ramsar Secretariat participated at the Fourth Meeting of the JLG and this forum has continually stated that it welcomes appropriate participation by other bodies outside the Rio network. In contrast, however, the BLG has traditionally been more restrictive, although more recent meetings have seen a concerted effort to invite observers from other bodies and institutions, notably at the UN level.

[75] Decision VI/26, para. 11.

work now concentrating on the revised 2020 strategy.[76] Throughout its tenure there has been a steady accretion of topics to the BLG agenda, derived both from the Group's individual meetings and through steering from the CBD.[77] Considerable attention has been focused on knowledge management, with the BLG keen to improve data-exchange and to share models of good practice. One practical BLG initiative that has borne fruit is the development of an interactive CD-ROM addressing implementation of the Addis Ababa Principles and Guidelines, encompassing guidance specific to each treaty, which has been welcomed by the parties as a helpful management tool.[78] In the mid-term future, there is also scope to develop closer thematic linkages with the JLG on issues of mutual interest.[79] There would appear to be little impediment to this, given that a number of the BRCs have fostered individual working relationships with both the UNCCD and UNFCCC.

The BLG has, however, traditionally been hamstrung by a vague and unfocused structure that has been compounded by a dearth of clear long-term aims and objectives. There is also some evidence to suggest that a degree of suspicion was accorded to this forum by particular regimes, at least initially, as primarily a vehicle for the advancement of the CBD's agenda.[80] The BLG has tended to meet on an ad hoc basis with a fluctuating cast of participants,[81] which prompted a review of its practices under the auspices of the CBD.[82] Operational problems have also undermined the JLG, which in April 2011 subsequently endorsed five guiding principles underpinning its common framework, acknowledging that it remains primarily a forum for information exchange, with particular emphasis on facilitating national implementation; respect for the differing mandates of the conventions; a commitment to decreasing bureaucracy; a need to reduce costs, especially for the parties; and for actions to be realistic in timescale and finance.[83] Mirroring these developments, the BLG has also sought 'a more formal *modus operandi*',[84] which was duly adopted in September 2011.[85] Nevertheless, it remains clear that this is likely to constitute the extent of executive harmonisation between the various conventions: the possibility of an ExCOP-style approach as

---

[76] For a concise review of this development *see* SR Harrop, '"Living in Harmony With Nature"? Outcomes of the 2010 Nagoya Conference of the Convention on Biological Diversity' (2011) 23 Journal of Environmental Law 117.

[77] *See* Decision IX/27: Cooperation among multilateral environmental agreements and other organizations and Decision X/20.

[78] *Report of the Sixth Meeting of the Liaison Group of Biodiversity-related Conventions*, 1.

[79] This has been advocated by the parties: *Report of the Special Meeting, ibid.*; moreover Decision X/20 directed the CBD Secretariat to explore potential linkages between the two Groups.

[80] Velázquez Gomar (n 15), 11.

[81] *Report of the Sixth Meeting*, 2.

[82] Decision IX/27.

[83] *Report of the Eleventh Meeting of the Joint Liaison Group of the Rio Conventions*, 1–2.

[84] *Report of the Special Meeting of the Liaison Group of Biodiversity-related Conventions*, 1.

[85] *Modus Operandi for the Liaison Group of the Biodiversity-related Conventions*, reproduced on-line at <http://www.cbd.int/cooperation/doc/blg-modus-operandi-en.pdf>.

advanced by their counterparts within the chemical cluster has been explicitly rejected both by the Rio treaties[86] and the BRCs.[87]

Since then the BLG has broadly sought to reorientate itself towards the development of synergies on a national level, implementing the CBD's Aichi targets, engaging efficiently with pertinent streamlining initiatives at the UN level and elaborating common activities.[88] Most recently, the BLG has sought to prioritise tactical interactions with the Global Environment Facility[89] and the Intergovernmental Panel on Biodiversity and Ecosystem Services (IPBES).[90] This is a sensible tactic and one that may ultimately serve to promote the biodiversity agenda more effectively within these key UN initiatives: there is already a degree of evidence to suggest that these vital fora are likely to be more receptive to combined approaches by treaty clusters rather than solicitations from individual regimes.[91]

3.2 Scientific and Technical Collaboration

In addition to executive arrangements, synergies have also been pioneered on a scientific and technical level between the BRCs. In 2007, a forum for the Chairs of the Scientific Advisory Boards of Biodiversity-Related Conventions (CSAB) was inaugurated.[92] There are signs that CSAB presents a clearer avenue for the development of meaningful collective practices, primarily because it engages matters with a unifying methodological resonance for the BRCs in a manner that is perhaps not so evident on an administrative and managerial level. Strategic priorities for CSAB have mainly constituted examining areas of cooperation and attempting to translate scientific considerations into clear policies, alongside identifying emerging issues with reference to problems and priorities within the individual scientific fora of the participating regimes.[93]

Thus far, such work has largely mirrored synergies between individual conventions in highlighting case-studies and examples of good practice. Although seemingly mundane, in the scientific context such endeavours have an understated importance in driving

---

[86] *Report of the Eleventh Meeting of the Joint Liaison Group of the Rio Conventions*, 4–5.
[87] *Report of the Eighth Meeting of the Liaison Group of Biodiversity-related Conventions*, at 3 (stating definitively that 'the model of synergies adopted among the chemicals Conventions could not be adapted to the biodiversity Conventions').
[88] *Report of the Ninth Meeting of the Liaison Group of Biodiversity-related Conventions*, 6.
[89] *Ibid.*, 2–6.
[90] *Report of the Eighth Meeting of the Liaison Group of Biodiversity-related Conventions*, 4.
[91] Indeed, the BLG has observed that 'a series of anticipated and coordinated requests from the six biodiversity-related Conventions would be very useful and give additional weight to the requests'; *ibid.* As noted below, joint lobbying has also proved useful in a scientific context.
[92] *Report of the Fifth Meeting of the Liaison Group of Biodiversity-related Conventions*, 5–6; *see* also Decision VIII/16, calling for enhanced cooperation between scientific and technical bodies in addition to Secretariats.
[93] *Report of the First Meeting of Chairs of Scientific Advisory Bodies of Biodiversity-related Conventions*, 2.

synergistic arrangements. They have contributed to a closer alignment on methodologies for environmental impact assessments[94] and have also proved important in framing emergent policies on issues such as climate change adaptation[95] and emergency responses. Likewise, the development of guidance on ecosystem restoration – a pressing concern for all the BRCs – will constitute a major area of future activity for CSAB.[96] Significantly, these meetings also promote an improved flow of data which, despite numerous commitments towards information-sharing outlined in individual MOUs and work programmes between regimes, has often lacked a formalised outlet for inter-treaty dissemination.

As with the BLG, it may be considered that CSAB has also been compromised by a lack of clearly defined targets and objectives. More recently, attempts have been made to streamline working arrangements, for which it is intended that 'CSAB should focus on very concrete activities, and potentially start with a small number of activities'[97] and, ultimately, to coordinate more closely and thematically with the BLG,[98] from which it has generally operated independently. While tangible results remain difficult to measure from this forum, there is some evidence that a more coordinated approach between the scientific and technical bodies of the BRCs has been of considerable assistance in lobbying for further leverage within the process towards the finalisation of IPBES. Indeed, this appears to have resulted in biodiversity matters being given a somewhat higher profile within the system than would seemingly have been the case had these conventions applied on a more individual basis.[99] Likewise, this development appears to have galvanised a more collaborative approach towards future interactions with IPBES for which, as noted above, the BLG has recommended collective support for potential applications to this forum.

While collective scientific alignment across each of the BRCs has proved elusive – and potentially counter-productive – a series of significant technical synergies have nonetheless been developed on an individual basis between particular regimes. A notable example in this context is the response of AEWA, Ramsar and the CMS to the avian influenza crisis.[100] In August 2005, following the H5N1 outbreak, the CMS established a Task Force on Avian Influenza and Wild Birds with the assistance of the

---

[94] *Report of the Third Meeting of Chairs of Scientific Advisory Bodies of Biodiversity-related Conventions*, 4.

[95] *See* A Trouwborst, 'International Nature Conservation Law and the Adaptation of Biodiversity to Climate Change: A Mismatch?' (2009) 21 Journal of Environmental Law 419, 430–437.

[96] H MacKay and S Alexander, *Towards a Multi-Convention Collaboration on Ecosystem Restoration*; Discussion Paper presented at the Fourth Meeting of CSAB.

[97] *Report of the Sixth Meeting of Chairs of Scientific Advisory Bodies of Biodiversity-related Conventions*, 4.

[98] *Report of the Fifth Meeting of Chairs of Scientific Advisory Bodies of Biodiversity-related Conventions*, 4. It is currently considered that back-to-back meetings should be convened, but as yet the schedules of these two fora have not yet aligned.

[99] *Ibid.*, 2.

[100] For a full appraisal of the international policy response to the outbreak of avian influenza *see* R Cromie and others, 'Responding to Emerging Challenges: Multilateral Environmental Agreements and Highly Pathogenic Avian Influenza H5N1' (2011) 14 Journal of International Wildlife Law 206.

Food and Agriculture Organization (FAO) and incorporating a number of pertinent actors, specifically including AEWA and the Ramsar Convention.[101] From this, the CMS, AEWA and Ramsar institutions formulated an integrated response to the crisis, framed by executive Resolutions adopted in close cooperation with their scientific bodies.[102] A series of common approaches have emerged from these Resolutions to unify future policies towards avian influenza. Both AEWA and Ramsar called for restraint, stating that culls of waterbirds and the destruction of habitats are misguided and do not constitute the 'wise use' of wetlands. All three regimes have observed the need for improved data collection and prompt analysis, integrated mitigation responses between all relevant stakeholders, the development of emergency responses and capacity-building programmes and a unified scientific response. The CMS Task Force convened a workshop in 2007 on practical lessons learned, the outcomes of which were also endorsed by these actors. In turn, the Ramsar Scientific and Technical Review Panel (STRP) has developed Practical Guidance on avian influenza responses, which has been endorsed by the CMS in framing future disease strategies.[103] Indeed, disease strategies will extend beyond the narrow confines of veterinary responses and quarantine policies, with the illegal trade in wildlife – which specifically engages the concern of CITES – considered to be a major factor in perpetuating and amplifying the impact of large-scale infection events and a basis for further cooperation.

3.3 Inter-treaty Cooperation and Work Programmes

Notwithstanding clustering practices, the elaboration of MOUs between individual treaties remains the most entrenched means of promoting integrative policies between the BRCs. These arrangements tend to follow a relatively predictable pattern, commencing with the negotiation of an instrument of cooperation, which identifies general areas of potential cooperation and may subsequently be bolstered by COP pronouncements. These are often accompanied by a Joint Work Programme – framed either within the MOU or developed subsequently – which evolves in conjunction with the experience of the treaties concerned and by ongoing cluster priorities. This general trend emerged in 1996 through the activities of the CBD, which has since concluded some 223 separate instruments towards cooperation with a vast array of disparate actors and institutions.[104] These are primarily adopted in the form of an MOU or a

---

[101] Resolution 9.8: Responding to the Challenge of Emerging and Re-emerging Diseases in Migratory Species, Including Highly Pathogenic Avian Influenza H5N1.

[102] Resolution 3.18: Avian Influenza and Resolution 4.15: Responding to the spread of Highly Pathogenic Avian Influenza H5N1 (AEWA); Resolution IX.23: Managing wetlands and waterbirds in response to highly pathogenic avian influenza and Resolution X.21: Guidance on responding to the continued spread of highly pathogenic avian influenza H5N1 (Ramsar); Resolution 8.27: Migratory Species and Highly Pathogenic Avian Influenza and Resolution 9.8: Responding to the Challenge of Emerging and Re-emerging Diseases in Migratory Species, Including Highly Pathogenic Avian Influenza H5N1 (CMS).

[103] Resolution 10.22: Wildlife Disease and Migratory Species (exhorting parties to 'use and promote the Ramsar Disease Manual' in managing diseases).

[104] Reproduced on the CBD institutional website at <http://www.cbd.int/agreements/>.

Memorandum of Cooperation (MOC) or, more occasionally, a Letter of Intent.[105] The varying nomenclature seems to have little impact upon the legal status of the document in question, nor in practice does there appear to be any discernible hierarchy between the various cooperative instruments. Many such documents are surprisingly complex and comprehensive; even the prosaic-sounding Letters of Intent lay down the expectations and duties of the parties in some considerable detail. Similar instruments of cooperation – if not necessarily on the same voluminous scale – have been concluded by other individual members of the BRCs with pertinent bodies.

As a general position, political impetus towards servicing these MOUs is provided by the formal pronouncements of the various COPs, alongside the periodic Strategic Plans that most of the BRCs routinely establish for themselves. The precise legal status of these documents is somewhat uncertain, although as Scott observes 'many such agreements appear to wear (if not flaunt) some of the trappings of a legally binding instrument'.[106] There is, however, little evidence to suggest that they are intended to have any legally enforceable effect and are seemingly envisaged as providing a platform for the pursuit of additional work programmes and a remit to attend meetings of other conventions, for which the individual terms of reference may be inadvertently hostile to participation by external operators.

Spatial constraints preclude a full appraisal of the progress made under all the various MOUs concluded between treaties to date.[107] The MOUs adopted to date have generally been concise documents pledging greater institutional cooperation, information-exchange and mutual representation at meetings of interest. Many MOUs also include more open-ended possibilities towards streamlining policies, the potential harmonisation of administrative practices and joint programmes of activity. This is particularly true in the case of certain members of the cluster, such as CITES and the CMS, which operate a 'listing' approach to their mandates, thereby periodically designating individual species and populations to their appendices, which may further facilitate clear lines of potential cooperation on species of mutual concern. Nevertheless, in and of themselves, such MOUs contribute relatively little more than vague statements of potential compatibility. As Matz observes in the context of the CMS, '[t]he general will to co-operate, even if repeated in decisions of the Conferences of States Parties, is not sufficient, but rather needs implementation'.[108] Indeed, although MOUs may exemplify endeavours towards a fuller implementation of broad commitments to cooperate, they have tended to be relatively ineffective unless accompanied by clear, targeted, realistic and partner-specific programmes of work.

---

[105] For the purposes of abbreviation, these arrangements between the BRCs are referred to collectively as 'MOUs', although as specified in particular instances the documentation in question has been concluded as an MOC.

[106] Scott (n 15), 194.

[107] For discussion of aspects of this practice *see* Caddell (n 15); Scott (n 15); Caddell (n 37); Bowman, Davies and Redgwell (n 11), 448–449, 478–481, 531–532, 578–582 and 624–626.

[108] N Matz, 'Chaos or Coherence? Implementing and Enforcing the Conservation of Migratory Species through Various Legal Instruments' (2005) 65 Zeitschrift für ausländisches öffentliches Recht und Völkerrecht 197, 211.

Since the earliest MOUs/MOCs were established in the mid-1990s, inter-treaty arrangements have evolved from being rather ambiguous documents into the instruments establishing a clearer basis for national implementation that have emerged in recent years. The MOU-based approach has been largely developed by the CBD, although it has also been replicated between other individual BRCs. In 1996, the CBD concluded MOCs with the Ramsar Convention, CITES and the CMS, which were swiftly followed by similar documents with UNESCO (1998) and, latterly, the ITPGRFA (2010). In 1997 a pioneering Joint Work Programme (JWP) was concluded between the Ramsar Bureau and the CBD Secretariat to seek to advance the broad cooperative themes outlined in the earlier MOC on a more practical basis. A similar initiative was developed between the CBD and its other 'lead partner' on biodiversity matters, the CMS, in 2002. The JWPs have been subsequently revised and refreshed in line with further guidance from the respective COPs. However, early incarnations of the JWPs tended to be extremely vague and, in retrospect, generated few tangible projects between treaties.

The first CBD-Ramsar JWP essentially sought to identify advisory roles and potential administrative synergies, albeit without particular targets and with relatively few specific projects in mind. A second JWP for 2000–01 introduced a degree of target-setting, primarily concerning the review of outputs and assisting in the practical implementation of convention commitments alongside a concerted effort to streamline reporting obligations. Meanwhile, in the context of the CMS, these arrangements were conversely undermined by an absence of targets and a resulting lack of operational coherence. Virtually all of the action points in the first CBD-CMS JWP were accorded a 'high' priority, requiring implementation before the end of 2003.[109] Indeed, the only objective considered a 'low' priority was the need to 'develop legislation for the protection and conservation of migratory species, as appropriate'.[110] Ironically, given that the CBD champions the development of coordinated national activity towards biodiversity conservation, this particular objective would seemingly constitute the most tangible and permanent benefit of synergy.

In the case of CITES, however, the JWP approach with the CBD proved to be relatively short-lived. An MOC was adopted in 1996, pledging institutional cooperation and information exchange, alongside the mooted coordination of work programmes and reporting requirements and encouraging joint conservation actions. Shortly afterwards, CITES adopted a rather ambiguous Resolution framing its relationship with the CBD, suggesting that parties streamline activities between national focal points and encouraging partnership opportunities between the conventions.[111] In 2000 the MOC was amended to include an option to develop joint work plans 'from time to time'. Ultimately, only one such instrument has ever been developed with the CBD, which was operational between 2000 and 2002, and which proposed cooperation on economic

---

[109] Document UNEP/CMS/Inf.7.13: CBD-CMS Joint Work Programme.
[110] Action 15.3.
[111] CITES Resolution Conf. 10.4, Cooperation and Synergy with the Convention on Biological Diversity (20 June 1997). On the early scope for cooperation, *see*: R Cooney, 'CITES and the CBD: Tensions and Synergies' (2001) 10 Review of European Community and International Environmental Law 259.

incentives, green-labelling, plant conservation and bushmeat concerns. This is not, however, to suggest that CITES and the CBD have failed to interact on a concerted basis; significant collaborative activities have been undertaken in the context of bushmeat, invasive species and issues affecting the trade in endangered plants.[112] Such activities have instead been pursued on a more ad hoc and opportunistic basis, as opposed to following a preordained programme of activity established through an overarching JWP.

If the initial JWPs focused primarily on executive and administrative synergies between treaties, more recent JWPs have demonstrated a subtle but significant shift in policy towards improving synergies at the national level. Indeed, the key CBD obligation for parties to establish National Biodiversity Strategies and Action Plans (NBSAPs)[113] presents further opportunities to streamline cumulative BRC commitments within domestic conservation strategies and remains the most tangible forum through which national parties to multiple biodiversity treaties can incorporate their various multilateral commitments in an efficient and coordinated manner. Indeed, this policy was reinforced by the CBD at its tenth COP as a key strategy for implementing the revised global biodiversity targets.[114] The third CBD-Ramsar JWP, operational between 2002 and 2006, was arguably the first such document to address this concern, coinciding as it did with a revised MOC that reinforced mutual commitments towards ensuring consistency between national wetlands policies under Ramsar and the NBSAPs developed under the CBD. This was put into practice in a fourth JWP developed for 2007–10, which allocated primary responsibility to the national focal points for promoting operational synergies. Under this initiative parties may identify particular national actions, based on domestic priorities, for which implementation strategies are to be pursued through 'proactive and flexible' cooperation between the focal points established under the respective Conventions. A fifth JWP was concluded for 2011–20, which further emphasises the need to facilitate 'developing and implementing National Biodiversity Strategies and Action Plans and National Wetland Policies in a consistent and mutually supportive way'.

A similar approach was followed in the second CBD-CMS JWP, which was operational between 2006 and 2008, and also prioritised linkages between national focal points. The CBD subsequently advocated reinforcing JWP arrangements, with particular emphasis on 'providing support and guidance to Parties on the integration of migratory species considerations in national biodiversity strategies and action plans'.[115] The CMS has developed a series of Guidelines to promote the implementation of

---

[112] On these initiatives *see* Caddell (n 37), 272.
[113] CBD, Article 6. This obligation has been reinforced through Aichi Biodiversity Target 17, which calls upon the parties to develop, adopt and begin implementing NBSAPs by 2015.
[114] The Executive Secretary of the CBD is required to '[c]ollaborate with the secretariats of the other biodiversity-related conventions to facilitate the participation of national focal points of these agreements, as appropriate, in the updating and implementation of national biodiversity strategies and action plans and related enabling activities': Decision X/5: Implementation of the Convention and the Strategic Plan.
[115] Decision X/20: Cooperation with other conventions and international organisations and initiatives.

Convention commitments within NBSAPs.[116] Significantly, they promote a clearer focus upon synergies on a national 'grassroots' basis, as opposed to the more abstract executive levels.[117] Similarly, CITES has sought to improve synergies on a national level by assisting mutual parties in integrating complementary activities within their NBSAPs[118] and has also developed distinct guidelines to this end.[119]

The scope for the BRCs to promote closer alignment within the various NBSAPs remains difficult to assess in the immediate short term. Thus far, although the development of indicative guidelines does represent a degree of success and endorsement of this policy, practical implementation remains mixed. Indeed, participants in the CBD-Ramsar initiative have reported relatively few joint actions between their focal points.[120] Meanwhile, there is little concerted evidence of greater treaty streamlining within the NBSAPs that have been revised and submitted to the CBD post-2010. Indeed, most national targets have been set with a view to implementing CBD requirements – unsurprisingly, perhaps, given that NBSAPs are fundamentally an obligation derived from the 1992 treaty with which states are seeking to demonstrate compliance as their primary objective – and there has been little movement towards coordinating Aichi considerations with other multilateral commitments. While there remains scope for national synergies to be promoted within such documents, notably in the context of addressing invasive species – which remains an issue of common concern of varying priority to each of the BRCs – such an approach has yet to be assertively or centrally articulated on a concerted basis within the NBSAPs developed in recent years.

If the NBSAP-centred approach to synergy arrangements has yet to yield tangible results, it is nonetheless apparent that targeted interactions between individual treaties can generate a basis for effective partnerships, conservation practices and a more efficient use of resources. This can have positive benefits both at an executive level and for treaty implementation by national agencies. Crucially, however, it is also clear that the conventions in question must demarcate explicit roles and responsibilities in implementing collaborative actions. As an example of good practice, instead of focusing on vague intimations towards executive and administrative streamlining, since 2004 CITES and the CMS have worked effectively to identify issues and species that are of most obvious direct interest between them.[121] Moreover, this process also establishes the degree of threat posed to that species and, by implication, its position

---

[116] Resolution 10.18: Guidelines on the Integration of Migratory Species into National Biodiversity Strategies and Action Plans (NBSAPs) and other Outcomes from CDB COP10.

[117] Indeed, the Guidelines pointedly criticised previous synergy initiates as being 'focused largely on processes at the international level': *ibid.*, 17.

[118] Notification to the Parties No. 2011/021.

[119] *Contributing to the Development, Review, Updating and Revision of National Biodiversity Strategies and Action Plans (NBSAPs): A Draft Guide for CITES Parties*; Notification to the Parties 2011/026.

[120] *The Joint Work Programme (JWP) between the CBD and the Ramsar Convention on Wetlands (Ramsar, Iran, 1971): Progress with Implementation and Development of the Fifth Joint Work Programme (2011 Onwards)*; Document UNEP/CBD/COP/10/INF/38, at 8.

[121] For a full account of collaboration between CITES and the CMS generally *see* Caddell (n 37), 273–279.

within the operational priorities of the treaty in question. In principle, this ought to facilitate a clearer understanding of the potential scope for synergies with allied bodies. To this end, CITES has identified saiga antelopes, snow leopards, African elephants, marine turtles, whale and great white sharks and sturgeons as priority species for synergistic activities,[122] which has provided an operational basis for cooperation with the CMS. To this end, a first suite of Joint Activities was agreed for 2005–07, from which a targeted List of Joint Activities was developed for 2008–10, subsequently extended into 2014, with a new CMS-CITES Joint Work Programme 2015–20 having been concluded at the time of writing. One of the primary areas of focus from these initiatives has been identifying commonalities in species coverage, which transpired to be an unexpectedly complicated task given increasingly apparent taxonomic and population discrepancies between the two conventions. Significant efforts have thus been expended in harmonising taxonomic designations between the two conventions, a problem first observed by CITES in 2000.[123] This is not merely a theoretical exercise, as attested to by the position of the saiga antelope, explicitly considered by CITES to be one of the main candidate species for collaborative activities with the CMS. Saiga antelopes have been one of the main beneficiaries of cooperative activities between the two conventions, but only after a rather inauspicious start where it was revealed that each convention applied to different sub-species, a position that took considerable efforts to streamline.[124] Partly as a result of these difficulties, the standardisation of nomenclature has been a significant issue in relations between the treaties, with the CMS having invested considerable time and resources in addressing this problem so as to improve linkages.[125]

In a similar manner, individual treaties have also taken advantage of overlapping remits to develop mutual processes and allow for the participation of other organisations in core outreach and inspection mechanisms developed under their specialised auspices. Perhaps the clearest example in this context has been the interaction between Ramsar and AEWA, both of which intersect in their purported regulation of wetland habitats and waterbirds respectively. The Wetlands Convention has developed an innovative implementation mechanism, the Ramsar Advisory Mission (RAM),[126] which mandates on-site inspection and remedial advice. In 2008 AEWA launched a parallel initiative, the Implementation Review Process (IRP), with its Standing Committee charged with ensuring that it operates 'in mutual cooperation with other relevant agreements to eliminate any possibility of duplication'.[127] Mutual reinforcement of the Ramsar implementation process is already conducted under AEWA, with its Standing Committee involved in follow-up activities for recommendations arising under RAMs in common areas.[128] While the IRP is still in its relative infancy, it has demonstrated a strong collaborative outlook, most recently advocating joint missions between itself,

---

[122] Resolution Conf. 13.3: Cooperation and synergy with the Convention on the Conservation of Migratory Species of Wild Animals (CMS).
[123] CITES Resolution Conf. 12.11, Standard Nomenclature.
[124] On the saiga saga *see* Caddell (n 37), 276.
[125] CMS Recommendation 9.4, Standardized Nomenclature for the CMS Appendices.
[126] Recommendation 4.7: Mechanisms for improved application of the Ramsar Convention.
[127] Resolution 4.6: Establishment of an Implementation Review Process.
[128] *Report of the Sixth Meeting of the AEWA Standing Committee* (UNEP, 2010), at 7.

Ramsar and the wider CMS in addressing potential implementation problems within mutual parties.[129] Meanwhile, the Ramsar regime has already exhibited a strong culture of collaborative actions in relation to wetland conservation. In 1999 an MOU was concluded between Ramsar and the World Heritage Centre, given that many protected natural heritage sites are also areas of significant wetland habitats.[130] A series of joint advisory missions have been conducted by experts on behalf of both treaties, which has not only facilitated the aggregation of vital expertise and the avoidance of duplicate or incomplete work, but has also ensured that a more comprehensive level of financial support has been available for the respective activities of the inspection teams of each convention than might otherwise have been the case. The Wetlands Convention has also established the Ramsar Cultural Network to further ensure that cultural values are addressed in the course of the work of the Convention and to provide a platform for further synergies to this end.

Furthermore, the context of illegal trade exemplifies perhaps one of the most unheralded yet clearest examples of successful inter-treaty synergy, which also illustrates the strong degree of inter-connectedness of the work of many of the BRCs. With the notable exception of CITES, the enforcement capabilities of the BRCs has proved to be decidedly limited. CITES has therefore led the way in facilitating the development of a series of operational toolkits to address particular instances of environmental crime,[131] as well as providing vital training under its auspices to national focal points charged with implementing multiple conventions, many of which will be undermined by these practices. In 2010, at the instigation of CITES, the International Consortium on Combating Wildlife Crime (ICCWC) was established, providing a forum for interaction between the major transnational investigative agencies.[132] In return, CITES has been the beneficiary of considerable volumes of specialist information from other members of the biodiversity cluster. For instance, the CMS cetacean agreements have commenced a process of maintaining detailed genetic data on captive marine mammals to specifically assist efforts under CITES to disrupt the illegal trade in live specimens,[133] while efforts to minimise the poaching of gorillas and elephants has long constituted a central feature of cooperation between the two conventions.[134] Moreover, as the CMS has observed, other members of the BRCs play a key role 'in creating a

---

[129] *Report of the Fifth Meeting of the Parties to AEWA* (UNEP, 2012), at 13. The mission in question concerned wetland habitats in Montenegro.

[130] The cultural remit of the Ramsar Convention is reinforced in its preamble, which notes that 'wetlands constitute a resource of great economic, cultural, scientific, and recreational value, the loss of which would be irreparable'.

[131] Notably UN Office on Drugs and Crime, *Wildlife and Forest Crime Analytic Toolkit* (United Nations, 2012).

[132] For an inside account of the establishment of ICCWC and its founding intentions see JM Sellar, *The UN's Lone Ranger: Combating International Wildlife Crime* (Whittles Publishing, 2014), at 159–163.

[133] *Identification of Cetaceans for the Needs of CITES*; Document UNEP/CMS/COP11/Inf.37. On the problems presented by the live capture of cetaceans and the deceptively fragile legal framework underpinning this practice *see* A Trouwborst, R Caddell and E Couzens, 'To Free or Not Free? State Obligations and the Rescue and Release of Marine Mammals: A Case-Study of Morgan the Orca' (2013) 2 Transnational Environmental Law 117.

[134] Caddell (n 37), 276–277.

462 *Research handbook on biodiversity and law*

platform for engaging all relevant stakeholders in addressing wildlife crime in concert with all other aspects of wildlife conservation and management'.[135] Indeed, in addition to sharing vital information, this mobilisation of effort remains of considerable value to CITES, not only in practical terms but by helping to establish wildlife crime as a higher priority for both global and regional instruments and national institutions.

4. FUTURE CHALLENGES TO REGIME INTERACTION

As noted above, a series of initiatives have been developed to foster improvements in the collaborative arrangements between the BRCs, with varying degrees of success. Nevertheless, considerable challenges remain to the formation and maintenance of effective linkages between these treaties. In this regard, four central concerns may be considered particularly pressing. First, challenges remain in establishing a prevailing ethos for the biodiversity cluster and objective priorities for synergy arrangements between its constituent members. Secondly, it has proved difficult to find common thematic and administrative ground, with many of the BRCs operating to different timescales and reporting requirements. Moreover, relatively few areas of core mutual interest may be apparent, even between seemingly similar conventions. Thirdly, and perhaps of greatest acuity, the spectre of resource constraints continues to undermine the capacity of individual treaties to maintain effective linkages. Finally, concerns may also be raised that the focus of synergy arrangements has continued to be centred upon abstract executive linkages rather than improving coordination at a national level.

4.1 Objectives and Ethos

One significant inhibitor to the potential success of synergy initiatives is that there has been relatively little agreement upon the ultimate aims and objectives of collaborative arrangements. On a basic level, the rationale for promoting executive synergies between the BRCs has been articulated by the CBD as being 'to enhance coherence and coordination in their implementation'.[136] Nevertheless, this relatively prosaic and self-evident mandate masks significant existential questions as to precisely the end to which the BRCs are intended to converge.

The BRCs remain a disparate group of treaties, each with individual mandates, guiding principles and philosophical approaches towards addressing the specific issues engaged under their auspices. Although further administrative streamlining and coordination is a broadly desirable outcome of synergy arrangements as espoused by the BLG – even if, as suggested below, it may not necessarily be an especially straightforward objective to achieve – a further convergence of approaches and ideals is unlikely to be attained without compromising the individual ethos of the BRCs to at least some degree. In recent years, this has manifested itself in a trend towards a closer alignment with the primary policies pursued under the CBD. Ultimately, however, this

---

[135] *Draft Resolution on Fighting Wildlife Crime Within and Beyond Borders*; Document UNEP/CMS/COP11/Doc.23.4.7/Rev.1.
[136] Decision VII/26 on Strategic Plan for the Convention on Biological Diversity.

has proved to be a more palatable prospect to some members of the cluster than for others. For the ITPGRFA, close coordination with the CBD remains a central treaty commitment.[137] The Ramsar Convention has ultimately been prepared to amend fundamental working principles to accommodate CBD concerns: interactions with the CBD have prompted a reformulation of key Ramsar commitments regarding 'wise use' so as to explicitly incorporate the ecosystem approach,[138] championed by the CBD, which the CBD Secretariat has subtly considered to be one of its greatest collaborative achievements.[139] Meanwhile, and equally strikingly, a series of subsidiary instruments developed by the CMS since 2003 have been primarily motivated by 'international responsibilities ... pursuant to *the Convention on Biological Diversity*' and the status of the Bonn Convention as its lead partner,[140] with specific CMS commitments rather less explicit. Moreover, the CMS has recently considered the need for its Scientific Council 'to adjust its expertise to reflect the evolving needs of the Convention' as the new Strategic Plan for the CMS 'draws heavily on the CBD Strategic Plan and the Aichi targets'.[141]

Nevertheless, synergy arrangements through the BLG and other collectivisation policies have been somewhat coloured as a means to attain essentially CBD-related commitments, to which other members of the cluster have at times proved resistant. CITES, which is perhaps the more confident of the other treaties due to its distinctive mandate, has repeatedly voiced concerns over this trajectory and has criticised the expectation by the CBD that other members of the BRCs will commit pressurised resources to measure performance towards meeting the various biodiversity-loss targets, since these are effectively CBD priorities that were ultimately formulated without the involvement of the other Secretariats.[142] The rather unsubtle CBD-orientated focus of the BLG's agenda has meant that a number of members of the cluster have treated this body with suspicion and a degree of wariness, initially only sending junior or temporary staff to participate at earlier meetings.[143] A perception has accordingly lingered among other treaty bodies that the BLG remains a tool of the CBD and is predicated on the advancement of a preordained agenda that favours that convention over others.[144] This is perhaps not such a difficult conclusion to draw: in its most recent pronouncement on cooperation with other bodies, the CBD noted with satisfaction 'the

---

[137] The objectives of the ITPGRFA encompass the conservation and sustainable use of plant genetic resources and the equitable sharing of benefits, which are to be attained by 'closely linking' it to the CBD: Article 1.1.
[138] Resolution IX.1: Additional Scientific and Technical Guidance for Implementing the Ramsar Wise Use Concept (at Annex A).
[139] B Siebenhüner, 'Administrator of Global Biodiversity: The Secretariat of the Convention on Biological Diversity' (2007) 16 Biodiversity Conservation 259, 267.
[140] Emphasis added; see the preambles to the Aquatic Warbler MOU, West African Elephants MOU, Saiga MOU, Pacific Islands Cetaceans MOU, Mediterranean Monk Seal MOU and Western Marine Mammals MOU.
[141] *Options for the Restructuring of the Scientific Council*; Document UNEP/CMS/COP11/Doc.17.1, para 18.
[142] See the comments of CITES at *Report of the Fifth Meeting of the Liaison Group of Biodiversity-related Conventions*, 2 and *Report of the Sixth Meeting*, 4.
[143] Velázquez Gomar (n 15), 11.
[144] *Ibid.*

464 *Research handbook on biodiversity and law*

progress made under the Convention on Migratory Species, the International Treaty on Plant Genetic Resources for Food and Agriculture, the World Heritage Convention, the Ramsar Convention on Wetlands and the Convention on International Trade in Endangered Species of Wild Fauna and Flora to reflect the Strategic Plan for Biodiversity 2011–2020 and the Aichi Biodiversity Targets in their work'.[145]

4.2 Commonality

Allied to these concerns, a key problem experienced thus far in synergy arrangements has been the establishment of common priorities. Unlike the chemical cluster, which may involve the same states addressing fundamentally the same range of materials presenting similar ecological and human dangers, obvious points of collective interaction are less conspicuous within the biodiversity grouping. Indeed, it has proved deceptively difficult to identify clear technical and policy issues that cut across the agendas of each of the BRCs.

At first glance, the application of a range of treaties to a broadly similar group of species ought to present clear opportunities for synergies between them. However, as noted above in the context of saiga antelope, many such regimes apply to a disparate range of sub-species and populations. Accordingly, while a group of treaties may engage a particular land mammal or bird, it is not axiomatic that they will correspond to the same distinct sweep of species, populations and groupings. Indeed, the divergence between the distinctive remits of the individual BRCs further militates against a strong degree of convergence in species coverage.[146] Moreover, even where such overlaps exist, conservation priorities may also diverge markedly. As a case in point, although some degree of commonality can be found in relation to species covered between CITES and the network of migratory species instruments, trade and crime considerations have tended to present comparatively limited threats to such animals as opposed to ecosystem-based concerns.[147] Forging specific synergies between these bodies may therefore carry little tactical advantage beyond a more abstract sense of inter-organisational solidarity, with the limited resources of these bodies accordingly better deployed on other priorities.

In addition to the identification of common priorities and issues between groups of eclectic specialist treaties, a significant challenge is posed by the working practices of the BRCs themselves. Each individual member of the cluster was developed and concluded in general isolation from the others and has developed its own institutional structures, procedures and cultures. Many bodies operate on disparate timescales, with different monitoring and reporting procedures. National agencies attempting to implement treaty commitments have long bemoaned a lack of alignment in monitoring requirements. That numerous secretariats demand largely similar information in different formats, with fluctuating reporting deadlines, creates additional pressure on

---

[145] Resolution XI/6: Cooperation with other conventions, international organizations, and initiatives.
[146] *See*, for example, Caddell (n 37), 277–278 (appraising the limited degree of species commonality in practice between the extensive annexes of CITES and AEWA respectively).
[147] *Ibid.*

ministries and environmental bodies at the national level. The harmonisation of administrative practices has accordingly constituted a long-standing operational priority for the United Nations Environment Programme (UNEP). Despite numerous meetings, working groups, case-studies, pilot projects, position papers and, ultimately, a consistent theme within the various MOUs, administrative arrangements between MEAs remain fragmentary.

The coalescence of reporting deadlines is therefore a key consideration in the effective interaction between allied regimes. Nevertheless, this has remained a singularly difficult task to achieve. Even within the chemical cluster, which appears best placed to advance the closest degree of institutional alignment between like MEAs, considerable complications remain. At present, the various secretariats of the cluster are developing proposals for a more coordinated reporting calendar, with a view to synchronising administrative processes in time for a mooted ExCOP3. Meanwhile, coordinated reporting between the Rio Conventions appears to be a practical impossibility: the UNCCD has observed that it is extremely difficult to identify thematic issues in a manner that would be consistent with the CBD approach, while the UNFCCC has stated that its reporting structure, which is focused on generating highly specific scientific data, is also unsuitable for collective adaptation.[148]

Similar difficulties have also been reported among the BRCs. Even where individual members of the cluster have demonstrated a clear overlap in function and application, clear administrative synergies have remained elusive. The closest possibility to date has been a series of initiatives advanced through AEWA, which had intended from May 2012 to schedule its subsequent Meetings of the Parties (MOPs) in close proximity to the Ramsar timetable of meetings, 'since it was vital that synergies between AEWA and Ramsar be maximised'.[149] Indeed, the AEWA MOP will be convened in the same year as the Ramsar COP, in 2015, albeit five months later than the meeting of the Wetlands Convention. The Fifth MOP to AEWA nevertheless yielded mixed results from the standpoint of administrative harmony. Despite numerous initiatives tabled to promote harmonised reporting cycles, both with the Ramsar regime and EU requirements, the proposal to move to a quadrennial cycle was ultimately withdrawn.[150] Anecdotally, this appears to have been attributable to financial pressures incumbent upon donor states in an era of global economic austerity, rather than a specific rejection of the principle of closer integration with the Ramsar regime.[151] Indeed, the Fifth MOP yielded the signing of a further JWP with the Wetlands Convention and a move towards the twinning of protected areas established under both conventions.[152] Nevertheless, there

---

[148] *Report of the Eleventh Meeting of the Joint Liaison Group of the Rio Conventions*, 6–7.
[149] *Report of the Fourth Meeting of the Parties to AEWA* (Nairobi: UNEP, 2008), 40. Although a significant incentive, administrative streamlining was not the purpose of this strategy; the cost implications of the increased membership of AEWA was also a strong motivating factor.
[150] *Report of the Fifth Meeting of the Parties to AEWA* (Nairobi: UNEP, 2012), 28–29.
[151] These sentiments were unreported in the official minutes, but see the contemporaneous bulletin updates of the International Institute for Sustainable Development available at <http://www.iisd.ca/cms/aewa-mop5/>.
[152] Resolution 5.19: Encouragement of Further Joint Implementation of AEWA and the Ramsar Convention; Resolution 5.20: Promote Twinning Schemes between the Natural Sites

has been relatively little subsequent momentum towards further administrative changes in this manner.

In addition to these difficulties, it is increasingly difficult to map the specific policies pursued by MEAs on particular issues, and this may further inhibit potential synergies. In recent years there has been a steady proliferation of working groups, committees, processes, institutions and actors, some of which may have only a transient or virtual existence. This increasing fragmentation 'is not only an obstacle for law makers involved in the implementation of the convention at the national level, but also for negotiators willing to ensure mutual supportiveness of the convention in other international processes'.[153] Tracking the precise outputs of subsidiary bodies, thematic strategies, work plans and allied initiatives across even a narrow range of specialist treaties is becoming an increasingly complex and time-consuming task in detective work. To this end, considerable faith has been placed in new mapping initiatives, notably the United Nations Information Portal on Multilateral Environmental Agreements (InforMEA) launched in June 2011, establishing a thematic database of policies from a range of MEAs,[154] which could streamline efforts to locate policy alignments if updated effectively and regularly.

4.3 Resource Implications

Ultimately, and perhaps most pervasively, where clear lines of commonality are identified, synergies have often been undermined by resource constraints. Wholesale underfunding has long bedevilled MEAs generally, with virtually every environmental treaty body having consistently reported financial and staffing pressures affecting their daily work. Cooperative strategies are, ultimately, expensive. The practices of inter-treaty cooperation outlined above generally entail the creation of working groups, additional collaborative meetings, and the development of joint working and management plans, alongside open-ended monitoring and evaluative exercises. Such strategies invariably involve additional burdens on stretched budgets and staffing complements. Inter-treaty cooperation is rarely listed within official budget and staff lists and remains a significant yet largely invisible expense for most regimes. Indeed, as the BLG has expressly stated, synergy work 'is facing limited resources in terms of time and financial means', accordingly generating 'a need to define clear priorities'.[155]

Relatively few MEAs maintain designated liaison officers – although a conspicuous example outside the context of the biodiversity cluster is a cohort of staff established to promote cooperation between the CBD and UNCCD under the terms of their initial MOC. In practice synergy tasks duly fall on the already over-burdened Secretariats, which remain chronically over-reliant on transient or co-opted staff and a constant

---

Covered by the AEWA and the Network of Sites Listed under the Ramsar Convention. No documentation was available on the scheduled Sixth MOP to AEWA at the time of writing.

[153] E Morgera and E Tsioumani, 'The Evolution of Benefit Sharing: Linking Biodiversity and Community Livelihoods' (2010) 19 Review of European Community and International Environmental Law 150, 173.

[154] *See* <http://www.informea.org>.

[155] *Report of the Eighth Meeting of the Liaison Group of Biodiversity-related Conventions*, 6.

stream of temporary interns to operate. While there have been examples of some successes at the executive levels – most recently, for instance, the BLG has mobilised communications strategies towards promoting a stronger degree of interaction between the various communications and outreach specialists within the BRCs[156] – it remains the case that, in many instances, wholesale synergy activities remain a luxury that is beyond the capacity of many administrative bodies to facilitate. This has been particularly acute, for instance, in the context of the WHC, which has long operated a very small unit charged with delivering natural heritage considerations and has little scope for undertaking regular liaison work.[157] The requirement for collegiality is clearly one that must be balanced against limited financial and human resources, and ultimately there is the need to prioritise the fulfilment of the extensive mandate of the treaty in question as a basic task over monitoring the activities of others. Indeed, as one recently retired senior member of CITES recalls:

> I have sometimes seen the same UN official attending nothing but workshops, seminars and conferences and there's a risk that networking becomes a be-all and end-all and one's core activities get left aside ... I was invited to many, many events each year that might have been interesting or semi-beneficial, for me and others, but would have ended up taking, to my mind, too much of my time and have diverted me away from what I saw as my priorities.[158]

Echoing this, the trade convention has also expressed concerns that cooperative arrangements should not be pursued at the expense of the individual duties of the treaties, noting that 'some CITES parties are cautious on the issue of synergies because they wish to avoid being distracted from the Convention's core mandate'.[159]

While parties to particular treaties do occasionally proffer generous financial support to underwriting their work, such funding is usually earmarked for specific – and highly visible – conservation projects rather than promoting paper-based efficiencies. Individual donations to support governance projects are therefore sporadic, although welcome and effective when forthcoming. For example, France contributed €95,000 to underwrite the costs of identifying future areas of cooperation between CITES and the CMS,[160] which has supported the development of a clearer programme of joint work. Nevertheless, there is no guarantee that such funds will be forthcoming in the future – from France or any other interested party – to support and promote synergy initiatives on a consistent basis. Meanwhile, the clearing-house mechanism joint work plan of the chemical cluster requires an estimated $360,000 to underwrite projected organisational costs, for which the parties have solicited voluntary contributions.[161] While parties may be more motivated to contribute towards the establishment of nascent institutions at the

---

[156] *Report of the Ninth Meeting of the Liaison Group of Biodiversity-related Conventions*, 6.
[157] Velázquez Gomar (n 15), 9.
[158] Personal communication with Mr John M. Sellar OBE, Former Chief of Enforcement of CITES (on file).
[159] *Report of the Eighth Meeting of the Liaison Group of Biodiversity-related Conventions*, 6.
[160] *Cooperation with Other Organisations: Convention on the Conservation of Migratory Species of Wild Animals*; Document SC61 Doc. 15.4 (Rev. 1), 2.
[161] *Joint Activities: Note by the Secretariats*; Document UNEP/FAO/CHW/RC/POPS/EXCOPS.1/2, 4.

commencement of this synergy initiative – at least initially – there are again no guarantees that continued funding will be available to support what may ultimately become a large joint structure in addition to the smaller frameworks of the individual treaties. Beyond the realms of state donations, MEAs have also struggled to generate the sustained private sector and corporate funding that might assist in advancing such projects further.

Likewise, the long-term success of effective synergy projects is frequently imperilled by ongoing financial constraints. Already the BRCs have observed that financial problems are affecting the proposed strategic alignments through the CBD's NBSAP programme, with the CMS recently reporting that 'sufficient funds have not been received to assist national implementation through capacity building and the effective participation of CMS focal points in the NBSAP regional and national processes'.[162] Resource problems were initially experienced in fully implementing participation in and access to the InforMEA portal, designed specifically to aid communication and synergy possibilities between treaties.[163] Meanwhile, examples of financial problems undermining synergy projects on an individual treaty basis remain legion. To give but one striking example, the 'Wings over Wetlands' programme, a much-heralded partnership between the CMS, AEWA and Ramsar to promote the long-term conservation of migratory birds, recorded a financial shortfall of €1 million – thereby leaving 13 out of its 15 international implementation priorities bereft of funding.[164] Perhaps of greater concern in this context, given that AEWA has arguably been at the centre of a number of the more promising initiatives towards effective coordination, the Agreement Secretariat was appealing in July 2014 for voluntary contributions to allow the scheduled Sixth MOP to take place.[165]

4.4 National Synergies

Finally, and as discussed in Section 3 above, the BRCs have tended to focus primarily on collaborations at an executive level between the various bodies established under their auspices, with relatively limited attention accorded to the issue of national implementation. This remains a significant issue for national nature conservation actors, which may be required to address species and ecosystem considerations for similar species under a wide range of potentially conflicting conventions, which has come at the expense of coherent implementation. As the Ramsar Strategic Plan

---

[162] *Ibid.*, 6.
[163] *Online National Reporting, Harmonization of Information and Knowledge Management for MEAs*; Document UNEP/CMS/Conf.10.10.
[164] *Report of the Fourth Meeting of the Parties to AEWA* (Nairobi: UNEP, 2008), 15.
[165] *See* <http://www.unep-aewa.org/en/node/2570>. This highly unsatisfactory state of affairs is attributable to Israel having been the only country to offer to host the MOP, with a number of delegations thereby unable to accept such an invitation for political reasons. This resulted in the MOP having to be hosted unexpectedly at the CMS headquarters in Bonn. Although unlikely to be replicated in future, this turn of events is nonetheless revealing as to the underlying scarcity of funds available to AEWA to absorb unexpected costs, which may have implications for participation in synergistic activities.

2009–15 presciently observes, a host of cooperative actions have been advanced under the auspices of the Wetlands Convention,

> [y]et much of this collaboration to date with the CBD, and with other biodiversity and environment conventions and agreements, such as the Convention on Migratory Species and the UN Convention to Combat Desertification, has been through global-scale mechanisms – secretariats, scientific bodies, etc. – and there is an urgent need for closer communication and cooperation between convention national focal points to achieve joint on-the-ground implementation.[166]

This disconnect between the executive and grassroots levels has also been recognised by treaty insiders: while the development of executive links is a central feature of the Resolutions framing interaction between treaties, 'there is much less consistency in the extent of national-scale implementation included between the Conventions' respective national focal points'.[167]

As noted above, there has been a series of initiatives to assist national focal points with streamlining multilateral commitments so that iterative documents, such as NBSAPs, can facilitate a more coherent implementation of national commitments. Ultimately, however, this is a long-term strategy to which the BLG and other fora have only recently addressed concerted efforts. Current NBSAPs have yet to reflect a more integrated approach, focusing primarily on meeting the revised CBD Aichi Targets. Treaties such as the CMS and CITES have developed draft guidance for further integrating the individual commitments and approaches prescribed under their auspices into national CBD documentation; monitoring whether such an approach has ultimately engendered meaningful change in this regard will remain a long-term project given that the CBD has as yet declined to make this policy a formal commitment under its auspices. Tellingly, commitments towards integrative practices within the CBD remain largely coordinated at the executive treaty level,[168] beyond a vague recognition that cooperation between focal points 'will provide a useful tool for such collaboration'.[169]

5. CONCLUDING REMARKS

Synergy arrangements and the elaboration of cooperative partnerships has become a key institutional priority for the BRCs since approximately the mid-1990s. It is clear that considerable energy and resources will continue to be expended on the pursuit of such arrangements, yet the precise benefits of these endeavours have proved to be

---

[166] Ramsar Strategic Plan 2009–15, para. 25.
[167] N Davidson and D Coates, 'The Ramsar Convention and Synergies for Operationalizing the Convention on Biological Diversity's Ecosystem Approach for Wetland Conservation and Wise Use' (2011) 14 Journal of International Wildlife Law and Policy 199, 203.
[168] See most recently Decision XI/2: Review of progress in implementation of national biodiversity strategies and action plans and related capacity-building support to Parties and Decision XI/21: Biodiversity and climate change: integrating biodiversity considerations into climate-change related activities.
[169] Decision XI/6: Cooperation with other conventions, international organizations, and initiatives.

decidedly mixed. Inter-treaty linkages, while seemingly prosaic, are nonetheless a significant element of the daily reality of MEA activities, yet remain decidedly understudied. There remains considerable scope for investigating the impact of multilateral synergies upon national implementation of international biodiversity commitments, as well as the ongoing success (or otherwise) of current and emerging initiatives.

A tentative degree of success is apparent from current collaborative practices. At the macro-level, where synergies have been most closely directed, clear lines of dialogue have been established. At the time of writing, new possibilities for UN funding were emerging, for which the process of clustering and mutual support for common objectives among the BRCs presents a greater scope for success than more individualised approaches. Particular scientific benefits have also been realised, especially from a methodological and taxonomical standpoint. Individual members of the BRCs have also aligned themselves to operate on a more collegiate basis. As a result, there is some evidence to suggest that at the individual species level, closer alignment between regimes has generated additional funding for conservation work, helped to ensure improvements in the quality and availability of data, led to enhanced training possibilities for national actors and exhibited potential for improving the conservation status of particular populations.

Considerable problems nevertheless remain in relation to synergy initiatives. Uncertainty still reigns as to the ultimate rationale for synergy and the extent to which disparate administrations and working practices are sufficiently malleable to accommodate more standardised approaches. Concerns are also raised that many such initiatives have been pursued in an aimless fashion and to the detriment of the 'day jobs' of the conventions in question. More worryingly, current arrangements demonstrate a tangible unease at the strategy of the CBD, not least that other bodies may lose their distinct identities and philosophies or, at least, face a concerted agenda creep from the 1992 Convention. Personal chemistry – or, perhaps more accurately, volatile reactions between key individuals – has also played an inhibiting role.[170] There has been little concerted engagement on a national level to promote synergies, which remain skewed towards collaboration at the more rarefied executive level.

Moreover, collaborative policies must be supported by significant financial and human resources if they are to be successful in the mid- to long term, an issue that is likely to pose the greatest single challenge to the future success of synergy initiatives. The BRCs have traditionally been chronically underfunded, with their resources and relatively small staffing complements having been increasingly stretched as, ironically, they have become more successful in attracting multilateral participants. Treaty secretariats are already under considerable political pressure to demonstrate that their budgets are generating meaningful conservation advances. Notwithstanding the professed support of the parties for more cooperative initiatives between treaties, they are nonetheless rarely enamoured by the prospect of investing in the invisible research initiatives and the expensive consultants and executives necessary to render synergy arrangements a coherent and productive reality. As MEAs are required to demonstrate

---

[170] Velázquez Gomar (n 15), 12.

greater efficiency in a straitened economic climate, the possibilities of closer coordination remain highly attractive in certain quarters. As the above evidence demonstrates, however, synergy is no panacea for underlying operational problems within the biodiversity cluster and poorly coordinated, under-funded and under-strategised approaches may ultimately present just as pressing an obstacle to the effective discharge of their mandates as the more pernicious impacts of treaty congestion.

SELECT BIBLIOGRAPHY

Caddell, R, 'The Integration of Multilateral Environmental Agreements: Lessons from the Biodiversity-Related Conventions (2012) 22 Yearbook of International Environmental Law 37

Churchill, R and G Ulfstein, 'Autonomous Institutional Arrangements in Multilateral Environmental Agreements: A Little-Noticed Phenomenon in International Law' (2000) 94 American Journal of International Law 623

Jóhannsdóttir, A, I Cresswell and P Bridgewater, 'The Current Framework for International Governance of Biodiversity: Is It Doing More Harm than Good?' (2010) 19 Review of European Community and International Environmental Law 177

Maes, F, A Cliquet, W du Plessis and H McLeod-Kilmurray (eds.), *Biodiversity and Climate Change: Linkages at International, National and Local Levels* (Edward Elgar Publishing, 2013)

Scott, KN, 'International Environmental Governance: Managing Fragmentation through Institutional Connection' (2011) 6 Melbourne Journal of International Law 177

Velázquez Gomar, JO, 'Environmental Policy Integration among Multilateral Environmental Agreements: The Case of Biodiversity' (2015) 15 International Environmental Agreements: Politics, Law and Economics (in press)

Young, MA, *Saving Fish, Trading Fish: The Interaction between Regimes in International Law* (Cambridge University Press, 2011)

Young, MA (ed.), *Regime Interaction in International Law: Facing Fragmentation* (Cambridge University Press, 2012)

Index

Aarhus Convention 79, 294
  Almaty Guidelines 104, 109, 110
  compliance committees 115–16
access to justice 80, 281, 294–5, 296
acidification, ocean 124, 125, 126, 128–9, 137–44, 150, 169–70
  cause 137–8
activism, environmental 81, 88
Addis Ababa Principles and Guidelines 290, 292, 371, 452
Afghanistan 248, 423
Africa 36, 114, 363, 367
  African-Eurasian Waterbirds Agreement (AEWA) 230, 393, 394, 454–5, 460–461, 465, 468
  African Nature Conservation Convention 60–61, 102, 132
  armed conflict 260–261, 265–6
  fragmentation of habitats 220, 234–5
  Eastern Africa regional seas regime 107
  elephants 460
  gorillas 50–51, 229–30, 247, 266, 394, 461
  Transboundary Ramsar Site (TRS) 225
  see also individual countries
agency and strategy 69–70, 72
Agenda 21 167, 173, 329
Agent Orange 250, 267
Agreement on the Conservation of Cetaceans of the Black Sea, Mediterranean Sea and Contiguous Atlantic Area (ACCOBAMS) 107, 445
Agreement on the Conservation of Small Cetaceans of the Baltic, North East Atlantic, Irish and North Seas (ASCOBANS) 107, 113, 328
agriculture 73, 112
Aichi Biodiversity Targets 176–7, 191, 232, 284, 405–6, 453, 463, 464, 469
albatrosses 393, 394
alien invasive species 3, 11, 124, 150, 184–218, 275–6, 389, 399
  climate change-induced migration 193–4
  definition 184–5
  developing legal response 190–192
  EU Regulation 185, 192–218
    border control and quarantine 209–10
    cooperation 211–12
    ecosystem approach 205–6
    education and public awareness 208–9
    eradication, containment and control 215–16
    information exchange 210–211
    intentional introduction 212–13
    mitigation 215
    national lists 203
    precautionary approach 195–201
    research and monitoring 207–8
    states, role of 206–7
    three-stage hierarchical approach 201–5
    unintentional introductions 213–15
  human health 189, 196, 198, 213
  impact of 188
    biodiversity 188
    economic activities 189–90
    ecosystem services 188–9
    human health 189
  marine environment 166, 171–4, 176
    ballast water 172, 173–4, 182, 187, 190, 214–15
    Law of the Sea Convention 130
  restoration and prevention 398
  risk assessment 194, 197, 198–201, 207–8, 211, 213, 216
  spread of
    intentional introduction 185–6
    unintentional introduction 186–8
  WHC 432
  WTO: SPS Agreement 190, 199–201, 218
Angola 229, 256
animal rights 50–51
animal welfare 213, 308
Antarctic Treaty (1959) 131, 336
  1991 Protocol on Environmental Protection 60, 347–8
Antarctic Whaling case 24, 25–6, 322–5, 326, 353, 435
antelopes, saiga 460
Anthropocene 15–16
anthropocentrism 12–14, 15, 48, 55, 56, 58–9, 60–61, 345
anthropomorphism 68, 69
anti-personnel mines 256

aquaculture 138, 182, 193
Arctic Council 143
Arctic Ocean 135
Argentina 50, 104, 315, 326
Aristotle 69
armed conflict 46, 245–69
  biodiversity/conflict nexus 246–9
  international humanitarian law 245, 246, 249, 263–4, 265
    environment-specific protections 249–51
    targeting 251–3
    weapons 254–7
  towards better protection
    continuity and 'co-application' of obligations 260–263
    enhancing protection 263–8
    'place-based' special status 257–60
Arnstein, S 81–2, 87, 99
Asia 114
  see also individual countries
Asian pangolin (or scaly anteater) 423
aspirin 363
Australia 22, 25, 325, 326
  alien invasive species 186, 187, 190, 209
  Great Barrier Reef 136, 158
  Kakadu National Park 433
Austria 225
autopoiesis 70, 71, 72, 76
autotrophs 44
avian influenza crisis 454–5

ballast water 172, 173, 182, 187–8
  Convention 173–4, 190, 214–15
Baltic Sea 103, 105, 107, 108
Bangladesh 314
Barbados Programme of Action (1994) 79
Barcelona Convention for the Mediterranean 105, 107, 108, 131, 191
  code of conduct 110, 111
Belgium 212
Belize Barrier Reef 136, 181, 432
Bequia 304
Bering Sea Fur Seals Arbitration 27, 58, 338–9, 364
Bern Convention on the Conservation of European Wildlife and Natural Habitats (1979) 13, 28, 132
  alien invasive species 191–2, 195, 196, 212, 218
  fragmentation of habitats 220, 235–8, 242
  intrinsic value 59
  non-compliance procedures 416, 419, 420, 425–8

  restoration 390, 395–6, 403, 408–9
biocentric approaches 68–70, 72, 75
Biodiversity Convention 13, 16, 22–3, 36, 43, 47, 152, 338
  alien invasive species 172, 184–5, 191, 275
    EU Regulation and CBD principles 195–218
  armed conflict 245–6, 261, 262, 263, 264, 265, 266, 267, 268
  bioprospecting 361–2, 369–77, 379, 381, 384, 385
  Clearing-House Mechanism (CHM) 211, 373
  common concern of humankind 24, 340, 342–5, 348–9, 356
  definition of biodiversity 130–131, 245–6, 273, 388
  ecosystem approach 44–5, 64, 163, 176–7, 205–6, 283–4, 285, 287–90, 291, 292–3
  fragmentation of habitats 220, 231–3
  interaction, regime 438, 455, 457, 458–9, 462–4, 465, 470
    Biodiversity Liaison Group (BLG) 451–3, 454, 463–4, 466, 467, 469
    conflict clause 444–6
    funding 466, 468
    hierarchy 446, 447
    Joint Liaison Group (JLG) 451
    Joint Work Programmes (JWPs) 457–8
    lead partners 447
    legal personality 442
    national synergies 469
  intrinsic value 60, 384
  legal personality 442
  mainstreaming 284–5
  marine environment 130–131, 137, 143–4, 160, 161, 343
    alien invasive species 172
    conflict clause 444–6
    ocean fertilisation 143
    VMEs: habitat protection 175–7
  National Biodiversity Strategies and Action Plans (NBSAPs) 175–7, 285, 286, 458–9, 468, 469
  objectives 175, 273
  participatory processes
    non-state actors 101, 102, 106, 107, 109–10, 111, 112–13
  precautionary principle 195–6, 197, 217–18, 280, 282, 284, 288, 291–2, 293
  Protocols 275, 282, 293
    Cartagena (2000) 190, 275, 281

Nagoya (2010) 275, 282, 293, 371–4, 385
restoration 390, 392, 398, 405–7, 410
Rules of Procedure 106
Strategic Plan 2011–2020 191, 232, 283–7, 290, 405, 406, 464
  Aichi Biodiversity Targets 176–7, 191, 232, 284, 405–6, 453, 463, 464, 469
sustainable development and 273–96
  concept 276–8, 293
  conclusions 292–5
  contents of CBD 279–82
  COP decisions 282–92
  further research 295–6
  two-track approach 274–6
sustainable use 290–2
biological and chemical weapons 255
bioprospecting and biopiracy 124, 385–6
  benefit-sharing contracts 370, 373, 374–7, 379, 385
  conceptualising (loss of) biodiversity 382–5
  environmental law and 368–71
    Nagoya Protocol 371–4
  ethics, politics and 362–5
    emerging controversies and framing of biopiracy 365–8
  IP, competing knowledge systems and 377–82
  prior informed consent 361, 370, 372–3, 381
BirdLife International and the Wildlife Conservation Society (WCS) 104
birds 11, 347, 393, 398, 403, 464
  African-Eurasian Waterbirds Agreement (AEWA) 230, 393, 394, 454–5, 460–461, 465, 468
  American mink 188
  armed conflict 247
  avian influenza crisis 454–5
  cane toads 186
  capercaillies 243
  Colombia 246–7
  EU Wild Birds Directive 238, 239, 240, 266
  instrumental value 57
  migration 227–9, 230, 236, 336, 468
  of prey 11–12, 52–3
  rose-ringed parakeets 189–90
  Waterbird Site Network 225
birth control pill 363
Black Sea 112, 173
Bolivia 23

Bonn Convention on the Conservation of Migratory Species of Wild Animals (1979) 29, 131, 136, 190, 281
  concerted actions procedure 50
  fragmentation of habitats 220, 226–31
  instrumental value 58
  interaction, regime 438–9, 456, 457, 458–60, 461, 463–4, 469
    avian influenza crisis 454–5
    Biodiversity Liaison Group (BLG) 451–3, 454, 463–4, 466, 467, 469
    environmental crime 461–2
    funding 467, 468
    Joint Work Programme (JWP) 457, 458
    legal aspects 442
    non-state actors 104, 107, 118
    restoration 390, 393, 396, 399, 403, 411
Brazil 50, 326
brown bears 243
Brundtland Commission 30–31, 276–7
Bukhara deer 230–231
Bulgaria 426

Cambodia 247, 256, 267
Cameroon 229, 423
Canada 22, 83, 225, 226
  alien invasive species 173
  whaling 300, 330, 331, 332
cane toad 186
capercaillies 243
capitalism 381–2, 383–4, 385
carbon capture and storage
  sub-seabed 140–141
Caribbean 104, 111
  alien invasive species 172
  case study: participatory resource management 78–93
  coral reefs 148, 172
  pollution from LBSP 168–9
Caspian Sea 102–3
causation
  environmental damage 67
Central African Republic 229, 425
chemical and biological weapons 255
chemical cluster: regime interaction 448–50, 464, 465, 467
Chile 104, 298
China 165, 174
Chytrid fungus 188
CITES (Convention on International Trade in Endangered Species of Wild Fauna and Flora) 9, 49, 347

interaction, regime 438, 455, 456, 457–8, 459–60, 467, 469
  Biodiversity Liaison Group (BLG) 451–3, 454, 463–4, 466, 467, 469
  commonality 464
  conflict clause 444
  environmental crime 461–2
  legal aspects 443, 444, 446–7
  non-compliance procedures 416, 419, 420, 421–5
  non-state actors 102, 106, 111, 118
  Rules of Procedure 106, 111
  whales 300, 446–7
civil society 294, 310, 365, 419
  organisations 93, 104
  see also non-governmental organisations
climate change 3, 22, 46, 150, 305, 342, 415
  alien invasive species 193–4
  connectivity conservation 219, 220, 222–3, 228, 230, 232–3, 237–8, 244
  inter-generational equity 312
  Intergovernmental Panel on (IPCC) 124, 126, 127–8, 129, 138, 145, 169, 171
  marine biodiversity, international law and 123–45
    effects of climate change 126–8
    high seas fisheries 132–7
    international legal framework 129–32
    ocean acidification 124, 125, 126, 128–9, 137–44, 169–70
    sea level rise 126
    'whole of ocean' strategy 129, 145
  restoration and prevention 389, 390–391, 399, 402, 403–5, 410
  UN Framework Convention on (UNFCCC) 66, 118, 170–171, 418
    common concern of humankind 340, 342–3, 345, 348–9, 356
    ecosystem services 64
    interaction, regime 443, 451, 452, 454, 465
    Kyoto Protocol 138, 139, 170, 421
    ocean acidification 138–9
    prevention and restoration 403–4, 407
    wetlands 430
cluster munitions 255–6, 268
clustering see interaction of regimes
Coen, E 46
collective interest and global environmental responsibility 335–7
  community of interest 339–40
  traditional legal order 337–9

Colombia 171, 246–7, 252
colonialism 363, 364, 381
common but differentiated responsibilities 277, 280, 305, 349–50
common concern of humankind 24, 47, 151, 262, 273, 280, 334–5, 339, 342–50
  normative aspects 350–351, 356–7
  environmental obligations *erga omnes* 352–4
  institutional coordination 351–2
  normative development 354–6
common heritage of mankind 153–4, 334–5, 339, 340–342, 351, 355–6, 357
common ragweed 189
community 45–6, 48–9
  of interest 339–40
  international community and the environment 21–3
  legal implications of community commitment 23–30
  nature of 18–21
  sharpening focus upon 43–5
conferences of the parties (COPs) 37, 48, 151, 389, 396, 397, 412, 440–441
  CBD 160, 176, 191, 232–3, 398, 405–6, 457, 458
    sustainable development 273, 274, 276, 281–92, 293, 294, 295, 296
  CITES 421–2, 423
  climate change 171
  CMS 227–9, 231, 456, 457
  common concern of humankind 351, 357
  interaction, regime 439, 451, 455, 456, 457, 458, 465
    combined COP 448–9, 450
    legal aspects 440–441, 442–3
  interpretation 221
  legal personality 442
  non-state actors 107
  Ramsar 178, 224, 392, 396, 407–8, 430, 457, 465
connectivity conservation *see* fragmentation of habitats
constructivism 96
Convention on Biological Diversity (CBD) *see* Biodiversity Convention
Convention for the Conservation of Antarctic Marine Living Resources (CCAMLR) (1980) 59, 131, 164
  Commission 133, 134, 137, 143
  non-state actors 102, 110

Convention on the Conservation and
    Management of High Seas Fisheries
    Resources in the North Pacific Ocean
    (2012) 164
  Commission 165–6
Convention on the Law of the Sea (UNCLOS)
    see Law of the Sea Convention (LOSC)
    (1982)
Convention on Migratory Species (CMS) 29,
    131, 136, 190, 281
  concerted actions procedure 50
  fragmentation of habitats 220, 226–31
  instrumental value 58
  interaction, regime 438–9, 456, 457,
    458–60, 461, 463–4, 469
    avian influenza crisis 454–5
    Biodiversity Liaison Group (BLG)
      451–3, 454, 463–4, 466, 467, 469
    environmental crime 461–2
    funding 467, 468
    Joint Work Programme (JWP) 457, 458
    legal aspects 442
  non-state actors 104, 107, 118
  restoration 390, 393, 396, 399, 403, 411
Convention for the Protection, Management
    and Development of the Marine and
    Coastal Environment of Eastern African
    Region (1985) 102, 103, 104, 107
  Protocol 103
Convention on the Protection of the Marine
    Environment of the Baltic Sea Area
    (1992) 103
Convention for the Protection of the Marine
    Environment of the North-East Atlantic
    (OSPAR) (1992) 111–12, 131, 355
  Commission 143
Convention for the Protection of the Natural
    Resources and Environment of the South
    Pacific Region (1986) 131
Coombe, R 361, 365, 381, 382, 384
cooperation, inter-treaty *see* interaction of
    regimes
coral reefs 124, 126, 133, 136, 143–4, 169
  alien invasive species 172
  Belize Barrier Reef 136, 181, 432
  biological diversity 147
  bottom trawling 165, 166, 167
  climate change 150, 171
  coastal fringing 155
  marine algae 167
  ocean acidification 129, 143–4, 150, 170
  recreation/tourism 148, 156
  territorial waters 151

vulnerability 149
wave energy 148
World Heritage Convention 180, 181, 432
Coral Triangle Initiative 105, 113
cost-benefit analysis 63
Costa Rica 23, 64
Côte d'Ivoire 226
Council of Europe 13, 61
courts *see* judicial process
coypu 189, 209
crimes, environmental 74, 461–2
crocodile, Nile 423
customary international law 220, 252, 255,
    261, 329, 334, 339
  common concern 350, 354–5
  common heritage 342
Cyprus 426, 427–8
Czech Republic 225

Danube River 246, 250, 426–7
databases
  alien invasive species 210
  traditional knowledge 380
  UN Information Portal on Multilateral
    Environmental Agreements
    (InforMEA) 466, 468
democracy
  deliberative democratic processes 96–7, 98,
    100, 101, 102
Democratic Republic of Congo (DRC) 229,
    246, 247, 248, 252, 255, 425
Denmark 310
depleted uranium ammunition 250, 254
Desertification Convention (UNCCD) (1994)
    345–7, 451, 452, 465, 466
developing countries 98, 279–80, 341, 364,
    369
  intra-generational equity 301–2, 327
  public participation 78
  small island 79
Djibouti 423
Dominican Republic 64

East African Community Treaty 28–9
East African Court of Justice 28
East Asia Seas Action Plan 105
East Asian-Australasian Flyway 225
East Rennell 181
ecocentric theories of intrinsic value 70–72,
    75
ecocide 74
ecological restoration *see* restoration in
    international biodiversity law

economic development 29–30, 248, 284, 396, 400, 412
  social and 278, 279–80, 286
  wetlands 178
economics 49, 63, 76
  environmental 63–5
  neo-classical 14
economy, global 62
ecosystem approach 44–5, 64, 131, 162–5
  CBD 44–5, 64, 163, 176–7, 205–6, 283–4, 285, 287–90, 291, 292–3
  CCAMLR 133
  climate change 64, 134
  EU Regulation: alien invasive species 205–6, 216
  Fish Stocks Agreement (FSA) 133
  Ramsar 178, 224
  regional fisheries agreements 133–4, 164, 331
ecosystem restoration 233, 285, 390, 398, 406–7, 454
ecosystem services and alien invasive species 188–9
Ecuador 51
  Galapagos Islands 158, 181
education 112, 266–7
  and public awareness 281
    alien invasive species 208–9
elephants 461
  African 460
Ellsworth, RE 6, 7
emergency responses: interaction of regimes 454, 455
empowerment
  participatory resource management 81, 82, 83, 88, 89, 91, 92
endangered status and intrinsic value 75
Enlightenment 13, 43, 47
Enola bean 367
environment impact statements (EISs) 308
environmental assessments (EAs) 308, 309
environmental impact assessments (EIAs) 212, 278, 280–281, 288, 294, 354–5, 391, 401, 402, 420, 426
  interaction of regimes 454
Equatorial Guinea 330
equity 47, 277, 278
  capacity for effective participation 91–2
  inter-generational see separate entry
  intra-generational see separate entry
erga omnes obligations 337, 342, 352–4, 414, 415
Estonia 315

ethics, politics and bioprospecting 362–8
European Convention on Human Rights (ECHR) 50
European Patent Office 367
European Union 22, 290, 402, 425
  alien invasive species 185, 191, 192–218
    available scientific evidence 197–8
    border control and quarantine 209–10
    cooperation 211–12
    ecosystem approach 205–6
    education and public awareness 208–9
    emergency action 194, 195, 196–7, 198, 200–201, 203, 207, 216, 217, 218
    eradication, containment and control 215–16
    information exchange 210–211
    intentional introduction 212–13
    mitigation 215
    national lists 203
    precautionary approach 195–201, 217–18
    research and monitoring 207–8
    states, role of 206–7
    three-stage hierarchical approach 201–5
    time limits 204, 205, 208, 210, 211, 214, 215
    unintentional introductions 213–15
  enforcement 217
  Habitats Directive 28, 192, 220, 238–44, 266
  interaction, regime 445, 465
  internal market 208, 210
  Natura 2000 sites 239, 240, 241–2, 243–4, 267
  subsidiarity 194
  TFEU
    environment 198
  Wild Birds Directive 238, 239, 240, 266
exclusive economic zones (EEZs) 135, 153, 156–9, 161–3, 338
  Australia 326
  whaling 299, 326, 329

factory farming 73
Faroe Islands 330–331
financial crash of 2008 14, 48
Fish Stocks Agreement (FSA) 132, 133, 135, 163, 164
fish/fisheries 11, 13–14, 36, 49, 123, 160–162, 285, 339, 402
  armed conflict 247
  bottom trawling 133–4, 165–6, 167, 182
  by-catch 162, 163, 165, 241, 445
  climate change, effects of 126–8, 150

Index 479

climate change-induced species migration: problems 134–5
highly migratory fish stocks (HMFS) 132, 134, 135, 154, 336
instrumental value 57
interaction, regime 445
international legal framework 129–32
  high seas fisheries regime and climate change 132–7
maximum sustainable yield (MSY) 57, 133, 162–3, 164
methods 164, 165–6
  bottom trawling 133–4, 165–6, 167, 182
ocean acidification 128–9, 137–44, 150
straddling fish stocks (SFS) 132, 135, 154, 164, 336, 338, 347
sturgeon, European 409, 423
total allowable catch (TAC) 162–5, 182, 332
vulnerable marine ecosystems *see separate entry*
Food and Agriculture Organization (FAO) 364, 369, 455
fisheries 137, 148, 159
  Code of Conduct 163, 164, 165, 190
vulnerability 149
see also International Treaty on Plant Genetic Resources for Food and Agriculture (ITPGRFA)
food security 128, 302
forests 285
  18th century Prussia 10–11
  armed conflict 247–8, 252, 253, 255, 267, 269
  carbon emissions from deforestation 64
  inter-generational equity 312
  interaction of regimes 451
  participatory approaches to management of 80–83
    forms of arrangements 81
    Jamaica 81, 83–93
  quinine: Amazonian 363
  US Forest Service 12
fragmentation of habitats 219–44
  African Nature Conservation Convention 234–5
  armed conflict 247
  Bern Convention 235–8
  Biodiversity Convention 231–3
  Bonn Convention 226–31
  connectivity and 221–3
  EU Habitats Directive 238–44
  Ramsar 224–5

  WHC 225–6
fragmentation of international law 29, 357, 466
Framework Convention for the Protection of the Marine Environment of the Caspian Sea 102–3
France 212, 316, 467
free-riding 64, 337
fur seals 11, 27, 57, 58
  Bering Sea Fur Seals Arbitration 27, 58, 338–9, 364

Gabon 229
Galapagos Islands 158, 181
Gambia 225
genetic resources 36, 59, 342, 384, 439
  Biodiversity Convention 152, 175, 273, 280, 369–71
    Nagoya Protocol 275, 371–4
  FAO 364, 369
    International Treaty on Plant Genetic Resources for Food and Agriculture (ITPGRFA) *see separate entry*
geographical indications 380
Germany 226, 298, 315, 434
giant hogweed 189
global constitutional order 18
  international community and the environment 21–3
  legal implications of community commitment 23–30
  reflections on nature of 'community' 18–21
Global Environment Facility (GEF)
  ballast water 187
  interaction of regimes 453
good governance 47, 277, 289–90, 295
good neighbourliness principle 63
Goodwin, P 82, 83, 92
Gordon, Seton 11–12
gorillas 50–51, 229–30, 247, 266, 394, 461
Greece 28, 174, 426, 428
Greenland 303, 304, 310, 330–331
Grotius, Hugo 152–3
Guinea 226, 422–3
Gulf Conflict (1991) 247, 248, 250

habeas corpus 50
habitat loss 9, 123
  Aichi Biodiversity Targets 176
  armed conflict 247
  marine environment 150
    fish 128, 129
    ocean acidification 129

property rights 49
habitat protection 3, 338, 345, 393, 394, 399, 404
  avian influenza crisis 455
  Bern Convention on the Conservation of European Wildlife and Natural Habitats (1979) *see separate entry*
  Biodiversity Convention 131, 231–3
  fragmentation 219–44
    African Nature Conservation Convention 234–5
    armed conflict 247
    Bern Convention 235–8
    Biodiversity Convention 231–3
    Bonn Convention 226–31
    connectivity and 221–3
    EU Habitats Directive 238–44
    Ramsar 224–5
    WHC 225–6
  Habitats Directive 28, 192, 220, 238–44, 266
  Law of the Sea Convention (LOSC) 130
  Particularly Sensitive Sea Areas (PSSAs) 158
  vulnerable marine ecosystems: capacity and 175
    CBD 175–7
    Ramsar 177–9
    WHC 179–81
  see also marine environment; wetlands
harmony 47, 284, 405
  UN's Harmony with Nature project 8, 23, 51, 59, 384
Hayden, C 361, 362, 363, 364, 365, 370, 371, 375, 376, 383, 384
hazardous chemicals and wastes
  chemical cluster: regime interaction 448–50, 464, 465, 467
health 112, 189, 196, 198, 213
hedgehogs 203, 209
heterotrophs 44
high seas 153, 159–60, 162–4, 165, 299, 329, 336, 337–9, 342
  erga omnes obligations 353
hippopotamus 423
Hobbes, T 57
Hoodia plant 367, 368, 376
human dignity 335
human rights 29–30, 32, 74, 263, 278, 293, 309, 335
  UDHR 344
Humane Trapping Standards Agreement (1997) 36

Hungary 316

ICBP (International Centre for Birds of Prey) 32–3
ICCROM (International Centre for the Study of the Preservation and Restoration of Cultural Property) 432
Iceland 165, 300, 306, 318, 320, 321–2, 324, 330–331, 429
ICOMOS (International Council on Monuments and Sites) 432
implied powers doctrine 441–2
India 23, 187, 189, 366–7, 368
indigenous peoples 103, 117
  Biodiversity Convention 101, 113, 275, 280, 281–2, 286, 289, 290, 293, 294, 296
  bioprospecting 369, 370, 372–3, 375–7
  ICCPR 309
  Johannesburg Declaration 281
  knowledge 363, 365–8, 370, 373, 375–6, 381
  Nairobi Convention 103
  New Zealand 51
  UN Declaration 281, 304, 310
  whaling 297, 299, 300, 302–11, 322, 324, 329–30, 331, 332
    definition of indigenous 303–4
Indonesia 330
information 102–3, 276, 277, 278, 281, 289, 294–5, 296
  alien invasive species: exchange and sharing of 210–212
  armed conflict 265–6, 268
  InforMEA (Information Portal on Multilateral Environmental Agreements) 466, 468
  interaction of regimes 452, 454, 456
  LAC Declaration on Principle 10 80
  non-compliance procedures 415, 420
  participation 79, 81, 98, 99, 106, 108, 109, 110, 112, 113, 115
    stumbling blocks 116
insecticide and traditional knowledge 367
instrumental value 55–9, 60, 61
  early treaties 57–8
  intrinsic value and 71, 72, 75–7
  limits of 62–7, 74
  rationale of 61–2
integration principle 277–8, 280, 285, 288, 291, 292, 293, 295, 296
intellectual property 47, 362, 363, 365–8, 385

bioprospecting, competing knowledge systems and 377–82
CBD 369–70, 374
FAO 369
TRIPS Agreement 366, 374, 379–80, 381
inter-generational equity 47, 278, 279, 288, 292, 293, 312–17
  courts 313–15, 317
  national constitutions 315
  RFMOs 331–2
  whaling 26, 297–301, 302, 317, 332–3
    conservation methods 317–20
    NAMMCO 330–331
    pirate 329–30
    sanctuaries 325–7
    scientific 301, 320–325, 327
    small and medium-size cetaceans 327–9
interaction of regimes 437–71
  future challenges 462
    commonality 450–451, 464–6
    national synergies 468–9
    objectives and ethos 462–4
    resource implications 466–8, 470
  legal aspects 440–447, 456
    conflict clauses 443–5
    hierarchy 446–7
    lead partners 447
  models of cooperation 447–8
    collective synergies and 'clustering' 448–53
    cooperation and work programmes 455–62
    scientific and technical collaboration 453–5
  nomenclature 460
  reporting deadlines 465
  whaling 300, 328–9, 444, 446–7
interdependence 48, 77
inter-governmental organisations (IGOs) 32, 440
  Caribbean 111
  CITES 102
  implied powers 441–2
  Mediterranean 105
  Whaling Convention Rules of Procedure 106
internal waters 153, 154–5, 161
International Committee of the Red Cross (ICRC) 267
International Consortium on Combating Wildlife Crime (ICCWC) 461
International Court of Justice (ICJ) 26–7, 245

Antarctic Whaling case 24, 25–6, 322–5, 326, 353, 435
Barcelona Traction 352
common concern of humankind 352, 353, 354–5
East Timor 353
Gabčíkovo-Nagyramos Dam 277, 314–15
inter-generational equity 26, 314–15
scientific whaling 301, 320, 322–5, 326
Legality of the Threat or Use of Nuclear Weapons, Advisory Opinion 314, 337
Pulp Mills 354, 355
International Covenant on Civil and Political Rights (ICCPR) 309
International Criminal Court (ICC) 74
international environmental movement 18, 21, 22, 26, 34
international humanitarian law 245, 246, 249, 263–4, 265
  environment-specific protections 249–51
  targeting 251–3
  weapons 254–7
International Labour Organization (ILO) 51–2, 310
International Law Association (ILA) 355
International Law Commission (ILC)
  armed conflict 245, 246, 248, 249, 256, 260, 261–2, 265, 268
  fragmentation 29
  state responsibility 352, 353–4
  treaty interpretation 220–221
International Maritime Organization (IMO) 141, 143, 156, 157, 158, 159
  ballast water 173, 187, 214–15
International Plant Protection Convention (IPPC) 437, 439
  Biodiversity Liaison Group (BLG) 451–3, 454, 463–4, 466, 467, 469
International Seabed Authority 153–4, 159
international trade
  alien invasive species 171, 172, 184, 186, 199
International Treaty on Plant Genetic Resources for Food and Agriculture (ITPGRFA) 441, 457, 463
  Biodiversity Liaison Group (BLG) 451–3, 454, 463–4, 466, 467, 469
International Waterfowl Research Bureau 32–3
International Whaling Commission (IWC) *see under* whaling
'interstitial headway' 37–8, 50

intra-generational equity 47, 278, 280, 282, 288, 291, 293
  RFMOs 331–2
  whaling 297–302, 327, 332–3
    definitions 303–4
    indigenous/subsistence 297, 299, 300, 302–11, 322, 324, 329–30, 331, 332
intrinsic value 13–14, 55–7, 59–61, 68, 76–7, 262, 344–5, 384
  biocentric approaches 68–70, 72, 75
  consequences of 72–6
  early treaties 58
  ecocentric theories of 70–72, 75
  Rio Declaration 59
  three meanings 56
  where is 68–72
invasive alien species *see* alien invasive species
IPBES (Intergovernmental Platform on Biodiversity and Ecosystem Services) 8, 30, 453, 454
Iran 33
Iraq 27–8, 66–7, 245, 247, 251, 260, 263
Ireland 240
Israel 254, 316
Italy 189, 209, 426
IUCN (International Union for the Conservation of Nature and Natural Resources) 32–3, 53, 102, 104, 260, 267, 389, 424, 432
ivory 424

Jamaica 79
  local forest management committees in 81, 83–93
    capacity 91–2
    co-management 86–90, 92
    empowerment 81, 82, 83, 88, 89, 91, 92
    functions of 84, 91, 92
    maintaining local community interest 90–91
    outreach programmes 88
    power sharing 86–90
    stakeholder representation 86
    sustainable livelihood projects 81, 83, 89, 90–91, 92, 93
    training 92, 93
Japan 64, 165, 187
  whaling 25, 298, 308, 311, 320–321, 322–4, 325–6, 328
Japanese Knotweed 190
judicial process 27, 28, 50, 51, 357
  ECJ 198, 217, 238, 240, 241, 243–4

environmental crimes 74
ICJ *see* International Court of Justice
inter-generational equity 313–15, 317
participation 79
patents 366
Permanent Court of International Justice 339
transboundary harm or damage 63
US courts 306, 308–9
justice, global 383

Kazakhstan 231, 423
Kenya 23
Kosovo 250, 253, 256
krill 124
Kuwait 27–8, 66–7, 248, 251

landmines 255–6, 268
Laos 256
Latin America 79–80
  see also individual countries
Latour, B 363, 383
Law of the Sea Convention (LOSC) (1982) 66, 130, 131, 132, 165, 338, 347
  armed conflict 261
  best scientific evidence 133
  exclusive economic zones 156–7, 158, 161–2, 163, 329
  high seas 160, 163, 164
  modern maritime zones 152–4
  ocean acidification 139–40, 141, 142, 143, 144
  pollution 130, 166–7, 168, 341
    alien invasive species 172–3, 182, 190
  regime interaction: CBD 445–6
  territorial sea 153, 155, 161
    sea lanes 155–6
  United States 342
  whaling 328–9
legal personality/personhood 32, 34, 50, 52, 98, 442
Legality of the Threat or Use of Nuclear Weapons, Advisory Opinion 314, 337
legitimacy 419
  non-state actors: participation 96, 97, 98–9, 100, 101, 102, 103, 105, 106, 108–9, 115–17, 118
Leopold, A 48, 72
Lesotho 423
Letters of Intent 456
lex posterior derogat legi priori 447
lex specialis derogat legi generali 447
Lionfish 171–2

Index 483

local communities 117
  CBD 113, 275, 280, 281, 282, 286, 289, 290, 293, 294, 296
  knowledge 370, 371, 372–3, 375–7
local forest management committees in Jamaica 81, 83–93
Locke, J 400
London Dumping Convention (LC) (1972) 130, 140, 141
  London Protocol (LP) 130, 140–142
lynx, Iberian 243
Lyster's International Wildlife Law 5, 11, 12, 13, 14, 16–18, 28, 30, 31, 32, 33–6, 38, 50, 61, 164, 177, 178, 179, 191, 221, 298, 342, 391, 418, 439

Madagascar 423
Makah 305–9, 311
mangroves 126, 147, 148, 150, 151, 155, 167, 177, 182
  Vietnam War 252, 255
margin of appreciation 323, 324
marine environment 338, 343, 404, 410
  armed conflict 247, 259–60
  CBD 130–131, 137, 143–4, 160, 161, 343
    alien invasive species 172
    conflict clause 444–6
    ocean fertilisation 143
    VMEs: habitat protection 175–7
  climate change, marine biodiversity and international law 123–45
    effects of climate change 126–8
    high seas fisheries 132–7
    international legal framework 129–32
    ocean acidification 124, 125, 126, 128–9, 137–44
    sea level rise 126
  fish/fisheries *see separate entry*
  fur seals 11, 27, 57, 58
  Global Programme of Action for the Protection of the Marine Environment from Land-Based Activities (GPA) 142–3, 167–8
  interaction, regime 460, 461
    CBD conflict clause 444–6
    CITES conflict clause 444
  non-state actors in treaty regimes *see separate entry*
  UN General Assembly 133, 137, 150, 159–60, 165, 182, 341, 446
  VMEs *see* vulnerable marine ecosystems
  whaling *see separate entry*

marine protected areas (MPAs) 133, 154–60, 176, 182
Marine Stewardship Council 109
MARPOL Convention (1973) 130, 143
Mauritania 423
Mediterranean 103, 108, 131, 133, 439, 463
  alien species 187, 191
  Barcelona Convention 105, 107, 108, 131, 191
  code of conduct for NGOs 110, 111
meeting of the parties (MOP) 351
  AEWA 230, 465, 468
Melbourne Principles for Sustainable Cities (2002) 73–4
memoranda of understanding/cooperation (MOUs/MOCs) 439, 442–3, 448, 454, 455–7, 461, 465, 466
Memorandum of Understanding for the Conservation of Cetaceans and Their Habitats in the Pacific Islands Region 108
mercury 348
Mexico 328, 376
Millennium Development Goals (MDGs) 22, 284–5, 286
Millennium Ecosystem Assessment 126, 147, 148, 388
Minamata Convention (2013) 348
mink 188
Moldova 427
mono-cultural practices 73
Mozambique 256

Nairobi Convention (1985) 102, 103, 104
  Protocol 103
national parks 12, 28, 181, 229, 433
  armed conflict 253
NATO 250, 253
Neem tree 366–7, 368
neo-classical economics 14
neoliberalism 382–3, 385
Nepal 252
Netherlands 33, 212, 226
networks
  fragmentation of habitats 222, 225, 226
    Bern Convention 236, 237
    CBD 232–3
    CMS 227, 228–9, 230, 231
    EU 239, 241, 242
  of interdependence 48
New International Economic Order 341
New Zealand 22, 25, 27, 187, 326, 434
  Wanganui River Treaty (2012) 51

Nile crocodile 423
no-harm principle 336–7
non-compliance procedures 351–2, 414–16, 418–21, 435–6
　Bern Convention 416, 419, 420, 425–8
　CITES 416, 419, 420, 421–5
　compliance 416–19
　dispute resolution and 435
　Ramsar Convention 420, 428–31
　WHC 416, 421, 431–4
non-governmental organisations (NGOs) 32–4, 49–50, 95, 117
　Aarhus Convention
　　compliance committees 116
　African Convention on the Conservation of Nature and Natural Resources 102
　alien invasive species 209
　Biodiversity Convention 101, 286, 289, 294, 296
　bioprospecting/biopiracy 361, 365, 367, 377
　Black Sea 112
　Caribbean 80, 85, 86, 111
　CCAMLR 110
　CITES 102, 118, 424
　climate change 118
　Coral Triangle Initiative 105
　decision-making areas 112
　development policy 104
　fisheries: VMEs 133–4
　legitimacy 97, 98–9, 100, 102
　Mediterranean 105, 108, 111
　Nairobi Convention 104
　non-compliance procedures 415, 419, 420, 424, 425, 427, 430, 432, 433, 434
　OSPAR Convention 111, 112
　Ramsar Convention 178, 430
　SASCEP 105
　subjecthood in international law 98
　threshold requirements for participatory rights 106, 109
　whaling 100, 106, 307–8, 311, 330
non-state actors in treaty regimes 95–118
　further research 117–18
　hierarchy 116–17
　justifications and limitations 96–9
　legitimacy and participatory rights 115–17
　non-compliance procedures 419, 420
　participation
　　meaning of 99–101
　　modes of 110–113
　　objectives 101–5
　　threshold requirements 106–10
　taxonomy of participatory rights 113–15

non-use value 56, 61–2, 64–5, 76
North Atlantic Marine Mammal Commission (NAMMCO) 330–331
North Korea 248
Norway 189, 300, 308, 318, 320, 328, 330–331
Noumea Convention (1986) 131
nuclear weapons 255
Nuclear Weapons Advisory Opinion 314, 337

Obama, Barack 306
ocean acidification 124, 125, 126, 128–9, 137–44, 150, 169–70
　cause 137–8
ocean fertilisation: sequestration of CO_2 139, 141–3
Oman 434
ombudsman and inter-generational equity 316
O'Neill, J 56, 67
organic farming 73
Organisation of American States (OAS)
　public participation 78
OSPAR Convention (1992) 111–12, 131, 143, 355
overexploitation 3, 123, 150, 160–166, 176, 240, 325, 387, 399, 402
Ozone Convention (1985) 66
　Montreal Protocol 421

Pacific Islands Region 108
pacta tertiis nec nocent nec prosunt 337
Pakistan 189, 314
pangolin, Asian (or scaly anteater) 423
participation 46, 47–8, 49
　alien invasive species 209
　approaches 80–83, 87–8
　CBD and sustainable development 277, 278, 281, 285, 286, 288–9, 290, 291, 292, 294, 296
　meaning of 99–101
　non-compliance procedures 415, 419, 420, 424, 425, 427, 430, 432, 433, 434
　resource management: Caribbean case study 78–93
　see also non-governmental organisations; non-state actors in treaty regimes
participatory resource management: Caribbean case study 78–93
　local forest management committees in Jamaica 83–93
　　barriers to participation 86–92
　　capacity 91–2

maintaining local community interest 90–91
power sharing and management role 86–90
stakeholder representation 86
participatory approaches 80–83, 87–8
Particularly Sensitive Sea Areas (PSSAs) 158–9, 182
Parvo, Arvid 341
patents 365–8, 374, 379–80
peace parks 268
permaculture 73
Permanent Court of International Justice 339
persistent organic pollutants 348
　chemical cluster: regime interaction 448–50, 464, 465, 467
Peru 367
petrels 393, 394
pharmaceutical substances 363, 364–5, 366–8, 369, 375
Philippines 147, 187, 313–14, 330
pirate whaling 329–30
plant breeders' rights 380–381
pluralism 43–4, 48
point of view 68
Poland 315
Polar Bears Agreement (1973) 132, 136
politics, ethics and bioprospecting 362–8
polluter-pays principle 215, 216, 280, 292, 295
pollution 3, 64, 66, 123–4, 126, 150, 399, 415
　armed conflict 247, 248, 249–53, 254
　inter-generational equity 312, 321
　international legal framework
　　marine environment 130, 139–40, 142, 143, 155, 156, 157, 158, 166–71, 182, 190, 341
　mercury 348
　persistent organic pollutants 348
　　chemical cluster: regime interaction 448–50, 464, 465, 467
population growth 62
positivism 98
poverty 302, 312
　eradication of 277, 279, 282, 284, 286–7
　reduction 284, 285, 286
precautionary approach/principle 66, 149, 163, 164, 165, 166
　alien invasive species: EU Regulation 195–201, 217–18
　armed conflict 253, 254
　CBD 195–6, 197, 217–18, 280, 282, 284, 288, 291–2, 293

common concern of humankind 350
non-compliance procedures 420
regional fisheries agreements 133–4, 165, 166, 331
sustainable development 277, 278
WTO 199–201, 218
prevention *see* restoration in international biodiversity law
prior informed consent 361, 370, 372–3, 381
private property 400–401
private sector 34, 38, 80, 103, 412, 468
property rights
　habitat loss 49
proportionality 252–3, 264, 373
Prussia 10–11
public awareness
　alien invasive species 208–9
public goods 62, 64
public participation *see* participation

Quammen, D 221–2
quinine 363

racial discrimination
　Committee on the Elimination of 310
Ramsar Convention 9, 16, 33, 37, 131, 152, 190, 253, 281, 347
　armed conflict 265–6
　habitat fragmentation 220, 224–5
　habitat protection and capacity 177–9, 180
　instrumental value 58
　interaction, regime 438, 457, 460–461, 463, 465
　　avian influenza crisis 454–5
　　Biodiversity Liaison Group (BLG) 451–3, 454, 463–4, 466, 467, 469
　　funding 468
　　Joint Work Programmes (JWP) 457, 458, 459
　　legal aspects 442, 443
　　national synergies 468–9
　Montreux Record 179, 429–30
　non-compliance procedures 420, 428–31
　non-state actors 102
　restoration 389–90, 392–3, 396–7, 403, 407–8, 410
　Small Grants Fund 429
　Transboundary Ramsar Sites (TRS) 224–5
rational use 59
rationality, instrumental 55
red deer 230–231
regime interaction 437–71
　future challenges 462

commonality 450–451, 464–6
national synergies 468–9
objectives and ethos 462–4
resource implications 466–8, 470
legal aspects 440–447, 456
hierarchy 446–7
lead partners 447
models of cooperation 447–8
collective synergies and 'clustering' 448–53
cooperation and work programmes 455–62
scientific and technical collaboration 453–5
nomenclature 460
reporting deadlines 465
whaling 300, 328–9, 444, 446–7
Regional Fisheries Management Organisations (RFMOs) 130, 131, 132–6, 159, 164, 165, 166, 338
inter- and intra-generational equity 331–2
res communis 152–3, 338, 347, 356
research and monitoring
alien invasive species: EU 207–8
resource management
Caribbean case study: participatory 78–93
participatory approaches 80–83, 87–8
restoration in international biodiversity law 387–412
ecosystem restoration 233, 285, 390, 398, 406–7, 454
failure to prevent 409–12
international nature protection law 392–6
prevention 389, 391
addressing weak 403–9
interrelationship between restoration and 396–8
weak side of 398–403
specialness 403
terminology 390–391
revisionism
vision, visionaries and risks of 8–16
rhododendron, pontic 189
rights of nature 51
Rio+20 Conference on Sustainable Development (2012) 51, 276, 286–7
interaction, regime 438
Rio Declaration (1992) 336
anthropocentrism 58–9
common but differentiated responsibilities 305
integration 29–30
participation 79, 294

precautionary approach 195, 198
risk assessment
alien invasive species 194, 197, 198–201, 207–8, 211, 213, 216
armed conflict 265
Romania 427
Romanticism 9–13
Ruddy duck 188, 212
ruddy-headed goose 104
Russia 165, 225
whaling 304, 307, 318, 328
see also Soviet Union
Rwanda 246, 248, 255, 266

saiga antelopes 460
St Vincent and the Grenadines 79, 304
sanctions
individuals: travel and financial 425
trade 419, 421
scholarship 38
vision and virtue in legal 16–18
Schwägerl, C 15, 16
Science and Technology Studies (STS) 362
Scotland 203
Sea Shepherd Society 308
sea turtles 426, 428, 460
seabed
International Seabed Authority 153–4, 159
Seabed Disputes Chamber 341–2, 351, 353, 354–6, 357
seahorse, common 423
second law of thermodynamics 20, 44
selfish behaviour 401
Senegal 225
Serbia 246, 250, 253
Seychelles 325
sharks, great white 460
shellfish poisoning 187–8
Shiva, Vandana 383
sic utere tuo ut alienum non laedas 63
Slovakia 225
slugs, Spanish 188–9
small island developing states 79
Smith, A 400–401
snow leopards 460
social capital 82
social inclusion 82
soft law 296, 313, 351, 397–8
alien invasive species 191, 201
anthropocentrism 58–9
indigenous peoples 310
marine environment 162, 163, 259–60

non-state actors 95, 104, 107–8, 114, 115, 117
Somalia 423
South Africa 22, 187, 315, 326
   Hoodia plant 367, 368, 376
South Asia Co-operative Environmental Programme (SASCEP) 105
South Korea 165, 248, 320
Southern Bluefin Tuna 134
sovereignty 335, 336, 337, 338, 339–40, 342
   Biodiversity Convention 369
   natural resources 47, 59, 176, 273, 329, 336, 337, 338, 343, 356, 364
Soviet Union 33, 298, 318, 319
   see also Russia
Spain 202, 243–4
specialness rhetoric 403
speciesism 67
Spinoza, B 69
squirrels, grey 75, 209, 426
Sri Lanka 252, 256
starch 34, 44
state responsibility 24, 25, 63, 352, 353–4, 435
Stockholm Declaration (1972) 15, 23, 30, 336–7, 387
   anthropocentrism 58
   participation 78
Stone, C 51
straddling fish stocks (SFS) 132, 135, 154, 164, 336, 338, 347
strategic environmental assessments (SEA) 288
sturgeon 460
   European 409, 423
subjecthood in international law 98
Suez Canal 187
sustainable development 21–2, 29, 47, 182, 301–2, 337, 355, 400
   Biodiversity Convention and 273–96
      conclusions 292–5
      contents of CBD 279–82
      COP decisions 282–92
      further research 295–6
      two-track approach 274–6
   concept of 276–8, 293
   Ramsar 224
   Sustainable Development Goals (SDGs) 22
Switzerland 315
synergies *see* regime interaction

Tajikistan 231
Tanzania 28

territorial sea 153, 155–6, 161, 299, 329
Thailand 187
tourism 138, 148, 156, 181, 186–7, 209, 256, 285, 319
   World Heritage List 434
traditional knowledge/medicine 363, 365–8, 369
   benefit-sharing 370, 373, 374–7, 379, 385
   bioprospecting, IP and competing knowledge systems 377–82
TRAFFIC 420, 424
tragedy of the commons 65, 153, 154, 338
Trail Smelter Arbitration 63
transboundary harm or damage 63
transparency 103
treaties 32, 34, 337, 352, 355–6
   compliance 416–19
   conferences of the parties (COPs) *see separate entry*
   evolutionary change 37
   international legal order: role of 35–8
   interpretation 29, 220–221, 296
   law of 24
   living instruments 45, 400
   meeting of the parties (MOP) 351
      AEWA 230, 465, 468
   non-compliance procedures *see separate entry*
   non-state actors in treaty regimes *see separate entry*
   participation 46
   regime interaction *see separate entry*
   secretariats and treaty-making powers 443
   see also individual treaties
Trinidad and Tobago 79
TRIPS Agreement 366, 374, 379–80, 381
tuna 134, 135
Turkey 426
Turkmenistan 231
turmeric powder 366, 368
turtles 426, 428, 460

Ukraine 426, 427
uncertainty 98, 133, 149, 151
   alien invasive species 174, 195, 198
   bioprospecting 376–7
   common heritage 342
   see also precautionary approach/principle
UNCLOS *see* Law of the Sea Convention (LOSC) (1982)
UNESCO (United Nations Educational, Scientific and Cultural Organization) 310, 457

United Kingdom 83, 212, 263
  fishing 148
  fur seals 58
  grey squirrels 75
  Liverpool Mercantile City 432
United Nations 286, 287, 453, 470
  Charter 446
  Compensation Commission (UNCC) 27–8, 66–7, 251, 267
  Economic Commission for Europe (UNECE) 79
  ECOSOC 117, 424
  Environment Programme (UNEP) 167
    armed conflict 254, 256
    ballast water 187
    compliance 416
    instrumental use values 64
    interaction, regime 465
    Iraq 247
    regional seas 131–2
  General Assembly 59
    marine environment 133, 137, 150, 159–60, 165, 182, 341, 446
  harmony with nature 8, 23, 51, 59, 384
  indigenous peoples 281, 304, 310
  Information Portal on Multilateral Environmental Agreements (InforMEA) 466, 468
  NGOs 112, 117
  Security Council 27–8, 424–5
United States 22, 64, 226, 263
  alien invasive species 171, 173
  Ballast Water Convention 174
  Florida Keys 158
  fur seals 58
  national parks 12
    Everglades 181
  patents 366–7, 380
  recreation 148
  UNCLOS 342
  USAID 86
  Vietnam War 249, 252, 255, 256, 267
  whaling 304, 306
    Makah 305–9, 311
Universal Declaration of Human Rights 344
Universal Postal Union 21
urban planning 73–4
Uzbekistan 231

values in international biodiversity law 55–77
  agenda for reconciliation 76–7
  defining the value of nature 55–7
  instrumental value 55–9, 60, 61
    early treaties 57–8
    intrinsic value and 71, 72, 75–7
    limits of 62–7, 74
    rationale of 61–2
  intrinsic value 13–14, 55–7, 59–61, 68, 76–7, 262, 344–5, 384
    biocentric approaches 68–70, 72, 75
    consequences of 72–6
    early treaties 58
    ecocentric theories of 70–72, 75
    Rio Declaration 59
    three meanings 56
    where is 68–72
  locating value 57–61
Veblen, T 401
vertebrate species 3, 4
Vienna Convention on the Law of Treaties (1969) 24, 337
  interaction, regime 446
  interpretation 29, 220, 296
  termination or suspension on breach 414–15, 435
Vietnam 423
Vietnam War 249, 252, 255, 256, 267
vision 5–8
  2020 vision and beyond 39–53
    blinkers of 'Enlightenment' worldview 40–43
    focus upon 'community' 43–5
    momentum 45–8
  envisioning of vision itself 30–31
    role of actors in international legal order 31–5
    role of treaties in international legal order 35–8
  and virtue in legal scholarship 16–18
  visionaries and risks of revisionism 8–16
vulnerable marine ecosystems (VMEs) 146–82
  alien invasive species 166, 171–4, 176
  ballast water 172, 173–4, 182, 187–8, 190, 214–15
  bottom trawling 133–4, 165–6, 167, 182
  habitat protection and capacity 175–81
    CBD 175–7
    Ramsar 177–9, 180
    WHC 179–81
  marine ecosystems 147–8
    role for public international law 151–2
    vulnerabilities 150–151
    vulnerability in 149–50
  marine protected areas (MPAs) 133, 154–60, 176, 182

exclusive economic zones 156–9
high seas 159–60
internal waters 154–5
territorial sea 155–6
modern maritime zones 152–4
Particularly Sensitive Sea Areas (PSSAs) 158–9, 182
pollution 166–71
greenhouse gases 169–71, 182
LBSP 166, 167–9, 182
unsustainable fisheries 160–166
archipelagic and coastal State authority 161–2
methods 164, 165–6, 182
TACs (total allowable catch) 162–5, 182

Wadden Sea 225, 226
Seal Agreement 104
Wandesforde-Smith, G 5, 18, 22–3, 30, 31, 33–6, 37
war *see* armed conflict
Western Hemisphere Convention (1940) 16, 57, 132
Western Indian Ocean Marine Science Association (WIOMSA) 104
wetlands 33, 345, 411
armed conflict 247, 252, 253, 265–6, 269
CBD 392
climate change 430
definition 177
Ramsar Convention 9, 16, 33, 37, 131, 152, 190, 253, 281, 347
armed conflict 265–6
habitat fragmentation 220, 224–5
habitat protection and capacity 177–9, 180
instrumental value 58
Montreux Record 179, 429–30
non-compliance procedures 420, 428–31
non-state actors 102
regime interaction 438, 442, 443, 454–5, 457, 458, 459, 460–461, 463, 465, 468–9
regime interaction: BLG 451–3, 454, 463–4, 466, 467, 469
restoration 389–90, 392–3, 396–7, 403, 407–8, 410
Small Grants Fund 429
Transboundary Ramsar Sites (TRS) 224–5
whale watching 148, 319
whales: CITES-CMS cooperation 460
whaling

Antarctic Whaling case 24, 25–6, 322–5, 326, 353, 435
instrumental value 57
inter-generational equity 26, 297–301, 302, 317, 332–3
conservation methods 317–20
NAMMCO 330–331
pirate 329–30
sanctuaries 325–7, 332
scientific 301, 320–325, 327
small and medium-size cetaceans 327–9
International Whaling Commission (IWC) 27, 112, 299–300
inter-generational equity 297, 301, 317–20, 321–2, 324, 325, 327–30, 331, 332
intra-generational equity 297, 301, 304–6, 307, 310, 311, 332
non-state actors and legitimacy 100
Scientific Committee 305, 318, 320, 321, 327
intra-generational equity 297–302, 327, 332–3
indigenous/subsistence 297, 299, 300, 302–11, 322, 324, 329–30, 331, 332
maximum sustainable yield 304, 318
pirate 329–30
small and medium-size cetaceans 300, 327–9, 330–331
small-type coastal 310–311
Whaling Convention (1937) 11, 57, 298
Whaling Convention (ICRW) (1946) 9, 24, 25–6, 27, 57, 130, 131, 137, 297–301, 338
CITES and 300, 446–7
common interest 345
indigenous whaling 301, 303, 306, 322
interaction, regime 328–9, 444, 446–7
moratorium 300–301, 302, 311, 318, 320, 322, 323, 325–6, 332
normative *lacunae* 299–301
objectives 298–9
opting-out 300, 318, 332
Rules of Procedure 106
sanctuaries 325–7, 332
scientific whaling 301, 320, 322–5, 327, 435
small and medium-size cetaceans 300, 327–8
territorial scope 299
UNCLOS and 328–9
women 101
World Charter for Nature (WCN) 13, 23, 30

intrinsic value 59–6
World Conservation Strategy 30, 49
World Heritage Convention (1972) 12, 24, 131, 136, 137, 152
  armed conflict 248, 258, 261, 262, 263–6
  Committee 28, 265, 432–4
  common concern 347
  danger list 180–181, 248, 395, 398, 431–3, 434
  definition of natural heritage 179–80
  fragmentation of habitats 220, 225–6
  interaction, regime 438, 461, 467
    Biodiversity Liaison Group (BLG) 451–3, 454, 463–4, 466, 467, 469
    conflict clause 443
  non-compliance procedures 416, 421, 431–4
  restoration 390, 394–5, 398
  transboundary sites 226
  VMEs: habitat protection and capacity 179–81
  World Heritage Fund 431, 432, 434
World Trade Organisation (WTO) 21
  alien invasive species 190, 199–201
  dispute settlement 29, 200–201
  SPS Agreement 190
    precautionary principle 199–201, 218
  TRIPS Agreement 366, 374, 379–80, 381
World Wide Fund for Nature (formerly World Wildlife Fund) (WWF) 4, 104

Yellowstone National Park 12

Zaire 23
Zebra mussels 173, 202